## STUDENT COMPANION TO ACCOMPA

# FUNDAMENTALS OF
# BIOCHEMISTRY
## LIFE AT THE MOLECULAR LEVEL
Fourth Edition

**Akif Uzman**
*University of Houston*

**Jerry Johnson**
*University of Houston*

**Joseph Eichberg**
*University of Houston*

**William Widger**
*University of Houston*

**Donald Voet**
*University of Pennsylvania*

**Judith G. Voet**
*Swarthmore College*

**Charlotte W. Pratt**
*Seattle, Washington*

**JOHN WILEY & SONS, INC**

Cover Designer Madelyn Lesure
Cover Illustration Norm Christiansen
Cover Photos
(Vitruvian Man): © Odysseus/Alamy and © Dennis Hallinan/Alamy
Molecular structures clockwise from top: based on X-ray structures that were respectively determined by Richard Dickerson and Horace Drew, Caltech; Gerard Bunick, University of Tennessee; Thomas Steitz, Yale University; Alfonso Mondragón, Northwestern University; Venki Ramakrishnan, MRC Laboratory of Molecular Biology, Cambridge, U.K.; Andrew Leslie and John Walker, MRC Laboratory of Molecular Biology, Cambridge, U.K.

Founded in 1807, John Wiley & Sons, Inc. has been a valued source of knowledge and understanding for more than 200 years, helping people around the world meet their needs and fulfill their aspirations. Our company is built on a foundation of principles that include responsibility to the communities we serve and where we live and work. In 2008, we launched a Corporate Citizenship Initiative, a global effort to address the environmental, social, economic, and ethical challenges we face in our business. Among the issues we are addressing are carbon impact, paper specifications and procurement, ethical conduct within our business and among our vendors, and community and charitable support. For more information, please visit our website: www.wiley.com/go/citizenship.

To order books or for customer service, please call 1-800-CALL WILEY (225-5945).

Student Companion (paperback)     ISBN-13   978-1-1182-1827-3

Printed in the United States of America

10  9  8  7  6  5  4  3  2

# Contents

Welcome to biochemistry! You are about to become acquainted with one of the most exciting scientific disciplines. The biotechnology industry, with its roots in molecular genetics, is one of the most visible manifestations of the explosion of biochemical knowledge that has occurred during our lifetime. Drug design and novel approaches such as gene therapy rely on the fundamental knowledge of the chemistry of biological molecules, particularly proteins. Our most common diseases (e.g., diabetes and heart disease) have pleiotropic multifaceted physiological effects that are best understood in terms of biochemistry. You will soon discover that biochemistry's impact on our lives cannot be over-emphasized. We are excited to bring you an ever-expanding understanding of this magnificent subject!

Learning biochemistry is not easy but it can be fun! Most students discover that biochemistry is a synthetic science, merging knowledge of general chemistry, organic chemistry, and biology. Hence, a more mature and creative kind of thinking is required to gain a deep understanding of biochemistry. In addition to a solid foundation in chemistry and biology, you will need to recognize and assimilate some general principles from other disciplines within biology, including physiology, genetics, and cell biology. In this respect, biochemistry is not all that different from nonscientific pursuits that require some degree of "lateral thinking" across disciplines.

This Student Companion accompanies *Fundamentals of Biochemistry Fourth Edition* by Donald Voet, Judith G. Voet, and Charlotte W. Pratt. It is designed to help you master the basic concepts and exercise your analytical skills as you work your way through the textbook. Each chapter of the Student Companion is divided into four parts, beginning with a general summary reminding you of the topics covered in that chapter. This is followed by a section called Essential Concepts, which provides an overview of the main facts and ideas that are essential for your understanding of biochemistry. This can be regarded as a set of brief notes for each chapter, alerting you to the key facts you need to commit to memory and to the concepts you need to master. You will soon notice that biochemical knowledge is cumulative: new concepts often rely on a solid understanding of previously presented concepts. Hence, one of the key goals of this Companion is to help you gain this understanding. The third and last section is the Questions. These are organized in a manner to help you gain a firm understanding of each section of a chapter. Some questions ask you to recall essential facts while others exercise your problem-solving skills. Answers to all of the questions are provided at the end of the Student Companion. However, you do yourself a great disservice by turning to them too soon. Don't know the answer right away? Keep trying! Go back to the text to find clues for yourself. Use the answers to check yours, not to fill in a temporary void in your understanding. The Solutions to all end of chapter problems in the text can be found after the Answers to Questions.

As a new addition to this edition of the Student Companion, we have added three new features to select chapters: Behind the Equations, Calculation Analogies and Play It Forward. Behind the Equations is a section that tackles key equations in Biochemistry. This section will provide insights and learning strategies for using and understanding what these equations are really telling us, and how to use them intuitively. The Calculation Analogies section will tackle key equations conceptually using simple analogies that will help students understand how to solve problems. The Play It Forward Section will take critical concepts from early chapters and integrate them into material in later chapters to help students see how the content continues to build upon itself, hopefully allowing students to synthesize Biochemistry as a discipline.

"How should I study biochemistry beyond reading the textbook and working in this Companion?" is a likely question from students. The phrase, "if you don't use it, you lose it" applies here. It is pointless to simply read your biochemistry textbook over and over. Unless you are actively engaged in working with the material, you become a passive reader. Active engagement includes using your hands to work problems, drawing pictures, or writing an outline or flow chart. As you read, ask yourself questions and seek answers from your text and your instructor. By doing these things, you use the material, and it becomes more efficiently transferred to your long-term memory.

"How often should I study biochemistry?" is also a common question. Most instructors agree that frequent short study sessions—even daily—will pay greater dividends than a single long session once a week. Because short-term memory lasts just a few minutes, take a few minutes after every class to review your notes. Similarly, stop reading your text, and review on paper with diagrams, word charts, flow diagrams, what you just read. Talk biochemistry with anyone who will listen. Form a study group to enrich your knowledge and test your memory. All these activities will result in the transfer of knowledge from short-term memory to long-term memory. In other words, the more you use biochemistry, the better you know it and the more fun you will have with it.

One of the truly most satisfying ways to learn biochemistry is to apply its principles and findings to problems that integrate your knowledge. To this end, Dr. Kathleen Cornely has developed numerous case studies that test your analytical skills. You will find these case studies on the *Fundamentals of Biochemistry 4e* Student Companion Site at **www.wiley.com/college/voet**. They can also be found in the *Fundamentals of Biochemistry 4e* WileyPLUS course **(www.wileyplus.com)**. Topics for the case studies were chosen to cover a range of interesting areas relevant to biochemistry. The cases themselves are based on data from research reports as well as clinical studies. The prerequisites include material that you are likely to be studying in biochemistry class, but occasionally include concepts from genetics and immunology on a level likely to be encountered in a first-year general biology course. The answers to the questions posed in these case studies can be obtained from your instructor.

We have many people to thank for helping us get this Companion to you. First and foremost, we would like to acknowledge Associate Editor, Ms. Aly Rentrop, and Production Editor, Ms. Sandra Dumas. In addition, we would like to acknowledge Ms. Petra Recter, Associate Publisher for Physics and Chemistry at John Wiley and Sons for her support of this project. We are also indebted to Caroline Breitenberger of The Ohio State University and Laura Mitchell of St. Joseph's University for their much-appreciated reviews of our original draft. We would also like to thank our students at Swarthmore College, the University of Houston, and the University of Houston-Downtown for pointing out errors and ambiguities in earlier drafts of this work. It is still possible that errors persist, and so we would greatly appreciate being alerted to them. Please forward you comments to Akif Uzman (uzmana@uhd.edu).

*Akif Uzman*

*Jerry Johnson*

*Joseph Eichberg*

*William Widger*

*Charlotte W. Pratt*

*Donald Voet*

*Judith G. Voet*

# Introduction to the Chemistry of Life

This chapter introduces you to life at the biochemical and cellular level. It begins with a discussion of the chemical origins of life and its early evolution. This discussion continues into ideas and theories about the evolution of organisms, followed by a brief introduction to taxonomy and phylogeny viewed from a molecular perspective. The chapter concludes with an introduction to the basic concepts of thermodynamics and its application to living systems. Biochemistry, like all other sciences, is a based on the measurement of observable phenomena. Hence, it is important to become familiar with the conventions used to measure energy and mass. Box 1-2 presents the essential biochemical conventions that we will encounter throughout *Fundamentals of Biochemistry*.

## *Essential Concepts*

### *The Origin of Life*

1.  Living matter consists of a relatively small number of elements, of which C, N, O, H, Ca, P, K, and S account for ~98% of the dry weight of most organisms (which are 70% water). These elements form a variety of reactive functional groups that participate in biological structure and biochemical reactions.

2.  The current model for the origin of life proposes that organisms arose from the polymerization of simple organic molecules to form more complex molecules, some of which were capable of self-replication.

3.  Most polymerization reactions involving the building of small organic molecules into larger more complex ones occur by the formation of water. This is called a condensation reaction.

4.  A key development in the origin of life was the formation of a membrane that could separate the critical molecules required for replication and energy capture from a potentially degradative environment.

5.  Complementary surfaces of molecules and macromolecules provide a template for biological specificity (e.g., macromolecular assembly, enzyme activity, and expression and replication of the genome).

6.  Modern cells can be classified as either prokaryotic or eukaryotic. Eukaryotic cells are distinguished by a variety of membrane-bounded organelles and an extensive cytoskeleton.

*Organismal Evolution*

7.   Prokaryotes show a limited range of morphologies but very diverse metabolic capabilities.

8.   Phylogenetic evidence based on comparisons of ribosomal RNA genes have been used by Woese and colleagues to group all organisms into three domains: archaea, bacteria, and eukarya.

9.   The evolution of sexual reproduction marks an important step of the evolution of organisms, because it allows for genetic exchanges that lead to an increase in the adaptability of a population of organisms to changing environments.

10.  Eukaryotes contain several membrane-bounded structures, such as mitochondria and chloroplasts, which may be descended from ancient symbionts.

11.  Archaea represent a third domain or branch of life in the three-domain system of classification. While they outwardly resemble bacteria, their genomes and the proteins encoded in them more closely resemble those of eukaryotes.

12.  Biological evolution is not goal-directed, requires some built-in sloppiness, is constrained by its past, and is ongoing.

13.  Natural selection directs the evolution of species.

*Thermodynamics*

14.  The first law of thermodynamics states that energy ($U$) is conserved; it can neither be created nor destroyed.

15.  Enthalpy ($H$) is a thermodynamic function that is a sum of the energy of the system and the product of the pressure and the volume ($PV$). Since biochemical processes occur at constant pressure and have negligible changes in volume, the change of energy of the system is nearly equivalent to the change in enthalpy ($\Delta U \approx \Delta H$).

16.  The second law of thermodynamics states that spontaneous processes are characterized by an increase in the entropy of the universe, that is, by the conversion of order to disorder.

17.  The spontaneity of a process is *determined by its free energy change ($\Delta G = \Delta H - T\Delta S$)*. Spontaneous reactions have $\Delta G < 0$ (exergonic) and nonspontaneous reactions have $\Delta G > 0$ (endergonic).

15.  For any process at equilibrium, the rate of the forward reaction is equal to the rate of the reverse reaction, and $\Delta G = 0$.

18.  Energy, enthalpy, entropy, and free energy are state functions; that is, they depend only on the state of the system, not its history. Hence, they can be measured by considering only the initial and final states of the system and ignoring the path by which the system reached its final state.

19. The entropy of a solute varies with concentration; therefore, so does its free energy. The free energy change of a chemical reaction depends on the concentration of both its reactants and its products.

20. For the general reaction

$$aA \ + \ bB \ \rightleftharpoons \ cC \ + \ dD$$

the free energy change is given by the following relationship:

$$\Delta G \ = \ \Delta G^\circ \ + \ RT \ln \left( \frac{[C]^c [D]^d}{[A]^a [B]^b} \right)$$

21. The equilibrium constant of a chemical reaction is related to the standard free energy of the reaction when the reaction is at equilibrium ($\Delta G = 0$) as follows:

$$\Delta G^\circ \ = \ -RT \ln K_{eq}$$

where $K_{eq}$ is the equilibrium constant of the reaction:

$$K_{eq} \ = \ \frac{[C]^c_{eq} [D]^d_{eq}}{[A]^a_{eq} [B]^b_{eq}}$$

The equilibrium constant can therefore be calculated from standard free energy data and *vice versa*.

22. The equilibrium constant varies with temperature by the relationship:

$$\ln K_{eq} \ = \ \frac{-\Delta H^\circ}{R} \left( \frac{1}{T} \right) + \frac{\Delta S^\circ}{R}$$

where $\Delta H^\circ$ and $\Delta S^\circ$ represent enthalpy and entropy in the standard state. A plot of $K_{eq}$ versus $1/T$, known as a van't Hoff plot, permits the values of $\Delta H^\circ$ and $\Delta S^\circ$ (and therefore $\Delta G^\circ$ at any temperature) to be determined from measurements of $K_{eq}$ at two (or more) temperatures.

23. The biochemical standard state is defined as follows: The temperature is 25°C, the pH is 7.0, and the pressure is 1 atm. The activities of reactants and products are taken to be the total activities of all their ionic species, except for water, which is assigned an activity of 1. $[H^+]$ is also assigned an activity of 1 at the physiologically relevant pH of 7. These conditions are different than the chemical standard state, so that the biochemical standard free energy is designated as $\Delta G^{\circ\prime}$ and in the chemical standard state is $\Delta G^\circ$. We assume that activity equals molarity for dilute solutions.

4 Chapter 1 Introduction to the Chemistry of Life

24. An isolated system cannot exchange matter or energy with its surroundings. A closed system can exchange only energy with its surroundings. A closed system inevitably reaches equilibrium. Open systems exchange both matter and energy with their surroundings and therefore cannot be at equilibrium. Living organisms must exchange both matter and energy with their surroundings and are thus open systems. Living organisms tend to maintain a constant flow of matter and energy, referred to as the steady state.

25. Living systems can respond to slight perturbations from the steady state to restore the system back to the steady state.

26. The recovery of free energy from a biochemical process is never total, and some energy is lost to the surroundings as heat. Hence, while the system becomes more ordered, the surroundings experience an increase in entropy.

27. Enzymes accelerate the rate at which a biochemical process reaches equilibrium. They accomplish this by interacting with reactants and products to provide a more energetically favorable pathway for the biochemical process to take place.

## Key Equations

Be sure to know the conditions for which the following thermodynamics equations apply and be able to interpret their meaning.

1. $\Delta U = q - w$

2. $H = U + PV$

3. $\Delta H = q_P - w + P\Delta V$

4. $S = k_B \ln W$

5. $\Delta S \geq \dfrac{q}{T} = \dfrac{\Delta H}{T}$

6. $\Delta G = \Delta H - T\Delta S$

7. $\Delta G = \Delta G° + RT \ln \left( \dfrac{[C]^c [D]^d}{[A]^a [B]^b} \right)$

8. $\Delta G° = -RT \ln K_{eq}$

9. $K_{eq} = \dfrac{[C]^c_{eq} [D]^d_{eq}}{[A]^a_{eq} [B]^b_{eq}}$

10. $\ln K_{eq} = \dfrac{-\Delta H°}{R} \left( \dfrac{1}{T} \right) + \dfrac{\Delta S°}{R}$

**Calculation Analogy for Determining Gibbs Free Energy for a Chemically Defined System**

The change in Gibbs Free Energy ($\Delta G$) is used to determine whether or not a process is spontaneous or non-spontaneous under a given set of conditions (state of a thermodynamic system).

For a thermodynamic system that is a chemical process such as:

$$aA + bB \rightleftharpoons cC + dD$$

The $\Delta G$ of this process can be determined using equation 7 (see *Key Equations* above):

$$\Delta G = \Delta G^\circ + RT \ln \left( \frac{[C]^c [D]^d}{[A]^a [B]^b} \right)$$

The typical student approaches a problem utilizing this equation with a 'plug and chug' mentality, without understanding exactly what the equation is trying to tell them.

Ultimately, this equation is informing us how far a thermodynamic system is from equilibrium ($\Delta G = 0$). A simple mathematical analogy can be used to demonstrate this concept and provide a visual perspective when solving problems requiring the use of this equation.

Consider a number line used in mathematics:

$$\underline{\hspace{2cm}|\hspace{2cm}}$$
*0*

Number lines are constructed with a particular reference in mind; here, the number zero. Any system located on the right-hand side of zero on the number line is characterized by a positive number and would represent a system that is non-spontaneous under the specified conditions.  Any system located on the left-hand side of zero on the number line is characterized by a negative number and would represent a system that is spontaneous under the specified conditions. Since a system at thermodynamic equilibrium has a $\Delta G = 0$, then any value of $\Delta G \neq 0$ would represent a thermodynamic system not at equilibrium.

We could use a number line to plot the position of any thermodynamic system not at equilibrium in reference to thermodynamic equilibrium ($\Delta G = 0$). In doing so, we could physically measure the distance between zero and the position of the thermodynamic system that is not at equilibrium using a ruler. You could plot the number $-30$ on the number line, and this position would correspond to a thermodynamic system with a $\Delta G = -30$ kJ•mol$^{-1}$:

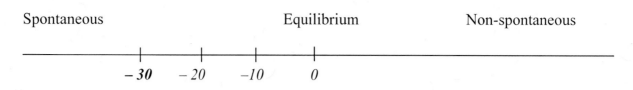

You could use a ruler to measure the distance between the system at that state ($\Delta G = -30$ kJ•mole$^{-1}$) and thermodynamic equilibrium ($\Delta G = 0$). Thus, it is convenient and accurate to say that $\Delta G$ is a measure of how far a thermodynamic system is from equilibrium. While we cannot use an actual ruler to gain an appreciation of this in real life, the further a thermodynamic system is from equilibrium, the more useful work that system is capable of performing.

Solving a $\Delta G$ equation for a chemical system using equation (1) is analogous to using a number line. In order to determine the $\Delta G$ of a process, we need to know two pieces of information about that system (or reference points).

Equation (1) provides you with that information. Ultimately, the $\Delta G$ equation can be divided into two mathematical terms: the "$\Delta G°$" term and the "$RT\ln([C]^c[D]^d/[A]^a[B]^b)$" term. These two mathematical terms are your reference points!

$\Delta G°$ is the $\Delta G$ of the system at the chemical standard state (25° C, 1 atm, and all reactants with an activity of 1). This represents the position of the reaction under the defined conditions. It is an arbitrary state that we can now use a reference point.

The second mathematical term of equation (1) is the current state of the thermodynamic system as defined by the variables of the problem you are trying to solve. Specifically, it provides the temperature and current molar concentrations of all reacting species in the thermodynamic system. This provides a second reference point. Using these two pieces of information, you now can calculate the $\Delta G$ of that system, and gain an appreciation for how far that system is from thermodynamic equilibrium!

### Questions

*The Origin of Life*

1. What are the elements that account for 98% of the dry weight of living cells?

2. In what critical ways was the atmosphere of the primitive earth different than the earth's current atmosphere?

3. What is the rationale and significance of the Miller and Urey experiments?

4. What was, and continues to be, the source of atmospheric oxygen?

5.    Examine the reaction shown below for the condensation and hydrolysis of lactose (two covalently linked sugars, a disaccharide). Circle the functional groups that will form water during the condensation.

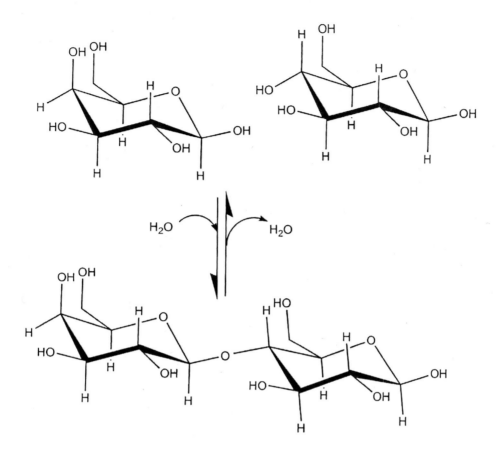

6.    The condensation of two functional groups can result in the formation of another common functional group, which can be referred to as a compound functional group. Examine the functional groups in Table 1-2. Which compound functional groups are the combination of two other functional groups found in that figure? Show how these compound functional groups form.

## Cellular Architecture

7.    Draw a schematic diagram of an eukaryotic cell showing its principal organelles.

8.    Give the principal distinguishing feature(s) of each pair of terms:
      (a) Prokaryote and eukaryote
      (b) Cytosol and cytoplasm
      (c) Endoplasmic reticulum and cytoskeleton

9.    Living organisms are classified into three domains: _____,

_____ , and _____ .

10.   What is the most compelling evidence that mitochondria and chloroplasts represent descendents of symbiotic bacteria that lived inside of ancient eukaryotic cells?

11.   What is the relationship between mutation and genetic variation in a population of organisms? Of what significance is it to evolution?

## Thermodynamics

12.   Distinguish between enthalpy and energy. Under what conditions are they equivalent?

13.   What does it mean when $q$ and $w$ are positive?

14.   When crystalline urea is dissolved in water, the temperature of the solution drops precipitously. Does the enthalpy of the system increase or decrease? Explain.

15.   List and define the four major thermodynamic state functions.

16.   Which of the following pairs of states has higher entropy?
      (a)   Two separate beakers of NaCl and KCH$_3$COO in solution and a beaker containing a solution of both salts.
      (b)   A set of dice in which all the dice show 6 dots on the top side and a set of dice in which the 6's show up on one of the side faces.
      (c)   A small symmetric molecule that can form a polymer through reaction at either end, and a small asymmetric molecule that can polymerize from only one end.

17.   Rationalize the temperature dependence of Gibbs free energy changes when both the enthalpy and entropy terms are positive values or when they are both negative values.

18.   Hydrogen gas combines spontaneously with oxygen gas to form water

$$2\,H_2 + O_2 \rightarrow 2\,H_2O$$

Which term, enthalpy or entropy, predominates in the equation for the Gibbs free energy? How are the surroundings affected by this reaction?

19.   Evaluate the following statement: Enzymes accelerate the rate of a reaction by increasing the spontaneity of the reaction.

20.   Based on your reading in this chapter, suggest simple criteria for a reasonable definition of life.

21.   The breakdown of glucose to pyruvate utilizes the free energy of 2 ATP molecules; however, the synthesis of glucose from pyruvate requires 4 ATP. Using the basic ideas of thermodynamics presented in this chapter, provide a thermodynamic explanation. Hint: Think about the relationship $\Delta G = \Delta H - T\Delta S$ in your argument.

# Water

This chapter introduces you to the unique properties of water and to acid–base reactions. The discussion of water begins with a look at its structure and how its polarity provides a basis for understanding its powers as a solvent. You are then introduced to the hydrophobic effect, osmosis, and diffusion. The chemical properties of water are then described, beginning with the ionization of water, which sets the stage for a discussion of acid–base chemistry and the behavior of weak acids and buffers. This discussion includes the Brønsted–Lowry definition for acids and bases, the definition of pH, and the derivation of the Henderson–Hasselbalch equation. As we shall see in subsequent chapters, a solid understanding of acid–base equilibria is fundamental to understanding key aspects of amino acid biochemistry, protein structure, enzyme catalysis, transport across membranes, energy metabolism, and other metabolic transformations.

## Essential Concepts

1.  Water is essential to biochemistry because:
    (a)  Biological macromolecules assume specific shapes in response to the chemical and physical properties of water.
    (b)  Biological molecules undergo chemical reactions in an aqueous environment.
    (c)  Water is a key reactant in many reactions, usually in the form of $H^+$ and $OH^-$ ions.
    (d)  Water is oxidized in photosynthesis to produce molecular oxygen, $O_2$, as part of the process that converts the sun's energy into usable chemical form. Expenditure of that energy under aerobic conditions leads to the reduction of $O_2$ back to water.

### Physical Properties of Water

2.  The structure of water closely approximates a tetrahedron with its two hydrogen atoms and the two lone pairs of electrons of its oxygen atom "occupying" the vertices of the tetrahedron.

3.  The high electronegativity of oxygen relative to hydrogen results in the establishment of a permanent dipole in water molecules.

4.  The polar nature of water results in negative portions of the molecule being attracted to the positive portions of neighboring water molecules by a largely electrostatic interaction known as the hydrogen bond.

5.  Hydrogen bonds are represented as D—H$\cdots$A, where D—H is a weakly acidic compound so that the hydrogen atom (H) has a partial positive charge, and A is a weakly basic group

that bears lone pairs of electrons. A is often an oxygen atom or a nitrogen atom (occasionally sulfur).

6.   Water is strongly hydrogen bonded, with each water molecule participating in four hydrogen bonds with its neighbors; two in which it donates and two in which it accepts. Hydrogen bonds commonly form between water molecules and the polar functional groups of biomolecules, or between the polar functional groups themselves.

7.   The strongly hydrogen bonded character of water is responsible for many of its characteristic properties, most notably:
   (a)   A high heat of fusion, which allows water to act as a heat sink, such that greater heat loss is necessary for the freezing of water compared to other substances of similar molecular mass.
   (b)   A high heat of vaporization, such that relatively more heat must be input to vaporize water compared to other substances of similar molecular mass.
   (c)   An ability to dissolve most polar compounds.
   (d)   An open structure makes ice less dense than liquid water, thereby making ice float, insulating the water beneath it, and inhibiting total freezing of large bodies of water.

8.   A variety of weak electrostatic interactions are critical to the structure and reactivity of biological molecules. These interactions include, in order of increasing strength, London dispersion forces, dipole–dipole interactions, hydrogen bonds, and ionic interactions (see Table 2-1).

9.   Water is an excellent solvent of polar and ionic substances due to its property of surrounding polar molecules and ions with oriented shells of water, thereby attenuating the electrostatic forces between these molecules and ions.

10.   The tendency of water to minimize its contact with nonpolar (hydrophobic) molecules is called the hydrophobic effect. This effect is largely driven by the increase in entropy caused by the necessity for water to order itself around nonpolar molecules. This causes the nonpolar molecules to aggregate, thereby reducing the surface area that water must order itself about. Consequently, nonpolar substances are poorly soluble in aqueous solution.

11.   Many biological molecules have both polar (or charged) and nonpolar functional groups and are therefore simultaneously hydrophilic and hydrophobic. These molecules are said to be amphiphilic or amphipathic.

12.   Osmosis is the movement of solvent across a semipermeable membrane from a region of lower concentration of solute to a region of higher concentration of solute. Osmotic pressure of a solution is the pressure that must be applied to the solution to prevent an inflow of solvent. Hence, an increase in solute concentration results in an increase in osmotic pressure.

13. Diffusion is the random movement of molecules in solution (or in the gas phase). It is responsible for the movement of solutes from a region of high concentration to a region of low concentration.

## Chemical Properties of Water

14. Water is a neutral, polar molecule that has a slight tendency to ionize into $H^+$ and $OH^-$. However, the proton is never free and binds to a water molecule to form $H_3O^+$ (hydronium ion).

15. The ionization of water is described as an equilibrium between the unionized water (reactant) and its ionized species (products)

$$H_2O \rightleftharpoons H^+ + OH^-$$

In which

$$K = \frac{[H^+][OH^-]}{[H_2O]}$$

Since in dilute aqueous solution, the concentration of water is essentially constant (55.5 M), the concentration of H2O is incorporated into the value of K, which is referred to as Kw, the ionization constant of water.

$$K_w = [H^+][OH^-]$$

16. The values of both $H^+$ and $K$ are inconveniently small; hence, their values are more conveniently expressed as negative logarithms, so that

$$pH = -\log [H^+]$$

$$pK = -\log K$$

17. According to the Brønsted–Lowry definition, an acid is a substance that can donate a proton, and a base is a substance that can accept a proton. The strength of a weak acid is proportional to its dissociation constant, which is expressed as

$$K = \frac{[H^+][A^-]}{[HA]}$$

18. The pH of a solution of a weak acid is determined by the relative concentrations of the acid and its conjugate base. The equilibrium expression for the dissociation of a weak acid can be rearranged to

$$pH = pK + \log \frac{[A^-]}{[HA]}$$

This relationship is known as the Henderson–Hasselbalch equation. When the concentration of a weak acid is equal to the concentration of its conjugate base, pH = p$K$. Hence, the stronger the acid, the lower its p$K$ (see Table 2-4).

19. Solutions of a weak acid at pH's near its p$K$ resist large changes in pH as OH$^-$ or H$^+$ is added. Added protons react with the weak acid's conjugate base to reform the weak acid; whereas added OH$^-$ combines with the acid to form its conjugate base and water. A solution of a weak acid and its conjugate base (in the form of a salt) is referred to as a buffer. Buffers are effective within 1 pH unit of the p$K$ of the component acid.

## Key Equations

Henderson–Hasselbalch equation:

$$pH = pK + \log \frac{[A^-]}{[HA]}$$

## Behind the Equations

### Which acids are defined by a pK value?

The key to understanding which acids are defined by a p$K$ value (here we use an abbreviation for p$K_a$, the dissociation constant for a weak acid) and which are not is understanding the behavior of strong and weak acid chemistry at equilibrium. Ultimately, only weak acids are defined by $K$ values and thus p$K$ values. This can be demonstrated both conceptually and mathematically. Conceptually, strong acids are defined as those that dissociate completely in aqueous solution:

$$HA \longrightarrow H^+ + A^-$$

The unidirectional arrow above implies that as the solution of strong acid approaches equilibrium, the molar concentration of HA becomes smaller and smaller, approaching zero. Once the strong acid has reached equilibrium, there is virtually no HA remaining in solution.

Mathematically, the equilibrium acid dissociation constant is defined as:

$$K = \frac{[H^+][A^-]}{[HA]} \qquad \text{(Text Equation 2–5)}$$

And the p$K$ is defined as:

$$pK = -\log K \qquad \text{(Text Equation 2–6)}$$

If a strong acid dissociates completely in aqueous solution, the concentration of HA, and thus the denominator of Equation 2 approaches the limit of zero. As the denominator of a fraction becomes smaller and smaller, the value of that fraction becomes larger and larger, approaching the limit of infinity. Since Equation 2-5 refers to concentrations at equilibrium, the denominator of Equation 2-5 would be 0, and the value of $K$ is therefore undefined. Thus, it is not appropriate to define a strong acid with a $K$ value.

Because $K$ approaches infinity, p$K$ would approach negative infinity. Again, this makes it inappropriate for describing the strength of a strong acid.

However, the situation is very different for weak acids. Weak acids are defined as those acids that are incompletely dissociated at equilibrium in aqueous solution:

$$HA \rightleftharpoons H^+ + A^-$$

The bidirectional arrow above implies that at equilibrium, both reactants and products will be present in measurable amounts Thus, the value of $K$ will not be undefined, and the value of p$K$ will not approach negative infinity. This allows weak acids to have measurable values of both $K$ and p$K$.

### *What does p$K$ really tell us about a weak acid?*

The value of p$K$ provides a basis for comparing the strengths of weak acids. Since $K$ is an equilibrium dissociation constant, it indicates the tendency and extent to which a weak acid will dissociate to a proton and the conjugate base of the weak acid at equilibrium. The larger the value of $K$, the larger the value of the numerator in Equation 2-5. The larger the value of $K$, the smaller the value of p$K$ (Equation 2–6).

In order to understand the concept of p$K$, one must realize that the p$K$ value is a relative measure of the affinity a weak acid has for its proton. By definition, affinity is the attraction between two things, in this case, the attractive force between the conjugate base of a weak acid and the proton. Consider the following analogy: two men are walking down the street. The first man has $100 in his wallet and is a miser with his money, while the second man has $100 in his wallet and is well known for his generosity.

Which of the two men has a higher affinity for (or desire to hold onto) his $100? Naturally, the first man will have a higher affinity for his $100, and is less likely to give up his money compared to the second man. The first man is analogous to a weaker acid while the second man is analogous to a stronger acid. A weaker acid holds on to a proton with higher affinity compared to a stronger acid.

In other words, a stronger acid has a lower proton affinity and a lower p$K$ value, whereas a weaker acid has a higher proton affinity and a higher p$K$ value. As a result, the numerator of Equation 2-5 and the ultimate value of $K$ for a stronger acid will be larger than the numerator of Equation 2-5 and value of $K$ for a weak acid. Finally, a weak acid must have a lower p$K$ value than a weaker, weak acid. Relatively stronger acids have low proton affinities and lower p$K$ values while weaker acids have higher proton affinities and higher p$K$ values.

This can be summarized as:

**The strength of a weak acid is inversely proportional to the affinity of that acid for its proton and inversely proportional to the p*K* of that acid.**

*How can we use this information when solving problems with the Henderson – Hasselbalch equation?*

The Henderson – Hasselbalch equation describes the relationship between the p*K* of a weak acid, the pH of a solution, and the ratio of the molar concentrations of the weak acid and conjugate base at equilibrium. Two important relationships become evident when one looks at two different scenarios with the same weak acid. Consider the **Sample Calculation 2-2** in the textbook:

| | |
|---|---|
| *Calculate the pH of a 2 L solution containing 10 mL of 5 M acetic acid and 10 mL of 1 M sodium acetate.* | |
| *First calculate the concentrations of the acid and conjugate base, expressing all concentrations in units of moles per liter.* | |
| *Acetic acid: (0.010 L)(5M) / (2L) = 0.025 M* | |
| *Sodium acetate: (0.010 L)(1 M) / (2L) = 0.0050M* | |
| *Substitute the concentrations of the acid and conjugate base into the Henderson-Hasselbalch equation. Find the pKa for acetic acid in Table 2-4.* | |
| *pH = pKa + log([acetate]/[acetic acid])* | |
| *pH = 4.76 + log (0.0050/0.025)* | |
| *pH = 4.76 – 0.70* | |
| *pH = 4.06* | |
| The first important relationship is revealed if we alter the problem slightly. What would the pH of the solution be if the concentration of acetate was equal to the concentration of acetic acid? | |
| *pH = 4.76 + log (0.025/0.025)* | |

In this scenario, the ratio of the conjugate base to acid is equal to 1.0. The log of 1.0 = 0, so this equation reduces to:

$$pH = pK$$

If the concentrations of the weak acid and conjugate base are equal, then the pH of the solution is equal to the $pK$ of the weak acid. One of the first steps you should employ when solving Henderson – Hasselbalch problems is to look at the molar ratio of the conjugate base to the weak acid: the closer these values are to each other, the closer the pH of the environment will be to the pKa of the weak acid. In fact, because the Henderson – Hasselbalch equation is logarithmic, **for every order of magnitude 10-fold difference between the concentrations of the conjugate base and weak acid at equilibrium, the pH will differ by 1 unit from the p$K$.** Notice that in the original Sample Calculation 2-2 from the textbook, the molar concentrations of the conjugate base and weak acid were not even 10-fold different from each other, and thus the pH of the solution at equilibrium was not even 1 pH unit different from the p$K$ value (pH = 4.06 vs. p$K$ = 4.76).

The second relationship has to do with the relative concentrations of acid and conjugate base depending on the values of pH and p$K$. For the original Sample Calculation 2-2, the pH of the solution is lower than the p$K$ of the acid. Notice that when the pH of the solution is lower than the value of the p$K$ of the weak acid, the ratio of conjugate base to acid favors the acid form (HA), or the denominator of the equation.

Now consider what would happen if the pH were instead equal to 5.50 in this scenario? What would the ratio of conjugate base to acid be then? Which molecular species predominates? Which term in Equation 2-5 is larger?

Application of the Henderson – Hasselbach equation yields (where X refers to ratio of the concentrations of base to acid)

$$5.50 = 4.76 + \log X$$

$$5.50 - 4.76 = \log X$$

$$10^{0.74} = X$$

$$X = 5.50$$

Notice now that when the pH of the solution is higher than the p$K$ of the acid, then the ratio of conjugate base-to-acid favors the conjugate base (A⁻), or the numerator of the equation.

Understanding this relationship between pH, p$K$ and the ratio of conjugate base to acid is key to predicting the correct outcomes of Henderson – Hasselbalch problems and can be summarized as:

**When the pH of the solution is lower than the p$K$ of the weak acid, then the weak acid is mostly protonated (in the HA form), and the denominator of the conjugate base to weak acid fraction in the Henderson – Hasselbalch equation predominates.**

**When the pH of the solution is higher than the p$K$ of the weak acid, then the weak acid will be mostly deprotonated (in the A⁻ form), and the numerator of the conjugate base to weak acid fraction in the Henderson – Hasselbalch equation predominates.**

## *Applying these concepts*
Consider two organic, weak, monoprotic acids: acetic acid and formic acid.

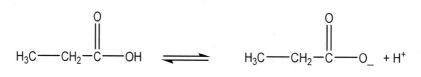

Acetic acid: p$K$ = 4.76

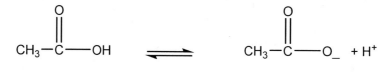

Formic acid: p$K$ = 3.75

1.  Which of these two acids is the stronger acid, and which has the higher affinity for protons?

2.  Consider a scenario where a solution of an unknown amount acetic acid and sodium acetate has a pH 1.75. Without using a calculator:
    (a)  Predict which form of the acetic acid will predominate in solution?
    (b)  Estimate the ratio of conjugate base-to-weak acid form be.
    (c)  Now use a calculator to prove that your answers to parts a and b are correct.

3.  Consider a scenario where 10 millimoles of formic acid and 250 millimoles of sodium formate are dissolved in 0.75 liters of water. Without using a calculator:
    (a)  Predict whether the pH of the resulting solution will have a pH value higher or lower than the pKa of formic acid.
    (b)  Approximate the pH value of the solution.
    (c)  Use a calculator to now prove that your answers to parts a and b are correct.

4.  For problem #12 in the textbook, predict whether the pH will be lower or higher than the pK of $KH_2PO_4$ and by approximately how much without using a calculator.

## Questions

### Physical Properties of Water

1.  Draw a 3-dimensional structure of a water molecule, including any lone pairs of electrons, and indicate the dipole moment. What is the name of the geometrical figure have you drawn?

2.  Which gas in each of the following pairs would you expect to be more soluble in water? Why?
    (a)  oxygen and carbon dioxide
    (b)  nitrogen and ammonia
    (c)  methane and hydrogen sulfide

3.  Why do bottles of beer break when placed in a freezer?

4.  Distinguish between hydrophilic, hydrophobic, and amphipathic substances.

5.  Mixing olive oil with vinegar creates a salad dressing that is an emulsion, a mixture of vinegar with many tiny oil droplets. However, in a few minutes, the olive oil separates entirely from the vinegar. Describe the changes in entropy that occur during the initial mixing and the subsequent separation of the olive oil and vinegar.

6.  Below is a beaker that contains a solution of 0.1 M NaCl. Floating inside is a cellulose bag that is permeable to water and small ions like $Na^+$ and $Cl^-$ but is impermeable to protein. The bag contains 1 M NaCl and 0.1 M protein. Describe the movement of solutes and solvent.

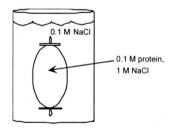

## Chemical Properties of Water

7.   In the most general definition of an acid and a base, a Lewis acid is a compound that can accept an electron pair, and a Lewis base is a compound that can donate an electron pair. Which of the following compounds can be classified as Brønsted–Lowry acids and bases and which as Lewis acids and bases?

$$CO_2BF_3, \quad NH_3, \quad Zn^{2+}, \quad HCOOH, \quad NH_4^+, \quad OH^-, \quad Cl^-, \quad H_2O$$

8.   Define p$K$ and pH, and write the Henderson–Hasselbalch equation that relates the two.

9.   The p$K$'s of trichloroacetic acid and acetic acid are 0.7 and 4.76, respectively. Which is the stronger acid? What is the dissociation constant of each?

10.   What is the $[OH^-]$ in a 0.05 M HCl solution? What is the pH?

11.   A solution of 0.1 M HCl has a pH of 1. A solution of 0.1 M acetic acid has a pH of 2.8. How much 1 M NaOH is needed to titrate a 100 mL sample of each acid to its respective equivalence point? Hint: It may be useful to review Le Chatelier's principle.

12.   A 0.01 M solution of a weak acid in water is 0.05% ionized at 25°C. What is its p$K$?

13.   What is a buffer? How does it work? What compounds act as buffers in cells?

14.   What is the pH of a 0.1 M solution of cacodylic acid (p$K$ = 6.27)?

15.   How would the pH of a 0.1 M solution of acetic acid be affected by the addition of a 4.5 M solution of sodium acetate (NaOAc)? NaOAc is a relatively strong base (weak acids have strong conjugate bases). Hint: Consider Le Chatelier's principle.

16.   A 0.02 M solution of lactic acid (p$K$ = 3.86) is mixed with an equal volume of a 0.05 M solution of sodium lactate. What is the pH of the final solution?

17.   You need a KOAc solution at pH 5, which is 3 M in $K^+$. Such a solution is used in bacterial plasmid DNA isolation. How many moles of KOAc and acetic acid (HOAc) do you need to make 500 mL of this solution?

18.   What are the predominant phosphate ions in a neutral pH phosphate buffer?

19.   A beaker of pure distilled water sitting out on your lab bench is slightly acidic. Explain. Hint: see Box 2-2.

20.   What is the pH of a solution of 10-12M of HCl? Explain.

21. Mammalian erythrocytes contain the enzyme carbonic anhydrase, which catalyzes the equilibrium CO2 + H2O ⇆ HCO3- + H+ with no carbonic acid intermediate. Hence, we have the following acid-base equilibrium:

$$CO_2 + H_2O \leftrightarrows HCO_3^- + H^+$$

(a) Describe how this equilibrium is affected in the lungs. What happens to blood pH in the lungs?

What process decreases pH in the peripheral tissues?

## *Graphical Analysis Questions*

Consider this modified version of Table 2–4 from the textbook containing two added columns of data:

| Acid | Ka | pKa | pH of environment | [A⁻]/[HA] |
|---|---|---|---|---|
| Oxalic acid | $5.37 \times 10^{-2}$ | 1.27 (pk$_1$) | 3.25 | |
| $H_3PO_4$ | $7.08 \times 10^{-3}$ | 2.15 (pk$_1$) | | 100/1 |
| Formic acid | $1.78 \times 10^{-4}$ | 3.75 | 3.75 | |
| Succinic acid | $6.17 \times 10^{-5}$ | 4.21 (pk$_1$) | 1.21 | |
| Oxalate⁻ | $5.37 \times 10^{-5}$ | 4.27 (pk$_2$) | | 10,000/1 |
| Acetic acid | $1.74 \times 10^{-5}$ | 4.76 | | 1/10 |
| Succinate⁻ | $2.29 \times 10^{-6}$ | 5.64 (pk$_2$) | 3.65 | |
| MES | $8.13 \times 10^{-7}$ | 6.09 | 7.1 | |
| $H_2CO_3$ | $4.47 \times 10^{-7}$ | 6.35 (pk$_1$) | | 1/100 |
| PIPES | $1.74 \times 10^{-7}$ | 6.76 | | 1/10,000 |
| $H_2PO_4^-$ | $1.51 \times 10^{-7}$ | 6.82 (pk$_2$) | 10.80 | |
| MOPS | $7.08 \times 10^{-8}$ | 7.15 | 1.15 | |
| HEPES | $3.39 \times 10^{-8}$ | 7.47 | | 1,000/1 |
| Tris | $8.32 \times 10^{-9}$ | 8.08 | 5.1 | |
| $NH_4^+$ | $5.62 \times 10^{-10}$ | 9.25 | | 1/100,000 |
| Glycine (amino group) | $1.66 \times 10^{-10}$ | 9.78 | 2.8 | |
| $HCO_3^-$ | $4.68 \times 10^{-11}$ | 10.33 (pk$_2$) | | 1/1 |
| Piperidine | $7.58 \times 10^{-12}$ | 11.12 | 5.1 | |
| $HPO_4^{2-}$ | $4.17 \times 10^{-13}$ | 12.38 (pk$_3$) | | 1/1,000 |

The fourth and fifth columns of the table indicate that the weak acid in a particular row is poised to the pH value listed in the fourth column, or alternatively, poised to an unknown pH value that results in the ratio of [A⁻]/[HA] listed in the fifth column.

1. Are the weak acids listed in this table in order from top to bottom as strongest (at the top) to weakest (at the bottom) or weakest to strongest? Why?

2. Which weak acid on the table has the highest affinity for protons? Which weak acid on the table has the lowest affinity for protons?

3.   If the range of physiological pH of the cell cytoplasm is 7.1 – 7.8, which of the weak acids in the above table, when formulated as a buffer, are the most appropriate buffers to use for experiments which need to control pH near the pH range of the cytoplasm? Why?

4.   The last two columns of the table (labeled 'pH of environment' and '[A–]/[HA]') have been partially completed for you. The 'pH of environment' column lists a hypothetical pH at which the weak acid is poised, while the '[A–]/[HA]' is the ratio of conjugate base-to-weak acid at that particular pH. Without using a calculator, fill in the remaining cells of the table.

### Play It Forward: Thermodynamics of Sucrose Solubility

Have you ever tried to dissolve table sugar (sucrose) in iced tea? It can take a significant amount of time. There is an interesting experiment that you can perform in your own home to illustrate how thermodynamic state functions ($\Delta G$, $\Delta H$, and $\Delta S$) influence the speed and extent at which sucrose dissolves in aqueous solutions.

To illustrate the point more clearly, try to find a very thin-walled glass from your kitchen. This would be a glass where the glass is so thin you can easily feel temperature changes of liquids inside of the glass. Do not use a plastic cup.  Fill the glass up approximately 60 – 70% full with tap or bottled water. Now add a significant (6 – 8 tablespoon full) amount of sucrose to the water and stir it with a spoon to try to get the table sugar to fully dissolve in the glass of water at room temperature. Two observations can be made while doing this:
   (a)   It takes a significant amount of time for all of the sugar granules to become completely solubilized.
   (b)   As the sugar goes into solution, the glass containing the sugar-water solution becomes cold to the touch, and may even form condensation on the outside.

Think about the content presented in chapters 1 and 2 of the textbook in order to answer these questions.

1.   Is the process of dissolving sucrose in water spontaneous or non-spontaneous under the described conditions and why?

2.   What purpose does stirring have in the process of dissolving sucrose in water? Is it absolutely necessary?

3.   What other strategy could one use other than stirring to increase the ability of the sucrose to go into solution?

4.   If the glass gets cold to the touch as the dissolving process occurs, what can we say about the enthalpy changes ($\Delta H$) as the process occurs?

5.   What could be happening at the molecular level to account for the endothermic nature of this process? Try to be as complete as possible in your answer.

6.   Given your answer to question #4, what must be true about the magnitude and value of $T\Delta S$ and $\Delta G$ for this process?

# 3

# Nucleotides, Nucleic Acids, and Genetic Information

This chapter introduces you to the structure and function of nucleotides and their polymers, ribonucleic acid (RNA) and deoxyribonucleic acid (DNA). The chapter begins with a discussion of the various kinds of nucleotides and the large variety of their functions in cellular processes. The nucleic acid polymers, RNA and DNA, are the primary players in the storage, transmission, and decoding of the genetic material. Scientists use a variety of powerful techniques to characterize and manipulate DNA from any organism. This chapter discusses the sequence-specific cleavage of DNA by restriction endonucleases; DNA sequencing; amplification of DNA by cloning in unicellular organisms such as bacteria and yeast; and the *in vitro* amplification of DNA by the polymerase chain reaction.

## Essential Concepts

### Nucleotides

1.  Of the four major classes of biological molecules (amino acids, sugars, lipids, and nucleotides), nucleotides are the most functionally diverse. They are involved in energy transfer, catalysis, and signaling within and between cells, and are essential for the storage, decoding, and transmission of genetic information.

2.  Nucleotides are composed of a nitrogenous base linked to a ribose sugar to which at least one phosphate group is attached. The eight common nucleotides, which are the monomeric units of RNA and DNA, contain the bases adenine, guanine, cytosine, thymine, and uracil.

3.  The best known nucleotide is the energy transmitter adenosine triphosphate (ATP), which is synthesized from adenosine diphosphate (ADP). Transfer of one or two of the phosphoryl groups of ATP is an exergonic process whose free energy can be used to drive an otherwise nonspontaneous process.

### Introduction to Nucleic Acid Structure

4.  Nucleic acids are polymers of nucleotides in which phosphate groups link the 3' and 5' positions of neighboring ribose residues. This linkage is called a phosphodiester bond because the phosphate is esterified to the two ribose groups. The phosphates are acidic at biological pH and so the polynucleotide is a polyanion.

5.  Nucleic acids are inherently asymmetric, so that one end (with a 5' phosphate) is different form the other end (with a 3' hydroxyl). This asymmetry, or polarity, is critical for the information storage function of nucleic acids. In fact, all linear biological molecules show this kind of polarity.

6. The structure of DNA was determined by Francis Crick and James Watson in 1953. Key information used to build their model included the following:
   (a) DNA has equal numbers of adenine and thymine residues and equal numbers of cytosine and guanine residues (called Chargaff's rules).
   (b) X-Ray diffraction studies (principally by Rosalind Franklin) revealed that the polymer is helical with a uniform width.
   (c) Structural studies had indicated that the nitrogenous bases should assume the keto tautomeric form.
   (d) Chemical evidence indicated that the polymer was linked by phosphodiester bonds between the 3' and the 5' carbons of adjoining ribose units.

7. The Watson–Crick model has the following major features:
   (a) Two polynucleotide chains wind around a central axis to form a right-handed helix.
   (b) The polynucleotide chains are antiparallel to each other.
   (c) The nitrogenous bases occupy the core of the double helix, while the sugar–phosphate chains are the backbones of DNA, running along the outside of the helical structure.
   (d) Adenine is hydrogen bonded to thymine to form a planar base pair (like the rung of a ladder). In a similar fashion, cytosine is hydrogen bonded to guanine.

8. The complementary strands of DNA immediately suggest that each strand of DNA can act as a template for the synthesis of its complementary strand and so transmit hereditary information across generations.

9. The DNA of an organism (its genome) is unique to each organism and, in general, increases in amount in rough proportion to the complexity of the organism. In eukaryotes, genomic DNA occurs in discrete pieces called chromosomes.

10. While single-stranded DNA is uncommon in cells, RNA (ribonucleic acid) is usually single-stranded. However, by complementary base pairing, RNA can form intrastrand double helical sections, which bend and fold these molecules into unique three-dimensional shapes.

*Overview of Nucleic Acid Function*

11. The link between DNA and proteins is RNA. DNA replication, its transcription into RNA, and the translation of messenger RNA into a protein constitute the central dogma of molecular biology: DNA makes RNA makes protein.

12. RNA has many functional forms. The product of DNA transcription that encodes a protein is called messenger RNA (mRNA) and is translated in the ribosome, an organelle containing ribosomal RNA (rRNA). In the ribosome, each set of three nucleotides in mRNA base pairs with a transfer RNA (tRNA) molecule that is covalently linked to a specific amino acid. The order of tRNA binding to the mRNA determines the sequence of amino acids in the protein. It follows that any alteration to the genetic material (mutation) manifests itself in a new nucleotide sequence in mRNA, which often results in the pairing of a different RNA and hence a new amino acid sequence.

13. Some RNAs have catalytic capabilities. In cells, the joining of two amino acids in the ribosome is catalyzed by rRNA.

## Nucleic Acid Sequencing

14. Nucleic acid sequences provide the following:
    (a) Information on the probable amino acid sequence of a protein (but not on any posttranslational modifications to these amino acids).
    (b) Clues about protein structure and function (based on comparisons of known proteins).
    (c) Information about the regulation of transcription (DNA sequences adjoining amino acid coding sequences are usually involved in regulation of transcription).
    (d) Discoveries of new genes and regulatory elements in DNA.

15. The overall strategy for sequencing DNA or any other polymer is
    (a) cleave the polymer into specific fragments that are small enough to be fully sequenced.
    (b) sequence the residues in each segment.
    (c) determine the order of the fragments in the original polymer by repeating the preceding steps using a different set of fragments obtained by a different degradation procedure.

16. Nucleic acid sequencing became relatively easy with the following:
    (a) The development of recombinant DNA techniques, which allows the isolation of large amounts of specific segments of DNA.
    (b) The discovery of restriction endonucleases, which cleave DNA at specific sequences.
    (c) The development of the chain-termination method of sequencing.

17. Gel electrophoresis is a technique in which charged molecules move through a gel-like matrix under the influence of an electric field. The rate a molecule moves is proportional to its charge density, its size, and its shape. Nucleic acids have a relatively uniform shape and charge density, so their movement through the gel is largely a function of their size.

18. The most common method used to sequence DNA is called the chain-terminator or dideoxy method. This technique relies on the interruption of DNA replication *in vitro* by small amounts of dideoxynucleotides. These lack the 3'-hydroxyl group to which another nucleotide would normally be added during polymerization. The DNA fragments, each ending at a position corresponding to one of the bases, are separated according to size by electrophoresis, thereby revealing the sequence of the DNA.

19. Automated DNA sequencing has allowed scientists to obtain over 500,000 base pairs of DNA sequence information per day. This technology utilizes a different fluorescent dye that is attached to a primer in each of four different reaction mixtures of the chain-termination procedure. The four reaction mixtures are combined for gel electrophoresis, wherein each lane of the gel is scanned by a spectrophotometer that records the fluorescence along the lane from top to bottom. Each band, which corresponds to an oligonucleotide terminated by either ddATP, ddCTP, ddGTP, or ddTTP, is identified by one of the four characteristic fluorescence emissions (indicated by the green, red, black, and blue curves in Fig. 3-20).

20. Phylogenetic relationships can be evaluated by comparing sequences of similar genes in different organisms. The number of nucleotide differences roughly corresponds to the extent of evolutionary divergence between the organisms.

21. The human genome appears to contain ~23,000 protein-encoding genes, compared to ~6000 in yeast, ~13,000 in *Drosophila*, ~18,000 in *C. elegans*, and ~26,000 in *Arabidopsis*. Most proteins encoded in the human genome occur in all the other organisms sequenced to date. Up to 60% of the human genome is transcribed, and 50% of it consists of various repeated sequences, a few of which are repeated >100,000 times.

*Manipulating DNA*

22. Recombinant DNA technology makes it possible to amplify and purify specific DNA sequences that may represent as little as one part per million of an organism's DNA.

23. Cloning refers to the production of multiple genetically identical organisms from a single ancestor. DNA cloning is the production of large amounts of a particular DNA segment through the cloning of cells harboring a vector containing this DNA. Cloned DNA can be purified, after which it can be sequenced. If the DNA of interest is flanked by the appropriate regulatory sequences (which themselves can be introduced into the vector), then the cells harboring the cloned DNA will transcribe and translate the cloned DNA to produce large amounts of protein. The protein can be purified and used for a variety of analytical or biomedical purposes.

24. Organisms such as *E. coli* and yeast are commonly used to carry cloning vectors, which are small, autonomously replicating DNA molecules. The most frequently used vectors are circular DNA molecules called plasmids, which occur as one or hundreds of copies in the host cell. Other important vectors for DNA cloning are bacteriophage λ in *E. coli* and yeast artificial chromosomes (YACs) and bacterial artificial chromosomes (BACs). Both YACs and BACs can accommodate much larger pieces of foreign DNA than can plasmids.

25. The following strategy is typically used to clone a segment of DNA:
    (a) A fragment of DNA is obtained using restriction endonucleases that generate sticky ends. The fragment is then isolated for subsequent ligation to a vector that has been cut with the same restriction endonuclease. The vector contains two selectable genes: (i) one that allows for the selection of transformed bacteria (usually via antibiotic resistance); and (ii) another that allows for the identification of recombinant vectors (vector plus inserted DNA) among a population of transformed colonies.
    (b) The fragment is ligated to the vector using an enzyme called DNA ligase, which catalyzes the formation of phosphodiester bonds.
    (c) The recombinant DNA molecule is introduced into cells by a method called transformation. Cells containing the recombinant DNA are usually selected for by their ability to resist specific antibiotics.
    (d) Cells containing the desired DNA are identified by a second selectable marker, often via the interruption of an easily assayable gene such as β-galactosidase. These identified recombinant clones are then isolated and stored or allowed to grow in order to obtain large quantities of the recombinant DNA.

26.  In most cases, the desired DNA exists in just a few copies per cell, so all the cell's DNA is cloned at once by a method called shotgun cloning to produce a genomic library. The challenge is then to use a screening technique to find the desired clone among thousands to millions in the library. Colony hybridization is a common screening technique in which cells harboring recombinant DNA are transferred from a master plate to a filter that traps the DNA after the cells have been lysed. The filters are then probed using a gene fragment or mRNA that binds specifically to the DNA of interest.

27.  If some sequence information is available, a specific DNA molecule can be amplified using a powerful technique called the polymerase chain reaction (PCR). In PCR, a DNA sample (often quite impure) is denatured into two strands and incubated with DNA polymerase, deoxynucleotides, and two oligonucleotides flanking the DNA of interest that serve as primers for DNA replication by DNA polymerase. Twenty rounds of DNA replication and DNA denaturation (all of which are automated in simple computerized incubators) can increase the amount of the target sequence by a million-fold.

28.  Individuality in humans and other organisms derives from their high degree of genetic polymorphism. For example, homologous human chromosomes differ in the number of tandemly repeated sequences of 2 to 7 bp known as short tandem repeats (STRs). A number of STR sites with multiple alleles in the human population are the basis of DNA fingerprinting. In this technique, the allelic forms at several sites are identified following electrophoresis of PCR-amplified DNA, with results that can identify an individual with high confidence.

29.  DNA cloning can be used to obtain large amounts of protein, referred to as recombinant protein, in bacteria or other organisms (although posttranslational modifications unique to eukaryotes cannot be obtained in bacteria). Another important application is the ability to mutate the amino acid sequence in the DNA of interest at specific sites by a technique called site-directed mutagenesis. Another exciting application of DNA cloning is the ability to introduce modified or novel genes into an intact animal or plant, creating a transgenic organism. The transplanted gene is called a transgene. When this technique produces a therapeutic effect, it is referred to as gene therapy. Gene therapy also refers to the introduction of recombinant DNA into selected cells to obtain a therapeutic effect.

30.  Recombinant DNA technology raises ethical considerations that must be evaluated before its use in medicine, forensics, and agriculture.

## Key Equation

$$P = 1 - (1 - f)^N$$

## Questions

### Nucleotides

1.  Indicate which of the following are purine or pyrimidine bases, nucleosides, or nucleotides:
    (a)  uracil
    (b)  deoxythymidine

(c)  guanosine monophosphate
(d)  adenosine
(e)  cytosine
(f)  guanylic acid
(g)  UMP
(h)  guanine

## Introduction to Nucleic Acid Structure

2.  What key sets of data were used to build the Watson–Crick model of DNA?

3.  Define or explain the following terms: (a) antiparallel; (b) complementary base pairing.

4.  The base compositions of samples of genomic DNA from several different animals are given below. Which samples are likely to come from the same species?
(a)  27.3% T
(b)  29.5% G
(c)  13.1% C
(d)  36.9% A
(e)  22.7% C
(f)  19.9% T

## Nucleic Acid Sequencing

5.  A certain gene in one strain of *E. coli* can be cleaved into fragments of 3 kb and 4 kb using the restriction endonuclease *Pst*I, but it is unaffected by *Pvu*II. In another strain of bacteria, the same gene cannot be cleaved with *Pst*I, but treatment with *Pvu*II yields fragments of 3 kb and 4 kb. Explain the probable genetic difference between the two strains.

6.  You have inserted a 5-kb piece of eukaryotic DNA into a 3.5-kb plasmid at the *Bam*HI site. When you digest the recombinant DNA with various restriction enzymes, you obtain the fragments listed below. Draw a circular map of the recombinant plasmid showing all the restriction sites. Is it possible to determine the orientation of the insert in the plasmid? If so, what information reveals this? If not, what restriction digest would you perform to determine the orientation of the insert?

| *Restriction enzyme* | *Size of DNA fragments (kb)* |
| --- | --- |
| *Eco*RI and *Bam*HI | 1, 1.25, 2.75, 3.5 |
| *Pst*I and *Bam*HI | 1.75, 3.25, 3.5 |
| *Eco*RI, *Pst*I, and *Bam*HI | 0.75, 1, 1.25, 2, 3.5 |
| *Sal*I | 8.5 |
| *Sal*I and *Bam*HI | 0.5, 3, 5 |
| *Bam*HI | 3.5, 5 |

7.   In order to separate different-sized fragments of purified DNA by agarose gel electrophoresis at pH 7.5, should one allow the DNA to migrate from the cathode (the – electrode) or from the anode (the + electrode)? Explain.

8.   T7 DNA polymerase is less sensitive to dideoxynucleotides than the *E. coli* DNA polymerase I fragment described in Section 3-4C.

(a)   For a given concentration of dideoxynucleotides, which enzyme is more likely to give a longer "ladder" on the sequencing gel?
(b) Which enzyme is best used for sequencing DNA close to the primer?

9.   You are interested in the sequence of a 15-base segment of a gene. Using the chain-terminator procedure, you obtain the following results after gel electrophoresis. What is the sequence of the gene segment?

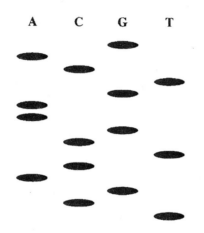

10.  An automated DNA sequencing system using chain-terminator sequencing technology can identify ~ 550,000 bases per day. A sequencing center has 15 such systems running simultaneously. How many days would it take to obtain a single read of human DNA on both strands (assuming no maintanence is required)?

11.  Assume that the average gene size in a prokaryotic genome (listed in Table 3-3) is about 1000 bases pairs in the smaller prokaryotic genomes and 1400 base pairs in the larger bacterial genomes. What is the average number of genes in the smallest and largest prokaryotic genomes shown in the table if all of the DNA in these organisms is part of a gene?

12.  Using the data in Table 3-3, compute the average chromosome length in humans. Assuming that the length of ten base pairs of DNA found in one turn of the helix is 34Å, what is the average length of a human chromosome in centimeters? Does your calculation lead to a surprising result?

*Manipulating DNA*

13.   Outline a procedure for identifying *E. coli* cells that contain a recombinant pUC18 plasmid in which a foreign gene has been inserted at the polylinker site.

14.   The size of the human haploid genome is about $3.2 \times 10^6$ kb. Using human sperm DNA and bacteriophage $\lambda$ as a cloning vector, you have made a human genome library containing $10^5$ clones. Each clone has a 10-kb insert on average. What is the probability that a particular 10-kb single-copy gene is present in your library? Are you confident that you can find your gene with this library? Does a library with 10 times more clones significantly improve your chances? If not, why not?

15.   The amplification of DNA by PCR is exponential for many rounds and then becomes linear. What factors are likely to become rate limiting so that rate of DNA accumulation becomes linear?

16.   During PCR, which cycle of DNA replication results in a double-stranded, unit-length DNA product? How does the rate of accumulation of this product differ from that of the DNA product wherein only one strand is of unit length?

17.   Many eukaryotic peptide hormones cloned in and isolated from bacteria have no biological activity. Why?

18.   DNA synthesized using RNA as a template is known as cDNA. In the technique called subtractive hybridization cloning, small amounts of cDNA from all the mRNA of tissue A is hybridized to all the RNA from tissue B. The unhybridized cDNA is then cloned. What has been cloned?

19.   During DNA replication *in vitro*, two populations of DNA will be produced. What are they? What problem is faced by the molecular biologist once she transforms bacteria with this DNA?

# 4
# Amino Acids

This chapter introduces you to the structure and chemistry of amino acids. The chapter begins by discussing the zwitterionic character of amino acids at physiological pH, followed by a brief introduction to the amide linkage known as the peptide bond. Amino acids are categorized by the chemistry of their unique R groups and by their mode of biosynthesis. The standard amino acids are specified by the genetic code; nonstandard amino acids include the D stereoisomers and modified forms of the standard amino acids. This chapter also introduces you to the conventions used to describe stereoisomers, namely the Fischer convention and the *RS* system. Amino acids contain ionizable groups, so this chapter discusses the acid–base properties of amino acids and introduces you to the concept of the isoelectric point. A clear understanding of the acid–base properties of amino acids is critical for appreciating the structural and catalytic behavior of proteins. The chapter closes with a brief discussion of the biological significance of some amino acid derivatives.

## *Essential Concepts*

### *Amino Acid Structure*

1. All proteins are composed of 20 standard amino acids, which are specified by the genetic code.

2. The standard amino acids are called α-amino acids because they have a primary amino group and a carboxyl group bound to the same carbon atom (the α carbon). Only proline has a secondary amino group attached to the α carbon, but it is still commonly referred to as an α-amino acid.

3. The generic structure of an amino acid at pH 7 is shown below.

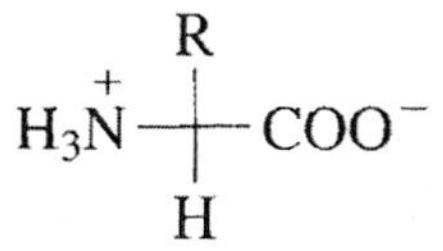

$$\overset{+}{H_3N}-\overset{\overset{\displaystyle R}{|}}{\underset{\underset{\displaystyle H}{|}}{C}}-COO^-$$

At pH 7, the amino acid is a zwitterion, or dipolar ion. A unique side chain, or R group, characterizes each amino acid.

4. Amino acids are polymerized by condensation reactions to form a chain called a polypeptide. Each polypeptide is polarized: One end has a free amino group and the other end has a free carboxyl group, referred to as the N-terminus and the C-terminus, respectively.

5.   The R groups of the standard 20 amino acids are classified into three categories based on their polarities and charge at pH 7: the nonpolar amino acids, the polar uncharged amino acids, and the charged amino acids (see Table 4-1).

6.   Among the nonpolar amino acids, glycine (shorthand Gly or G) has a hydrogen atom as its R group. Alanine (Ala; A), valine (Val; V), leucine (Leu; L), isoleucine (Ile; I), and methionine (Met; M) have aliphatic chains as R groups (Met has a sulfur rather than a methylene group). Tryptophan (Trp; W) and phenylalanine (Phe; F) contain bulky indole and phenyl groups, respectively.

7.   The polar uncharged amino acids include asparagine (Asn; N), glutamine (Gln; Q), serine (Ser; S), threonine (Thr; T), tyrosine (Tyr; Y), and cysteine (Cys; C). Amide functional groups occur in Gln and Asn. Alcoholic functional groups occur in Ser and Thr. Tyrosine and cysteine are characterized by a phenolic group and a thiol group, respectively.

8.   Among the charged amino acids, aspartate (Asp; D) and glutamate (Glu; E) contain carboxylic groups in their R groups. Lysine (Lys; K), arginine (Arg; R), and histidine (His; H) contain a butylammonium group, a guanidino group, and an imidazole group, respectively.

9.   The pK values of ionizable groups depend on the electrostatic influences of nearby groups. In proteins, the pK values of ionizable R groups may shift by several pH units from their values in the free amino acids.

*Stereochemistry*

10.   Except for glycine, the standard amino acids have asymmetric structures and rotate the plane of polarized light; thus, they are optically active. These molecules cannot be superimposed on their mirror images. Such nonsuperimposable pairs of molecules are called enantiomers. The asymmetric atom of an optically active molecule is called the chiral center and the molecule is said to have the property of chirality.

11.   Fischer projections are used to represent the absolute configuration of substituents around a chiral center. In the Fischer convention, horizontal lines extend above the plane of the paper, while vertical lines extend below the surface of the paper. The α-amino acid shown above is a Fischer projection of an L-amino acid. The specific arrangement of substituents around the chiral carbon is related to that of L-glyceraldehyde.

12.   A molecule with $n$ chiral centers has $2^n$ different possible stereoisomers. Molecules with two or more chiral centers are better described by the *RS* system, in which each substituent bound to the chiral center is prioritized according to its atomic number. Hence, the exact molecular arrangement of a molecule can be unambiguously described.

13.   Biochemical reactions almost invariably produce pure stereoisomers, in large part due to the precise arrangement of chiral groups of enzymes, which restricts the geometry of the reactants.

*Amino Acid Derivatives*

14. Nonstandard amino acids in polypeptides arise from posttranslational modifications of the R groups of amino acid residues of the polypeptide. These modifications have critical roles in the structure and function of proteins. Many unpolymerized nonstandard amino acids are synthesized by chemical modifications of one of the standard amino acids. Cells use many of these amino acids as signaling molecules, particularly in the central nervous system. Among the nonstandard amino acids are also the D isomers of the standard amino acids; many of these occur in bacterial cell walls and bacterially produced antibiotics. The tripeptide glutathione is a cellular reducing agent.

## Key Equation

$$pI = \tfrac{1}{2}(pK_i + pK_j)$$

## Play It Forward: pH, pK and Predominant Forms of Amino Acids in Solution

In Chapter 2 Behind the Equations: Weak Acids and the Henderson-Hasselbalch Equation, we outlined a key relationship between the pH of the environment, the p$K$ of a weak acid functional group, and the predominance of the protonated (HA) vs. deprotonated (A$^-$) form of a weak acid. This can be summarized as:

**If the pH of the environment is lower than the p$K$ of the weak acid, then the weak acid will be predominantly protonated (in the HA form), and the denominator of the conjugate base-to-weak acid fraction in the Henderson-Hasselbalch equation will predominate.**

**If the pH of the environment is higher than the p$K$ of the weak acid, then the weak acid will be predominantly deprotonated (in the A$^-$ form), and the numerator of the conjugate base-to-weak acid fraction in the Henderson-Hasselbalch equation will predominate.**

Isolated amino acids contain at least two weak acid functional groups, the primary amino group attached to the alpha carbon (C$_\alpha$) and the carboxylic acid group attached to C$_\alpha$. Each is defined by a unique p$K$: p$K_1$ for the carboxylic acid and p$K_2$ for the amino group. The amino acids D, E, C, Y, H, K and R also have weak acid functional groups in their side chains, each of which is also defined by a unique p$K$. As with weak acids that were encountered in the text and problem sets associated with Chapter 2, the protonation states of the weak acid functional groups of the amino acids are sensitive to the pH of the environment and obey the rules outlined in Chapter 2 and repeated above. Specifically, each of the weak acid functional groups present on an amino acid can exist predominantly in a protonated or deprotonated form depending upon the pH of the environment in relation to the p$K$ of that functional group.

As an example, consider the isolated L-amino acid alanine (Ala, A). p$K_1$ for alanine is reported as 2.35 whereas p$K_2$ is reported as 9.87. We can use the rules above to predict what the predominant structure of alanine will be at any particular pH of the environment by applying these rules.

Assume that a solution of alanine is set to pH 1.0. What is the predominant structure of alanine under these conditions? Simply apply the rules above to each functional group.

The pH of the environment (1.0) is less than $pK_1$ (2.35). The rules state that *"If the pH of the environment is lower than the pK of the weak acid, then the weak acid will be predominantly protonated (in the HA form)"*. Thus we predict that the carboxylic acid group of alanine will be predominantly in the protonated form in solution at pH = 1.0. Carboxylic acids can exist in solution in only two forms, the protonated (HA) form (R-COOH) and the deprotonated (A⁻) form (R-COO⁻):

$$R\text{--}COOH \rightleftharpoons R\text{-}COO^- + H+$$

Thus, we expect that the carboxylic acid functional group of alanine will be predominantly in the R-COOH form at pH = 1.0.

Moreover, the pH of the environment (1.0) is less than $pK_2$ (9.87). Applying the rule, we predict that the amino group of alanine will be predominantly in the protonated form at pH = 1.0. Primary amines can exist in solution in only two forms, the protonated (HA) form (R-NH₃⁺) and the deprotonated (A⁻) form (R-H₂N:):

$$R\text{-}NH_3^+ \rightleftharpoons R\text{-}H_2N\text{:} + H^+$$

In summary, because the pH of the environment is lower than both $pK_1$ and $pK_2$, then both weak acid functional groups of alanine will be predominantly in the protonated form, and the predominant structure would be:

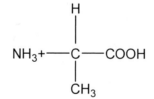

This should also be the predominant structure of alanine at pH = 2.0 because the pH of the environment is still lower than both $pK_1$ and $pK_2$. The net charge on alanine at pH = 2.0 is +1.

However, consider how the structure changes when the pH increases above the value of $pK_1$ but remains lower than $pK_2$.

At a pH = 3.0, pH > $pK_1$. The rule states that *"If the pH of the environment is higher than the pK of the weak acid, then the weak acid will be predominantly deprotonated (in the A⁻ form)"*. Thus, we anticipate that the carboxylic acid functional group will be predominantly in the deprotonated (A⁻) form (R-COO⁻). Simultaneously, the pH < $pK_2$, so the amino group should still be in the protonated (R-NH₃⁺) form. Thus, the predominant structure of alanine at pH 3.0 would be:

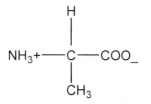

Notice now that we have a structure of alanine at pH 3.0 that is different than the structure of alanine at pH 1.0; and we have a net charge of 0. This will also be the predominant structure of alanine at pH = 4.0, 5.0, 6.0, etc… until we reach a pH value that is greater than $pK_2$.

At pH = 10.0, pH > $pK_1$, and is now greater than $pK_2$ also. Applying the rules, we predict that both weak acid functional groups will be predominantly in the deprotonated form, giving a predominant structure for alanine at pH = 10.0 of:

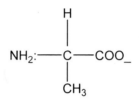

Notice that the net charge on this molecule is –1.

What we have done thus far is perform an imaginary base titration of alanine. We imagined that alanine began the titration at a low pH (pH = 1.0), and had a particular predominant structure and net charge at that pH value. As base was added to increase the pH of the environment, the pH eventually moved to the other side of $pK_1$, and caused a change in the structure to a new predominant form with a different net charge. Continuing the base titration eventually raised the pH to the other side of $pK_2$, causing a second change in the structure, which generated a new predominant form with a different net charge.

Examining this closely, a new relationship becomes evident, and a very important principle is revealed:

1.  For a molecule with weak acid functional groups, each defined by a unique p$K$, the number of possible structures is the sum of the number of functional groups plus one.

    Alanine has two weak acid functional groups (defined by $pK_1$ and $pK_2$), so there are 2 + 1 = 3 possible structures for alanine in solution. The pH of the solution will determine which possible structure predominates.

2   Every amino acid will carry a charge somewhere on the molecule at every pH value. Note the structures for alanine above. While the net charge of alanine can vary between –1, 0, or +1, there is at least one formal charge present on alanine at any possible pH value.

What about amino acids where the side chain also contains a weak acid functional group and has a $pK_R$?

Consider the amino acid Lysine (Lys, K). Lysine has 3 weak acid functional groups ($pK_1 = 2.16$, $pK_2 = 9.06$, and $pK_R = 10.54$).
How many possible structures for Lysine are there?

Therefore, for every weak acid functional group present on a molecule (each defined by a unique $pK$), there will be that many unique possible structures, + 1, that can exist in solution. Thus, for Lysine, there will be $3 + 1 = 4$ possible structures.

Draw the predominant structure for Lysine at pH = 1.0, 5.0, 10.0, and 12.0

At pH 1, pH $< pK_1$, $pK_2$, and $pK_R$; net charge $= + 2$

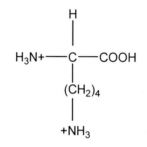

At pH = 5.0, pH $> pK_1$ but $< pK_2$ and $pK_R$; net charge = +1

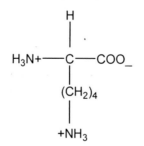

At pH = 10.0, pH $> pK_1$ and $pK_2$ but $< pK_R$; net charge = 0

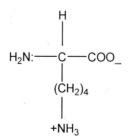

At pH = 12.0, pH > p$K_1$, p$K_2$, and p$K_R$; net charge = −1

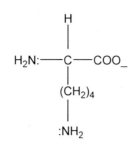

## Questions

### Amino Acid Structure

1. Without consulting the text, draw a generic L-α-amino acid using the Fischer convention.

2. Examine the amino acids below. Assume that the p$K$ of the carboxyl group attached to the α carbon is 2.0 and that of the primary amine attached to the α carbon is 9.5 The p$K$'s of ionizable R groups are shown below each structure. (a) Categorize these amino acids as nonpolar, uncharged polar, or charged polar at pH 7. (b) Which of the structures cannot exist as shown at any pH in aqueous solution? (c) Name each of the amino acids.

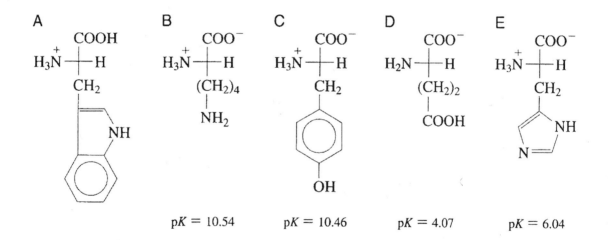

3. Which of the amino acids in Table 4-1 can be converted to another amino acid by mild hydrolysis that liberates ammonia?

4. Which of the amino acids in Table 4-1 can generate a new amino acid by the addition of a hydroxyl group?

5. The ionic characters of which amino acids are likely to be sensitive to pH changes in the physiological range? Explain.

6.   Circle the functional groups that are eliminated in the formation of a peptide bond between the amino acids shown below. Draw the structure of the dipeptide.

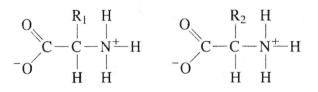

7.   What percentage of the histidine imidazole group is protonated at pH 7.2?

8.   What percentage of the cysteine sulfhydryl group is deprotonated at pH 7.6?

9.   In proteins, the imidazole groups of histidine play key roles in catalysis involving reversible protonation/deprotonation events. The $pK$ of the imidazole group is influenced by surrounding amino acids and in many enzymes is near 7.
     (a)  What is the significance of this apparent $pK$ in catalysis?
     (b)  Some unprotonated imidazole groups participate in hydrogen bonding a substrate to an enzyme. Do you expect the $pK$ of these imidazole groups also to be near 7?

10.  Using the $pK$ values in Table 4-1, identify the amino acids that have the titration curves shown below.

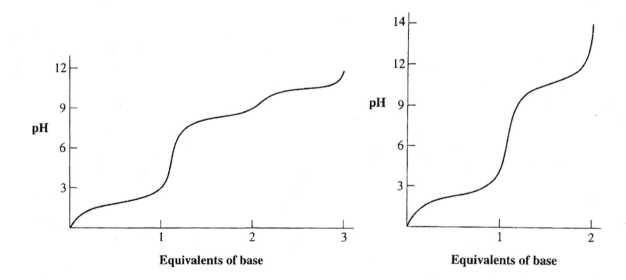

11.  Which of the following tripeptides contain peptide bonds NOT commonly found in proteins? Provide a name for tripeptide B.

A

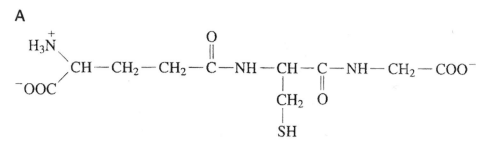

**glutathione**

B

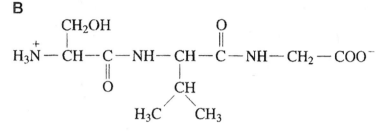

C

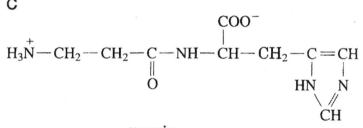

**carnosine**

D

H$_3$N$^+$—CH—CH$_2$—CH$_2$—CH$_2$—NH—CH—CH$_2$—SH
      |                                |
    COO$^-$                          C=O
                                   |
                                 NH
                                 |
                               CH$_2$
                                 |
                               COO$^-$

12.   Why is the p$K$ of the carboxyl group of glycine (p$K$ = 2.3) less than that for acetic acid (p$K$ = 4.76)?

13.   Calculate the p$I$'s of aspartate, lysine and serine.

14.   Calculate the p$I$ of the tripeptide Asp-Lys-Ser. Is it simply the average of the p$I$'s of the individual amino acids?

15.   In proteins, the p$K$ of the C-terminus is about 3.8, while that of the N-terminus is about 7.8. Rationalize these differences from the p$K$ values of the $\alpha$-carboxyl and the $\alpha$-amino groups in free amino acids.

16.   Is a protein as good a *cellular* buffer at physiological pH as its constituent amino acids would be if they were present as free amino acids in proportional concentrations in the cell? Explain.

17.   Based on your rationale in the previous question, describe the difference in the dissociation constants of the $\alpha$-COOH of GABA (see page 88) and glutamate (Table 4-1), and the differences in the dissociation constants of the amino group in each of these amino acids.

18.   100 mg of anhydrous powder of lysine is not completely soluble in 10 mL of water but dissolves completely when base is added. Explain.

19.   Examine the paper by McCoy, Meyer, and Rose (*J. Biol. Chem.* **112,** 283–302, 1935) at the Web site for the Journal of Biological Chemistry (http://www.jbc.org).
   (a)  What was the name of Unknown II? What is its name today?
   (b)  What remained to be solved about its structure?

20.   Shown below is a titration curve for glutamic acid. Draw the structure of the species that predominates at each labeled point.

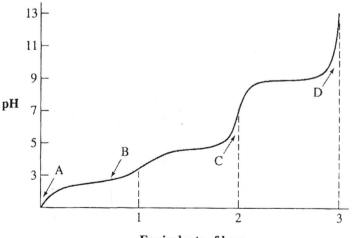

**Equivalents of base**

*Stereochemistry*

21.   Which amino acid found in proteins has no optical activity?

22.   Draw Fischer projections of L-aspartate and L-cysteine and show why L-Asp is (*S*)-Asp and L-Cys is (*R*)-Cys. (Refer to Box 4-2.)

23.   Which amino acids in Table 4-1 have two or more prochiral centers? A prochiral center can be made chiral by substituting a different group for one of the two identical groups attached to it.

*Amino Acid Derivatives*

24.   Which of the nonstandard amino acids below cannot occur in the interior of a polypeptide chain? Explain.

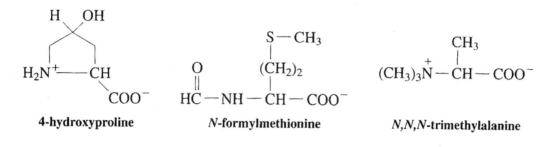

4-hydroxyproline          *N*-formylmethionine          *N,N,N*-trimethylalanine

25.   From which of the standard amino acids are the following physiologically active amines derived? What modifications gave rise to these products? (See Figure 4-15 for structures.)
(a) GABA
(b) histamine
(c) thyroxine
(d) dopamine

26.   How many conjugated double bonds occur in the fluorescent tyrosine derivative of green fluorescent protein?

27.   Study the functional groups of the amino acids in Table 4-1 and devise a Venn diagram showing amino acids in the following categories: nonpolar, polar, negatively charged, positively charged, aliphatic, small R groups, and bulky R groups. (In a Venn diagram, objects with a given common property are enclosed by an oval. A set of objects may have several fully or partially overlapping ovals that represent different properties or the ovals may not overlap.)

# 5
# Proteins: Primary Structure

This chapter covers protein purification and primary structure. In this chapter you will learn how a protein's size, charge, and general shape can be analyzed and used to develop procedures to purify the protein. Many of the principles you learned in the preceding chapters about thermodynamics, aqueous solutions, and acid–base chemistry can be applied to the isolation of proteins. This chapter also includes a discussion of the strategies and chemical methods for determining the primary structure of proteins, focusing on Edman degradation and introducing mass spectrometry as a powerful technique for determining polypeptide mass and sequence. The chapter concludes with a section discussing protein evolution. Here you will see how comparisons of primary structures have led biochemists to categorize proteins into families and identify specific modules or motifs involved in specific functions.

## *Essential Concepts*

### *Polypeptide Diversity*

1.  The primary structure of a protein is the amino acid sequence of its polypeptide chain (or chains if the protein contains more than one polypeptide).

2.  Proteins are synthesized in cells by the stepwise polymerization of amino acids in the order specified by the sequence of nucleotides in its gene. The 5′ end of the messenger RNA corresponds to the amino terminus of the polypeptide, which is the end that contains a free amino group bound to the α carbon.

3.  Although the theoretical possibilities for polypeptide composition and length are unlimited, polypeptides found in nature are limited somewhat in size and composition. Most polypeptides contain between 100 and 1000 residues and do not necessarily include all 20 genetically encoded amino acids. Leu, Ala, Gly, Ser, Val, and Glu are the most abundant amino acids in proteins, while Trp, Cys, Met, and His are the least common.

### *Protein Purification and Analysis*

4.  A general approach to protein purification requires the following:
    (a) A rapid and efficient method to disrupt cells so that the contents of the lysed cells (lysate) can be quickly stabilized in a buffer of appropriate pH and ionic strength.
    (b) Consideration of factors that affect the stability of the desired protein, such as pH, temperature, presence of degradative enzymes, adsorption to surfaces, and solvent conditions for long term storage.
    (c) Appropriate separation techniques to effectively purify the desired protein from the total proteins of the lysate.
    (d) A test or assay to easily assess the activity of the desired protein and hence measure its concentration at each step in the purification.

5. A protein contains multiple charged groups, so its solubility in aqueous solution varies with the concentrations of dissolved salts (ionic strength), pH, and temperature. When salt is added to a protein at low ionic strength, the protein's solubility increases with increasing ionic strength (called salting in). At higher salt concentrations, the solubility of the protein decreases (called salting out). This variable solubility can be exploited to selectively precipitate desired or undesired proteins.

6. Chromatography, which is used for analysis as well as purification, relies on the ability of a substance dissolved in a solvent (mobile phase) to interact with a solid matrix (stationary phase) as the solvent percolates through it. The chromatographic matrix is often in a column so that the solvent that elutes (exits) from it can be collected and assayed for the desired substance.

7. Ion exchange chromatography takes advantage of a molecule's ionizable groups. The binding affinity of a substance for a cation or anion exchanger depends on its net charge and the pH and ionic strength of the solvent. The bound substance is eluted by increasing the salt concentration or changing the pH of the solvent.

8. Hydrophobic interaction chromatography, which uses a matrix substituted with nonpolar groups, takes advantage of the exposed hydrophobic regions of a protein. The bound protein is eluted with a solvent that weakens the hydrophobic effects that cause the protein to bind to the matrix.

9. Gel filtration chromatography separates molecules according to their size and shape. The stationary phase consists of beads containing pores that span a relatively narrow size range. Smaller molecules spend more time inside the beads than larger molecules and therefore elute later (after a larger volume of mobile phase has passed through the column). Within the molecular mass range that is fractionated by the specific stationary phase used, there is a linear relationship between a substance's elution time and the logarithm of its molecular mass.

10. Affinity chromatography exploits a protein's specific ligand-binding behavior. The protein binds to a ligand that is covalently attached to the stationary phase and can be eluted with a solution containing a high salt concentration or excess ligand, which competes with the bead-bound ligand for binding sites on the protein.

11. Electrophoresis separates molecules according to size and net charge. The molecules move through a polyacrylamide or agarose gel under the influence of an electric field. Proteins can be visualized by soaking the gel in a dye that binds to proteins or by electroblotting the proteins to nitrocellulose paper and probing with an antibody that is specific for the desired protein (Western blotting).

12. In SDS polyacrylamide gel electrophoresis (SDS-PAGE), proteins are denatured with the detergent sodium dodecyl sulfate (SDS), which binds to a polypeptide and imparts a large negative charge by virtue of its negatively charged sulfate group. The net charge of an SDS-polypeptide is proportional to its length, giving all polypeptides the same charge density, so SDS-PAGE separates polypeptides almost entirely by sieving effects. Smaller polypeptides migrate faster since they are less impeded by the cross-linked polyacrylamide and hence polypeptides are separated according to their molecular mass.

*Protein Sequencing*

13. Knowledge of a protein's amino acid sequence (its primary structure) is important for:
    (a) Determining its three-dimensional structure and elucidating its molecular mechanism of action.
    (b) Comparing sequences of analogous proteins from different species to gain insights into protein function and evolutionary relationships among the proteins and the organisms that produce them.
    (c) Developing diagnostic tests and effective therapies for inherited diseases that are caused by single amino acid changes in a protein.

14. Determining a protein's sequence may require several steps:
    (a) End group analysis reveals the number of different polypeptides or subunits in a protein. Either dansyl chloride or phenylisothiocyanate (Edman degradation) can be used to identify the N-terminal residue of a polypeptide.
    (b) Disulfide bonds within and between polypeptide chains are cleaved by reducing agents such as 2-mercaptoethanol, and the free sulfhydryl groups are then alkylated, e.g., with iodoacetate, to prevent re-formation of the disulfide bonds.
    (c) The different polypeptides of a multisubunit protein are separated so that each can be sequenced.
    (d) Polypeptides longer than 100 residues cannot be sequenced directly and must therefore be cleaved into smaller fragments using endopeptidases such as trypsin or chymotrypsin or chemical agents such as cyanogen bromide.
    (e) Each fragment is isolated and then sequenced using repeated cycles of the Edman degradation.
    (f) The order of sequenced fragments in the intact polypeptide is established by cleaving the polypeptide with a different reagent to obtain a second set of fragments whose sequences overlap those of the first.
    (g) The positions of disulfide bonds are identified by cleaving the polypeptide before it has been reduced. Each fragment containing a disulfide bond is reduced, and the resulting two peptides are separated and sequenced.

15. In Edman degradation, phenylisothiocyanate (PITC) reacts exclusively with the N-terminal amino group of a polypeptide chain. Treating the resulting PTC polypeptide with trifluoroacetic acid releases the N-terminal residue as a thiazolinone derivative, which is converted to a phenylthiohydantoin–amino acid that can be identified by chromatography. The procedure is then repeated for the newly exposed N-terminal residue of the polypeptide. This procedure has been automated and refined to the point where up to 100 residues can be sequenced with as little as 5 to 10 picomoles of protein ($< 0.1$ μg)!

16. Mass spectrometry is a powerful tool for polypeptide sequencing. In particular, electrospray ionization (ESI) allows for direct sequencing of short polypeptides (< 25 amino acid residues) by measuring the mass-to-charge ratio of ionic gas-phase peptides.

17. Protein sequences are deposited in a number of public databases, most notably at the National Center for Biotechnology Information (NCBI; http://www.ncbi.nih.gov) or at the European site Expert Protein Analysis System (ExPASy: http://www.expasy.org).

*Protein Evolution*

18. Mutations that alter a protein's primary structure may or may not affect the protein's function. Mutations that are deleterious to the organism are less likely to be passed on to the next generation. Mutations that enhance an organism's ability to survive and reproduce tend to propagate quickly. This is the essence of Darwinian evolution.

19. Because related species have evolved from a common ancestor, the genes specifying each of their proteins have also evolved from a corresponding gene in that ancestor. Hence, the more closely related the two species, the more similar are the primary structures of their proteins. A phylogenetic tree can therefore be constructed by comparing the sequence divergence of proteins in different organisms.

20. Comparisons of the primary structure of evolutionarily related proteins (homologous proteins) may indicate which amino acid residues are essential to the protein's function: Essential positions tend to contain invariant residues; less essential positions tend to contain chemically similar (conserved) residues; and positions that do not significantly affect protein structure or function can accommodate a variety of residues.

21. The rates of evolution of proteins vary considerably. This reflects the ability of a given protein to accept amino acid changes without compromising its function.

22. Large families of proteins such as the immunoglobulins or globins arose by gene duplication and subsequent primary structure divergence among the copies. Hence, the new copies are capable of evolving new functions.

23. Many proteins contain amino acid sequence modules of about 40–100 residues that may have unique functions. Some modules are repeated many times in a single polypeptide. Hence, shuffling of modules is potentially a mechanism for rapidly generating new proteins with new functions.

24. Most primary structures of proteins are known from sequencing the genes that encode them. However, DNA sequencing cannot provide information about the locations of disulfide bonds or posttranslational modifications of amino acids. In addition, some newly synthesized polypeptides are trimmed by site-specific cleavage to generate the mature functional protein. In practice, protein sequences are used to confirm nucleic acid sequences and *vice versa.*

## Questions

*Protein Purification and Analysis*

1.  Discuss the disadvantages of each of the following in the handling of proteins:
    (a)  Low concentrations of protein
    (b)  Sudsing of protein solutions
    (c)  Lack of sterile conditions
    (d)  Absence of protease inhibitors
    (e)  Warm environments
    (f)  pH's far from neutrality

2.  List possible ways to track the presence of a protein during its purification.

3.  You have several cell lysates and want to determine which of them contain protein X in its phosphorylated form (with phosphorylation occurring on a tyrosine residue). You have two antibodies: One recognizes phosphotyrosine and the other recognizes protein X. Describe how you would analyze the cell lysates for the presence of phosphorylated protein X using just these two antibodies.

4.  The salting out procedure outlined in Fig. 5-5 results in significant protein purification by eliminating many less- and more-soluble proteins. What is its other advantage?

5.  You wish to separate two proteins in solution by selectively precipitating one of them. What method could you use other than salting out?

6.  The p$I$ of pepsin is < 1.0, whereas that of lysozyme is 11.0. What amino acids must predominate in each protein to generate such p$I$'s?

7.  Evaluate the solubility of the following peptides at the given pH. For each pair, which peptide is more soluble and why? *Hint:* Evaluate the ionization states of the amino acid functional groups.
    (a)  pH 7: $Gly_{20}$ and $(Glu\text{-}Asp)_{10}$
    (b)  pH 5: $(Cys\text{-}Ser\text{-}Ala)_5$ and $(Pro\text{-}Ile\text{-}Leu)_5$
    (c)  pH 7: $Leu_3$ and $Leu_{10}$

8.  Would you use an anion or cation exchange column to purify bovine histone at pH 7.0? (See Table 5-3.)

9.  Why is a DEAE ion exchange column ineffective above pH 9?

10.  When immunoaffinity chromatography is used to purify a protein, the cell lysate is often subjected to one or more purification steps before the material is applied to the immunoaffinity column. Why is this necessary?

11.  Describe two methods for determining the molecular mass of a polypeptide. Which is more accurate and why?

12.   Shown below are the biochemical fractionation and purification of an enzyme. Fill in the table on the next page with the values for specific activity and fold purification, and answer the accompanying questions. (nkat, nanokatal; one katal is the amount of enzyme that catalyzes the transformation of one mol of substrate per second.)
(a) Which step provides the greatest purification?
(b) Is the enzyme pure? How can purity be assessed?

| | *Total Protein (mg)* | *Activity (nkat)* | *Specific Activity (nkat/mg)* | *Fold Purification* |
|---|---|---|---|---|
| Crude extract | 50,000 | 10,000,000 | | |
| Ammonium sulfate precipitation | 5,000 | 3,750,000 | | |
| DEAE-cellulose, stepwise KCl gradient | 500 | 500,000 | | |
| DEAE-cellulose, KCl linear gradient | 250 | 500,000 | | |
| Gel filtration | 25 | 250,000 | | |
| Substrate affinity chromatography | 1 | 100,000 | | |

## Protein Sequencing

13.   Match the terms below with the accompanying statements. (More than one term may apply to a statement.)

A. dansyl chloride
B. phenylisothiocyanate
C. iodoacetate
D. 2-mercaptoethanol
E. cyanogen bromide

_____   An alkylating agent that reacts with Cys residues.
_____   A reducing agent that carries out reductive cleavage of disulfide bonds.
_____   A reagent used to identify N-terminal residues.
_____   A reagent that cleaves polypeptides into smaller fragments.
_____   A reagent used in sequencing polypeptides from the N-terminus.

14.   How many fragments can be obtained by treating bovine insulin (Fig. 5-1) with endopeptidase V8? Does this number depend on whether the protein is first reduced and alkylated?

15. You have a protein whose molecular mass is about 6.8 kD according to SDS-PAGE. However, SDS-PAGE of the protein in the presence of 2-mercaptoethanol reveals two bands of 2.3 kD and 4.5 kD. Are these polypeptides too large for direct sequencing by an automated sequencer using Edman degradation? What steps must you take to determine the complete sequence of this protein?

16. You have a polypeptide that has been degraded with cyanogen bromide and trypsin. The sequences of the fragments are listed below (using one-letter codes). Is there enough information to derive the complete amino acid sequence of the polypeptide? What further treatments might support the deduced sequence?

*Cyanogen bromide treatment:*
(1) **K**
(2) **RFM**
(3) **MLYCRGM**
(4) **NIKGLM**

*Trypsin treatment:*
(5) **FMK**
(6) **GMNIK**
(7) **GLMR**
(8) **MLYCR**

17. You want to know the amino acid sequence of a purified polypeptide (1341 D). Your new technician has performed the analyses outlined below. Do you have enough data to determine the amino acid sequence?
    (a) Partial acid hydrolysis yields tripeptides with the following amino acid sequences as determined by manual Edman degradation: **HSE, EGT, DYS, TSD, FTS, YSK.**
    (b) Dansyl chloride treatment yields **H** and **K**.
    (c) Brief carboxypeptidase A treatment yields one **K** per mole peptide (a carboxypeptidase removes one residue at a time from the C-terminus of a polypeptide).
    (d) Chymotrypsin treatment yields two peptides (A and B) with molecular masses of 659 D and 467 D, respectively, and a dipeptide composed of **S** and **K**.
        Peptide A
            (1) Dansyl chloride treatment yields **H**.
            (2) Brief carboxypeptidase A treatment yields **F** and **T**.
        Peptide B
            (3) Dansyl chloride treatment yields **T**.
            (4) Brief carboxypeptidase A treatment yields **Y** and **D**.
    (e) Trypsin treatment results in no fragmentation of the polypeptide.

18. Compare the molecular mass ($M$) of horse heart apomyoglobin calculated from the two adjacent peaks at 1884.7 and 1696.3, and 893.3 and 848.7 (Figure 5-16b). How do these results compare with those in Sample Calculation 5-1?

*Protein Evolution*

19. Explain why single-nucleotide mutations in DNA occur at a constant rate, but the rate of protein evolution varies.

20. Rank the following residue positions in cytochrome $c$ in order from least to most conserved: 56, 61, 74, 85, and 89.

21. Which protein(s) shown in Figure 5-23 would be most useful for assessing the phylogenetic relationships between mammalian species (note that mammals diverged from reptiles about 300 million years ago)? Explain.

22. The sequences of two cAMP regulatory transcription factors were compared using the BLAST alignment program from NCBI. The results below show the sequences from *Escherichia coli* (the Query sequence) and *Methylobacterium extorquens* (the Sbjct sequence). The BLAST "hit" compares residues 190–369 of the *E. coli* protein and residues 64–242 of the *M. extorquens* protein. Amino acid sequence identity is revealed by letters in the middle row; mismatches are spaces between the rows; and substitutions of similar amino acids are shown as a "+" in the middle row. Take a look at these "positives" as they are called in the BLAST output. Rationalize these positives based on your knowledge of R-group chemistry (recall the Venn diagram you made in Chapter 4).

```
Query   190   SGFACRHKLRASGARQINAYLLPGDLCDLDVALLDEMDHTITTLSTCTVVRLAPELIADL   249
              GFACR+KLR +GARQI AYL+PGD+CDLD    L+ MDHT+ TLSTC V R+ PE  A
Sbjct   64    EGFACRYKLRENGARQIMAYLVPGDVCDLDNGALNRMDHTVGTLSTCRVARIMPE-TARE   122

Query   250   LTHHPQIARALRKNTLVDEATLREWLMNVGRRSSVERMAHLFCELLLRFRAVGLAPEDSY   309
              L  HP +AR LRK  L+DEATLREWLMN+GRRS+VER+AHLFCELL+R RAVGL   +SY
Sbjct   123   LRQHPALARGLRKAALIDEATLREWLMNIGRRSAVERLAHLFCELLVRLRAVGLTSGNSY   182

Query   310   PLPLTQADLADTTGITSVHVSRSLKELRQGGLIELQGGRLKILDYARLRALAEFRANYLH   369
              LP+TQ DLADTTG+TSVHV+RSL EL++ GLIE +  RL + D ARL  +AEFR +YLH
Sbjct   183   ELPITQLDLADTTGMTSVHVNRSLGELKREGLIERKSKRLTLCDPARLAEVAEFRPDYLH   242
```

23. Comparisons between amino acid sequences can be used to understand which areas of a polypeptide are important for its function. A large family of proteins in bacteria includes a transcriptional regulator called the cAMP regulatory or activator protein (CRP or CAP).

    (i)   Go to the NCBI site at *http://www.ncbi.nlm.nih.gov/*

    (ii)  Retrieve the amino acid sequences for the following CRP/CAP family proteins by selecting the Protein database in the Search menu and typing the underlined accession number into the adjacent box:
       - *E. coli* catabolite gene activator (cAMP receptor protein), P0ACJ8
       - *Yersinia pestis* cyclic AMP receptor protein, AAM87500
       - *Klebsiella pneumoniae* cAMP receptor protein, A44903
       - *Shewanella* sp. MR-4 Crp/Fnr family transcriptional regulator, YP_734144
       - *Aeromonas hydrophila* fumarate and nitrate reduction regulatory protein YP_856826

    (iii) For each retrieval, select "FASTA" from the display menu then copy and paste the amino acid sequence into a Word document, giving each sequence a recognizable byline in FASTA format: ">byline".

    (iv)  Remove the numbers from the amino acid sequence

(v) Go to the EMBL Clustal W site (http://www.ebi.ac.uk/Tools/clustalw/index.html), which will allow you to compare up to 10 different amino acid sequences.

(vi) Copy and paste the five sequences in FASTA format consecutively into the open box and hit RUN.

(vii) In a few moments, you will see an output that shows all five sequences lined up according to the best matches between them.

(a) Which parts of this protein show the most amino acid divergence?

(b) Which parts of this protein appear to be the most conserved?

(c) Go to http://www.ncbi.nlm.nih.gov/Structure/cdd/wrpsb.cgi at NCBI. Paste in the *E. coli* sequence to see what domains appear. Do the same with the *A. hydrophila* sequence. What areas of the polypeptide are identified with the two smaller, common domains (colored in red and blue)? Comparing these two domains between these species, which domain appears more divergent? Does it make sense relative to the information given about each domain (scroll the arrow over each domain to retrieve information)? The e-value is a measure of similarity; similar sequences yield low e-values.

# 6

# Proteins: Three-Dimensional Structure

This chapter introduces you to an area of biochemistry that has shown an incredible growth of information in recent years: the three-dimensional structures of proteins. The chapter first discusses in some detail the geometric and physical properties of the planar peptide group, which contains the peptide bond. Here you see that the number of polypeptide conformations, while quite large, is finite due to restrictions imposed by colliding van der Waals spheres of the atoms of the polypeptide. The chapter then examines the major forms of secondary or local three-dimensional structure in polypeptides, in particular, the $\alpha$ helix and the $\beta$ sheet. Keratin and collagen are the most common and best-characterized fibrous proteins and are largely composed of one type of secondary structure. The structural properties and biology of these proteins are discussed at this point.

The chapter then turns its attention to the methods by which overall three-dimensional structure, or tertiary structure, is determined: X-ray diffraction of protein crystals and nuclear magnetic resonance (NMR) of proteins in solution. You are then introduced to the various kinds of supersecondary structures (motifs) and domains that have been identified in many proteins, including a brief discussion of the symmetries observed in the quaternary structures of proteins that consist of more than one subunit. This chapter also introduces structural bioinformatics, the branch of biochemistry that is involved in the visualization, analysis, and classification of protein three-dimensional structure.

The factors that affect protein stability are then explored, followed by a discussion of how newly synthesized proteins fold. In this presentation, you are introduced to proteins that facilitate protein folding *in vivo*: protein disulfide isomerase and the molecular chaperones (the Hsp70 family and the chaperonins). The section closes with a discussion of diseases related to protein misfolding.

Box 6-1 provides a brief biography of one of the 20th century's greatest chemists and scientists, Linus Pauling, and his major contributions to protein structure. Box 6-2 provides an example of how changes in primary structure affect changes in tertiary and quaternary structures leading to pathology (collagen diseases). Boxes 6-3 and 6-4 discuss the discovery of thermostable proteins and recent progress in predicting the structures of proteins, respectively.

## *Essential Concepts*

1.  There are four levels of organization in protein structure:
    (a) Primary structure, which is the amino acid sequence.
    (b) Secondary structure, which is the local three-dimensional structure of a polypeptide without regard to the conformations of its side chains.
    (c) Tertiary structure, which refers to the overall three-dimensional structure of an entire polypeptide.
    (d) Quaternary structure, which refers to the three-dimensional arrangement of polypeptides in a protein composed of multiple polypeptides.

*Secondary Structure*

2.	Amino acids are joined by the condensation of the amino group of one amino acid with the carboxyl group of another, which results in a linkage called a peptide bond. The peptide group has a resonance structure such that the bond connecting the carboxyl carbon and the amino nitrogen has ~ 40% double-bond character.

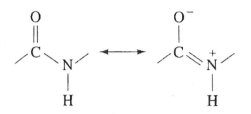

	With rare exceptions, the *trans* conformation shown above occurs in proteins.

3.	The backbone of a polypeptide chain can be viewed as a linked series of planar peptide groups that can rotate about the single covalent bonds involving the α carbon. When the peptide groups are fully extended and all lie in a plane, the torsion angles of the $C_\alpha$—N bond (φ) and the $C_\alpha$—C bond (ψ) are defined as 180° (Figure 6-4). As these planar groups rotate, the van der Waals spheres of their various atoms collide; hence the rotational freedom of the bonds is restricted. The allowed conformations, given as φ and ψ values, are summarized in the Ramachandran diagram (Figure 6-6).

4.	In regular secondary structures, successive residues have similar backbone conformations, that is, repeating values of φ and ψ (e.g., the α helix and the β sheet).

5.	The α helix is a right-handed helix with 3.6 residues per turn and a pitch of 5.4 Å. Hydrogen bonds connect the peptide C = O group of the *n*th residue with the peptide N—H group of the (*n* + 4)th residue. The core of the helix is tightly packed, and the amino acid side chains project outward and downward from the helix.

6.	β sheets result from hydrogen bonding between segments of extended polypeptide chains. Antiparallel β sheets consist of polypeptide strands that extend in alternate directions. Parallel β sheets consist of polypeptide chains that all extend in the same direction. Parallel β sheets are slightly less stable than antiparallel sheets because the hydrogen bonds connecting the polypeptides are distorted. Mixed β sheets, consisting of both parallel and antiparallel polypeptide chains, are common.

7.	Proteins are typically classified as fibrous or globular. A single type of secondary structure dominates fibrous proteins. Two of the best best-characterized fibrous proteins are α keratin and collagen, both of which form higher order structures that are insoluble in water.
	(a) α Keratin consists of two α-helical chains that wrap around each other in a left-handed coiled coil. Each polypeptide chain has a 7-residue pseudorepeat, *a-b-c-d-e-f-g,* such that hydrophobic residues at positions *a* and *d* form the contacts between the helices.

This dimer is assembled further into higher order structures that are less well characterized.

(b) Collagen contains long stretches of the three-residue repeating unit Gly-X-Y, in which X is often Pro, and Y is often 4-hydroxyPro (Hyp; hydroxylation of Pro residues requires ascorbic acid). Each polypeptide strand forms a left-handed helix with around three residues per turn. Three such parallel chains then wrap around each other in a gentle right-handed coil (Figure 6-17). Interchain hydrogen bonds connect all three strands, which are staggered so that a Gly residue occurs at every position along the triple helix (Figure 6-18). Covalent cross-links between modified Lys residues (allysine) and His residues bind together aggregates of collagen fibers.

8. Most proteins contain irregular secondary structures such as β bulges, reverse turns (β bends), and other structures whose residues have nonrepeating backbone conformations.

9. Certain amino acid residues occur more often in α helices or β sheets. Moreover, certain residues tend to disrupt or break secondary structures. These tendencies provide a foundation for predicting protein structure from amino acid sequences and for designing proteins with particular structures.

10. Linus Pauling (1901–1994) laid the foundations for structural biochemistry, culminating with a description of α helices, β sheets, and a precise structural basis for sickle-cell hemoglobin involving a single amino acid substitution.

## Tertiary Structure

11. The tertiary structure of a protein can be determined by X-ray diffraction or nuclear magnetic resonance (NMR).
   (a) The interaction of a beam of X-rays with the electrons of a crystallized protein yields a diffraction pattern that can be mathematically analyzed to reconstruct a three-dimensional electron density map of the atoms of the protein. Knowledge of the protein's primary structure is necessary to identify the amino acid residues in a three-dimensional model of the protein.
   (b) NMR spectroscopy can be used to obtain information about the tertiary structure of a protein in solution. This technique is limited to proteins with ~350 residues but may soon increase to ~900 residues.

12. Protein crystals are typically 40–60% water by volume, giving them a jelly-like consistency. For most crystals, the resolution ranges from 1.5 to 3.0 Å. Several lines of evidence suggest that molecules of crystalline protein have nearly the same structure that they have in solution.

13. In globular proteins, nonpolar residues occur most often in the protein interior. Charged residues most commonly occur on the protein surface. Polar residues that are buried in the protein interior are almost always "neutralized" by forming hydrogen bonds to other internal groups.

14. Groups of secondary structural elements form motifs, the most common of which are the βαβ motif, β hairpin, αα motif, Greek key motif, and β barrel.

15. Although nearly 50,000 protein structures are known, about 200 folding patterns account for nearly half of all known protein structures. Hence, it appears that evolution tends to conserve protein structures, motifs and domains in particular, rather than amino acid sequences. This probably reflects biochemical constraints based on the ability of these domains to
    (a) form stable folding patterns.
    (b) tolerate amino acid deletions, substitutions, and insertions without significant loss of biological activity.
    (c) support essential biological functions.

16. Larger globular domains form in polypeptides containing more than ~200 amino acids. These domains are often associated with a specific biochemical function (e.g., $NAD^+$ binding). There appears to be a limited number of domains, which allows biochemists to group proteins into families.

17. The Protein Data Bank archives the atomic coordinates of proteins and nucleic acids. These can be interpreted by a number of computer programs, many of which are freely available on the internet, to generate three-dimensional, manipulatable images. Several internet sites also use this information to compare the structures of proteins, providing investigators powerful tools to analyze potential structure-function relationships and the evolution of protein structure. One should keep in mind that all such bioinformatic tools help scientists *generate hypotheses* that must ultimately be tested in the laboratory.

*Quaternary Structure and Symmetry*

18. Most proteins with molecular masses >100 kD consist of more than one polypeptide chain. The arrangement of the chains is the protein's quaternary structure. The contact regions between subunits of a protein resemble the interior of a polypeptide.

19. The advantages of having multisubunit proteins include the following:
    (a) A defective region of a protein (e.g., caused by mistranslation or improper folding) can be easily repaired by replacing the defective subunit.
    (b) The only genetic information necessary to specify a large protein is that specifying its few different self-assembling subunits.
    (c) It provides a structural basis for the regulation of enzymatic activity.

20. Proteins with more than one subunit are called oligomers, and their identical subunits (which may contain more than one polypeptide) are called protomers. In most proteins, protomers occupy geometrically equivalent positions, displaying rotational symmetry. The most common are cyclic symmetry (most often involving two protomers) and dihedral symmetry (in which an *n*-fold rotation axis intersects a two-fold rotation axis at right angles).

*Protein Stability*

21. Native proteins are only marginally stable under physiological conditions. The hydrophobic effect is the principal force that stabilizes protein structures. Electrostatic interactions, especially hydrogen bonds and ion pairs, contribute little to the overall stability of tertiary structures because water interacts similarly with fully unfolded (denatured) proteins. However, these weak interactions play important roles in aligning residues in specific secondary structures and domains.

22. Disulfide bonds cross-link structures in extracellular proteins, which occur in a relatively oxidizing environment. Disulfide bonds are rare in intracellular proteins, presumably because the reducing environment of the cytosol weakens them. In some proteins, metal ions such as $Zn^{2+}$ cross-link small structures that would otherwise be unstable.

23. Theoretical calculations suggest that native protein structure consists of rapidly interconverting conformations (breathing), allowing small molecules to move into and out of the interior of proteins.

24. Most proteins unfold or denature at temperatures well below 100°C, with a sharp transition indicating that denaturation is a cooperative process. Conditions or substances that denature proteins include heat, variations in pH, detergents, and chaotropic agents (e.g., guanidinium ion and urea).

*Protein Folding*

25. Under proper conditions, most unfolded proteins will renature spontaneously, thereby indicating that tertiary structure is dictated by primary structure.

26. Protein folding is an ordered process rather than a random search for a stable conformation. During folding, secondary structures form first, followed by motifs and domains. A key driving force is a hydrophobic collapse that buries hydrophobic regions in the interior of the protein, out of contact with the aqueous medium (to yield a state referred to as a molten globule).

27. Certain proteins facilitate protein folding *in vivo,* which is often considerably faster than folding *in vitro.* Protein disulfide isomerase mediates the formation of disulfide bonds. Molecular chaperones prevent improper associations between exposed hydrophobic segments of unfolded polypeptides that could lead to non-native folding as well as nonspecific aggregation of the hydrophobic regions of different unfolded polypeptides.

28. The molecular chaperones include several groups of proteins: (1) the Hsp70 family of proteins; (2) chaperonins; and (3) the Hsp90 family of proteins. The Hsp70 chaperones are a family of 70 kD proteins that prevent premature folding of polypeptides as they are being synthesized in the ribosome.

29.   The chaperonins, which form barrel-shaped structures, consist of two types of proteins, Hsp60 (GroEL in *E. coli*) and Hsp10 (GroES in *E. coli*). A cycle of conformational changes in GroEL/ES requiring the hydrolysis of 7 ATP molecules coordinates protein folding in the interior of the GroEL/ES complex. A polypeptide engages in an average of 24 cycles of binding to the GroEL/ES complex before acquiring its native structure.

30.   Several neurological diseases are characterized by the intracellular accumulation of proteins in aggregates called amyloid plaques. The most notable of these are Alzheimer's disease, prion diseases, and an array of amyloidoses. The aggregates form fibrils largely comprised of β structures. In prion diseases (e.g., scrapie, "mad cow disease," Creutzfeldt–Jakob disease), an α-helical prion changes its conformation to a mixture of α helices and β sheets and forms insoluble fibers. It has been proposed that prions act as infectious agents by catalyzing the misfolding of other prion proteins. The formation of amyloid fibrils in other diseases is less clear; however, once formed they appear to be kinetically resistant to destruction by cellular proteases.

## Exercise

Use a molecular model set to satisfy you that glycine at the center of a tripeptide can assume conformations not shaded in the Ramachandran diagram (i.e., ϕ values between 60° and 180° and all ψ values except those between −45° and 45°).

## Questions

*Secondary Structure*

1.   Why does the peptide bond result in a planar configuration for its adjoining functional groups?

2.   In the dipeptide below, indicate which bonds are described by ϕ and ψ.

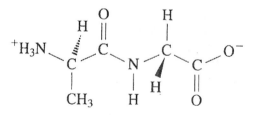

3.     Shown below is a portion of poly-β-alanine₁₀. Comment on the secondary structure of this peptide.

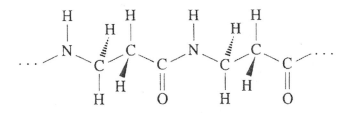

4.     When ϕ = 180° and ψ = 0°, what groups come into closer-than-van-der-Waals contact? (See Figure 6-4.)

5.     How do R groups constrain the potential conformations of a protein? (See Figure 6-4 and Kinemage 3-1.)

6.     Study Figure 6-7 and describe the orientation of the groups that contribute to the dipole moment of the α helix and the overall direction of this dipole moment. *Hint:* The sum of local dipole moments determines the overall dipole moment of the helix.

7.     Why are antiparallel β sheets more stable than parallel β sheets?

8.     How does the 7-residue repeat of α keratin promote formation of a coiled coil?

9.     Wool from sheep is steamed to generate long fibers for knitting. Wool sweaters shrink after subsequent drying with heat. Explain what is happening at the molecular level.

10.    What is the repeating sequence of collagen and how is it essential to the structure of the triple coils of collagen fibers?

11.    Scurvy is caused by a deficiency of _____, which is necessary for the activity of _____.

12.    Organic compounds containing radioactive atoms can be used to follow the biosynthesis of molecules in cells and organisms (see Section 14-4A). In these experiments, the amount of radioisotope in proteins derived from the cells or organisms is monitored. The administration of ¹⁴C-labeled 4-hydroxyproline to mice results in no appearance of ¹⁴C in the animals' collagen fibers. However, the administration of ¹⁴C-labeled proline does result in the appearance of ¹⁴C in the collagen fibers of mice. Explain these results.

13.    List the three amino acids that are least likely to occur in a β sheet.

*Tertiary Structure*

14. Examine Table 4-1 (pp. 76–77) and determine which amino acids might exhibit identical electron densities at 2.0-Å resolution. *Hint:* The electron density of a hydrogen atom is too small to be apparent at this resolution.

15. A resolution of ~3.5 Å is necessary to clearly reveal the course of the polypeptide backbone in X-ray crystallography. Can any useful information about protein structure be obtained at 6 Å resolution?

16. What advantages does NMR provide over X-ray crystallography in characterizing protein structure? What is the major limitation of NMR analysis?

17. What specific role does knowledge of the primary amino acid sequence play in the determination of tertiary structure by X-ray crystallography?

18. How does the polarity or charge of an amino acid affect its likely location within a protein?

19. How does the ratio of hydrophilic to hydrophobic amino acid residues change in a series of globular proteins ranging from 10 to 50 kD?

20. Go to the National Center for Biotechnology Information (NCBI: http://www.ncbi.nlm.nih.gov/) and click on the word Structure along the menu line (last word on the right). If you choose to download the Cn3D program, you will be able to view the three-dimensional structures directly from the links after you have done a search for a protein structure. While Cn3D is not as powerful as some of the programs discussed in the text, it is a very handy tool. In the search box, type in "cam kinase" for calmodulin-dependent kinase and observe the output. Protein kinases are enzymes (transferases) that transfer a phosphate from ATP to another protein. This covalent modification results in structural changes that either activate or inhibit enzyme activity. Hence, protein kinases are regulatory enzymes in the cell (see Chapter 13).
    (a) Click on 2V7O. This brings up the MMDB Structure Summary page. What kinds of information can you find here?
    (b) This site recognizes two structural domains in calcium-calmodulin-dependent kinase. What structural features seem to dominate each domain? *Hint:* Click on the PDB file name again, and then the folder tab for "Sequence Details."
    (c) Go back to the Structure Summary page, and click on VAST (Vector Alignment Search Tool). Click on each domain and examine the proteins that are retrieved, which show similar structures (this information is based on available 3D structural information). Are any hits for either domain not protein kinases? (See the Structural Genomics Consortium at http://www.sgc.utoronto.ca/SGC-WebPages/StructureDescription/ 2V7O.php for more information about the biochemical functions of each domain.)

*Quaternary Structure and Symmetry*

21.   What are the possible symmetries of (a) a tetrameric protein and (b) a pentameric protein? Assume all the subunits are identical.

22.   The subunit composition of an oligomeric protein can be determined by treating the protein with a cross-linking agent (a bifunctional molecule that reacts with and links groups in two different portions of the polypeptide), denaturing the protein, and analyzing the products by SDS-PAGE (Section 5-2D).

(a)   An oligomer analyzed by SDS-PAGE shows a single 40 kD polypeptide. Brief treatment with a cross-linking agent yields the SDS-PAGE banding pattern shown below. What is the protein's probable subunit composition?

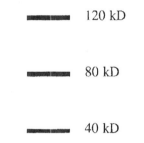

120 kD

80 kD

40 kD

(b)   Not shown in the electrophoretogram above are faint bands at 240 kD and 360 kD. What do these bands represent?

(c)   Another protein analyzed by SDS-PAGE shows two polypeptides at 20 kD and 50 kD. Chemical cross-linking of the native protein yields the results shown below in SDS-PAGE. Not shown is a very faint band at 540 kD. What is the subunit composition of this protein?

270 kD

180 kD

90 kD

50 kD

20 kD

23.   Consider the equilibrium of an oligomer:

    2 monomers ⇌ dimer

The $\Delta H°$ for this reaction is positive and varies little between 4°C and 45°C. $T\Delta S°$, in contrast, varies widely over the same temperature range. The behaviors of $\Delta H°$ and $T\Delta S°$ are shown in the graph below. How does the position of equilibrium of this oligomer differ at 4°C versus 45°C?

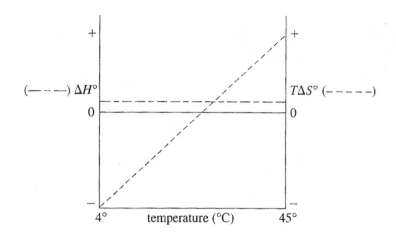

## Protein Stability

24.   What thermodynamic consideration prevents ion pairs from contributing significantly to protein stability?

25.   List four denaturants of proteins.

26.   What is the probability of recovering the native conformation (as judged by biological activity) of a dimeric protein, each of whose subunits contains three disulfide bonds? Assume that the disulfide linkages form randomly.

27.   What amino acid differences and molecular forces distinguish proteins found in hyperthermophiles from those found in mesophiles?

## Protein Folding

28.   Calculate the maximum length of a polypeptide that could recover its native conformation in about one day if the polypeptide explored all possible conformations (see p. 161).

29. To appreciate the role of hierarchical assembly in the rapid renaturation of proteins, consider the equation $t = 10^n/10^{13}$ as a reasonable approximation of renaturation time for a globular protein in which $n$ represents the number of nucleating structures ($\alpha$ helices and $\beta$ sheets). How does this improve the rate of renaturation of a 200-residue globular protein with eight $\alpha$ helices and six $\beta$ sheets?

30. Assume that the average molecular mass of an amino acid is 112, and that 4 ATP equivalents are expended to form a peptide bond. Using the current model for protein folding in GroEL/ES, compare the cost of protein synthesis with the cost of protein folding for a 70-kD polypeptide in *E. coli*.

31. What statement(s) in the text suggests that native protein structure is highly labile?

# 7

# Protein Function: Myoglobin and Hemoglobin, Muscle Contraction, and Antibodies

A fundamental question in biochemistry is how protein structure influences the functional properties of proteins. Chapter 7 begins this exploration by closely examining the structure and function of some of the best studied proteins in biochemistry: myoglobin and hemoglobin, actin and myosin, and antibodies. Myoglobin and hemoglobin are globular proteins involved in oxygen storage and transport. Myoglobin, whose primary role is to facilitate oxygen transport in muscle (and, in aquatic mammals, to store oxygen), binds oxygen with a simple equilibrium constant. Hemoglobin transports oxygen, first binding oxygen in the capillaries of the lungs (or gills or skin), where oxygen concentrations are high, and then releasing the oxygen in the capillaries of the tissues, where the oxygen concentrations are lower. The binding of oxygen to hemoglobin is not as simple as that for myoglobin. Several factors influence oxygen binding to hemoglobin, including the partial pressure of oxygen itself, pH, the concentration of $CO_2$, and 2,3-bisphosphoglycerate (BPG). The study of mutant hemoglobin molecules with amino acid substitutions that affect hemoglobin function provides important insights into the roles of individual amino acids.

Muscle fibers are largely bundles of myofibrils, consisting of interdigitated thick and thin filaments. The thick filaments are made of hundreds of myosin molecules, whose two heavy chains and four light chains form a long rod-like segment with two globular heads. Thin filaments contain three proteins: actin, tropomyosin, and troponin. Interactions between the thick and thin filaments allow the proteins to move past each other in a process that is driven by the hydrolysis of ATP. The myosin molecules undergo large conformational changes that generate easily measurable force and are the canonical examples of force-generating proteins called motor proteins, which include proteins involved in the movement of chromosomes and organelles (kinesins and dyneins), as well as DNA and RNA polymerases.

Antibodies are the major line of defense against extracellular pathogens such as bacteria and viruses. Cellular immunity is mediated by T lymphocytes, which are formed in the thymus. Humoral immunity is mediated by antibodies (immunoglobulins) that are produced by B lymphocytes, which mature in the bone marrow. A B cell makes only one type of antibody. Antigen binding to a specific antibody located on the surface of a B cell triggers an immune response in which the B cells secreting that antibody proliferate. B cells usually live for only a few days; however, memory B cells that recognize specific antigens remain and proliferate when more antigen is present. The specificity of an antibody–antigen interaction arises from variable amino sequences in the immunoglobulin, which create a unique antigen-binding site.

## *Essential Concepts*

### *Oxygen Binding to Myoglobin and Hemoglobin*

1. Myoglobin is a single polypeptide containing a heme group, which consists of a porphyrin ring whose coordinated Fe(II) atom binds molecular oxygen. The protein prevents oxidation of the heme iron to Fe(III), which does not bind oxygen. Oxidized myoglobin is called metmyoglobin.

2. Myoglobin facilitates oxygen diffusion in muscle, which under conditions of high exertion have a particularly high demand for oxygen. It also functions as an oxygen-storage protein in aquatic mammals, where it is in ten-fold higher concentration than in terrestrial animals. A simple equilibrium equation describes $O_2$ binding to myoglobin (Mb): Mb + $O_2$ $\leftrightarrows$ $MbO_2$. The fractional saturation of myoglobin is defined as $Y_{O_2} = pO_2/(K + pO_2)$, where $pO_2$ is the partial pressure of oxygen and $K$ is its dissociation constant. A plot of $Y_{O_2}$ versus $pO_2$ is a hyperbolic curve. It is convenient to define $K$ as $p_{50}$, the partial pressure of $O_2$ at which 50% of myoglobin has bound oxygen. The $p_{50}$ for myoglobin is 2.8 torr, which is much lower than the $pO_2$ in venous blood (30 torr), so that myoglobin is nearly saturated with oxygen under physiological conditions.

3. Hemoglobin binds oxygen in the lungs ($pO_2$ = 100 torr) and releases it in the capillaries ($pO_2$ = 30 torr). The efficiency of oxygen transport is greater than expected if oxygen binding were hyperbolic, as in myoglobin. Hemoglobin, which is an $\alpha_2\beta_2$ tetramer, each of whose subunits contains a heme group, binds $O_2$ cooperatively and thus has a sigmoidal oxygen binding curve. Deoxyhemoglobin is bluish (the color of venous blood), whereas oxyhemoglobin is bright red (the color of arterial blood).

4. The $p_{50}$ of hemoglobin is about 26 torr, which is nearly 10 times greater than that of myoglobin. Because hemoglobin exhibits a sigmoidal oxygen-binding curve, it releases a much greater fraction of its bound $O_2$ in passing from the lungs to the tissues than if I had a hyperbolic binding curve with the same $p_{50}$ (see Figure 7-6). The Hill equation describes the cooperative nature of oxygen binding, and the Hill constant, $n$, describes the degree of cooperativity. The Hill constant, which is not necessarily an integer, is obtained experimentally. The binding of $O_2$ to hemoglobin is said to be cooperative because the binding of $O_2$ to one subunit increases the $O_2$ affinity of the other subunits. The fourth oxygen to bind to hemoglobin does so with a 100-fold greater affinity than the first.

5. Hemoglobin and myoglobin are two examples of numerous oxygen-transport proteins that have been identified, among which are leghemoglobins from leguminous plants, chlorocruorins from some annelids, and hemerythrin and hemocyanin from invertebrates. Hemerythrin and hemocyanin are unique as oxygen-transporters since they do not contain heme groups.

6. Hemoglobin has only two stable conformational states, the T state (the conformation of deoxyhemoglobin) and the R state (the conformation of oxyhemoglobin). Oxygen binding causes the T state to shift to the R state, which has greater affinity for oxygen. The T to R shift is triggered by oxygen binding to the heme iron, which pulls the heme iron atom into the heme plane. This movement is transmitted to the F helix through His F8, which ligands the iron atom. Conformational changes in one subunit are transmitted across the $\alpha_1$–$\beta_2$ and $\alpha_2$–$\beta_1$ interfaces. Due to the conformational constraints at these interfaces, the conformational shift of one subunit must be accompanied by the conformational shift of all subunits, thereby increasing the oxygen affinity of the unoccupied subunits.

7. Decreases in pH promote the release of oxygen from hemoglobin. This effect, called the Bohr effect, is driven by dissolved carbon dioxide, which forms bicarbonate ion and a hydrogen ion. The hydrogen ion protonates hemoglobin, thereby stabilizing its T (deoxy) state. In the lungs, the reaction is reversed: Oxygen binding causes a switch to the R (oxy) state. The released hydrogen ions shift the equilibrium between bicarbonate and carbon dioxide, thereby forming carbon dioxide for expulsion from the lungs. Due to the Bohr effect, the low pH in highly active muscles causes the amount of oxygen delivered by hemoglobin to increase by nearly 10%. Carbon dioxide also binds preferentially to the N-terminal amino groups of T-state hemoglobin as carbamates and hence is released by R-state hemoglobin. This accounts for about half of the carbon dioxide released from the blood in the lungs.

8. Hemoglobin stripped of 2,3-bisphosphoglycerate (BPG) binds oxygen more tightly than hemoglobin in the blood. BPG binds to the T state but not the R state, thereby decreasing hemoglobin's oxygen affinity. This allows nearly 40% of the oxygen to be unloaded in venous blood. Fetal hemoglobin does not bind BPG as tightly and therefore has a higher affinity for oxygen than adult hemoglobin.

9. Humans use two mechanisms to adapt to the lower oxygen partial pressure at higher altitudes: (1) erythrocyte BPG concentration increases to promote oxygen delivery to tissues, and (2) the hemoglobin concentration rises. Other mammals that live at high altitude have isoforms of hemoglobin with increased affinity for oxygen.

10. Allosteric proteins are oligomers with multiple ligand-binding sites, in which ligand binding at one site alters the protein's binding affinity for a ligand at another site. Two models for cooperativity have been proposed: (1) the symmetry model and (2) the sequential model.

11. The symmetry model proposes that all the subunits in the protein exist in either the T state or the R state, and ligand binding to one subunit favors the conversion of all subunits to the R state. In contrast, the sequential model proposes that ligand binding induces a conformational change in one subunit (to the R state), which causes progressive changes in conformation in adjacent subunits (compare Fig. 7-15 with Fig. 7-16). Current studies suggest that hemoglobin shows features of both models, neither of which can fully explain the complexity of protein dynamics upon ligand binding.

12. There appear to be nearly 950 allelic variants of hemoglobin in humans, >90% of which are due to single amino acid substitutions. Most do not result in any discernible clinical symptoms. Hemoglobin variants that have demonstrated clinical symptoms are most commonly correlated with (1) decreased ability of globins to bind heme; (2) decreased cooperativity in which the protein is "locked" in either the T- or R-state; (3) increased propensity for oxidation of the ferrous iron to the ferric state; and (4) decreased protein stability resulting in protein degradation.

13. The analysis of mutant hemoglobins with altered functions has conveyed considerable knowledge about protein structure–function relationships. One variant, hemoglobin S, has a substitution of valine for glutamic acid in the β subunit. In the deoxy state, hemoglobin S aggregates, causing erythrocytes to sickle and block the capillaries. Individuals who are homozygous for the gene specifying hemoglobin S have sickle cell anemia, a debilitating and often fatal disease.

*Muscle Contraction*

14. Striated muscle is made of parallel bundles of myofibrils. As seen in the electron microscope, I bands of lesser electron density alternate with A bands of greater density. The repeating unit, the sarcomere, is bounded by Z disks at the centers of adjacent I bands and includes the A band, which is centered on the M disk. Within the sarcomere, thick filaments are linked to thin filaments by cross-bridges. Muscle contraction occurs when the filaments slide past each other, bringing the Z disks closer together.

15. Thick filaments are made of myosin, which consists of six subunits: two heavy chains and two pairs of light chains. A heavy chain consists of a globular head and a long α-helical tail. Two such tails associate in a 1600-Å-long left-handed coiled coil, yielding a rodlike molecule that has two globular heads. The light chains bind near the globular heads. Thin filaments are composed of actin, tropomyosin, and troponin. Tropomyosin is a coiled coil that winds in the actin polymer's helical grove so as to contact seven successive actin monomers. Troponin, a heterotrimer , interacts with tropomyosin and serves as a sensor molecule, binding $Ca^{2+}$ released from the sarcoplasmic reticulum upon stimulation of muscle contraction. Additional muscle proteins help define sarcomere structure.

16. Dystrophin, a critical protein link between F-actin filaments and a transmembrane glycoprotein, helps prevent large membrane ruptures from the mechanical stress of muscle contraction. Sufferers of Duchenne Muscular Dystrophy (DMD) lack this protein, while patients with milder Becker Muscular Dystrophy (BMD) have reduced levels.

17. The sarcomere contraction cycle involves conformational changes in myosin that drive the movement of actin filaments towards each other. (See Figure 7-32.) This model was born out of the sliding filament model proposed by Hugh Huxley and Andrew Huxley in 1954 (Box 7-4).
    (a) The myosin heads of the thick filaments bind to the thin filaments. When ATP binds to the myosin head, myosin dissociates from the actin filament.
    (b) Subsequent ATP hydrolysis cocks the head of myosin and allows it to rebind weakly to actin.

(c) The release of inorganic phosphate increases the strength of binding and causes the head of myosin to snap back in the power stroke. This pulls the filaments past each other. Each myosin head acts in this manner to cause muscle contraction.

(d) ADP is released, completing the cycle.

18. $Ca^{2+}$, which binds to troponin C, triggers muscle contraction. Nerve impulses induce the release of $Ca^{2+}$ from intracellular stores. This $Ca^{2+}$ binds to troponin and causes a conformational change that exposes additional myosin binding sites on the thin filament. At lower $Ca^{2+}$ concentrations, the myosin head is blocked from binding actin and the muscle is relaxed.

19. Nonmuscle microfilaments (~70-Å-diameter fibers), which consist of actin, occur in all eukaryotic cells, where their assembly and disassembly drives such cellular processes as ameboid locomotion and cytokinesis.

20. In most eukaryotic cells, actin is the most abundant cytosolic protein. The globular polypeptide has an ATP-binding site on one side of the protein. When actin polymerizes, this binding site marks the (−) end or pointed end of the polymer; the opposite end is referred to as the (+) end or barbed end. Actin also has a divalent cation–binding site that has a greater affinity for $Ca^{2+}$ than $Mg^{2+}$; however, most G-actin *in vivo* is bound to $Mg^{2+}$ due to its greater concentration in cells.

(a) Polymerization of G-actin into F-actin differs at each end of the growing filament: The rate of subunit addition at the (+) is 5- to 10-fold greater than that at the (−) end, so new growth tends to occur at the (+) end (hence its name).

(b) The rate of addition of G-actin monomers may be matched by the rate of G-actin monomer dissociation. This steady state process, called treadmilling, plays an important role in the amoeboid crawling by cells.

## Antibodies

21. Antibodies (immunoglobulins) are proteins produced by the immune system of higher organisms to protect them against pathogens such as viruses and bacteria. Antibodies are produced by B lymphocytes, which recognize foreign macromolecules (antigens). The primary response to an antigen requires several days for B cells to generate the required antibodies. If the organism subsequently encounters the same antigen, a secondary response results, in which large amounts of the antibody are produced more rapidly.

22. Antibodies contain at least four subunits, two identical light chains and two identical heavy chains, which are held together in part by interchain disulfide bonds to form a Y-shaped molecule. Of the five classes of antibodies, IgG is the most abundant. The classes are distinguished by the type of heavy chain ($\alpha$, $\delta$, $\varepsilon$, $\gamma$, and $\mu$). There are two types of light chain ($\kappa$ and $\lambda$). IgG heavy chains each consist of three domains of constant sequence, designated $C_H 1$, $C_H 2$, and $C_H 3$, and one variable domain, designated $V_H$. The light chains each consist of one constant domain, $C_L$, and one variable domain, $V_L$. The $V_H$ and $V_L$ are located at the two split ends of the Y-shaped protein.

23. The variable regions contain the antigen-binding sites in which hypervariable sequences determine the exquisite specificity of antibody–antigen interactions.

24. The binding of an antibody to its antigen is highly specific and has a dissociation constant ranging form $10^{-4}$ to $10^{-10}$ M. Because antibodies each have two identical antigen-binding sites, a population of antibodies can form large antigen–antibody aggregates that hasten the removal of the antigen and induce B cell proliferation.

25. A B cell produces one kind of antibody. Antigens stimulate the proliferation of a population of pre-existing B cells whose diversity arises from genetic changes that occur in B cell development, including somatic recombination and somatic hypermutation. Sequence variation allows the synthesis of potentially billions of immunoglobulins with different antigen-binding specificities.

26. In autoimmune diseases, the organism loses its self-tolerance and produces antibodies against its own tissues. This process is sometimes triggered by trauma or infection and may result from the resemblance of a self-antigen to some foreign antigen. Autoimmune diseases have variable symptoms ranging from mild to lethal.

30. Monoclonal antibodies, developed by César Milstein and Georges Köhler, are derived from clones of fused cells that make antibodies specific for one antigen (Box 7-5). Monoclonal antibodies can be used as the stationary phase of affinity chromatography columns for the purification of proteins, they can be used to identify specific infectious agents and test for the presence of drugs and other diagnostic molecules in body tissues, and they can be used in chemotherapy against specific cancers.

### Key Equations

$$Y_{O_2} = \frac{pO_2}{K + pO_2}$$

$$Y_{O_2} = \frac{(pO_2)^n}{(p_{50})^n + (pO_2)^n}$$

### Behind the Equations: What Does $p_{50}$ Tell Us? Insights into the Effects of Allosteric Effectors on Hemoglobin Structure and Function

Oxygen binding by myoglobin and hemoglobin is often described and characterized by a saturation curve. The saturation curve depicts the percent of total myoglobin or hemoglobin that binds oxygen as a function of the partial pressure of oxygen ($pO_2$) that the protein is exposed to. The fractional saturation curves for myoglobin and hemoglobin are noticeably different, and can be modeled by a mathematical equation. The saturation curves for myoglobin and hemoglobin are noticeably different: The curve for myoglobin is a rectangular hyperbola, and the curve for hemoglobin is sigmoidal (S-shaped). Despite the different shapes of the curves for each system, the equation that describes the fractional saturation for each protein is the same:

$$Y_{O_2} = \frac{pO_2^n}{(K + pO_2)^n}$$

The main difference between the two systems is the value of $n$, or Hill constant. For myoglobin, the value of n is equal to 1, indicating that myoglobin is a non-cooperative system. For hemoglobin the value of n is greater than one, and is often reported as $\approx 3$, indicating that hemoglobin is a positively cooperative system. The difference between Hill constant for myoglobin and hemoglobin is sufficient to produce differently-shaped fractional saturation curves.

The variable $K$ in the fractional saturation equation represents the equilibrium dissociation constant for either system. Equilibrium dissociation constants are inversely proportional to affinity.

Thus, a protein that binds a ligand tightly (has high affinity) has a low value of $K$, whereas a system with a high value of $K$ will have a low affinity (bind a ligand relatively loosely).

The binding of oxygen to myoglobin can be described chemically as: $Mb + O_2 \rightleftharpoons MbO_2$

The equilibrium dissociation constant is simply the concentration of reactants divided by the concentration of products:

$$K = \frac{[Mb][O_2]}{[MbO_2]}$$

The binding of oxygen to hemoglobin can be described chemically as: $Hb + nO_2 \rightleftharpoons Hb(O_2)_n$

The equilibrium dissociation constant for hemoglobin is:

$$K = \frac{[Hb][O_2]^n}{[HbO_2]^n}$$

The variable $K$ is often substituted by the variable $p50$ in the fractional saturation equations for myoglobin and hemoglobin. $p50$ is defined as the $pO_2$ required to achieve 50% saturation of the oxygen-binding sites in a myoglobin or hemoglobin population. This occurs when the value of $pO_2$ on the fractional saturation curve for oxygen produces a $Y = 0.5$. Thus, the fractional saturation equation can be rewritten as:

$$Y_{O_2} = \frac{pO_2^n}{(p50 + pO_2)^n}$$

What useful information does $p50$ tell us? The value of $p50$ reveals a wealth of information concerning the binding affinity of myoglobin/hemoglobin for oxygen. If $K = p50$, then $p50$ is inversely proportional to affinity.

In the case of hemoglobin, it also serves as a powerful predictive tool about the conformational state of the protein.

Consider Figure 7-13 from the textbook.

This figure depicts multiple hemoglobin fractional saturation curves, each generated in the presence of one or more allosteric effectors of hemoglobin. These curves are compared to a hemoglobin fractional saturation curve in the absence of allosteric effectors (hemoglobin has been stripped). The p50 value for stripped hemoglobin can be deduced by finding the point on the stripped hemoglobin fractional saturation curve of oxygen at $Y = 0.5$, and projecting down to the $x$-axis from that point. We can see that the $p50$ for stripped hemoglobin is between $10 - 20$ torr ($\approx 15$ torr).

Now compare this to the $p50$ value of hemoglobin in the presence of 2, 3-bisphophosphoglycerate (BPG) in the same figure. The $p50$ value of hemoglobin with BPG is between $20 - 30$ torr ($\approx 25$ torr). What is this telling us? The relationship between $p50$ and affinity (p50 $\propto$ 1/affinity) indicates that stripped hemoglobin ($p50 \approx 15$ torr) has a higher affinity compared to hemoglobin in the presence of BPG ($p50 \approx 25$ torr).

Structurally, we know that the allosteric effectors (BPG, $CO_2$, $Cl^-$, $H^+$) preferentially bind to and stabilize T-state (deoxy) hemoglobin. The combined action of the allosteric effectors in whole blood enables hemoglobin to effectively release oxygen to the peripheral tissues by lowering the affinity of hemoglobin for oxygen? As oxygenated (R-state) hemoglobin binds to allosteric effectors in the systemic circulation, the conformation of the protein changes from R-state to T-state. As described by the Perutz mechanism, the conformational change from R-state to T-state is accompanied by a decrease in the affinity of hemoglobin for oxygen. Thus the change from R-state to T-state should be accompanied by an increase in the p50 value of the system. If we could watch a hemoglobin fractional saturation curve change in real time as R-state hemoglobin binds to allosteric effectors, we would see a gradual shift in the curve along the x-axis towards the right as the p50 value increases. All of the allosteric effectors of hemoglobin have essentially the same structural and functional effects on the protein. They preferentially bind and stabilize T-state hemoglobin, shifting the fractional saturation curves to the right. In other words, these substances lower oxygen affinity, increasing p50 values, The actions of various allosteric effectors are synergistic, so hemoglobin structure and function can be fine tuned.

Fractional saturation and p50 changes as a function of ligands and allosteric effectors

| Change in Ligand/Effector | Change in p50 | Affinity for $O_2$ | Position of Fractional Saturation Curve |
|---|---|---|---|
| $\uparrow O_2$ | $\downarrow$ | $\uparrow$ | Shifts to Left |
| $\uparrow CO_2$ | $\uparrow$ | $\downarrow$ | Shifts to Right |
| $\uparrow H^+$ | $\uparrow$ | $\downarrow$ | Shifts to Right |
| $\uparrow Cl^-$ | $\uparrow$ | $\downarrow$ | Shifts to Right |
| $\uparrow BPG$ | $\uparrow$ | $\downarrow$ | Shifts to Right |

Fractional saturation curves and p50 values are valuable diagnostic tools for biochemists and physicians, because the value of p50 provides insights into hemoglobin structure and function.

*Critical Thinking Problem*: Carbon monoxide is a poisonous gas that competes with $O_2$ for the heme iron in hemoglobin. However, CO binds with much higher affinity to the heme iron than does $O_2$. With sufficient exposure, people can suffocate in the presence of CO even if $O_2$ is plentiful. Individuals suffering from CO poisoning often appear 'bright cherry red', which is an important visual clue for paramedics and physicians.

Imagine that you have a population of stripped hemoglobin that is in the T-state. There are no allosteric effectors present to influence the hemoglobin structure and affinity for $O_2$, and there is no $O_2$ currently bound to the protein. This completely deoxygenated, T-state hemoglobin is divided into two groups. The first group (control) is exposed to approximately 30 torr of $O_2$ only. The second group (experimental) is exposed to a small amount of CO, enough to bind only one of the four heme groups in a hemoglobin molecule, before being exposed to 30 torr $O_2$. The fractional saturation curves show that the second group (treated with CO) has a lower *p50* value than the first group (control). Rationalize this observation and propose a mechanism by which CO can lower the p50 of hemoglobin by binding to one out of 4 heme groups. Why are victims of CO poisoning bright cherry red?

Therefore, the lower *p50* of hemoglobin in the presence of CO indicates that the Hb-CO complex has a higher affinity for $O_2$ at the sites not occupied by CO. The Perutz mechanism suggests that the binding of ligand ($O_2$ or CO) to one heme group of T-state hemoglobin causes a conformational change that results in higher affinity for $O_2$ binding to the remaining heme groups on the same molecule. By shifting hemoglobin to the high-affinity R-state, CO boosts the amount of $O_2$ that can bind, even though $O_2$ cannot take the place of the CO that has already bound to one heme group (CO has a much higher affinity than $O_2$). The bright cherry red appearance is due to the hemoglobin being in the R-state, in which the electronic orbital configuration of the heme iron reflects red light. The electronic orbital configuration of T-state hemoglobin, such as in venous blood, absorbs red light and reflects more blue-purple light

## Questions

*Oxygen Binding to Myoglobin and Hemoglobin*

1.  Match the descriptions on the left with the terms on the right.

    _____ A component of cytochromes      A. methemoglobin
    _____ Binds $O_2$      B. myoglobin
    _____ Contains iron in the Fe(III) state      C. hemoglobin
    _____ Found in muscle only      D. hemoglobin S
    _____ Forms filaments in the deoxy state      E. heme

2.  The heme moiety by itself can bind oxygen. What physiological function does the globin serve?

3.  How do tissues with high metabolic activity facilitate oxygen delivery?

4.  How would a lower $p_{50}$ affect hemoglobin's oxygen acquisition in the lungs and oxygen delivery to the peripheral tissues?

5.  Describe, on the molecular level, the role of myoglobin in $O_2$ transport in rapidly respiring muscle tissue.

6.  What function(s) does carbamate formation in hemoglobin serve?

7.  Which of the following modulators of $O_2$ binding to hemoglobin counteract each other? $CO_2$, $H^+$, BPG.

8.  Explain why individuals with severe carbon monoxide poisoning are often given transfusions instead of oxygen-rich gas.

9.  Match each of the structural elements of myoglobin or hemoglobin with its function below.

    A.  His F8      F. $\alpha_1$–$\beta_1$ interface/$\alpha_2$–$\beta_2$ interface
    B.  His E7      G. oxymyoglobin/deoxymyoglobin
    C.  E and F helices      H. oxyhemoglobin/deoxyhemoglobin
    D.  $O_2$–Fe(II)      I. $\alpha_1$–$\beta_2$ interface/$\alpha_2$–$\beta_1$ interface
    E.  Val E11      J. C-terminal salt bridges

    _____ Forms a coordination bond with Fe(II)
    _____ Involved in the binding of heme
    _____ Partially occludes the $O_2$-binding site
    _____ Conformations are nearly superimposable
    _____ Forms a hydrogen bond with $O_2$
    _____ Is associated with a rotational shift of the hemoglobin $\alpha_1\beta_1$ dimer with respect to the $\alpha_2\beta_2$ dimer.
    _____ Disrupts the N- and C-terminal salt bridges in hemoglobin

_____ Stabilizes the T state
_____ Involved in the interactions of hemoglobin subunits

10. Oxygen binding to hemoglobin decreases the p$K$ of the imidazole group of His 146$\beta$ from 8.0 to 7.1. How does this contribute to the Bohr effect?

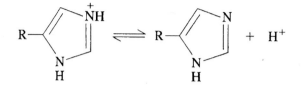

11. Describe how BPG decreases the $O_2$-binding affinity of hemoglobin in terms of the T $\leftrightarrow$ R equilibrium.

12. What kind of allosteric effect is inconsistent with the symmetry model?

13. How many conformational states are possible for a trimeric binding protein, whose three subunits each contain a ligand-binding site, when binding follows the (a) symmetry model or (b) the sequential model of allosterism?

14. Examine the hemoglobin variants in Table 7-1 and provide plausible answers to the following questions.
    (a) Which variant might be advantageous for an individual living in the Andes?
    (b) Which variants are likely to be the least stable, resulting in hemolytic anemia?
    (c) Which variants are likely to result in increased hematocrit (elevated erythrocyte content)?
    (d) Which variants are most like to affect the Hill constant?

15. What physiological condition leads to hemoglobin S fiber formation in the capillaries?

16. Some hemoglobin S individuals have significant levels of fetal hemoglobin in their erythrocytes. Why is this an advantage?

*Muscle Contraction*

17. Describe how ATP hydrolysis is involved in muscle contraction.

18. How does calcium regulate muscle contraction?

19. The graph below, the time course of microfilament polymerization *in vitro*, shows three phases of growth. (a) Describe the biochemical events taking place during each phase. Assume that ATP is in excess. (b) How might this graph look if ATP concentration becomes limiting as ATP is hydrolyzed by actin?

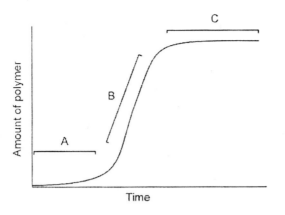

*Antibodies*

20. In a typical protocol for preparing antibodies for laboratory use, an animal is injected with an antigen several times over a period of weeks or months. Why are multiple injections useful?

21. Indicate which class of immunoglobulin (IgA, IgD, IgE, IgG, or IgM) best corresponds to each characteristic listed below (more than one may be implicated by each description).
    (a) First to be secreted in response to an antigen
    (b) Implicated in allergic reactions
    (c) Occurs in the intestinal tract
    (d) Contains J chains
    (e) The most abundant antibody

22. The following questions refer to the production of monoclonal antibodies.
    (a) Why are myeloma cells used?
    (b) Do all cells growing in the selective medium produce antibodies to antigen X?
    (c) Why do only hybrid cells grow in the selective medium?

23. You wish to isolate a large amount of Fab fragments in order to examine their binding to protein X by X-ray crystallography. At first, you inject a large rabbit with protein X to obtain antibody. On further reflection, you decide to inject the antigen into a mouse in order to produce monoclonal antibodies.
    (a) How can you purify protein X–specific antibodies from rabbit serum or the hybridoma medium?
    (b) How do the rabbit and mouse antibody preparations differ?
    (c) Would the rabbit or mouse antibodies be more suitable for X-ray crystallography?

24. Why are the hypervariable sequences of immunoglobulins located in loops?

# 8

# Carbohydrates

This chapter is concerned with the structures and properties of carbohydrates. These molecules contain just three elements, namely carbon, hydrogen and oxygen (although their derivatives may also contain nitrogen and sulfur). Carbohydrates are not only important metabolic energy sources, as detailed in Chapters 15 and 16, but they also have key functions in molecular and cellular recognition events. You will first learn the structures and chemical characteristics of monosaccharides and some of their derivatives, and then of oligosaccharides and polysaccharides. This is followed by a presentation of the composition, structure, and function of molecules in which carbohydrate are covalently linked to polypeptides, including proteoglycans, bacterial cell wall components, and glycoproteins. It is important to realize that a large proportion of proteins contain covalently attached carbohydrates and that the structures of these carbohydrate chains can vary enormously.

## *Essential Concepts*

### *Monosaccharides*

1.   Monosaccharides can be defined as aldehyde or ketone derivatives of straight-chain polyhydroxy alcohols containing a minimum of three carbon atoms. Sugars that have aldehyde groups are called aldoses, whereas those with ketone moieties are termed ketoses. Depending on the number of carbon atoms, monosaccharides are referred to as trioses, tetroses, pentoses, hexoses, etc.

2.   D-Glucose has four chiral centers and is therefore one of 16 possible aldohexose stereoisomers. D Sugars have the same absolute configuration at the asymmetric center most distant from the carbonyl group as does D-glyceraldehyde. Epimers are sugars in which the configuration around one carbon atom differs. Ketoses, which have a ketone function at C2, have one less chiral center than aldoses with the same number of carbons. Therefore, a ketohexose has only 8 possible stereoisomers.

3.   Sugars can be represented in their cyclic hemiacetal and hemiketal forms as planar Haworth projections. Sugars that form six-membered rings are known as pyranoses, whereas those with five-membered rings are known as furanoses. A cyclic monosaccharide exists as either an $\alpha$ or a $\beta$ anomer. Anomers freely interconvert in aqueous solution via the linear (open chain) form.

4.   Five- and six-membered sugar rings are most the abundant, probably because of their stability. The tetrahedral bonding angles of carbon prevent the rings from being truly planar. The pyranose ring prefers the chair conformation and exists predominantly in the form that minimizes steric interactions among bulky ring substituents (i.e., bulky groups tend to occupy equatorial rather than axial positions).

5. Sugars can undergo reactions characteristic of aldehydes and ketones:
    (a) Oxidation of the aldehyde group of an aldose yields an aldonic acid. Thus, D-glucose oxidation results in D-gluconic acid.
    (b) Oxidation of the primary hydroxyl group produces a uronic acid, such as D-glucuronic acid from D-glucose.
    (c) Reduction of aldoses and ketoses yields polyhydroxy alcohols called alditols. Thus, D-glucose reduction gives glucitol (also known as sorbitol).

6. Monosaccharides in which an OH group is replaced by an H are called deoxy sugars. The most important of these is β-D-2-deoxyribose, a component of DNA.

7. In amino sugars, an OH group is replaced by an amino group that is usually acetylated. A common amino sugar is *N*-acetylglucosamine. *N*-acetylneuraminic acid, an important constituent of glycoproteins (see below), is composed of an acetylated amino sugar, *N*-acetylmannosamine, covalently linked to pyruvic acid.

8. The anomeric carbon of a sugar can form a covalent bond with an alcohol to form an α or β glycoside. The bond is called a glycosidic bond. An *N*-glycosidic bond links an anomeric carbon and a nitrogen atom, as in the covalent bond between ribose and a purine or pyrimidine.

9. Reducing sugars are saccharides with anomeric carbons that have not formed glycosides. The free aldehyde of these sugars can reduce mild oxidizing agents. Hence, the identification of a nonreducing sugar represents evidence that the saccharide is a glycoside.

*Polysaccharides*

10. Polysaccharides are made up of monosaccharides covalently linked by glycosidic bonds. A homopolysaccharide contains one type of monosaccharide, whereas a heteropolysaccharide can contain diverse monosaccharides. Polysaccharides may be linear or branched, because glycosidic bonds can form between an anomeric carbon and any of the hydroxyl groups of another monosaccharide. Naturally occurring polysaccharides incorporate only a few types of monosaccharides and glycosidic linkages.

11. Disaccharides consist of two sugars linked by a glycosidic bond. One example is lactose, in which C1 of galactose is linked to C4 of glucose by an α-glycosidic bond. Lactose is a reducing sugar because the free anomeric carbon on the glucose residue can reduce a mild oxidizing agent. Another disaccharide is sucrose, common table sugar, in which the C1 anomeric carbons of glucose and fructose are joined by an α-glycosidic bond. Sucrose is therefore a nonreducing sugar.

12. Cellulose, the most abundant polysaccharide, is a large, linear polymer in which glucose units are linked by β(1→4) glycosidic bonds. Cellulose forms a highly hydrogen-bonded structure of enormous strength that contributes to the rigidity of plant cell walls. Cows and other herbivores can utilize cellulose as a nutrient because they harbor microbes that produce cellulases which cleave the glycosidic bonds. Another widely distributed polysaccharide is chitin, which comprises the exoskeletons of many invertebrates. It is a

homopolymer of *N*-acetylglucosamine residues linked in β(1→4) fashion. Another key component of plant cell walls are pectins, heterogeneous polysaccharides (similar to glycosaminoglycans described below).

13. Starch is the principal food reserve in plants. It has two components: α-amylose, a linear polymer of α(1→4) linked glucose units, and amylopectin, an α(1→4) linked glucose polymer bearing periodic branches linked by α(1→6) bonds. Digestion of starch by animals begins with the action of salivary amylase, an enzyme that cleaves α(1→4) glycosidic bonds, and is continued in the small intestine by pancreatic amylase, α-glucosidase, and a debranching enzyme that can cleave α(1→6) links. The resulting oligosaccharides are eventually converted to glucose, which can be absorbed by the intestine.

14. Glycogen is a polysaccharide that is synthesized and stored by animals, primarily in skeletal muscle and liver. The structure of glycogen resembles that of amylopectin but is more highly branched. When needed for metabolic energy, glycogen is broken down through the combined action of glycogen phosphorylase, which cleaves α(1→4) bonds, and glycogen debranching enzyme.

15. Glycosaminoglycans are major constituents of the extracellular matrix in eukaryotes and in bacterial biofilms. Most of these rigid, linear polysaccharides are composed of alternating uronic acid and hexosamine residues. For example, hyaluronic acid is composed of D-glucuronate linked β(1→3) to *N*-acetylglucosamine which in turn is linked β(1→4) to the next glucuronate residue. Because of its polyanionic nature, hyaluronic acid forms viscoelastic solutions, a property that makes it an effective biological shock absorber and lubricant.

16. Other types of glycosaminoglycans, all of which are composed of sulfated disaccharide units, include chondroitin sulfates, dermatan sulfate, keratin sulfate, and heparin. The last-named substance is found in mast cells and inhibits blood clotting.

*Glycoproteins*

17. A large proportion of all proteins have covalently bound carbohydrates and are therefore glycoproteins. The polypeptide chains of glycoproteins are encoded by nucleic acids, whereas the attached oligosaccharide chains are products of enzymatic reactions. This is the source of microheterogeneity, the variability in composition of the carbohydrate component in a population of glycoprotein molecules that all have the same polypeptide chain.

18. Proteoglycans occur mainly in the extracellular matrix and are combinations of proteins and glycosaminoglycans that associate by both covalent and noncovalent bonds. These molecules have a bottlebrush-like structure (e.g., Figure 8-15) in which up to 100 core proteins with attached glycosaminoglycans and both *N*-linked and *O*-linked oligosaccharides are linked to hyaluronate. This assembly of protein and carbohydrate is a

huge, space-filling macromolecule. Proteoglycans are highly hydrated, so that, in combination with collagen, they account for the high resilience of cartilage.

19.   Bacteria possess rigid cell walls that are responsible in part for their virulence. In gram-positive bacteria, the cell wall consists of polysaccharide and polypeptide chains that are covalently attached to form a baglike structure called peptidoglycan that completely envelops the cell. Gram-negative bacteria have a relatively thin peptidoglycan cell wall surrounded by a complex outer membrane.

20.   The polysaccharide of some bacterial cell walls consists of alternating residues of $\beta(1{\to}4)$-linked *N*-acetylmuramic acid and *N*-acetylglucosamine in which the *N*-acetylmuramic acid residues are linked via an amide bond to a tetrapeptide containing D-amino acids. A continuous meshlike framework is formed by cross-linking adjacent peptidoglycan chains through their tetrapeptide side chains (Figure 8-17). The enzyme lysozyme can degrade peptidoglycan by cleaving the glycosidic bond between *N*-acetylmuramic acid and *N*-acetylglucosamine. The antibiotic action of penicillin rests on its ability to inhibit the formation of cross-links in peptidoglycan.

21.   Glycoproteins include nearly all membrane-bound and secreted eukaryotic proteins. The oligosaccharide chains are attached to the proteins by either *N*-glycosidic or *O*-glycosidic bonds. In an *N*-glycosidic bond, the amide group of an asparagine in the sequence Asn-X-Ser/Thr is linked to *N*-acetylglucosamine. *N*-glycosylation occurs in stages:
(a)   An oligosaccharide rich in mannose and containing glucose and *N*-acetylglucosamine is attached cotranslationally, that is, while the polypeptide is being synthesized on ribosomes bound to the endoplasmic reticulum.
(b)   The oligosaccharide undergoes trimming, the enzymatic removal of some sugars, as the glycoprotein moves from the endoplasmic reticulum to the Golgi apparatus.
(c)   Further processing occurs in the Golgi apparatus, where monosaccharides such as *N*-acetylglucosamine, galactose, L-fucose, and *N*-acetylneuraminic acid are enzymatically added to the trimmed chain by glycosyltransferases. *N*-linked glycoproteins exhibit great diversity in their oligosaccharide chains due to differences in the extent of processing.

22.   *O*-glycosidically linked oligosaccharide chains are covalently linked to a Ser or Thr side chain in a protein and vary considerably in structure. *O*-linked oligosaccharides are added to completed polypeptides in the Golgi apparatus and are built up by stepwise addition of monosaccharides in reactions catalyzed by glycosyltransferases.

23.   The number and structure of *N*- and *O*-linked oligosaccharides attached to a given polypeptide chain can vary, giving rise to glycoprotein variants called glycoforms.

24.   Particular functions can sometimes be attributed to the presence of oligosaccharides. Oligosaccharides attached to proteins may modulate the conformational freedom of the polypeptide. The hydrophilic oligosaccharide chains may take up considerable volume and thereby tend to protect the protein from enzymatic attack or modify its activity.

25. The enormous number of oligosaccharide structures suggests that they contain biological information and are important in molecular recognition. All cells have a coat of glycoconjugates (a mixture of glycoproteins and glycolipids). Proteins called lectins specifically recognize and bind to individual monosaccharides or small oligosaccharides in discrete glycosidic linkages at the cell surface. For example, the leukocyte selectins bind to cell-surface carbohydrates on endothelial cells, an interaction that helps direct leukocytes to a site of blood vessel injury.

26. Different oligosaccharide components of certain glycolipids distinguish the ABO blood group antigens.

27. Oligosaccharide chains mediate a variety of biological functions that depend on molecular recognition between proteins and carbohydrates. Among these are delivery of proteins to appropriate destinations within cells and the regulation of cell growth.

## Questions

*Monosaccharides*

1. Indicate which of the following is an aldose, a ketose, a pentose, a hexose, a uronic acid, an alditol, a deoxy sugar, or a reducing sugar.

| (a) | (b) | (c) | (d) | (e) | (f) |
|---|---|---|---|---|---|
| CHO | CHO | $CH_2OH$ | CHO | $CH_2OH$ | CHO |
| HOCH | HCOH | C=O | HCOH | HCOH | HCOH |
| HCOH | HCOH | HCOH | HCOH | HOCH | HOCH |
| HCOH | HOCH | HCOH | HCOH | HCOH | HOCH |
| $CH_2OH$ | HCOH | HCOH | HCOH | HCOH | HCOH |
| | $CH_2OH$ | $CH_2OH$ | COOH | $CH_2OH$ | $CH_3$ |

2. Draw the following monosaccharides as Haworth projections:
   (a) α anomer of D-ribose
   (b) β anomer of D-glucose
   (c) β anomer of D-fructose
   (d) methyl-β-D-galactose.

   Which of these compounds contains a glycosidic bond?

3. An equilibrium mixture of D-glucose contains approximately 63% β-D-glucopyranose and 36% α-D-glucopyranose. There are trace amounts of three other forms. What are they?

4. Although β-D-glucopyranose is the predominant form of glucose in solution, crystalline glucose consists almost exclusively of α-D-glucopyranose. What accounts for this difference?

5. Draw the most stable chair conformation of α-D-galactose.

*Polysaccharides*

6. Match each term at the top with its definition below.
   A. Alditol
   B. Epimer
   C. Glycan
   D. Anomer
   E. Glycoside
   _____ Differs in configuration at the anomeric carbon
   _____ Polyhydroxy alcohol
   _____ Product of condensation of anomeric carbon with an alcohol
   _____ Differs in configuration at one carbon atom.
   _____ Polymer of monosaccharides

7. Which of the following di- and trisaccharides contains fructose, contains an α anomeric bond, and/or is a reducing sugar?

(a)  sucrose

(b)  cellobiose

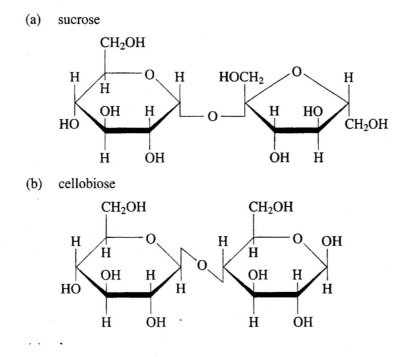

(c)  lactose

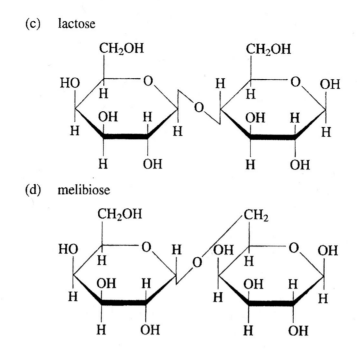

(d)  melibiose

8.  An unknown trisaccharide was treated with methanol in HCl (to methylate its free OH groups) then subjected to acid hydrolysis (to break glycosidic bonds). The products were 2,3,4,6-tetra-*O*-methylgalactose, 2,3,4-tri-*O*-methylglucose, and 2,3,6-tri-*O*-methylglucose. Treatment of the intact trisaccharide with β(1→6)-galactosidase yielded D-galactose and a disaccharide. Treatment of this disaccharide with α(1→4)-glucosidase yielded D-glucose. Draw the structure of this trisaccharide and give its systematic name.

9.  How do the chemical differences between starch and cellulose result in their very different polymeric structures?

10. Glycogen and starch are extensively branched high-molecular-weight polymers.  Suggest three reasons why such a structure is advantageous for a fuel-storage molecule.

## *Glycoproteins*

11. How could lectins be used to purify polysaccharides and glycoproteins?

12. Figure 8-19 presents structures of typical *N*-linked oligosaccharides. (a) Which sugars make up the core oligosaccharide? (b) What other sugars are typically found at the termini of branched oligosaccharides?

13. In some regions of the Great Plains, livestock grazing on a native plant began acting strangely and eventually died. The animals were considered to be "loco" and the plant was therefore called locoweed. Examination of the stomach contents of the animals revealed a buildup of glycoproteins in which *N*-linked oligosaccharides had an unusually high amount of mannose. Analysis of locoweed led to the isolation of a mannosidase inhibitor, which was named swainsonine. At what point in the synthesis of *N*-linked oligosaccharides does swainsonine exert its inhibitory effect? (Hint: see Figure 8-19)

14. The dissolution of blood clots involves hydrolysis of fibrin by the protease plasmin. Plasmin is activated by cleavage of inactive plasminogen by another protease called tissue plasminogen activator (tPA). tPA, in turn, must also be activated by proteolysis at a single site. Once plasmin is activated, it cleaves and activates more tPA in a positive feedback loop. tPA is a glycoprotein that can have either two or three *N*-linked oligosaccharide chains. Measurements indicate that tPA bearing two such chains is cleaved by plasmin much more rapidly than the molecule with three chains. Propose an explanation for the difference.

15. A major protein of saliva contains several hundred identical covalently attached disaccharides containing *N*-acetylneuraminic acid in α(2→6) linkage to *N*-acetylgalactosamine that is linked to a Ser residue. Solutions of the intact protein are extremely viscous. However, when the protein is treated with sialidase, the viscosity decreases markedly. What features of the glycoprotein structure give rise to the high viscosity and why does sialidase treatment bring about the observed change?

16. Explain why a type AB individual can receive a transfusion of type A or type B blood, but a type A or type B individual cannot receive a transfusion of type AB blood.

# 9
# Lipids and Biological Membranes

Lipids are vital components of all cells. Unlike nucleic acids, proteins, and carbohydrates, lipids do not have unifying structural features. They are defined operationally as a diverse group of nonpolar substances. They are therefore poorly or not at all water-soluble but dissolve readily in many organic solvents. Lipids are essential constituents of biological membranes. In addition, the acyl chains of lipids serve as energy sources. Finally, a number of cell signaling processes involve lipids. This chapter begins by focusing on the structures and physical characteristics of lipids and on the properties of the lipid bilayer. The chapter then turns to biological membranes, which are composed of a great variety of proteins as well as lipids. Membrane proteins act as enzymes, facilitate transmembrane transport, and use information concerning the extracellular milieu to prompt the correct intracellular response. This chapter describes the features of membrane proteins that enable them to interact with lipids, and describes the overall arrangement of lipids and proteins in membranes. The chapter concludes with discussions of the synthesis of membrane and secreted proteins (a process in which a polypeptide is translocated through the endoplasmic reticulum membrane), intracellular trafficking of vesicles during exocytosis and endocytosis, and membrane fusion.

## Essential Concepts

### Lipid Classification

1.  Fatty acids are common components of other lipids, where they occur in ester or amide linkages. They are straight-chain carboxylic acids, usually having between 14 and 22 carbons. Eukaryotic fatty acids usually possess an even number of carbon atoms and may contain one or more double bonds. Although the systematic names of fatty acids reveal their structures, many fatty acids also have common names.

2.  Saturated fatty acids, of which the most common are palmitic acid (C16; hexadecanoic acid) and stearic acid (C18; octadecanoic acid), contain no double bonds and are highly flexible molecules that tend to assume a fully extended conformation. In the pure compound, neighboring saturated fatty acyl chains pack together tightly. The resulting van der Waals interactions cause fatty acid melting points to increase as chain length increases.

3.  The double bonds of unsaturated fatty acids nearly always have the cis configuration, which introduces a 30° bend in the acyl chain. This prevents unsaturated fatty acids from packing together as closely as saturated fatty acids. As a result, the melting point of an unsaturated fatty acid is always lower than the melting point for a saturated fatty acid with the same number of carbons.

4.  Most fatty acid groups in eukaryotic lipids are unsaturated. The most common fatty acid with one double bond is oleic acid (C18; 9-octadecenoic acid). Fatty acids with two or more double bonds are termed polyunsaturated.

5.  Triacylglycerols, which contain three fatty acids esterified to the hydroxyl groups of d-glycerol, serve as energy reserves in plants and animals. A triacylglycerol generally contains more than one type of fatty acyl group. Mixtures of triacylglycerols may be fats (which are solid at room temperature) or oils (which are liquid at room temperature), depending on the properties of their component fatty acyl groups.

6.  The highly reduced nature of triacylglycerols makes them an efficient metabolic energy store. Adipocytes are cells specialized for the biosynthesis, storage, and breakdown of triacylglycerols. Adipose tissue also provides thermal insulation.

7.  The major lipid constituents of biological membranes are glycerophospholipids. These substances are composed of d-glycerol with fatty acids esterified to C1 and C2 and a phosphate group esterified to C3. The phosphate is almost always also esterified to a hydrophilic moiety. Thus, glycerophospholipids are amphipathic molecules with a polar head group and a nonpolar tail. Glycerophospholipids frequently contain saturated or monounsaturated acyl groups at C1 of glycerol and more highly unsaturated acyl moieties at C2.

8.  Phospholipases are enzymes that hydrolyze glycerophospholipids. For example, phospholipase A2, which occurs in bee and snake venoms, specifically cleaves the C2 ester linkage to yield a free fatty acid and a lysophospholipid. Several phospholipase-catalyzed hydrolysis products of glycerolipids play roles in cell signaling processes. These include 1,2-diacylglycerol and lysophosphatidic acid.

9.  Plasmalogens are glycerophospholipids in which a fatty acyl group is linked to C1 via an α,β-unsaturated ether bond.

10.  Sphingolipids all contain a long-chain nitrogen-containing alcohol named sphingosine. Ceramides have a fatty acyl group linked to the primary amino group of sphingosine and can be regarded as the basic building block of more complex sphingolipids. Among these are (a) sphingomyelin, in which ceramide is esterified to a phosphocholine or phosphoethanolamine head group; (b) cerebrosides, in which ceramide forms a glycosidic bond with either glucose or galactose; and (c) gangliosides, in which oligosaccharides containing one or more N-acetylneuraminic acid groups form the ceramide head group. Cerebrosides and gangliosides are therefore glycosphingolipids. On hydrolysis, sphingolipids, like glycerophospholipids, give rise to products that have signaling activity.

11.  Steroids are derivatives of the cyclopentaneperhydrophenanthrene fused-ring system. Cholesterol, the most abundant animal steroid, is weakly amphiphilic because it has a hydroxyl group at C3 (it is therefore classified as a sterol). It is a major component of biological membranes in animals. In a cholesteryl ester, the hydroxyl group of cholesterol is esterified to a fatty acid.

12. Cholesterol is the precursor of steroid hormones, which include (a) glucocorticoids (e.g., cortisol), which modulate metabolic processes, inflammatory reactions, and stress responses; (b) mineralocorticoids (e.g., aldosterone), which regulate salt and water excretion; and (c) androgens and estrogens, which influence sexual development and reproductive functions.

13. Vitamins D2 and D3 are produced from steroid precursors through photolysis by ultraviolet light. These two vitamins are then converted to active forms by enzymatic hydroxylation. Active vitamin D promotes absorption of dietary $Ca^{2+}$ and enhances the deposition of $Ca^{2+}$ from the blood into bone.

14. Isoprenoids are a large class of lipids that serve multiple functions in animals and plants. This group includes ubiquinone and plastoquinone (essential components of the electron transport chains of mitochondria and chloroplasts, respectively) and vitamins A, K, and E.

15. Eicosanoids are derived from the highly unsaturated C20 lipid arachidonic acid after its release from membrane glycerophospholipids by phospholipase A2. The eicosanoids include prostaglandins, thromboxanes, leukotrienes, and lipoxins, all of which are biologically active at extremely low concentrations. They are produced in a tissue-specific manner and play roles in inflammatory reactions, the regulation of the cardiovascular system, and reproduction.

## Lipid Bilayers

16. The physical properties of lipids in aqueous solution cause them to aggregate. Water tends to exclude the hydrophobic portions of amphiphilic lipids, whereas the polar head groups remain in contact with the aqueous environment. Amphiphiles with a single nonpolar tail, such as soaps, many detergents, and lysophospholipids, form globular micelles. Amphiphiles with two hydrocarbon tails, such as glycerophospholipids, instead form disk-like micelles that are actually lipid bilayers. A bilayer of phospholipids in an aqueous milieu may form a vesicle with a solvent-filled interior, called a liposome. Liposomes are useful as models of biological membranes and as water-soluble drug delivery vehicles.

17. The movement of an amphiphilic lipid across a lipid bilayer, a process called transverse diffusion or a flip-flop, occurs infrequently, because the passage of a polar head group through the hydrophobic interior of the bilayer is thermodynamically unfavorable. In contrast, lipids readily diffuse laterally in the plane of the bilayer, indicating that the hydrocarbon core of the bilayer is highly fluid. The motion of the acyl chains is greatest in the center of the bilayer and decreases markedly near the polar head groups.

18. Bilayer fluidity is a function of temperature. At high temperatures, the lipids form a highly mobile liquid-crystal state, and at low temperatures, they form a gel-like solid as the hydrocarbon tails tightly associate. The transition temperature, the temperature at which the phase changes, depends on bilayer composition and increases with increasing acyl chain length and decreasing unsaturation of the component lipids. The presence of cholesterol also reduces a bilayer's transition temperature. Living organisms alter their membrane composition in order to maintain constant fluidity as the ambient temperature varies.

19. Cholesterol stabilizes the bilayer over a range of temperatures: At high temperatures, its rigid ring system interferes with fatty acyl chain mobility and thereby decreases membrane fluidity. At low temperatures, cholesterol prevents tight packing of adjacent hydrocarbon tails and thereby promotes membrane fluidity.

*Membrane Proteins*

20. The types and relative amounts of membrane proteins and lipids vary among biological membranes. Membrane proteins, although extremely diverse, can be classified as integral, lipid-linked, or peripheral, depending on how they are associated with the membrane.

21. Integral proteins (also known as an intrinsic proteins) are amphiphilic in that some regions of each molecule strongly associate with nonpolar membrane components by means of hydrophobic effects, whereas hydrophilic portions extend into the aqueous surroundings. Integral proteins can be solubilized and extracted from membranes only by using relatively harsh reagents, such as detergents and organic solvents, to disrupt the membrane structure.

22. Membrane proteins contribute to the asymmetry of a biological membrane because they either are present on one side of a membrane or, if they extend through it, are oriented in only one direction. The transmembrane domain of an integral protein may consist of one $\alpha$ helix (as in glycophorin A), a bundle of helices (as in bacteriorhodopsin), or a $\beta$ barrel (as in porin). The backbones of both $\alpha$ helices and $\beta$ barrels can form all their possible hydrogen bonds and are therefore stable in the interior of the lipid bilayer. In all cases, hydrophobic residues contact the hydrocarbon chains in the membrane core, whereas charged polar groups tend to predominate at the membrane surface.

23. Certain membrane proteins are covalently linked to a lipid moiety, which provides the protein with a hydrophobic anchor in the membrane. There are three main types of lipid-linked proteins:
    (a) Prenylated proteins have an attached isoprenoid moiety, a polymer built of $C_5$ isoprene units. The most abundant of these are farnesyl ($C_{15}$) and geranylgeranyl ($C_{20}$) groups. As a rule, prenylation occurs at the C-terminus of a protein.
    (b) Fatty acylated proteins have a covalently attached myristoyl or palmitoyl group. Myristoylation is a stable modification that occurs through an amide bond to the $\alpha$ amino group of an N-terminal glycine. Palmitoylation occurs via a thioester linkage to a cysteine and is reversible.
    (c) Glycosylphosphatidylinositol (GPI)-linked proteins are located at the external surface of the cell and are anchored to the membrane by GPI, a glycolipid that is covalently attached to the C-terminus of the protein via an amide bond.

24. Peripheral membrane proteins (also known as extrinsic proteins) associate with membrane surfaces via electrostatic and hydrogen bonding interactions. As a result, these proteins can be removed from membranes by relatively gentle methods, such as extraction with salt solutions or variations in pH, that do not greatly disturb native membrane structure. Once dissociated, they exhibit properties typical of water-soluble proteins.

*Membrane Structure and Assembly*

25. The widely accepted concept of biological membrane structure, for which there is much evidence, is the fluid mosaic model. The model envisions integral proteins that float in a sea of lipid and can diffuse laterally unless restrained by other cell constituents. The rate of membrane protein diffusion can be assessed by fluorescent recovery after photobleaching, which measures the rate at which a fluorescent-labeled membrane component diffuses into an area of membrane that has been previously bleached by a laser beam.

26. The structure and function of the erythrocyte membrane have been extensively investigated, largely through the use of erythrocyte ghosts, which are cells from which hemoglobin and other soluble cell constituents have been removed by osmotic lysis. The erythrocyte's biconcave disklike shape, which is essential for its $O_2$-carrying properties, is maintained by the membrane skeleton, a network of proteins located just beneath the membrane. The chief element in the skeleton is spectrin, a fibrous heterotetrameric protein that is cross-linked to other skeletal and membrane proteins, including actin and ankyrin. Proteins similar to those of the erythrocyte membrane skeleton occur in many other cell types.

27. The lipid and protein components of biological membranes are unevenly distributed. The cytoskeleton apparently restricts the movements of many membrane proteins and hence contributes to this heterogeneity. Localized membrane domains that have different lipid compositions may result from the association of certain lipids with membrane proteins. Lipid rafts are microdomains of closely packed lipids that have a more ordered consistency than the bulk of membrane lipids, associate with certain proteins, and may have distinct functions, especially in cellular signaling.

28. Membrane constituents are asymmetrically distributed with respect to the outer and inner leaflets of the bilayer in natural membranes. For example, the carbohydrate components of plasma membrane glycoproteins and glycolipids are always located on the extracellular membrane surface. The inner and outer leaflets also contain lipids in different proportions.

29. Lipid asymmetry, which can be assessed through the use of specific phospholipases, results in part from the asymmetric synthesis of lipids on the cytoplasmic face of the plasma membrane (in prokaryotes) or the endoplasmic reticulum (in eukaryotes). Lipids are subsequently redistributed by two mechanisms: (a) Enzymes called flippases facilitate the flip-flop of specific phospholipids from one leaflet to the other; and (b) ATP-dependent translocases establish a nonequilibrium distribution of lipids. In eukaryotes, lipids are transported as components of vesicles that bud off the endoplasmic reticulum and ultimately fuse with other cellular membranes.

30. Protein synthesis starts at the N-terminus and proceeds to the C-terminus of a polypeptide chain. Synthesis takes place either on free ribosomes (for soluble proteins) or on ribosomes bound to the endoplasmic reticulum (for integral membrane and secretory proteins). The secretory pathway describes how polypeptides traverse the endoplasmic reticulum membrane:

(a)  Proteins synthesized on ER-bound ribosomes have an N-terminal amino acid sequence known as a signal peptide.

(b)  After an ~40-residue polypeptide chain has been synthesized, the signal peptide binds to the signal recognition particle (SRP), concomitantly with an exchange of GTP for GDP in the SRP. This event stops further polypeptide synthesis until the SRP–ribosome complex docks with the ER membrane.

(c)  The SRP–ribosome complex, bearing the nascent polypeptide chain, binds to a docking protein (the SRP receptor; SR) on the ER surface. The SRP receptor is associated with a pore structure, the translocon, a transmembrane channel, via which the growing polypeptide passes through the ER membrane into the lumen of the ER.

(d)  The SR and SRP–ribosome complex dissociate by a process that requires a conformational change, the energy for which is provided by the hydrolysis of GTP to GDP.
Protein synthesis can then resume, and the N-terminal sequence is translocated to the lumen of the ER through the translocon.

(e)  Once in the ER lumen, the signal peptide is removed from the polypeptide by hydrolysis catalyzed by a signal peptidase.

(f)  The growing polypeptide undergoes folding, facilitated by Hsp70, and posttranslational modification, such as addition of core N-linked oligosaccharides.

(g)  When synthesis is complete, secretory proteins have passed entirely through the ER membrane into the lumen, whereas transmembrane proteins remain embedded in the membrane with their C-terminus on the cytoplasmic face.

31.  The translocon is a multifunctional and highly conserved transmembrane protein that either allows soluble polypeptides to traverse the ER membrane or mediates the insertion into the bilayer of membrane-spanning polypeptide segments.

32.  After synthesis in the ER, immature transmembrane, secretory and lysosomal proteins pass through the series of Golgi apparatus cisternae, the site of posttranslational modification, especially the addition of oligosaccharide chains by glycosylation.

33.  Secreted proteins, lysosomal (and other vesicular) proteins, and membrane proteins are transported between cellular compartments in coated vesicles. The "coat" of these vesicles is composed of one of three classes of proteins:

(a)  A complex of tripartite clathrin proteins called triskelions that form a polyhedral network. These vesicles are usually bound for the plasma membrane or participate in endocytosis;

(b)  COPI proteins, which mediate transport within the Golgi apparatus;

(c)  COPII proteins, which mediate transport of vesicles from the ER to the Golgi apparatus.

34.  Proteins bound for the ER or lysosomes have specific signals that target their final localization. Mannose-6-phosphate residues covalently attached to lysosomal hydrolases target these proteins to lysosomes. Similarly, most proteins localized to the ER contain the amino acid sequence KDEL at the C-terminus. It is currently hypothesized that other carbohydrate moieties or specific peptide sequences are essential for targeting proteins to other cellular compartments.

35. Membrane fusion is critical for the trafficking of membrane components, secretion (exocytosis), and viral entry into cells. Membrane fusion is mediated by complementary pairs of proteins called SNAREs. Both vesicular SNAREs (R-SNAREs) and target membrane SNAREs (Q-SNAREs) comprise a large family of proteins, wherein they provide specificity for vesicular fusion with target membranes.

36. Bacterial toxins from anaerobic bacteria (eg., Clostridium tetani, Clostridium botulinum) cleave SNAREs, thereby interfering with neuromuscular and neuronal synaptic exocytosis (Box 9-3). Animal viruses such as influenza virus contain membrane fusion proteins (e.g., hemagglutinin) that stimulate receptor-mediated endocytosis, followed by membrane fusion, which allows the virus access to the host cell cytoplasm. In the low pH environment of endosomes, hemagglutinin undergoes dramatic conformational changes that turn it from a membrane receptor into a membrane fusion protein.

## *Questions*

### *Lipid Classification*

1. Draw structures for the following fatty acids:
   (a) 18:0 octadecanoic acid
   (b) 20:4 5,8,11,14-eicosatetraenoic acid
   (c) 22:6 4,7,10,13,16,19-docosahexaenoic acid
   (d) 13-(2-cyclopentenyl)-tridecanoic acid (a cyclic fatty acid)

2. How do the following alterations in the structures of fatty acids affect their physical properties?
   (a) Increasing the chain lengths of saturated fatty acids
   (b) Increasing the number of double bonds in unsaturated fatty acids
   (c) Changing a cis double bond in a fatty acid to a trans double bond

3. Draw and name a typical triacylglycerol.

4. List the two major functions of triacylglycerols.

5. Why are glycerophospholipids considered amphiphilic molecules?

6. Draw the structures of the following lipids:
   (a) 1-palmitoyl-2-oleoyl-3-glycerophosphatidylethanolamine
   (b) 1-stearoyl-2-arachidonoyl-3-glycerophosphatidylinositol

7. What distinguishes a plasmalogen from other glycerophospholipids?

8. What chemical moieties do sphingomyelin and gangliosides have in common and which are unique to each?

9. The compound shown below has been advertised as a dietary supplement that purportedly prevents obesity, heart disease, and the ill effects of aging. Based on its structure, what physiological function is it most likely to actually perform?

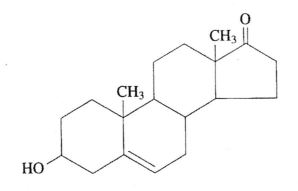

10. Why are eicosanoids called local mediators rather than hormones?

11. Arachidonic acid can arise from phospholipid precursors by the action of:
    (a) Phospholipase $A_1$
    (b) Phospholipase $A_2$
    (c) Phospholipase C
    (d) Phospholipase D

12. Summarize the functions of Vitamins A, D, E and K.

*Lipid Bilayers*

13. Explain why single-tailed amphiphiles tend to form micelles whereas two-tailed amphiphiles tend to form bilayers.

14. Why is a polar solute unlikely to penetrate a lipid bilayer?

15. Describe the structural changes that occur when a pure phospholipid bilayer is warmed and passes through its transition temperature. Explain what would happen if the bilayer contained a significant amount of cholesterol.

*Membrane Proteins*

16. Summarize the physical properties of a typical integral protein, including the forces involved in its interaction with the membrane, its orientation, and its movement within the membrane.

17. Lipids linked covalently to proteins often _____ the proteins in the membrane.

18. The three main types of lipid moieties that can be covalently attached to proteins are _____ , _____ , and _____ .

19. List the principal lipids that are found in lipid rafts. Would a phospholipid with a high content of docosahexaenoic acid be a raft component?

20. Mutated forms of the protein $p21^{c\text{-}ras}$ are associated with the development of a substantial proportion of tumors. However, a second mutation that changes the Cys residue in the C-terminal Cys-X-X-Y sequence of $p21^{c\text{-}ras}$ to some other residue inhibits its cancer-causing potential. What does this suggest about the molecular process leading to tumor formation?

### Membrane Structure and Assembly

21. In verifying the fluid mosaic model of biological membranes, Michael Edidin observed that fusion of mouse and human cell membranes was temperature dependent, such that fusion did not occur below 15°C but proceeded readily at 37°C. Explain this difference in terms of membrane properties (see Figure 9-26).

22. Explain why an erythrocyte membrane that synthesizes too little spectrin is a sphere rather than a biconcave disk.

23. How did Kennedy and Rothman demonstrate that newly synthesized bacterial membrane phospholipids rapidly cross from the cytoplasmic side to the external side of the membrane (see Fig. 9-35)?

24. Explain why the flip-flop rate of phospholipids in biological membranes is far greater than in artificial lipid membranes.

25. In the secretory pathway, match each of the following terms with its description below.
    _____ free ribosome
    _____ signal peptide
    _____ signal recognition particle (SRP)
    _____ core glycosylation
    _____ membrane anchor
    _____ translocon
    _____ signal peptidase
    _____ membrane-bound ribosomes
    _____ SRP receptor

    A. A hydrophobic sequence in a transmembrane protein that arrests movement of the protein through the membrane.
    B. A posttranslational modification that occurs while peptide bond formation continues.
    C. Cleaves the signal peptide.
    D. An N-terminal hydrophobic sequence flanked by hydrophilic sequences.
    E. The site of membrane and secretory protein synthesis.
    F. Binds the SRP and reinitiates ribosomal polypeptide synthesis.
    G. The site of soluble protein synthesis.

H.  Binds to a ribosome bearing an extruded signal sequence and stops polypeptide synthesis.

I.  Pore that provides conduit between the cytosol and the lumen of the ER.

26.  A membrane glycoprotein–synthesizing system was devised that contained endoplasmic reticulum membranes and all the necessary ingredients for the *in vitro* synthesis of a protein. The complete synthesis of the protein molecule required 40 minutes. When the detergent Triton X-100 was added within a few minutes after the initiation of polypeptide synthesis, the resulting protein was devoid of glycosyl groups. (a) Explain the action of the detergent. (b) Would the protein have been glycosylated if the detergent had been added 35 minutes after initiation of protein synthesis?

27.  The antibiotic monensin inhibits some steps of posttranslational protein modification in the Golgi apparatus. What effect would monensin have on protein targeting?

28.  Besides providing specificity to membrane fusion, SNAREs (and probably hemagglutinnin, too) are thought to play a role in the energetics of membrane fusion. Indeed, these proteins, beyond simple recognition, induce large conformation changes in the fusing membranes. Can you suggest an energetic barrier that these proteins may help traverse?

# 10
# Membrane Transport

Lipid bilayers are impermeable or weakly permeable to a variety of polar and nonpolar compounds that are essential to cellular metabolism. Hence, these substances must enter (and exit) cells by way of specialized membrane proteins that facilitate their transport across lipid bilayers. In order to function normally, cells must not only facilitate diffusion of ions and small molecules across their membranes (passive-mediated transport), they must also concentrate some substances on one side of the membrane (active transport). This chapter introduces the thermodynamics of transport and then discusses passive-mediated transport and active transport. Examples of passive-mediated transporters presented in this chapter include ionophores, porins, ion channels, aquaporins, and transport proteins such as the glucose and oxalate transporters. Many ion channels are gated so that they allow the movement of ions only in response to certain stimuli. Examples of active transporters discussed in this chapter include the ABC transporters, $(Na^+–K^+)$–ATPase, $Ca^{2+}$–ATPase, and ion gradient–driven active transport systems.

## *Essential Concepts*

### *Thermodynamics of Transport*

1. The thermodynamics of diffusion across a membrane can be expressed in terms of a chemical equilibrium. The difference in concentration of substance A on two sides of a membrane generates a chemical potential difference, $\Delta \bar{G}_A$. When A is ionic, an electrical potential difference ($\Delta \Psi$) may also develop. Thus, the equation for the electrochemical potential difference contains terms for the concentration and the charge of substance A:

$$\Delta \bar{G}_A = RT \ln ([A]_{in}/[A]_{out}) + Z_A \mathcal{F} \Delta \Psi$$

   where $Z_A$ is the ionic charge of A and $\mathcal{F}$ is the Faraday constant (96,485 $J \cdot V^{-1} \cdot mol^{-1}$). A negative value for $\Delta \bar{G}_A$ indicates spontaneous transport of substance A from the outside to the inside.

2. Transport may be classified as either mediated or nonmediated. Diffusion accounts for nonmediated transport. The chemical potential difference determines the direction of nonmediated transport, and the substance moves in the direction that will eliminate the concentration difference and at a rate proportional to the size of the gradient. Mediated transport occurs in two forms, based on the thermodynamics of the solutes traversing the membrane:
   (a) Passive-mediated transport or facilitated diffusion, in which solutes travel down a concentration gradient.
   (b) Active transport, in which solutes are transported against their concentration gradient via coupling to a sufficiently exergonic process.

3.   Mediated transport makes use of carrier molecules, called permeases, transporters, or translocases. In passive-mediated transport (also called facilitated diffusion), a molecule is transported from a high to a low concentration. In active transport, an energy-yielding process must be coupled to the movement of a substance from a lower to a higher concentration.

## *Passive-Mediated Transport*

4.   Ionophores are carrier molecules that greatly increase the permeability of a membrane to certain ions. Some of these organic compounds (such as valinomycin) bind the ion on one side of the membrane and carry it to the other side; others form transmembrane channels through which the ion can pass.

5.   Porins are β-barrel transmembrane proteins that display varying levels of solute selectivity. Some porins are weakly selective for cations or anions, while others, such as maltoporin, show a high degree of solute selectivity.

6.   All cells contain ion-specific channels, the best characterized of which is the $K^+$ channel from *Streptomyces lividans* (KcsA).

7.   The KscA $K^+$ channel is a homotetrameric transmembrane protein. The extracellular surface has a selectivity filter that preferentially allows $K^+$ to be dehydrated and enter the central cavity wherein $K^+$ is rehydrated. The geometry of the selectivity filter does not allow the entry of smaller $Na^+$ ions; hence, the channel is specific for $K^+$.

8.   Most ion channels are gated, meaning that they open or close in response to a variety of stimuli including (a) local deformations in the membrane (mechanosensitive channels); (b) binding of an extracellular signal such as a neurotransmitter (ligand-gated channels); (c) binding of an intracellular signal (signal-gated channels); and (d) changes in voltage across the membrane (voltage-gated channels).   .

9.   Voltage-gated channels propagate action potentials in nerve cells. Voltage gating in voltage-gated $K^+$ ($K_V$) channels is mediated by α-helical "paddles" that respond to critical voltage changes across the plasma membrane. The $K_V$ channel has two voltage-sensitive gates: one to open the channels and another to close it. Closure is mediated by the N-terminal domain, which swings into place 2–3 milliseconds after channel opening.

10.   Aquaporins are ubiquitous transmembrane channels that allow selective movement of $H_2O$ molecules across otherwise impermeable lipid bilayers.

11.   Integral membrane proteins that mediate passive transport cycle between conformational states in which binding sites for the molecule to be transported are alternately accessible on one side of the membrane and then on the other. For example, the glucose transporter binds glucose at the external cell surface then changes conformation so as to release glucose at the cytoplasmic surface. It then reverts to its previous conformation.

12.   Gap junctions are pairs of juxtaposed channels between two cells that allow non-selective movement of a variety of solutes such as ions, amino acids, sugars, and nucleotides. Gap junctions provide electrical coupling to cells, which is critical to heart function in vertebrates. The core protein is connexin, which forms a hexagonal ring in each membrane (Box 10-1).

13.   Mediated transport exhibits four properties common to enzymes: (a) speed and specificity, (b) saturability, (c) competition from structurally similar compounds, and (d) inactivation by compounds that chemically modify proteins. Indeed, mediated transporters can be viewed as "diffusion enzymes" (Box 10-2).

14.   Proteins that mediate passive transport can operate in either direction, depending on the concentrations of the transported substance on both sides of the membrane. A uniport mechanism transfers a single molecule across the membrane; symport involves two substances moving in the same direction; and antiport is the movement of two substances in opposite directions.

## Active Transport

15.   Active transport is coupled to the hydrolysis of ATP. A well-studied example is the plasma membrane ($Na^+$–$K^+$)–ATPase, which pumps 3 $Na^+$ out and 2 $K^+$ into the cell, thereby moving both ions against their concentration gradients, with every ATP hydrolyzed. The resulting electrochemical gradient of $Na^+$ and $K^+$ is the basis for the electrical excitability of nerve cells.

16.   The mechanism of the ($Na^+$–$K^+$)–ATPase involves a series of reaction steps that operate in only one direction because ATP breakdown and ion movement are coupled vectorial processes. The $Ca^{2+}$–ATPase, which pumps cytosolic $Ca^{2+}$ to the cell exterior against a large concentration gradient, functions in a similar manner to the ($Na^+$–$K^+$) pump. Cardiac glycosides (sugar components covalently linked to a steroid component) modulate cardiac function by blocking ATP hydrolysis in the ($Na^+$–$K^+$)–ATPase reaction sequence (Box 10-3).

17.   ABC (ATP-binding cassette) transporters are a large class of transporters that pump ions and a wide variety of polar and nonpolar molecules. These proteins include two transmembrane domains and two ATP-binding domains.
      (a)   In prokaryotes, a so-called ligand-binding domain is sometimes present, which concentrates the solute near the transporting transmembrane domains. Also, the three components (ligand-binding domain, transmembrane permease domain, ATP-binding domain) often exist as separate polypeptides in prokaryotes.
      (b)   In eukaryotes, these domains are found as a single polypeptide. The multidrug resistance (MDR) transporter or P-glycoprotein transports drugs and other substances out of eukaryotic cells.

(c) CFTR, the cystic fibrosis transmembrane regulator (Box 3-1), is an unusual ABC transporter in that ATP hydrolysis regulates the opening of a gate for $Cl^-$ to diffuse through, instead of actively transporting $Cl^-$ up a concentration gradient. CFTR is not a highly selective transporter of $Cl^-$, ions suggesting that it probably lacks a selectivity filter.

18. In secondary active transport, the energy generated by an electrochemical gradient is utilized to drive another, endergonic transport process. One example is the uptake and concentration of glucose by the intestinal epithelial cells. This task is accomplished by using the $Na^+$ gradient produced by the ($Na^+$–$K^+$) pump. Another example is the bacterial lactose permease, which utilizes a proton gradient across the bacterial membrane to power the transport of lactose.

## Key Equations

$$\Delta \bar{G}_A = RT \ln ([A]_{in}/[A]_{out}) + Z_A \mathcal{F} \Delta \Psi$$

## Questions

### Thermodynamics of Transport

1. Suppose the concentration differences across a membrane for glucose and sodium are both $10^4$ (that is, the concentrations are $10^4$ greater on one side of the membrane). Is the free energy change for the transmembrane movement of these solutes the same? Explain.

2. What is the electrochemical potential difference when the intracellular $[Ca^{2+}] = 1$ $\mu M$ and the extracellular $[Ca^{2+}] = 1$ mM? Assume $\Delta \Psi = -100$ mV (inside negative) and $T = 25°C$.

3. Rank the molecules below from lowest to highest according to their ability to diffuse across a lipid bilayer. Explain your rationale.

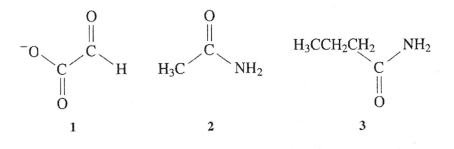

4. The following data (using arbitrary units) were obtained for the transmembrane movements of compounds A and B from outside to inside a cell:

| *A* | | *B* | |
|---|---|---|---|
| *Extracellular concentration* | *Flux into cell* | *Extracellular concentration* | *Flux into cell* |
| 2 | 1.2 | 2 | 4.5 |
| 5 | 3.4 | 5 | 6.2 |
| 10 | 6.5 | 10 | 7.5 |

Which compound enters the cell by mediated transport? Explain.

5. Which graph below shows the expected relationship between temperature and flux ($J$, rate of flow) of an ion transported across a biological membrane via a carrier ionophore? Explain.

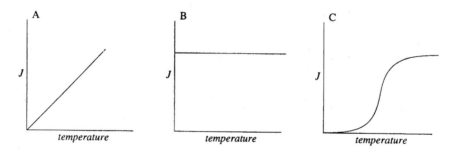

## Passive-Mediated Transport

6. $K^+$ channels have openings that are much wider than $Na^+$ ions, yet sodium ions cannot pass through efficiently. Explain.

7. The eggs of freshwater invertebrates and vertebrates are hypertonic (higher solute concentration) to their surrounding medium (pond water). What protein is likely to be absent from the egg plasma membranes?

8. The model for glucose transport shown in Figure 10-13 shows two conformational states of the transporter when it is not bound to glucose. Is one of these conformational states likely to be preferred?

## Active Transport

9. Does the activity of the $(Na^+–K^+)$–ATPase tend to make the cell interior electrically more negative or more positive with respect to the outside?

10. Determine whether each of the following transport systems is uniport, symport, or antiport. Which systems are active transport systems?

    (a) glucose transporter in erythrocytes
    (b) valinomycin
    (c) plasma membrane $(Na^+-K^+)$–ATPase
    (d) $Na^+$–glucose transporter of intestinal epithelium
    (e) *E. coli* lactose permease
    (f) $(H^+-K^+)$–ATPase of gastric mucosa

11. Fill in the diagram below showing glucose transport across a brush border cell from the intestinal lumen to the capillary bed. Also, draw arrows to show the flow of solutes and to indicate which transporter requires ATP hydrolysis. *Note: not all blanks and circles may be relevant.*

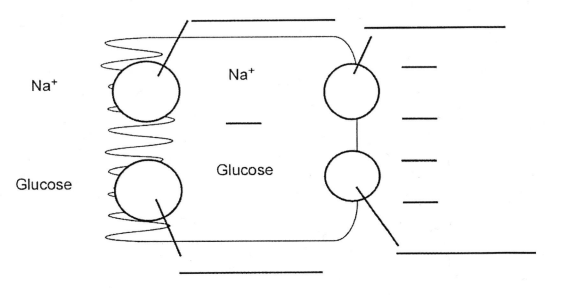

# 11

# Enzymatic Catalysis

Enzymes are biological catalysts that increase the rates of biochemical reactions. In some cases, the enzyme-catalyzed reaction is nearly $10^{15}$ times faster than the uncatalyzed reaction. Enzymes are proteins (or in some cases, RNA) with very specific functions and are active under mild conditions. Enzymes function by lowering the free energy of the reaction's transition state. This chapter first discusses the various types of enzymes, their substrate specificity, the roles of coenzymes and cofactors, and the reaction coordinate. Next, the five modes of catalysis are described in detail. The chapter ends with full descriptions of the catalytic activity of lysozyme and serine proteases.

## *Essential Concepts*

### *General Properties of Enzymes*

1. Enzymes differ from ordinary catalysts in their higher reaction rates, their action under milder reaction conditions, their greater reaction specificities, and their capacity for regulation.

2. The IUBMB enzyme classification system divides enzymes into six groups (each with subgroups and sub-subgroups) based on the type of reaction catalyzed. A set of four numbers identifies each enzyme.

3. Enzymes are highly specific for their substrates and reaction products. Hence, the enzyme and its substrate(s) must have geometric, electronic, and stereospecific complementarity. Enzymes, for example, aconitase, can distinguish prochiral groups.

4. Many enzymes require cofactors for activity. Cofactors may be metal ions or organic molecules known as coenzymes. Many vitamins are coenzyme precursors. Coenzymes may be cosubstrates, which must be regenerated in a separate reaction, or prosthetic groups, which are permanently associated with the enzyme. An enzyme without its cofactor(s) is called an apoenzyme and is inactive, and the enzyme with its cofactor(s) is a called a holoenzyme and is active.

### *Activation Energy and the Reaction Coordinate*

5. According to transition state theory, the reactants of a reaction pass through a short-lived high-energy state that is structurally intermediate to the reactants and products. This so-called transition state is the point of highest free energy in the reaction coordinate diagram. The free energy difference between the reactants and the transition state is the free energy

of activation, $\Delta G^{\ddagger}$. The reaction rate decreases exponentially with the value of $\Delta G^{\ddagger}$; that is, the greater the value of $\Delta G^{\ddagger}$, the slower the reaction. A catalyst provides a reaction pathway whose $\Delta G^{\ddagger}$ is less than that of the uncatalyzed reaction and hence increases the rate that the reaction achieves equilibrium.

## Catalytic Mechanisms

6.  Enzymes use several types of catalytic mechanisms, including acid–base catalysis, covalent catalysis, metal ion catalysis, catalysis by proximity and orientation effects, and catalysis by preferential binding of the transition state.

7.  Acid–base catalysis occurs when partial proton transfer from an acid and/or partial proton abstraction by a base lowers the free energy of a reaction's transition state. The catalytic rates of enzymes that use acid–base catalysis are pH-dependent. RNase A has two catalytic His residues, which act as general acid and general base catalysts.

8.  In covalent catalysis, the reversible formation of a covalent bond permits the stabilization of the transition state of the reaction through electron delocalization. Nucleophilic attack on the substrate by the enzyme to form a Schiff base intermediate capable of stabilizing (lowering the free energy of) a developing negative charge is an example of covalent catalysis. Common nucleophiles, which are negatively charged or contain unshared electron pairs, include imidazole and sulfhydryl groups.

9.  Metalloenzymes tightly bind catalytically essential transition metal ions. Metal-activated enzymes loosely bind metal ions such as $Na^+$, $K^+$, or $Ca^{2+}$ that play a structural role. Metal ions may orient substrates for reaction, mediate oxidation–reduction reactions, and electrostatically stabilize or shield negative charges. For example, the $Zn^{2+}$ ion of carbonic anhydrase makes a bound water molecule more acidic, thereby increasing the concentration of the nucleophile $OH^-$.

10. Enzymes lower the activation energies of the reactions they catalyze by bringing their reactants into proximity, by properly orienting them for reaction, by using charged groups to stabilize the transition state (electrostatic catalysis), and, most importantly, by freezing out the relative motions of the reactants and the enzyme's catalytic groups.

11. An enzyme's preferential binding of the transition state lowers $\Delta G^{\ddagger}$ and thereby increases the rate of the reaction. For this reason, an unreactive compound that mimics the transition state may be an effective enzyme inhibitor.

## Lysozyme

12. Hen egg white lysozyme, an enzyme that cleaves the glycosidic linkage between NAG and NAM residues in bacterial cell walls, has a substrate-binding cleft that accommodates six sugar residues such that the cleavage occurs between the residues in subsites D and E.

13. In the mechanism for lysozyme, the NAM residue that binds in the D site is distorted toward the half-chair conformation. Glu 35 (a general acid catalyst) then transfers a proton to O1 to cleave the C1—O1 bond of the substrate, generating an oxonium ion transition state. The nucleophilic carboxylate group of Asp 52 simultaneously attacks C1 to form a covalent intermediate (covalent catalysis). Hydrolysis of this bond completes the catalytic cycle. Because the substrate is distorted toward the transition-state conformation on binding, transition state stabilization is an important catalytic mechanism.

*Serine Proteases*

14. Serine proteases are a widespread family of enzymes that have a common mechanism. The active-site Ser (identified through its inactivation by diisopropylphosphofluoridate), His (identified through affinity labeling with a chloromethylketone substrate analog), and Asp (identified by X-ray crystallography) form a hydrogen-bonded catalytic triad. Nonhomologous serine proteases have developed the same catalytic triad through convergent evolution.

15. The differing substrate specificities of trypsin, chymotrypsin, and elastase depend in part on the shapes and charge distribution of the substrate-binding pocket near the active site.

16. Catalysis by serine proteases is a multistep process in which nucleophilic attack on the scissile bond by Ser 195 (using the chymotrypsinogen numbering system) results in a tetrahedral intermediate that decomposes to an acyl–enzyme intermediate. The replacement of the amine product with water is necessary for the formation of a second tetrahedral intermediate, which yields the carboxyl product and regenerated enzyme.

17. Serine proteases use acid–base catalysis (involving the Ser-His-Asp triad), covalent catalysis (formation of the tetrahedral intermediates), and catalysis through binding of the transition state in the oxyanion hole. The tight binding of bovine pancreatic trypsin inhibitor to trypsin inhibits the enzyme by preventing full attainment of the tetrahedral intermediate, as well as the entry of water into the active site.

18. Low-barrier hydrogen bonds (LBHBs) have been proposed to play important roles in the stabilization of transition states. LBHB's are strong hydrogen bonds, isolated from other water molecules, which have predicted bond energies –4 times those of normal hydrogen bonds in aqueous environments.

### *Play It Forward: A Primer on Understanding Catalytic Mechanisms Through a Common Set of Principles–*

Chapter 11 of the text presents some basic theory of enzymology, as well as the proposed catalytic mechanisms of three enzymes with varying detail: RNase A, lysozyme, and chymotrypsin. Understanding these and other enzyme mechanisms requires a synthesis and application of a large amount of material from previous chapters, including weak acid chemistry, amino acid structure and chemistry, protein structure, and protein conformational change. However, a common set of predictable principles often underlies many enzymatic mechanisms that can make it easier to analyze and understand them. These principles are outlined below, and will be used to partially analyze the three mechanisms presented in Chapter 11. This is then followed by critical thinking and application questions related to these mechanisms.

**Principle #1: Enzyme mechanisms are similar to a two act theatrical play.**

Imagine that you were required to verbally describe a two act theatrical play to a friend or family member who has not seen the play. What information would you need to convey to ensure that they understood the play from your description? Minimally, you would need to describe the main characters of the play, the motivations and roles of those characters in the story being told, as well as the plot and sequence of events of the story in the play. Understanding and describing enzyme mechanisms is very similar to understanding a two act theatrical play. The main characters represent the catalytic or important residues in the enzyme active site required for substrate binding and catalysis. The roles and motivations of the characters represent the chemistry and behavior of these active site residues. The plot and sequence of events represent the step-by-step chemical changes that occur to the substrate over time.

Consider the proposed mechanism of RNase A from the text. RNase A has a relatively simple two-step mechanism. Two primary catalytic residues are present (His 12 and His 119) and perform a majority of the action in the mechanism, while water also plays a role in the mechanism. Both His residues act as reversible acid-base catalysts (their motivation in the plot) to reversibly protonate and deprotonate key functional groups on the substrate to cleave the phosphodiester bond of the substrate (their role in the plot). The temporal sequence of reversible protonation and deprotonation steps, cleavage of the phosphodiester bond, as well as the formation and degradation of the 2′, 3′-cyclic intermediate represents the plot of the mechanism.

How is a mechanism similar to a two act play? Recall that enzymes are biological catalysts. In order to be qualified as a catalyst, the enzyme must not be permanently altered or consumed in the reaction that it catalyzes. Thus, any changes that occur to the enzyme structure during the course of the mechanism must be reversible and undone by the conclusion of the mechanism. With this in mind, it is convenient to analyze enzyme mechanisms with two main goals assigned to the catalytic steps: conversion of the substrate to product (catalytic goal) and undoing any changes that occur to the enzyme during the course of the mechanism (regeneration of the original state of the enzyme).

Consider the lysozyme mechanism presented in the text. While the text outlines 5 primary steps in the mechanism, it can be divided into two halves, where the catalytic goal of the enzyme is a good landmark for analysis. The catalytic goal of the enzyme is to cleave the glycosidic bond between the D and E sugar of the peptidoglycan substrate. Thus, we can use the breaking of this glycosidic bond as a landmark in the mechanism. Moreover, two significant changes occur to the catalytic residues during the breaking of the glycosidic bond of the substrate: Glu35 is deprotonated and Asp52 forms a covalent bond to stabilize the oxonium ion transition state to form the glycosyl-enzyme intermediate. These two changes must be reversed by the end of the mechanism, and this occurs only after the substrate glycosidic bond is cleaved. In summary, we can divide the Lysozyme mechanism into two halves (acts) with the dividing point between the two acts occurring at the conclusion of the third step in the mechanism. The first half of the mechanism is dedicated to the conversion of the substrate into product, and results in changes to the enzyme. The second half of the mechanism is dedicated to regenerating the original state of the enzyme by reversing changes that occurred to key catalytic residues.

**Principle #2: Enzyme active sites create microenvironments that differ from the bulk solvent.**

Enzymes catalyze a wide variety of reactions within their active sites, and some of the chemistry required in a particular mechanism may require a special set of conditions in order to proceed. For example, consider a scenario in which an enzyme (such as Lysozyme) requires a particular active site glutamic acid (Glu) residue to act as an acid catalyst. Under biochemical standard conditions, the side chain of the isolated Glu has a pKa value of 4.07 while the pH of the bulk solvent is 7.0. In Chapter 4 Play It Forward: pH, pKa and Predominant Forms of Amino Acids in Solution, we outlined how the predominant form of an amino acid in solution could be determined based upon the pKa values of the ionizable groups present on amino acids, and the pH of the environment. Under these conditions, we would expect the predominant form of the Glu side chain to be deprotonated because the pH of the solution is 3 pH units above the pKa of the side chain. Under these conditions, a population of Glu would exist such that approximately 1 out of 1000 would still be in the protonated form. If the active an enzyme requires that a particular Glu residue begin a mechanism in the protonated state, then this would not be very feasible if the active site has a pH of 7.0. Enzymes must have strategies to fine tune the catalytic environment of their active sites so that particular amino acid residues can function as needed.

We have seen in Chapter 4 of the text (Section 1-D) that the pKa of chemical groups can change depending upon their local environments. This is exemplified in how amino acid side chains can have pKa values that are different than that seen in the isolated amino acid when they are incorporated into a polypeptide chain. Many enzymes take advantage of this in their three-dimensional folding to create an active site that fine tunes the pKa of critical amino acid side chains by incorporating them into the polypeptide. However, enzymes also create an environment that is often radically different than the bulk solvent outside of the active site. This also helps to fine tune amino acid side chain chemistry to support the chemistry that is needed for catalysis. Together, we can infer that the enzyme active sites represent a unique microenvironment that is created to optimize the chemistry of catalysis, which would not often be possible in the environment of the bulk solvent.

Consider again the proposed mechanism of RNase A from the text. While the two-step mechanism presented is relatively simple, a number of unstated assumptions about the active site must be made to fully understand the catalytic strategy of RNase A. At the beginning of the mechanism the two catalytic His residues have slightly different pKa values and begin in different protonation states. His 12 begins the mechanism deprotonated while His 119 begins protonated. The text mentions that the two catalytic residues have pKa values of 5.4 and 6.4, both of which are different than the pKa of the His side chain in the isolated state (6.04). Based upon what we know from Chapter 4 Play It Forward: pH, pKa and Predominant Forms of Amino Acids in Solution, which pKa corresponds to which residue? What must the pH of the active site be (approximately) in order to facilitate the protonation states of these two His residues at the beginning of the mechanism? Apply the principles learned in Chapter 4:

**If the pH of the environment is lower than the pKa of the weak acid, then the weak acid will be predominantly protonated.**

**If the pH of the environment is higher than the pKa of the weak acid, then the weak acid will be predominantly deprotonated.**

If the two His residues have pKa values that differ by only 1 unit, but are beginning the mechanism in different protonation states, then we can predict that the apparent pH of the active site microenvironment should be somewhere between the two pKa values of these residues. In order for His 12 to begin the mechanism deprotonated, it would likely have a pKa of 6.4 while His 119 would likely have a pKa of 5.4 to begin the mechanism protonated.

Other features of active sites may also contribute to the formation of the active site microenvironment. For example, the oxyanion hole of the chymotrypsin active site represents a unique microenvironment feature that is created as a result of the three-dimensional folding of the enzyme, and it has an explicit role in catalysis.

**Principle #3: Enzyme active sites often control the access of water in their mechanisms to facilitate catalysis and the formation of microenvironments.**

The wide variety of chemistry that enzymes must facilitate includes some mechanisms that would be hindered or made impossible by the presence of water. Consider the proposed reaction mechanism of chymotrypsin. The transition state of the chymotrypsin mechanism is proposed to be stabilized, partially due to the presence of a low-barrier hydrogen bond (LBHB) in the chymotrypsin active site. The text notes that LBHBs are unlikely to exist in the presence of water, yet there is a proposed LBHB in the chymotrypsin active site. This implies that the chymotrypsin active site must be water deficient for this LBHB to exist. It is only in step 3 of the mechanism that chymotrypsin allows the controlled entry of water into the active site. Presumably, the controlled access of water in this mechanism is coupled to enzyme conformational changes that occur as the mechanism proceeds. Even as water enters the active site in step three, it is used to reprotonate His 57 and form a new tetrahedral intermediate from the acyl-enzyme intermediate.

Many enzymes create active site microenvironments that are devoid of water for at least a part of their reaction mechanisms. This represents a significant, but surmountable challenge for enzymes. How can an enzyme bind a substrate that is fully solvated, yet maintain an active site with limited access to water? The act of substrate binding must desolvate substrates by exchanging the interactions the substrates have with water with interactions that form with key enzyme amino acid residues. Moreover, the presence of substrate must sterically prevent water from entering the active site during catalysis. Consider Figure 11-19 of the text: the oligosaccharide substrate contains many hydrogen bond donor and acceptor groups that would be expected to engage in hydrogen bonding with water during solvation. However, these water molecules are replaced with peptide hydrogen bonding partners in the active site of Lysozyme upon substrate binding. Similar to the chymotrypsin mechanism, lysozyme does not allow water to enter the active site until late in the mechanism, after one part of the cleaved substrate is released from the active site. In both instances the enzyme is promoting the reactivity of water by facilitating the ionization of water, which was discussed in Chapter 2 of the text. While this is not universal for all enzymes, it is a common theme that appears in many enzyme mechanisms.

While many of the enzyme mechanisms presented in later chapters of the text are more complicated than those presented in Chapter 11, analyzing mechanisms with these three principles in mind can help organize and deconvolute the enzymatic chemistry. As practice, analyze the mechanisms of RNase A, lysozyme, and chymotrypsin as presented in the text according to these principles.

**Applying these concepts:**

1. Identify the catalytic goal of RNase A, lysozyme, and chymotrypsin, identify at which step this goal is completed in their mechanisms, and what intermediate/transition state is produced as a result of the completion of this goal.

   Therefore, the catalytic goal of RNAse A is to cleave the phosphodiester bond of RNA. This occurs by the end of the first step of the mechanism and results in the formation of the 2′, 3′-cyclic intermediate.

   The catalytic goal of the lysozyme mechanism is the cleavage of the glycosidic bond between the D and E sugars of the peptidoglycan substrate. This occurs by the end of the second step of the mechanism and results in the formation of the oxonium ion intermediate, which is subsequently stabilized by the formation of the covalent acyl-enzyme intermediate.

   The catalytic goal of the chymotrypsin mechanism is the cleavage of the peptide bond following W, Y or F residues in a peptide substrate (See chapter 5 of the text for the substrate specificity of chymotrypsin). This occurs by the end of the second step of the mechanism and results in the formation of the first tetrahedral intermediate of the mechanism.

2. The text describes the microenvironment around Glu35 as being different than the microenvironment around Asp52 in the Lysozyme active site. Considering the protonation states of Glu35 and Asp52 at the beginning of the mechanism, what general trend is revealed about the polarity of a microenvironment and the affinity an ionizable amino acid residue will have for a proton in active sites? Is the pKa of Glu35 higher or lower than isolated Glu at the biochemical standard state?

Therefore, the microenvironment around Glu35 is relatively nonpolar while the microenvironment around Asp52 is polar. Glu35 begins the mechanism protonated, indicating that it has a higher affinity for its proton than we would expect from the isolated amino acid. Thus, we may expect that Glu35 has a higher pKa than we would expect (See Chapter 2: Behind the Equations: Weak Acids and the Henderson-Hasselbalch Equation). In general, the more hydrophobic the environment is, the higher the affinity an amino acid will have for a proton and be less likely to deprotonate and ionize in a nonpolar environment. Compared to the isolated amino acid, an amino acid residue in a nonpolar microenvironment of an active site may have a higher pKa.

3. Chapter 4 does not describe the serine side chain as being ionizable, and Table 4-1 does not list a pKa for the Ser side chain. However, Ser195 reversibly deprotonates with the aid of His57 in the chymotrypsin mechanism. At the time of the deprotonation of Ser195 in the chymotrypsin mechanism, would you expect His57 to have a higher or lower pKa value compared to Ser195? Explain your answer.

Therefore, if His57 is acting as a base catalyst to deprotonate Ser195, then His57 is acting as a base while Ser195 is acting as an acid. His57 must have a higher affinity for the proton than Ser195 at that moment, and thus His57 should have a higher pKa than Ser195 at the time of the deprotonation.

4. Water appears in and performs critical roles in the mechanisms of RNase A, lysozyme and chymotrypsin. Describe three similarities for the role of water in all three mechanisms. There may be more than three.

   i) Water enters the active sites only after the catalytic goal of each enzyme has been accomplished.
   ii) Water is deprotonated by a basic catalytic residue that must be reprotonated to regenerate the enzyme.
   iii) Deprotonation of water generates a powerful nucleophile (⁻OH), which is used to destroy an intermediate/transition state.

## Questions

### General Properties of Enzymes

1. What is an enzyme's EC number?

2. Explain why enzymes are stereospecific.

3. What is an apoenzyme and how does it differ from a holoenzyme? Which form is active?

4. What is the relationship between vitamins and coenzymes?

5.  Proteins can be chemically modified by a variety of reagents that react with specific amino acid residues. How can such reagents be used to identify residues involved in an enzyme's activity? What are the shortcomings of this method?

*Activation Energy and the Reaction Coordinate*

6.  What is the rate-determining step of an enzyme-catalyzed reaction?

7.  Answer *yes* or *no* to the following questions and explain your *answer*.
    (a) Can the absolute value of $\Delta G$ for a reaction be larger than $\Delta G^{\ddagger}$?
    (b) Can $\Delta G^{\ddagger}$ for an enzyme-catalyzed reaction be greater than $\Delta G^{\ddagger}$ for the nonenzymatic reaction?
    (c) In a two-step reaction, such as the one diagrammed in Figure 11-6, must the intermediate (I) have less free energy than the reactant (A)?
    (d) In a multistep reaction, does the transition state with the highest free energy always correspond to the rate-determining step?

8.  An increase in temperature increases the rate of a reaction. How does the temperature affect $\Delta G^{\ddagger}$?

9.  $\Delta \Delta G^{\ddagger}$ for an enzymatic reaction at 25°C is 13 kJ·mole$^{-1}$. (a) Calculate the rate enhancement. (b) What is $\Delta \Delta G^{\ddagger}$ when the rate enhancement is $10^5$?

*Catalytic Mechanisms*

10. List the amino acid residues that are most likely to participate in general acid–base catalysis. Why isn't glycine among those listed?

11. If a protonated His residue acts as the proton donor in an acid-catalyzed enzymatic reaction, what happens to the enzyme's activity as the pH increases to a value that exceeds the p$K_R$ of that residue?

12. The p$K$ values of two essential catalytic residues in RNase A are 5.4 and 6.4. (a) Which corresponds to His 12 and which to His 119 (see Figure 11-10)? (b) Draw a titration curve for these two residues.

13. What is the difference between nucleophilic catalysis and general base catalysis?

14. A good covalent catalyst is highly nucleophilic and can form a good leaving group. What structural properties support these seemingly opposite characteristics?

15. Classify each of the following groups as an electrophile or nucleophile:
    (a) amine
    (b) carbonyl
    (c) cationic imine
    (d) hydroxyl
    (e) imidazole

16. For each of the following reactions, indicate the nucleophilic center and the electrophilic center. Draw curved arrows to indicate the movement of electrons and draw the reaction products.

    (a) carbonyl phosphate + ammonia $\rightleftharpoons$ carbamic acid + $P_i$

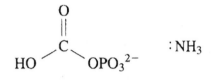

    (b) NADH + acetaldehyde $\rightleftharpoons$ ethanol + $NAD^+$

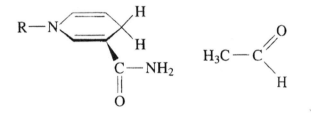

    (c) 2 amino acid $\rightleftharpoons$ dipeptide + $H_2O$

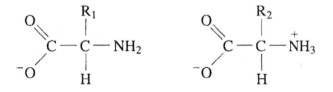

17. What is the role of $Zn^{2+}$ in carbonic anhydrase? *Hint*: See Interactive Exercise 7.

## Lysozyme

✦ *See Interactive Exercise 8 before answering the following questions.*

18. What are the roles of Glu 35 and and Asp 52? What is the role of water in catalysis?

19. Why does lysozyme appear to bind only NAG residues at subsites C and E?

20. Would $(NAG)_6$ or $(NAM)_6$ be a better substrate for lysozyme and why?

21. Hydrogen bonding of substrates to enzymes often involves the polypeptide backbone rather than amino acid side chains. What backbone–substrate hydrogen bond helps distort NAM in the D subsite of lysozyme? Can this hydrogen bond form when *N*-acetylxylosamine is in the active site?

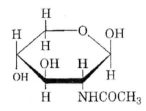

**N-Acetylxylosamine**

22. Draw the resonance forms of the half-chair conformation of the oxonium ion transition state of the lysozyme reaction.

*Serine Proteases*

✦ *See Guided Exploration 10 before answering the following questions.*

23. What is the role of the following amino acids in the active site of chymotrypsin?
    (a) Asp 102
    (b) His 57
    (c) Ser 195
    (d) Gly 193
    (e) Which three constitute the "catalytic triad?"

24. What two catalytic residues in chymotrypsin were identified by chemical modification? Could the same reagents used to identify these residues be used to label the catalytic residues of other serine proteases?

25. The different substrate specificities of chymotrypsin and trypsin have been attributed to the presence of different amino acid residues in the binding pocket. What problems can arise when site-directed mutagenesis is used to test predictions about the roles of such residues in substrate specificity?

26. The cleavage of the ester *p*-nitrophenylacetate (shown below) by chymotrypsin occurs in two stages. In the first stage, the product *p*-nitrophenolate is released in a burst, in amounts equivalent to the amount of active enzyme present. In the second stage, *p*-nitrophenolate is generated at a steady but much reduced rate. Explain this phenomenon in terms of the catalytic mechanism presented in Figure 11-29.

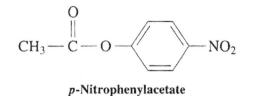

**p-Nitrophenylacetate**

27. What catalytic mechanism contributes the most to chymotrypsin's rate enhancement?

28. Lys 15 of bovine pancreatic trypsin inhibitor binds to the active site of trypsin but is not cleaved. Explain why the proteolytic reaction cannot proceed.

29. Since trypsin activation is autocatalytic, what is the role of enteropeptidase in activating trypsin?

# 12

# Enzyme Kinetics, Inhibition, and Control

This chapter introduces chemical kinetics—the study of reaction rates—followed by the kinetics of enzymatic reactions. An enzyme-catalyzed reaction can be described by the Michaelis–Menten equation, which expresses the reaction velocity in terms of its Michaelis constant, $K_M$, and its maximum velocity, $V_{max}$. Detailed knowledge of the kinetics of a reaction can contribute to the understanding of its step-by-step reaction mechanism. The effects of different substrates, inhibitors, and other factors may also reveal an enzyme's physiological function. This knowledge can be exploited to develop drugs that are enzyme inhibitors. In this chapter, the three types of reversible enzyme inhibition and the equations that describe them are presented. The chapter also describes the control of enzymes, using aspartate transcarbamoylase as an example of allosteric control, and glycogen phosphorylase as an example of control by covalent modification. The chapter concludes with a discussion of how drugs are developed and tested, and the basis of adverse drug reactions.

## Essential Concepts

### Reaction Kinetics

1.  A chemical reaction may proceed through several simple steps, called elementary reactions. The overall reaction pathway may therefore involve several short-lived intermediates.

2.  The rate, or velocity ($v$), at which a reactant is consumed or a reaction product appears can be mathematically described. Thus, for the conversion of reactant A to product P,

$$v = \frac{d[P]}{dt} = -\frac{d[A]}{dt} = k[A]$$

3.  The rate of an elementary reaction varies with the concentration(s) of the reacting molecule(s). For example, for a single-reactant reaction (a unimolecular or first-order reaction), the rate is directly proportional to the concentration of the reactant. For a two-reactant reaction (a bimolecular or second-order reaction), the rate is directly proportional to the product of the concentrations of the reactants or to the square of the concentration of a reactant that reacts with itself.

4.  The proportionality constant in the equation above, which is known as the rate constant, $k$, can be determined graphically. The rate equation for a first-order reaction is

$$\ln[A] = \ln[A]_o - kt$$

and that for a second-order reaction is

$$\frac{1}{[A]} = \frac{1}{[A]_o} + kt$$

where $[A]_o$ is the initial concentration of the reactant and $t$ is time. Consequently, if a plot of ln $[A]$ versus $t$ yields a straight line, the reaction is first-order, and if a plot of $1/[A]$ versus $t$ yields a straight line, the reaction is second-order. The slope of the line reveals the value of the corresponding rate constant, $k$.

5.  The kinetics of enzyme-catalyzed reactions are more complicated because the enzyme and substrate (reactant) combine to form a complex that then decomposes to product and free enzyme. For reactions involving a single substrate, S, the reaction velocity is typically measured under conditions where $[S] > [E]$. At very high substrate concentrations, the velocity is independent of $[S]$ and the enzyme is said to be saturated with substrate.

6.  The Michaelis–Menten equation, which describes an enzymatic reaction, is based on the assumption that the enzyme–substrate complex maintains a steady state; that is, its concentration does not change. This assumption is valid over most of the course of a typical enzymatic reaction.

7.  The Michaelis–Menten equation is

$$v_o = \frac{V_{max}[S]}{K_M + [S]}$$

where $v_o$ is the initial velocity of the reaction (before more than ~10% of the substrate has been consumed), $V_{max}$ is the maximum rate of the reaction, and $K_M$ is the Michaelis constant. This equation describes a rectangular hyperbola (the shape of the curve generated by a plot of $v_o$ versus $[S]$) whose asymptote is $V_{max}$.

8.  The Michaelis constant, $K_M$, is unique to each enzyme–substrate pair. Its value is the substrate concentration at which the reaction velocity is half-maximal. It is therefore a measure of the affinity of the enzyme for its substrate.

9.  The catalytic constant ($k_{cat}$), or turnover number, of an enzyme can be derived from $V_{max}$:

$$k_{cat} = \frac{V_{max}}{[E]_T}$$

where $[E]_T$ is the total enzyme concentration. The overall catalytic efficiency of an enzyme can be expressed as $k_{cat}/K_M$, which is an apparent second-order rate constant for the enzymatic reaction.

10. The kinetic parameters for an enzymatic reaction can be determined by taking the reciprocal of the Michaelis–Menten equation:

$$\frac{1}{v_o} = \left(\frac{K_M}{V_{max}}\right)\frac{1}{[S]} + \frac{1}{V_{max}}$$

A plot of $1/v_o$ versus $1/[S]$, a so-called Lineweaver–Burk or double-reciprocal plot, yields a straight line whose slope and intercepts yield the values of $K_M$ and $V_{max}$.

11. Steady state kinetics cannot unambiguously establish a reaction mechanism because there are an infinite number of mechanisms that are consistent with a given set of kinetic data. However, mechanisms that are not consistent with the kinetic data can be ruled out.

12. Many enzymes have multiple (usually two) substrates and products. For example, transferase reactions are bisubstrate reactions. In a sequential reaction, all the substrates bind to the enzyme before products are formed. In sequential reactions, a particular order of substrate addition may be obligatory (an Ordered mechanism) or not (a Random mechanism).

13. In a Ping Pong reaction, one or more products of a transferase reaction are released before all substrates bind.

*Enzyme Inhibition*

14. Many substances, often pharmaceutically active compounds, alter the activities of specific enzymes. Detailed information about the mechanism of an enzyme can aid in the design of drugs with the desired properties. Studies of the kinetics of enzymes in the presence of specific inhibitors help reveal how the inhibitors act and provide a way to quantitatively compare the effects of different inhibitors. An inhibitor may reversibly interact with an enzyme to interfere with its substrate binding, its catalytic activity, or both.

15. Three modes of reversible enzyme inhibition can be distinguished by their effects on the kinetic behavior of enzymes: competitive, uncompetitive, and mixed (noncompetitive) inhibition. Double-reciprocal plots of data collected in the presence of different concentrations of an inhibitor reveal the value of $K_I$, the dissociation constant of the inhibitor from the enzyme.

16. A competitive inhibitor competes with a normal substrate for binding to the enzyme. It therefore reduces the apparent affinity of the enzyme for its substrate (increases $K_M$). A large excess of substrate can overcome the effect of the inhibitor, so $V_{max}$ is not affected.

17. An uncompetitive inhibitor binds only to the enzyme–substrate complex and apparently distorts the active site. It decreases the apparent $K_M$ and $V_{max}$.

18. A mixed inhibitor binds to both free and substrate-bound enzyme and may interfere with both substrate binding and catalysis. As a result, the apparent $V_{max}$ decreases, and the apparent $K_M$ may increase or decrease. When only $V_{max}$ is affected, the inhibition is said to be noncompetitive.

## Control of Enzyme Activity

19. Enzyme activity can be controlled either by altering the amount of enzyme available for reaction or by modifying its catalytic activity through allosteric effects or by covalent modification. For example, in feedback regulation, the end product of a metabolic pathway inhibits the first committed step in the pathway. This ensures that the pathway is active when product concentrations are low but is inactive when product concentrations exceed the levels needed by the cell.

20. *E. coli.*aspartate transcarbamoylase (ATCase) provides an example of allosteric regulation that includes feedback inhibition by CTP (the ultimate product of the pyrimidine synthesis pathway, which begins with the ATCase reaction) and activation by ATP (which ensures that the concentrations of pyrimidine nucleotides keep pace with those of purines). The binding of either effector molecule (ATP or CTP) to the regulatory subunits of ATCase induces changes in the enzyme's quaternary structure that alter the activity of the catalytic subunits. ATP stabilizes the R (high activity) state, whereas CTP stabilizes the T (low activity) state of this allosteric enzyme.

21. Enzymes controlled by covalent modification usually undergo phosphorylation and dephosphorylation of a Ser, Thr, or Tyr residue, such that the enzyme shifts to a more or less active state. The modifications are catalyzed by kinases and phosphatases that are often part of signaling cascades.

## Drug Design

22. The majority of drugs act by modifying the activity of a receptor protein. The second largest class of drugs is enzyme inhibitors.

23. Methods of screening drugs to develop a potential candidate drug, called a lead compound, have evolved dramatically in the past thirty years, but in most cases, initial screening begins with an assessment of $K_I$. Rational or structure-based drug design takes advantage of knowledge of the three-dimensional structure of the target protein. Combinatorial approaches coupled with robotic systems have also quickened the pace of drug development.

24. Once a drug candidate has been developed, scientists must determine whether it can be delivered in sufficiently high concentrations to the target organ and protein. All drugs must ultimately be tested in humans, which involves expensive, federally regulated clinical trials. Currently, the Food and Drug Administration approves only 1 drug candidate in 5000.

25. Differences in patients' reactions to drugs arise from differences in genetic makeup as well disease states. The cytochrome P450 superfamily of heme-containing proteins oxidizes most drugs (and other xenobiotics). Humans express 57 isozymes of cytochrome P450, many of which are highly polymorphic. The resulting variation among individuals in the rates that specific drugs are metabolized are, in part, responsible for the variation in patients' responses to the same drug. Similarly, drug–drug interactions may result from cytochrome P450 activity on simultaneously administered drugs.

### Key Equations

$$\ln [A] = \ln [A]_0 - kt$$

$$\frac{1}{[A]} = \frac{1}{[A]_0} + kt$$

$$v_0 = \frac{V_{max}[S]}{K_M + [S]}$$

$$k_{cat} = \frac{V_{max}}{[E]_T}$$

$$\frac{1}{v_0} = \left(\frac{K_M}{V_{max}}\right)\frac{1}{[S]} + \frac{1}{V_{max}}$$

### Behind the Equations/Calculation Analogy: A Conceptual Approach to Michaelis-Menten Enzyme Kinetics

The text of Chapter 12 takes a systematic approach to defining (operationally and mathematically) the four kinetic parameters that we seek to solve for when performing Michaelis-Menten enzyme kinetic analysis: Vmax, $K_M$, $k_{cat}$ and $k_{cat}/K_M$. If a friend or family member asked you to describe what these four kinetic parameters really mean, could you explain them in such a way that the scientific layperson could understand? Can you look at kinetic data and interpret what the data are telling you about an enzyme? In order to reach this level of understanding, one must approach Michaelis-Menten kinetics conceptually as well as mathematically. Two simple analogies are often sufficient to reach this level of understanding: 1) automobiles and 2) imagining a large room of full of students with pencils.

### $V_{max}$: A simple concept that is often misused

$V_{max}$ is an easy kinetic parameter to grasp on the surface. However, students often misrepresent and over-extend what $V_{max}$ is really telling them. $V_{max}$ is operationally defined as the maximal velocity of an enzyme-catalyzed reaction. $V_{max}$ occurs only when the enzyme is saturated with substrate. To arrive at this conclusion, one simply needs to apply logic.

Consider an enzyme population in which there are multiple molecules of enzyme present along with substrate. At the most basic level, one can assume that the enzyme exists in one of two states: the enzyme without the substrate in the active site ([E]) and the enzyme with the substrate in the active site ([ES]). The total amount of enzyme ([E]$_T$) is simply the sum of both forms of the enzyme: [E]$_T$ = [E] + [ES]. One would expect that the maximal rate of the reaction catalyzed by that population of enzymes will only occur if substrate occupies every active site. An enzyme without substrate does not contribute to the rate of product formation at any instant in time. Thus, Vmax should occur when [E]$_T$ = [ES]. The only way that [E]$_T$ = [ES] continuously is if [S] is extremely high, or saturating for the enzyme. Thus, $V_{max}$ is expected to occur only at saturating [S], when the concentration of S is so high that the reaction rate becomes independent of S (adding more S will not increase the rate of reaction any further). Kinetically, we say that $V_{max}$ occurs when the reaction becomes zero-order with respect to substrate.

To arrive conceptually at the mathematical definition of $V_{max}$, one must arrive at one other logical conclusion. The maximal rate of any process is set by the slowest, or rate-limiting, step in that process. For enzyme-catalyzed reactions, it is assumed that the rate-limiting step is one of the steps in the actual catalytic mechanism, not the act of substrate binding to, or product release from, the active site. This rate-limiting step is described by the kinetic rate constant $k_2$. Students often question whether this assumption is valid because they perceive enzymes to be very fast at converting substrate to product. However, students should recall that the rate of diffusion is very fast, and that substrate binding to enzyme is often quite rapid.

To see the validity of this assumption, imagine that you were tasked with designing a new enzyme from scratch. Presumably, the reaction that you wish to catalyze with this new enzyme is already thermodynamically spontaneous, just slow in the absence of the enzyme. The reaction already has a mechanism in the absence of the enzyme. The enzyme that you design should not create a new mechanism, just speed up the existing mechanism. However, the enzyme must add one additional step to the mechanism: the act of substrate binding to the enzyme active site. After the substrate occupies the active site, the enzyme simply enhances the rate of one or more steps (at least the rate-limiting step) in the mechanism through one or a combination of the catalytic strategies discussed in Chapter 11 of the text. From a design standpoint, if you must add one step to a mechanism to enhance the rate of a reaction, the newly added step had better not be slower than every other step. If the newly added binding step is slower than every other existing step, you only succeeded in slowing the reaction rate. We know that enzymes enhance their reaction rates, thus we must conclude that the new substrate binding step that they add to the reaction mechanism must be faster than the slowest step in the existing mechanism.

From these two points of logic, one can now explain the mathematical definition of Vmax:

$$V_{max} = k_2[E]_T$$

Conceptually, this makes sense when you look at an analogy. Imagine that you have a very large classroom that contains only 30 students, where each student represents one molecule of the same enzyme. The students are spread out around the room such that there is a significant distance between each student. The catalytic goal of these enzymes is to break pencils in half as fast as possible. In order to accomplish this task each person must pick up a pencil (analogous to substrate binding), break the pencil in half (catalysis; represented by the rate constant $k_2$), and put down the broken pencil halves (product release).

First, notice that $V_{max}$ is dependent upon the total amount of enzyme present ($[E]_T$). For our example, 30 students would be able to break pencils at a particular speed. If we were to double the number of students in the room, then theoretically, they would be able to break pencils twice as fast. Now, assuming that breaking the pencil is the rate-limiting step, then the value of $k_2$ should be the inherent limiting factor in determining the maximal rate at which each person can break pencils. However, we must also consider how many pencils the students have access to. If the large room contains only 2 pencils, then the students must wander around the room trying to find a pencil (approximates a random encounter of an enzyme and substrate by diffusion), and then only 2 of the 30 students will grab the pencils and break them. Finding the pencils requires some time (but is not the rate-limiting step), all 30 students were not recruited to break pencils at all of their maximal rates continuously ($[E]_T \neq [ES]$, and thus the system has not achieved $V_{max}$.

Now imagine that we fill the room with pencils such that every student has continual access to so many pencils all of the time, that there is no chance that a student will not have almost instantaneous access to pencils. Under this condition, the fraction of time that each student spends with a pencil in hand (as ES) compared to not having a pencil in hand (as E) is extremely high, such that we can assume that at any instant in time, $E_T = ES$. At this point, the rate at which each student can break pencils is equivalent to the speed of the rate-limiting step ($k_2$). The total rate of reaction in the room then becomes the product of $k_2$ times the average speed of each person, or $V_{max} = k_2[E]_T$.

This raises a very important summary point: $\underline{V_{max} \text{ is a population parameter}}$. It is the measure of the maximal rate of a population of enzymes under conditions of saturating S. $V_{max}$ is not the speed of an individual enzyme active site. As a result, $V_{max}$ has limited value when assessing the speed and efficiency of one enzyme versus another. The more useful kinetic parameter to know is $k_{cat}$.

### $V_{max}$ vs. $k_{cat}$: A population vs. individual measure

To illustrate the difference between $V_{max}$ and $k_{cat}$, consider the following scenario:

A man walks into a car dealership looking to buy a very fast automobile. He asks the car salesperson "How fast can one million Ferraris go if you push the all their gas pedals all the way to the floor?" Obviously, this question misses point of what the man really wants to know. The more appropriate question is "How fast can one Ferrari go if you push the gas pedal all the way to the floor?" This is the difference between $V_{max}$ and $k_{cat}$. As mentioned previously, $V_{max}$ is a population parameter that measures the maximal speed of a population under conditions of saturating S. $V_{max}$ is essentially the answer to the question concerning the speed of one million Ferraris. The more practical information is how fast can one enzyme active site convert (or turn over) substrate to product under conditions of saturating substrate. This would be the true measure of how fast one enzyme is at catalyzing a reaction.

Mathematically, $k_{cat}$ is easy to derive. Consider our original analogy of students breaking pencils under conditions of saturating S. As students continuously break pencils at their maximal rate, each is contributing to an overall $V_{max}$. Imagine that the $V_{max}$ of this process is 900 pencils broken per minute by the 30 students. What is the average contribution of an individual student to this $V_{max}$? To arrive at this mathematically, we would simply need to normalize the $V_{max}$ set by 30 students to one individual student. To do this, we would simply divide the total rate by the number of students to determine the average contribution of each student: 900 pencils per minute / 30 students = 30 pencils per minute per student. This is the $k_{cat}$, or turnover number of an individual student (active site). Thus, to determine $k_{cat}$, we simply need to normalize the $V_{max}$ of the system to the number of active sites. To do this we simply divide the $V_{max}$ by $[E_T]$.

$$k_{cat} = \frac{k_2[E_T]}{[E_T]} = k_2$$

From the conceptual and mathematical definition of $k_{cat}$, we can see that kcat is equal to the kinetic rate constant of catalysis, $k_2$, and has the units of $s^{-1}$. The value of $k_2$ is an important piece of data for assessing enzymes. Imagine that a scientist is studying the same enzyme from multiple species, from bacteria to humans. Simply knowing the value of $V_{max}$ for the enzyme for each species is useless information for comparative purposes. However, knowing the value of $k_{cat}$ for the enzyme for each species is valuable for comparison. Knowing this is like comparing the maximal speed of a variety of individual automobiles of different makes and models.

## $K_M$: An approximation of enzyme affinity for substrate

The Michaelis constant, $K_M$, can be defined in three different ways: operationally, mathematically, and practically. Operationally, $K_M$ is a substrate concentration at which the reaction is at ½ of $V_{max}$. Thus, $K_M$ has the same units as substrate concentration, usually expressed in molar, micromolar or nanomolar. One can use a comparative analogy of two different automobiles to see the significance of this operational definition. Consider two different automobiles: A Ferrari and a hybrid Prius. The Ferrari is capable of achieving speeds of up to 200 miles per hour (mph), and is analogous to an enzyme that has a very high $k_{cat}$. A Prius is not capable of reaching such speeds, maybe only 120 mph, analogous to an enzyme with a lower $k_{cat}$ value. However, this is not the only difference between these two automobiles. The Prius is immensely fuel efficient, requiring very little gasoline to achieve and maintain ½ of the maximal speed. This is equivalent to an enzyme that requires very little S to achieve ½ of $V_{max}$. Conversely, a Ferrari requires a large continuous amount of gasoline to achieve and maintain ½ of the maximal speed, which is analogous to an enzyme that requires a great deal of S to achieve ½ $V_{max}$. The comparison of the Prius and the Ferrari is an illustration of fuel efficiency vs. fuel inefficiency, respectively. If substrate can be considered fuel for an enzyme, then the value of $K_M$, the substrate concentration required to achieve ½ $V_{max}$, is also one metric, but not a complete picture of enzyme efficiency. The lower an enzyme's $K_M$ value, the more efficiently it uses substrate to achieve its maximal velocity. But a relatively fast enzyme may have a high or a low $K_M$ value, and a relatively slow enzyme may also have a high or a low $K_M$ value. Consequently, we need additional information to assess enzyme efficiency.

The mathematical definition of $K_M$ leads to the practical definition of KM. Consider Equation 12–12 from the text:

$$E + S \rightleftharpoons ES \rightarrow P + E$$

where $k_1$ is the kinetic rate constant describing the arrow pointing towards the formation of ES from E + S

$k_{-1}$ is the kinetic rate constant describing the arrow pointing towards E + S from ES without catalysis having occurred

$k_2$ is the kinetic rate constant describing the arrow pointing towards the formation and release of product from ES

Mathematically, $K_M$ is defined as a ratio of these rate constants:

$$K_M = \frac{k_{-1} + k_2}{k_1}$$

Careful scrutiny reveals that the numerator is the sum of all rate constants that are associated with arrows pointing away from ES, while the denominator is comprised of the only rate constant that is associated with an arrow pointing towards ES. In conceptual terms, the numerator is the sum of all influences that act to lower the concentration of ES, while the denominator is the influence that acts to increase the concentration of ES. Thus, the mathematical definition represents a ratio of kinetic rates that determine the equilibrium or steady state concentration of ES. If the value of $K_M$ is low, then this implies that the rate of formation of ES is very rapid compared to the rate at which ES dissociates, either productively to produce a product, or simply back to the substrate. Conversely, if the value of $K_M$ is high, then this implies that the rates at which ES dissociates are significantly fast compared to the rate at which ES forms.

In order to derive the forward rate of equation 12 – 12 from the text, one of two simplifying assumptions were necessary. The assumption of equilibrium stated that ES was in equilibrium with E + S because $k_{-1} \gg k_2$. The assumption of steady state stated that ES achieved a steady state concentration through most of the reaction if [S] $\gg$ [E]. In order for ES to achieve a steady state, the values of $k_2$ and $k_{-1}$ must be very small compared to $k_1$. If the values of $k_2$ and $k_{-1}$ are very large compared to the value of $k_1$, then ES would dissociate as fast as it formed, and the steady state concentration of ES would be negligible. If the equilibrium assumption is valid, even to a degree, and recalling that $k2$ is the rate-limiting step of the enzymatic process, then we can assume the order of the kinetic rate constants, from fastest to slowest is: $k_1 \gg k_{-1} \gg k_2$. If the value of $k_2$ is significantly small compared to $k_{-1}$, then mathematically it does not significantly contribute to the numerator of the $K_M$ equation, and this equation reduces to:

$$K_M = \frac{k_{-1}}{k_1} = K_S$$

Thus, $K_M$ is an approximation of the equilibrium dissociation constant for the ES complex. Thus, we can say that $K_M$ is a reflection of the affinity E has for S as long as these assumptions are valid. This is the most useful interpretation of $K_M$, as it comprises one metric of enzyme efficiency and provides a basis for comparing enzymes.

**Catalytic Efficiency and the Diffusion-Controlled Limit**

Conceptualizing catalytic efficiency of enzymes is very similar to evaluating the efficiency of an automobile. Automobile performance is most often evaluated using speed and gas mileage. Enzymes are also evaluated on two similar criteria, kcat and $K_M$. Recall that the kcat value quantifies the turnover of substrate to product by a single enzyme active site, and thus is a reflection of speed. As mentioned in the previous section, $K_M$ is a reflection of the affinity of E for S. Catalytic efficiency is quantified by dividing kcat by $K_M$:

$$\text{Catalytic Efficiency} = k_{cat}/K_M$$

This equation represents the turnover of the enzyme normalized to the relative affinity E has for S. The resulting quantity of kcat/$K_M$ has units consistent with a diffusion-controlled reaction, $M^{-1} s^{-1}$. The significance of this is that it provides insight as to how efficient the enzyme is with a particular reference in mind. If we consider again the simplest scheme for an enzymatically catalyzed reaction and the three rate constants of $k_1$, $k_{-1}$ and $k_2$, we can see that an enzyme may become more efficient through mutations that decrease the $K_M$ or increase the $k_{cat}$ value. The upper limit of $k_1$ is set by the laws of physics and chemistry and cannot be altered. Thus, if enzyme efficiency has evolved to the point where $k_2$ is no longer rate-limiting, and $k_{-1}$ has been minimized by improving the affinity of E for S, then $k_1$ becomes rate-limiting. The value of $k_1$ is set by the physical limit of diffusion in an aqueous environment, and dictates the rate at which E and S come into contact. This limit of diffusion is often reported as $10^8 - 10^9 \ M^{-1} s^{-1}$. When enzymes reach this 'diffusion-controlled limit', they are said to have reached catalytic perfection.

**Applying these concepts:**

Consider Table 12–1 of the text to answer the following questions:

1. Which of the enzymes listed in Table 12-1 has the highest affinity for its substrate? Which has the lowest? Explain.

   Therefore, fumarase has the highest apparent affinity for the substrate fumarate while chymotrypsin has the lowest apparent affinity for the substrate N-acetylglycine ethyl ester because they have the lowest and highest $K_M$ values for these substrates, respectively.

2.  Which of the enzymes listed in Table 12–1 has the highest turnover number Which as the lowest turnover number? Explain.

    Therefore, catalase has the highest turnover number while chymotrypsin has the lowest turnover number (for the substrate N-acetylglycine ethyl ester) because they have the highest and lowest values of $k_{cat}$, respectively.

3.  Which of the enzymes listed in Table 12–1 has the highest catalytic efficiency? Which has the highest? Explain.

    Therefore, catalase has the highest catalytic efficiency while chymotrypsin has the lowest catalytic efficiency because they have the highest and lowest $k_{cat}/K_M$ values, respectively.

4.  Which of the enzymes listed in Table 12–1 have reached the diffusion-controlled limit of catalytic perfection? Explain.

    Therefore, acetylcholinesterase, catalase and fumarase have all achieved the diffusion-controlled limit of catalytic perfection because they have $k_{cat}/K_M$ values in the range of $10^8 - 10^9 \ M^{-1} \ s^{-1}$.

## Questions

### Reaction Kinetics

1.  For each of the following reactions, write a rate equation and determine the reaction order.
    (a)  $A \rightarrow P$
    (b)  $A + B \rightarrow P + Q$
    (c)  $2A \rightarrow P$

2.  List two different ways to measure the progress of a chemical reaction.

3.  A first-order reaction has a $t_{1/2}$ of 20 minutes.
    (a)  What is the rate constant $k$?
    (b)  What time is required to form 20% of the product?
    (c)  What time is required to form 80% of the product?
    (d)  How much starting material remains after 15 min?
    (e)  Compare the rate constant for this reaction to that of the decay of $^{32}P$, which has a half-life of 14 days.

4.  The energy of binding a transition state complex ($X^{\ddagger}$) can be determined by writing an equilibrium expression for the formation of the complex.
    (a)  For the reaction $A \leftrightarrows X^{\ddagger}$, what is the equilibrium expression?
    (b)  What is the expression for the free energy of binding to the transition state ($\Delta G^{\ddagger}$)?
    (c)  What factors are needed to equate the rate constant $k$ to the free energy of activation?

5.  For the following reaction:

$$E + S \underset{k_{-1}}{\overset{k_1}{\rightleftharpoons}} ES \overset{k_2}{\rightleftharpoons} P + E$$

(a)  What is meant by the term "enzyme–substrate complex"?
(b)  Write a rate equation for the production of ES.
(c)  What is the rate of product formation from ES?
(d)  If all the enzyme is bound to substrate, what is the effect of adding more substrate on the forward rate of the reaction?

6.  What is meant by (a) the steady state assumption, (b) $K_M$, (c) $k_{cat}$, (d) turnover number, (e) catalytic efficiency, and (f) diffusion-controlled limit?

7.  The following data were obtained for the reaction A $\rightleftharpoons$ B, catalyzed by the enzyme Aase. The reaction volume was 1 mL and the stock concentration of A was 5.0 mM. Seven separate reactions were examined, each containing a different amount of A. The reactions were initiated by adding 2.0 μL of a 10 μM solution of Aase. After 5 minutes, the amount of B was measured.

| Reaction | Volume of A added (μL) | Amount of B present at 5 minutes (nmoles) |
|---|---|---|
| 1 | 8 | 26 |
| 2 | 10 | 29 |
| 3 | 15 | 39 |
| 4 | 20 | 43 |
| 5 | 40 | 56 |
| 6 | 60 | 62 |
| 7 | 100 | 71 |

(a)  Calculate the initial velocity of each reaction (in units of $\mu M \cdot min^{-1}$)
(b)  Determine the $K_M$ and $V_{max}$ of Aase from a Lineweaver–Burk plot.
(c)  Calculate $k_{cat}$.

8.  Can you use kinetic data to prove that a particular model for an enzymatic reaction mechanism is correct? Explain.

9.  Why is it possible for sequential bisubstrate reactions to be Ordered or Random, whereas a Ping Pong reaction always has an invariant order of substrate addition and product release?

10. Assume that $k_2 = k_{cat}$ for a highly purified enzyme for which you seek to determine $\Delta G^{\ddagger}$. Using the Arrhenius equation, $k = Ae^{-\Delta G^{\ddagger}/RT}$, where $k$ is a reaction rate constant and A is a constant, you can determine the activation energy to a good approximation.

    (a) Derive a mathematical relationship that will allow you to calculate $\Delta G^{\ddagger}$ and show a graph that can used with experimental data. *Hint*: Convert the Arrhenius equation into a linear form, in which the constant A need not be known beforehand.

    (b) What are the easily measurable parameters that you can obtain?

    (c) What assumptions do you need to make in order to obtain a good approximation of the activation energy?

## Enzyme Inhibition

11. There are three general mechanisms for the reversible inhibition of enzymes that follow the Michaelis–Menten model. How does the mode of inhibitor–enzyme binding differ among the three mechanisms?

12. The catalytic behavior of an enzyme may depend on ionizable amino acids. Therefore, a change in pH may influence an enzyme's catalytic behavior. How can you tell whether pH affects substrate binding or catalytic activity?

13. The movement of glucose across the erythrocyte membrane is "catalyzed" by a transport protein (Section 10-2E and Box 10-2).

    (a) What is the kinetic behavior of this process?

    (b) Can glucose transport be subject to competitive, uncompetitive, or mixed inhibition? Explain.

14. In hen egg white lysozyme (Section 11-4), the substitution of Ala for Asn at position 37 or for Trp at position 62 may alter the enzyme's kinetics. What changes would you predict and why?

## Control of Enzyme Activity

15. Draw velocity versus [Asp] curves for the reaction catalyzed by the ATCase catalytic trimer and by the intact enzyme. Explain why the curves differ.

16. How do carbamoyl phosphate, aspartate, ATP, and CTP affect the T $\leftrightarrows$ R equilibrium of ATCase?

## Drug Design

17. What biochemical parameters are likely to vary among different allelic isozymes of cytochrome P450?

18. Draw a flow diagram that summarizes the steps involved in developing a drug from the generation of a lead compound through FDA approval. Which is the most time-consuming step?

# 13
# Biochemical Signaling

This chapter introduces the fundamental biochemical signaling pathways that allow the cell to respond to chemical and physical signals in its environment. The focus of this chapter is on biochemical signaling pathways that respond to intercellular signals called endocrine hormones and growth factors. The binding of a hormone to its receptor induces a conformational change that transduces the signal into new biochemical activity inside the cell. Three signal transduction pathways are discussed in detail: signaling pathways activated by receptor tyrosine kinases; the adenylate cyclase signaling system activated by G protein–coupled receptors; and the phosphoinositide pathway, which involves hydrolysis of a phospholipid. Alterations of signaling pathways, which often result from mutations in genes (oncogenes) that encode signaling proteins, are important to the development of cancers, as indicated in Box 13-3. Signal transduction pathways are the target of a variety of drugs and toxins, as discussed in Box 13-4. The chapter also highlights two key biochemical technologies used in the quantitative analysis of biochemical signaling: the radioimmunoassay developed by the late Rosalyn Yalow (Box 13-1) and receptor–ligand binding assays using Scatchard plots (Box 13-2).

## *Essential Concepts*

### *Hormones*

1.  In general, every biochemical signaling pathway consists of a receptor protein that binds a hormone or some other signaling molecule. This binding results in conformational changes that either allow the receptor–hormone complex to directly affect gene expression or transduce the extracellular signal through a series or cascade of reactions, often involving protein kinases and phosphatases. The latter mechanism may result in dramatic exponential amplification of the original signal molecule's interaction with a single receptor.

2.  Endocrine hormones are chemical signals that reach their target cells via the bloodstream. Each hormone binds to a specific receptor, and its binding results in the transduction of the hormone signal into a biochemical response inside the cell (often referred to as signal transduction). Endocrine hormones include soluble compounds such as amino acid derivative and small polypeptides, and insoluble lipids such as steroids. Many factors may modulate the specific response of a cell to a particular hormone, such as the presence or number of receptors, the synergistic or antagonistic activities of other hormone-bound receptors, and the specific isoforms of the kinases and phosphatases of the activated enzyme cascades.

3.  The pancreatic islet and adrenal gland hormones represent some of the best studied hormone signaling systems. Energy metabolism is regulated largely by these hormones. The pancreas secretes the polypeptide hormones, glucagon and insulin, which act antagonistically to each other to regulate blood glucose concentration. The adrenal medulla

secretes the catecholamines, epinephrine and norepinephrine, which bind to α- and β-adrenergic receptors. These receptors often respond to the same ligand in opposite ways in different tissues.

4.  Steroids are carried in the blood via the glycoprotein transcortin and albumin. As transcortins interact with cellular membranes, steroids pass across the membrane and bind to their cognate steroid receptors. Steroid–receptor complexes migrate to the nucleus, where they function as regulators of DNA transcription (see Chapter 28).

5.  The receptor for growth hormone (a polypeptide hormone) shows features characteristic on many membrane-bound receptors. The binding of growth hormone to the extracellular domain of its receptor results in dimerization of the receptor. This allows it to bind to protein kinases that phosphorylate Tyr residues on target proteins (via transfer of the γ-phosphate from ATP), and are therefore called tyrosine kinases. Growth hormone and insulin were among the first polypeptides to be produced in industrial quantities by recombinant DNA technology.

## Receptor Tyrosine Kinases

6.  A large family of cell surface receptors, known as receptor tyrosine kinases (RTKs), possess protein kinase activity that specifically phosphorylates tyrosine residues on target proteins. The insulin receptor is a receptor tyrosine kinase. RTKs usually dimerize upon binding ligand, which leads to autophosphorylation of the receptor protein. This autophosphorylation in turn may lead to the phosphorylation of target proteins. These proteins may bind directly to the RTK via a protein domain called the Src homology 2 (SH2) domain, which recognizes the phospho-Tyr residues, or they may bind to integrator proteins such as the insulin receptor substrate proteins (IRS-1 and IRS-2), which are directly phosphorylated by the RTK and then serve as docking proteins to which other proteins bind.

7.  The human genome encodes at least 90 protein Tyr kinases (PTKs) and 388 Ser/Thr kinases (representing >2% of human genes). PTK domains undergo dramatic conformational changes close to the active site, presumably bringing substrates in close juxtaposition, as well as excluding water from the active site, which could result in wasteful ATP hydrolysis.

8.  Many RTKs interact with proteins that activate the monomeric G protein Ras, which in turn activates a protein kinase cascade. In the best-known pathway, the activation of Ras leads to the activation of Raf (MAPKKK), which is a Ser/Thr protein kinase that phosphorylates and activates MEK (MAPKK). MEK is both a Tyr kinase and a Ser/Thr kinase that dually phosphorylates MAP kinase (MAPK), another Ser/Thr kinase with many targets including other kinases and transcription factors. Many MAPK cascade proteins are organized into complexes by scaffold proteins that are essential for the specificity and rapid amplification of signal transduction in these signaling pathways.

9. G proteins are an important family of regulatory proteins that bind and hydrolyze GTP, resulting in a cycle of conformational changes reminiscent of the ATP-binding-and-hydrolysis cycle of myosin.

    (a) Monomeric Ras interacts with a guanine nucleotide exchange factor (GEF; Sos in some cases), which stimulates Ras to exchange its bound GDP for GTP. Ras also interacts with a GTPase activating protein (GAP), which stimulates the minimal GTPase of Ras by a factor of $10^5$.

    (b) An activated RTK binds to an adapter protein (such as Grb2)–GEF complex, which then interacts with inactive Ras·GDP, resulting in a rapid exchange of GDP for GTP to produce active Ras·GTP. Activated Ras·GTP binds to and stimulates other effectors, such as Raf, becoming inactivated by the hydrolysis of GTP.

10. Receptor tyrosine kinase family members also bind proteins called growth factors that regulate cell growth and differentiation. Many genes that encode elements of growth-factor signaling pathways are known as proto-oncogenes, because their mutation may yield an oncogene. The product of an oncogene may disrupt normal cell growth and lead to cancer. Many cancer-causing viruses bear oncogenes. For example, the v-*erbB* oncogene, first identified in an animal virus, codes for the epidermal growth factor receptor without its ligand-binding domain, which results in constitutively active phosphorylation of protein substrates by this RTK. Other oncogenes include mutated forms of Ras and the transcription factors Fos and Jun (see Box 13-3).

11. Many receptors do not have intrinsic kinase activity (e.g., the growth hormone receptor and many cytokine receptors). Instead, dimerization or trimerization of the receptor usually leads to complexing and activation of nonreceptor tyrosine kinases (NRTKs).. One of the best studied NRTKs is Src (named after the sarcoma virus wherein it was first discovered), which is one member of a large family of NRTKs. Many of these have common structural domains that allow the protein to dock or interact with other signaling proteins, such as SH2 (Src homology-2) domain, which binds to receptor phospho-Tyrs, and the SH3 domain, which binds to proline-rich domains in downstream effector proteins.

12. Protein kinases that phosphorylate serine and threonine residues make up another large class of signaling enzymes, the best known of which are the mitogen-activated protein kinases (MAPKs). The MAPK cascade is often tethered by scaffold proteins that organize and spatially position combinations of kinase isoforms. These signaling pathways initiate a variety of cellular responses including cell growth and differentiation, proliferation, apoptosis (programmed cell death), and cell survival.

13. The activity of protein kinases is antagonized by the activity of protein phosphatases that hydrolyze the phosphoryl groups added to proteins by protein kinases, and thereby inhibit or reverse the effects of the signaling pathway that originally activated the protein kinase. The phosphatases are generally specific for phospho-Tyr or phospho-Ser/Thr, although dual specificity protein phosphatases are known Protein phosphatases in general have broader specificity than the protein kinases and so they form a smaller family of proteins.

14. Receptor-ligand binding interactions can be quantified using a variety of technologies, including radioimmunoassays (Box 13-1). This quantification can provide insights into the physiological hypersensitivity of some cancers to endogenous growth factors, or help distinguish the binding affinities of different isoforms of receptors (e.g., adrenergic receptor isoforms). This quantification is based on the dissociation constant for the receptor and its ligand at equilibrium, which can be graphically represented (Scatchard plot) to reveal both the dissociation constant and number of receptors (see Box 13-2).

*Heterotrimeric G Proteins*

15. Some signal transduction systems that involve heterotrimeric G proteins include a G protein–coupled receptor (GPCR) and adenylate cyclase, which produces the second messenger cAMP. Production of the second messenger leads to activation of many target proteins. Thus, the initial hormone signal is amplified, although its duration is usually limited.

16. GPCRs transmit the binding of an extracellular signal (or ligand) to the cell interior by alternating between two discrete conformations, one with the ligand bound and another without. The glucagon receptor and the α- and β-adrenergic receptors are GPCRs. All GPCRs contain seven transmembrane α helices, but have widely variable N- and C-termini and intrahelical loops

17. GPCR and other signaling systems respond to *changes* in stimulation rather than to specific values of extracellular signals, such that they reduce their response to long-term stimuli in a process called desensitization. Desensitization involves phosphorylation of the activated receptor and often its removal from the cell surface by endocytosis.

18. G proteins form a large class of GTP-hydrolyzing proteins, of which there are two families involved in signal transduction: heterotrimeric G proteins and small monomeric G proteins (e.g., the Ras family). Heterotrimeric G proteins consist of a $G_\alpha$, $G_\beta$, and $G_\gamma$ subunit, in which GTP binds to and is hydrolyzed in the $G_\alpha$ subunit. The inactive state consists of the trimeric holo G protein, in which GDP is bound to the $G_\alpha$ subunit. The binding of this complex to a ligand-bound receptor results in a conformational change that stimulates the exchange of GDP for GTP. The γ phosphate of GTP in turn induces a conformation change in the three switch regions of the $G_\alpha$ subunit, which causes the dissociation of the $G_\alpha$ subunit from the $G_{\beta\gamma}$ dimer. Both the dissociated $G_\alpha$ and the $G_{\beta\gamma}$ subunits can then act to allosterically activate downstream protein targets, although targets of the $G_\alpha$ subunit are the best characterized. A $G_\alpha$ subunit self-inactivates through the slow hydrolysis of its bound GTP to GDP and $P_i$.

19. The exchange of GTP for GDP and GTP hydrolysis in G proteins can be augmented by accessory proteins.
    (a) GTPase-activating proteins (GAP) stimulate GTP hydrolysis.
    (b) Guanine nucleotide exchange factors (GEF) stimulate the replacement of GDP with GTP in G proteins.
    These accessory proteins are most commonly associated with the small monomeric G proteins (e.g., Ras).

20. Signal transduction via the adenylate cyclase system involves activation of a heterotrimeric G protein. The $G_\alpha$ subunit of a stimulatory G protein ($G_{s\alpha}$) activates adenylate cyclase, an integral membrane protein, whereas the $G_\alpha$ subunit of an inhibitory G protein ($G_{i\alpha}$) inhibits the enzyme.

21. The second messenger 3',5'-cyclic AMP (cAMP), the product of the adenylate cyclase reaction, activates protein kinase A (PKA). cAMP binds to the regulatory subunits of PKA so as to release the catalytic subunits in active form. The effects of cAMP are attenuated by the action of a phosphodiesterase that hydrolyzes cAMP to AMP, and by protein phosphatases that reverse the phosphorylation catalyzed by PKA. Many drugs and toxins interfere with components of the adenylate cyclase signaling system (see Boxes 13-4 and 13-5).

*The Phosphoinositide Pathway*

22. The phosphoinositide signaling pathway produces two second messengers: inositol trisphosphate ($IP_3$) and diacyglycerol (DAG), which are hydrolysis products of the plasma membrane phospholipid, phosphatidylinositol-4,5-bisphosphate ($PIP_2$). Activation of the heterotrimeric G protein, $G_q$, by a ligand–GPCR complex activates phospholipase C (PLC), which hydrolyzes $PIP_2$. The resulting $IP_3$ binds to $IP_3$-gated $Ca^{2+}$ channels in the endoplasmic reticulum, thereby releasing $Ca^{2+}$ into the cytosol. This $Ca^{2+}$ can then bind to calmodulin and activate calmodulin-sensitive kinases. DAG activates protein kinase C (PKC), which phosphorylates other kinases and proteins. Other phospholipids, including sphingolipids, may be hydrolyzed in response to GPCR activation. In addition, some RTKs bind and activate a PLC isoform that contains SH2 domains.

23. $Ca^{2+}$-calmodulin dependent kinases (CaM-dependent kinases) are autoinhibited kinases wherein the N- or C- terminal sequence serves as a "pseudosubstrate" that occludes the kinase active site. Calmodulin, by binding to the inhibitory "pseudosubstrate" sequence of the CaM-dependent kinase, causes a conformational change that bends the pseudosubstrate domain away from the active site, thereby relieving the intrinsic autoinhibition of the enzyme. The use of autoinhibition is a widespread means of controlling protein kinase activity (e.g. src is regulated in a similar way)

24. Different signal transduction pathways may antagonize each other or act synergistically via cross talk. For instance, some RTKs activate a PKC isoform that contains SH2 domains. When the insulin receptor phosphorylates the docking proteins IRS-1 and IRS-2, other proteins from multiple signaling pathways can be recruited to form a complex Moreover, other phosphoTyr sites on the insulin receptor can bind proteins that activate other signaling pathways; hence, the insulin receptor is capable of stimulating numerous signaling pathways depending on which are present in the stimulated cell.

25. Nearly every step of a signal transduction pathway is subject to regulation such that the magnitude of the ultimate biochemical response reflects the presence and degree of activation or inhibition of multiple signaling components. This complexity gives rise to emergent properties, which may not be apparent from consideration of individual component steps.

26. Insulin signaling is an example of a complex signaling system in which cellular responses to insulin depend on the biochemical phenotype, which requires a larger systems-level approach to understand the physiological response to insulin in different cell types. Similarly, some drugs or toxins (e.g., anthrax toxin; discussed in Box 13-5) may affect several signaling pathways.

## Questions

### Hormones

1. The α- and β-adreonoreceptors stimulate different cellular effects in response to their ligands or agonists; however, the ultimate physiological response is the same. What is the net result of the activation of both kinds of receptors?

2. In what way are steroid receptors unique in their action once they have bound their cognate ligand?

### Receptor Tyrosine Kinases

3. How are SH2 and SH3 domains important in the activation of Ras by receptor tyrosine kinases?

4. How does the binding of ligand to a receptor tyrosine kinase lead to changes in gene expression?

5. How do scaffold proteins help to regulate signal transduction pathways?

### Heterotrimeric G Proteins

6. Compare the overall structure and mode of signal transduction by G protein–coupled receptors and receptor tyrosine kinases.

7. During G protein activation, which subunit exchanges GDP for GTP? What is the immediate consequence of this exchange?

8. Aluminum fluoride ($AlF_3$) has been widely used in studies of the molecular mechanism of G protein transformations between the GDP- and GTP-bound forms. Suggest why this inorganic compound has proved useful in these investigations..

9. Suggest three mechanisms by which the β-adrenergic signaling pathway utilizing the second messenger cAMP can be "turned off."

10. You are studying protein kinase A (PKA)–mediated cell signaling in a mammalian cell line, in which PKA levels are sensitive to two different hormones, XGF and YGF. Examine the results shown below and suggest a mechanism for how XGF and YGF affect PKA.

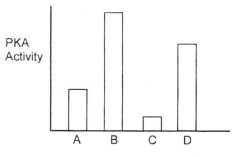

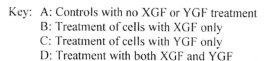

Key:  A: Controls with no XGF or YGF treatment
      B: Treatment of cells with XGF only
      C: Treatment of cells with YGF only
      D: Treatment with both XGF and YGF

11. The actions of caffeine, theobromine, and theophylline result in relative decreases in cAMP. Suggest a biochemical rationale based on the information given in Box 13-4.

*The Phosphoinositide Pathway*

12. Describe the intracellular signaling cascade mediated by phospholipase C activity. How is phospholipase C activated?

13. The insulin receptor can interact with several proteins. What are these proteins and what signaling pathways are they linked to?

14. $Li^+$ ions inhibit the recycling of $PIP_2$. Is $Li^+$ likely to counteract or enhance the effects of phorbol-13-acetate?

15. How is the regulation of the steady-state level of $IP_3$ analogous to that of cAMP?

# 14

# Introduction to Metabolism

This chapter provides a brief overview of the biological strategies and thermodynamics of metabolism. Metabolism is the overall biochemical processes that living systems use to acquire and use free energy. Organisms break down macromolecules to a common set of smaller molecules, or metabolites, which then serve as precursors for new biosynthesis. This chapter introduces some basic thermodynamic features of metabolic pathways and the mechanisms that have evolved to allow an organism to control the flow of a few common metabolites through different metabolic pathways.

Organisms harness the free energy from the degradation of macromolecules by trapping it in certain nucleotides (ATP, $NAD^+$, and FAD) and certain thioesters, which then make free energy available to energy-requiring pathways. The chapter describes these energy transmitters in terms of their thermodynamic features and their chemical properties. Some phosphorylated compounds have significant negative free energies of hydrolysis, which are described as their phosphoryl group-transfer potentials, their tendency to transfer their phosphoryl group to another compound. ATP, with its intermediate phosphoryl group-transfer potential, is the principal energy currency of life. Thioesters are also "high-energy" compounds. One of these, coenzyme A, shuttles acyl groups in metabolic processes.

Oxidation–reduction reactions are the most important process through which living organisms acquire and use free energy. The chapter reviews the principles of redox reactions, including the Nernst equation, and describes the mathematical relationship between free energy and reduction potentials. The chapter concludes with a brief look at the methods used to map the labyrinth of biochemical pathways in living cells and to understand the regulation of biochemical pathways.

## Essential Concepts

1.  Metabolism, the network of all biochemical reactions in cells, can be divided into two parts:
    (a)  Catabolism (degradation), in which free energy is released as organic molecules are broken down into smaller constituents.
    (b)  Anabolism (biosynthesis), in which biological molecules are synthesized from smaller, simpler molecules.

2.  In general, catabolic reactions release free energy, which can then be used to drive endergonic synthetic reactions. This coupling requires free energy transmitters including nucleotides (ATP, NADH, NADPH, and $FADH_2$) and thioesters (e.g., coenzyme A).

*Overview of Metabolism*

3.   Organisms use different strategies for capturing free energy from their environment:
     (a)  Autotrophs synthesize all their macromolecules from simple molecules obtained from their environment. The chemolithotrophs oxidize inorganic compounds, and the photoautotrophs use light to drive synthetic reactions.
     (b)  Heterotrophs obtain free energy from the oxidation of organic compounds (usually produced by autotrophs).

4.   Organisms can be further classified by their requirements for oxygen. Obligate aerobes require oxygen for the oxidation of nutrients. Obligate anaerobes are poisoned by oxygen and must use another electron acceptor for the oxidation of nutrients. Facultative anaerobes can oxidize nutrients both in the absence and presence of oxygen.

5.   Some oxidation–reduction reactions do not involve oxygen and are most easily recognized by examining the oxidation state of the carbon atom(s) involved in the chemical reaction. The oxidation state of a carbon atom does not reflect its atomic charge (which is close to neutral) but is a convenient tool for following oxidation–reduction reactions.

6.   Vitamins and many mineral nutrients participate in metabolic reactions as coenzymes (though most must first be chemically modified from the ingested form) and enzyme cofactors, respectively.  The water-soluble vitamins are key chemical participants in the active sites of enzymes involved in energy metabolism and biosynthesis. Fat-soluble vitamins are involved in light reception (vitamin A), $Ca^{2+}$ absorption (vitamin D), blood clotting reactions (vitamin K), and protection from oxidation (vitamin E). Minerals serve not only as enzyme cofactors (e.g., magnesium and zinc) but as key components of small molecules (e.g., sulfur in amino acids), and structural elements (e.g., calcium phosphate in bones and teeth).

7.   Metabolic pathways are compartmentalized in the cytosol of prokaryotic and eukaryotic cells. Eukaryotic cells further compartmentalize metabolic pathways in membrane-bounded organelles. In multicellular organisms, many metabolic pathways are compartmentalized in different tissues.

8.   In any given metabolic pathway, most reactions are near equilibrium, such that the law of mass action largely dictates the flow rate (flux) of metabolites.

9.   In each metabolic pathway, there is at least one reaction that is far from equilibrium, in which the reactants accumulate above their equilibrium values and $\Delta G \lll 0$. Such reactions are referred to as rate-determining steps since they control flux in the pathway. The flux changes only with a change in the enzyme's ability to increase the reaction rate.

10.   Metabolic pathways have three key characteristics:
      (a)  They are irreversible.
      (b)  They have an exergonic step that serves as the first committed step and ensures irreversibility.

(c) Catabolic and anabolic pathways involving the interconversion of two metabolites differ in key exergonic reactions.

11. Control of flux in the rate-determining step requires control of the enzyme catalyzing it by one or more of the following mechanisms:
    (a) Allosteric control by feedback regulation from an end product of the pathway.
    (b) Covalent modification of the enzyme, which may increase or decrease its ability to accelerate a reaction.
    (c) Substrate cycles in which interconversion of two substrates utilizes different rate-determining enzymes.
    (d) Genetic control, which regulates the steady-state levels of the enzyme.
       Mechanisms (a)–(c) respond quickly to changes in physiological states (seconds to minutes), whereas mechanism (d) involves slower, long-term adaptive changes to new physiological states (minutes to days).

## "High-Energy" Compounds

12. The free energy derived from the degradation of organic compounds is transiently captured in "high-energy" compounds whose subsequent breakdown provides the free energy to drive otherwise endergonic reactions.

13. ATP is the primary energy currency of cells. Its energy resides in the thermodynamic instability of its two phosphoanhydride bonds. The free energy of hydrolysis of ATP, in which the phosphate group is transferred to water, is called its phosphoryl group-transfer potential. ATP has an intermediate phosphoryl group-transfer potential, making it a conduit for the transfer of free energy from higher-energy compounds to lower-energy compounds.

14. The high-energy character of the phosphoanhydride bonds results from:
    (a) increased resonance stabilization of the hydrolysis products.
    (b) the destabilizing effect of electrostatic repulsions between the charged phosphates at neutral pH.
    (c) increased solvation energy of the hydrolysis products.

15. Many endergonic reactions in cells are coupled to the hydrolysis of ATP or pyrophosphate ($PP_i$) so that the net reaction is exergonic. In some biosynthetic reactions, the transfer of a nucleotidyl group "activates" the substrate for further reaction (e.g., the polymerization reactions of polysaccharides and the formation of aminoacyl–tRNA for protein synthesis).

16. ATP can be replenished by transfer of a phosphoryl group to ADP from a compound with a higher phosphoryl group-transfer potential. Such a transfer is called substrate-level phosphorylation. The concentrations of ATP and other nucleotides are maintained in part by the activity of kinases.

17. The transfer of acyl groups requires their "activation" by the formation of a thioester bond to a sulfur-containing compound such as coenzyme A. The hydrolysis of thioesters is about as exergonic as the hydrolysis of ATP. Hence, thioester cleavage drives the otherwise endergonic transfer of the acyl group.

*Oxidation–Reduction Reactions*

18.  Oxidation–reduction reactions are the principal source of free energy for life. The oxidation of organic compounds is coupled to the reduction of the nucleotide cofactors $NAD^+$ (and $NADP^+$) and FAD.

19.  A measure of the potential electrical energy (electromotive force or reduction potential) in an electrochemical cell is described by the Nernst equation:

$$\Delta \mathscr{E} = \Delta \mathscr{E}^{\circ\prime} - \frac{RT}{n\mathscr{F}} \ln \left( \frac{\left[ A_{red} \right]\left[ B_{ox}^{n+} \right]}{\left[ A_{ox}^{n+} \right]\left[ B_{red} \right]} \right)$$

Here, $\Delta \mathscr{E}$ is the reduction potential, $\Delta \mathscr{E}^{\circ\prime}$ is the reduction potential when all the components are in their biochemical st2andard states, $\mathscr{F}$ is the faraday (96,485 $J \cdot V^{-1} \cdot mol^{-1}$), $n$ is the number of moles of electrons transferred per mole of reactants reduced, and $R$ is the gas constant.

20.  $\Delta \mathscr{E}$ is related to the free energy change in a redox reaction by the following relationship:

$$\Delta G = -n\mathscr{F}\Delta \mathscr{E}$$

Electrons flow spontaneously from a compound with the lower reduction potential to a compound with the higher reduction potential.

*Experimental Approaches to the Study of Metabolism*

21.  Metabolites can be traced by labeling them with isotopes of certain atoms that can be detected by their radioactivity or through nuclear magnetic resonance spectroscopy. Such labels are useful for establishing precursor–product relationships and for examining the rates of biochemical transformation in living cells and tissues.

22.  Other methods for analyzing a metabolic pathway include the use of metabolic inhibitors, which inhibit specific enzymes, and genetic mutations in enzymes involved in the pathway. In recent years, genetic engineering has become a powerful new tool for studying metabolism, as the gene for a specific enzyme can be added, deleted, or specifically altered.

23.  An organism's metabolic capabilities can be assessed through the approaches of systems biology, which may include genomic sequencing, DNA microarray analysis, and techniques for protein identification and quantitation.

24.  A picture of gene expression, as represented by all of the different species of a cell's mRNA (transcriptome), can be obtained using DNA microarrays. These "DNA chips" can be used to determine allelic differences between individuals or to assess how

transcriptional activity of various genes differs between tissues or at different developmental stages.

25. An organism's overall metabolism can also be assessed through proteomics, the identification and quantitation of all of a cell's proteins, including their posttranslational modifications, in a given tissue at a given time.

## Key Equations

$$\Delta G = -n\mathscr{F}\Delta\mathscr{E}$$

$$\Delta\mathscr{E} = \Delta\mathscr{E}^{\circ\prime} - \frac{RT}{n\mathscr{F}} \ln\left( \frac{[A_{red}][B_{ox}^{n+}]}{[A_{ox}^{n+}][B_{red}]} \right)$$

$$\Delta\mathscr{E}^{\circ} = \mathscr{E}^{\circ}_{(e^- \text{ acceptor})} - \mathscr{E}^{\circ}_{(e^- \text{ donor})}$$

### Behind the Equations for the Nernst Equation: Similarities with Calculating Gibbs Free Energy

The Nernst Equation describes the change in the reduction potential (electromotive force) of a system, and can be used to determine whether or not a process involving reduction and oxidation (redox) is spontaneous or non-spontaneous under a given set of conditions (state of a thermodynamic system).

For a general redox reaction between an oxidized (ox) and reduced (red) species:

$$A^{n+}_{(ox)} + B_{(red)} \rightleftharpoons A_{(red)} + B^{n+}_{(ox)}$$

One would assume it would be possible to calculate the $\Delta G$ of the process using the $\Delta G$ equations reviewed in Chapter 1 (Calculation Analogy for Determining Gibbs Free Energy for A Chemically Defined System). However, scientists often use the Nernst Equation to calculate the degree of spontaneity of a redox process instead. In many ways, the Nernst Equation is similar to the $\Delta G$ equation reviewed previously:

$$\Delta\mathscr{E} = \Delta\mathscr{E}^{\circ\prime} - RT/n\mathscr{F}\ln\left([A_{red}][B_{ox}]/[A_{ox}][B_{red}]\right) \quad (1)$$

$$\Delta G = \Delta G^{\circ} + RT\ln([C]^c[D]^d/[A]^a[B]^b) \quad (2)$$

There are many similarities between these two equations. First, both equations tell us how far from equilibrium a thermodynamic system is ($\Delta G = 0$). As such, the number line analogy is also appropriate to use when thinking about the Nernst equation. Moreover, the Nernst equation is formulated to provide two reference points about the thermodynamic system by having two mathematical terms: the "$\Delta\mathscr{E}^{\circ\prime}$" term and the "$-RT/n\mathscr{F}\ln([A_{red}][B_{ox}]/[A_{ox}][B_{red}])$" term.

$\Delta\mathscr{E}^{\circ\prime}$ represents the $\Delta\mathscr{E}$ of the system at the biochemical standard state (25° C, 1 atm, pH = 7.0, etc…). As with the $\Delta G^\circ$ term of the Gibbs Free Energy equation, the $\Delta\mathscr{E}^{\circ\prime}$ term represents the position of the reaction under the defined conditions (state) of the system in relation to thermodynamic equilibrium, where $\Delta\mathscr{E} = 0$. Similar to $\Delta G^\circ$ or $\Delta G^{\circ\prime}$, $\Delta\mathscr{E}^{\circ\prime}$ is a convenient arbitrary state that we can use as a reference point.

The second term in the Nernst equation is the current state of the redox reaction as defined by the temperature and the concentrations of the reaction's components. However, one must now pay attention to the oxidized and reduced states of the chemical species exchanging the electrons.

In the Nernst equation, the $RT$ term is modified in two significant ways: it is divided by $n\mathscr{F}$ and is modified by a negative sign. As discussed in the text, $n$ is the number of moles of electrons transferred per mole of reactant converted, and is often = 1. The Faraday constant is merely a conversion factor that allows one to interconvert the units of an electrical potential (Volts) to that of $\Delta G$ (Joules per mole). The negative sign mathematically describes the tendency of electron movement. Electrons carry negative charge and will be attracted to positively charged species. Similarly, electrons will want to move towards more positive potentials. Thus, incorporation of the negative sign in the Nernst equation allows for spontaneous redox processes to be described by positive values of $\Delta\mathscr{E}$. With the exception of these two modifications, solving problems involving the Nernst equation is similar conceptually to solving Gibbs Free Energy equations.

## Questions

### Overview of Metabolism

1. In searching for life on Mars in the 1970s, the Viking spacecraft tested the Martian soil for rapid oxidation–reduction reactions. Explain why such reactions might indicate the presence of life.

2. Use the following terms to fill in the diagram below: ADP, $P_i$, $NADP^+$, ATP, NADPH, carbohydrates, proteins, lipids, acetyl-CoA, catabolism, anabolism.

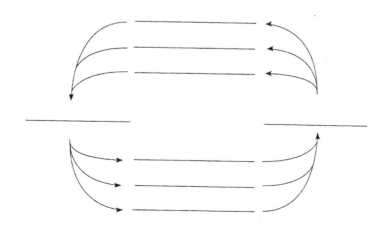

3.    Phosphofructokinase (PFK) catalyzes the reaction

ATP + fructose-6-phosphate (F6P) $\leftrightarrows$ fructose-1,6-bisphosphate (FBP) + ADP

Explain why the reaction rate is relatively insensitive to changes in the concentrations of F6P or FBP. What does this tell you about PFK?

*"High-Energy" Compounds*

4.    In the cell, divalent cations such as $Mg^{2+}$ bind to the anionic phosphate groups of ATP. How would the removal of the divalent cations affect the $\Delta G$ for the hydrolysis of ATP?

5.    What processes maintain the cellular concentration of ATP?

6.    Why don't "high-energy" compounds such as phosphoenolpyruvate and phosphocreatine (Figure 14-6) break down quickly under physiological conditions?

7.    Explain how phosphocreatine acts as an ATP "buffer."

8.    Many metabolic reactions are actually coupled reactions. A common coupled reaction is substrate phosphorylation by ATP. For example, the oxidation of glucose begins with its phosphorylation to glucose-6-phosphate (G6P).

(a)    For the reaction

Glucose + $P_i$ $\rightleftharpoons$ glucose-6-phosphate + $H_2O$ $\qquad \Delta G^{\circ\prime} = 14 \ kJ \cdot mol^{-1}$

Calculate the ratio of $[G6P]/[glucose][P_i]$ at equilibrium at 25°C.

(b)    In muscle cells at 37°C, the steady-state ratio of $[ATP]/[ADP]$ is 12. Assuming that glucose and G6P achieve equilibrium values in muscle, what is the ratio of $[G6P]$ to $[glucose]$?

9.    Aldolase catalyzes the reaction

Fructose-1,6-bisphosphate (FBP) $\rightleftharpoons$
Glyceraldehyde-3-phosphate (GAP) + Dihydroxyacetone phosphate (DHAP)

$\Delta G^{\circ\prime}$ for this reaction is 22.8 $kJ \cdot mol^{-1}$. In the cell at 37°C, the $\Delta G$ for this reaction is –5.9 $kJ \cdot mol^{-1}$. What is the ratio $[GAP][DHAP]/[FBP]$?

*Oxidation–Reduction Reactions*

10.    Using the curved-arrow convention, show the transfer of electrons in the reduction of pyruvate to lactate in the presence of NADH + $H^+$.

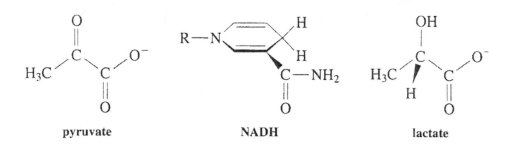

pyruvate               NADH               lactate

11.  In the diagram below, indicate the direction of flow of electrons and the voltage (on the meter) for the following half-reactions under standard conditions:

$$Zn^{2+} + 2e^- \rightarrow Zn \qquad\qquad E° = -0.763\ V$$
$$Ni^{2+} + 2e^- \rightarrow Ni \qquad\qquad E° = -0.250\ V$$

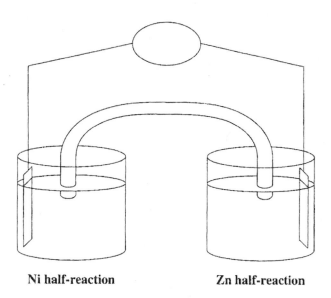

Ni half-reaction               Zn half-reaction

12.  At pH 0,

$$1/2\ O_2 + 2\ H^+ + 2\ e^- \rightleftharpoons H_2O \qquad\qquad \mathscr{E}° = 1.23\ V$$

Is oxygen reduction more favored at pH 0 or at pH 7? Explain in electrochemical terms as well as in terms of chemical equilibria.

13.  Consider the reaction in which acetoacetate is reduced by NADH to β-hydroxybutyrate:

$$\text{Acetoacetate} + \text{NADH} + \text{H}^+ \rightleftharpoons \text{β-hydroxybutyrate} + \text{NAD}^+$$

Calculate $\Delta G$ for this reaction at 25°C when [acetoacetate] and [NADH] are 0.01 M, and [β-hydroxybutyrate] and [NAD$^+$] are 0.001M.

## Experimental Approaches to the Study of Metabolism

14.  Radioactive isotope tracers and metabolic inhibitors have been essential to the elucidation of metabolic pathways. Can both kinds of agents be used to determine the order of metabolites in a metabolic pathway?

15.  Many biosynthetic pathways have been elucidated by the analysis of genetic mutations in organisms such as *Neurospora crassa* (a mold) and *Escherichia coli*. How would you elucidate the steps of a hypothetical biosynthetic pathway in *N. crassa* in which compound A leads to compound Z?

16.  You have isolated four mutants in amino acid metabolism of the mold *N. crassa*. Mutant 1 requires two compounds for growth, X and Z. Mutant 2 requires only X. Mutants 3 and 4 require only Z. Mutant 3 accumulates a compound, W, that supports the growth of Mutant 4 but not that of Mutants 1 or 2. Mutant 4 accumulates a compound, Y, that alone supports the growth of Mutant 1.
    (a)  Diagram the biosynthetic pathway connecting compounds W, X, Y, and Z, indicating the step at which each mutant is blocked.
    (b)  According to the diagram in (a), what is the first committed step in the synthesis of Z?

17.  Evaluate the accuracy of the following two statements on the interpretation of DNA chip results such as those presented in Figures 14-18 and 14-19.
    (a)  DNA chip data allow one to determine the number of mRNA copies present in the cell.
    (b)  DNA chip data provide an accurate assessment of the relative protein diversity and content of cells under different physiological states.

# 15
# Glucose Catabolism

Glycolysis, or the biochemical conversion of glucose to pyruvate, is one of the best understood metabolic pathways. The mechanisms and structures of the ten glycolytic enzymes are known in some detail, and they serve as models for the study of other enzymes with similar reaction mechanisms. The regulation of metabolic pathways, a topic introduced in Chapter 14, is illustrated here. This chapter also describes the fate of pyruvate (the end product of glycolysis), the entry of other sugars into the glycolytic pathway, and the pentose phosphate pathway that also catabolizes glucose.

## *Essential Concepts*

### *Overview of Glycolysis*

1    Glucose is a major source of metabolic energy in many cells. The energy released during its conversion (oxidation) to pyruvate is conserved in the form of ATP and the reduced coenzyme NADH. Ten enzymes catalyze the glycolytic pathway, which occurs in both prokaryotes and eukaryotes and is almost universal.

2.   Stage I of glycolysis is a preparatory stage in which glucose is "activated" by phosphorylation by ATP and broken down into two $C_3$ sugars. Stage II produces "high-energy" intermediates that phosphorylate ADP to form ATP, for a net gain of 2 ATP. One glucose molecule yields two pyruvate molecules and requires the oxidizing power of two $NAD^+$. The overall equation for glycolysis is

$$\text{Glucose} + 2\,NAD^+ + 2\,ADP + 2\,P_i \rightarrow$$
$$2\,\text{pyruvate} + 2\,NADH + 2\,ATP + 2\,H_2O + 4\,H^+$$

### *The Reactions of Glycolysis*

3.   Hexokinase, the first enzyme in the pathway, transfers a phosphoryl group from ATP to the C6-OH group of glucose to produce glucose-6-phosphate (G6P). Hexokinase undergoes a large conformational change on binding glucose, which excludes water from the active site and promotes the specific transfer of the phosphoryl group from ATP to glucose.

4.   Phosphoglucose isomerase catalyzes the conversion of glucose-6-phosphate to fructose-6-phosphate (F6P). This reaction proceeds via an enediolate intermediate.

5.   Phosphofructokinase (PFK) converts fructose-6-phosphate to fructose-1-6-bisphosphate (FBP) by another phosphoryl-group transfer from ATP. This reaction, which is irreversible, is the first committed step of the pathway. The phosphofructokinase reaction is the rate-determining step of glycolysis and the principal regulatory point.

6.  Aldolase cleaves fructose-1,6-bisphosphate to two $C_3$ compounds: glyceraldehyde-3-phosphate (GAP) and dihydroxyacetone phosphate (DHAP). In animals and plants, aldolase operates via Schiff base and enamine intermediates.

7.  Triose phosphate isomerase catalyzes the interconversion of glyceraldehyde-3-phosphate and dihydroxyacetone phosphate through an enediolate intermediate. This $\alpha/\beta$ barrel enzyme has achieved catalytic perfection. Its activity allows the products of the first stage of glycolysis to proceed through the second stage as glyceraldehyde-3-phosphate.

8.  Glyceraldehyde-3-phosphate dehydrogenase (GAPDH) catalyzes the formation of the first glycolytic intermediate that has sufficient free energy to synthesize ATP from ADP. Glyceraldehyde-3-phosphate is converted to 1,3-bisphosphoglycerate (1,3-BPG) by the reduction of $NAD^+$ and the addition of inorganic phosphate. The sulfhydryl group of an enzyme Cys residue attacks GAP, which is then reduced by $NAD^+$ to form an acyl thioester intermediate that is attacked by $P_i$.

9.  1,3-Bisphosphoglycerate, a "high-energy" mixed anhydride of a phosphate and a carboxylic acid, is the substrate for phosphoglycerate kinase, which transfers the phosphoryl group at C1 to ADP to generate 3-phosphoglycerate (3PG) and the first ATP product of glycolysis.

10. 3-Phosphoglycerate is converted to 2-phosphoglycerate (2PG) by phosphoglycerate mutase. The active form of this enzyme contains a phosphohistidine residue whose phosphoryl group is transferred to 3PG to produce 2,3-bisphosphoglycerate (2,3-BPG), which then transfers the phosphoryl group at the 3 position back to the histidine, yielding 2PG.

11. Enolase dehydrates (removes water from) 2-phosphoglycerate to form phosphoenolpyruvate (PEP). This reaction produces the second "high-energy" intermediate of glycolysis.

12. The free energy of phosphoenolpyruvate is released in the reaction catalyzed by pyruvate kinase. The transfer of the phosphoryl group to ADP produces the second ATP product of glycolysis and an enol product whose tautomerization to the keto form yields pyruvate. Most of the free energy of the pyruvate kinase reaction is supplied by this tautomerization step. Keep in mind that because the initial substrate of glycolysis is a $C_6$ compound that is converted to two $C_3$ compounds, the first stage of glycolysis consumes 2 ATP but the second stage generates 4 ATP (2 for each GAP), for a net yield of 2 ATP.

*Fermentation: The Anaerobic Fate of Pyruvate*

13. The NADH produced by glycolysis must be converted back to $NAD^+$ in order for the glyceraldehyde-3-phosphate dehydrogenase reaction to proceed. Consequently, pyruvate is not the end product of glucose metabolism but can undergo one of three processes to regenerate $NAD^+$: homolactic fermentation, alcoholic fermentation, or oxidative metabolism. In oxidative metabolism, pyruvate is oxidized to $CO_2$ via the citric acid cycle.

14. In muscle cells, when $O_2$ is in short supply, lactate dehydrogenase reduces pyruvate to lactate, with the concomitant oxidation of NADH to $NAD^+$. The lactate that builds up in muscles upon strenuous exertion can either be reconverted to pyruvate later or carried by the blood to the liver, where it can be converted back to glucose by a process called gluconeogenesis.

15. Under anaerobic conditions, yeast carry out alcoholic fermentation. In the first reaction of this process, pyruvate is decarboxylated to acetaldehyde and carbon dioxide. Pyruvate decarboxylase catalyzes this reaction with the aid of the coenzyme thiamine pyrophosphate, which stabilizes the reaction's carbanion intermediate. In the second reaction, acetaldehyde is reduced with NADH to form ethanol and $NAD^+$, as catalyzed by alcohol dehydrogenase. These two reactions have been known for thousands of years: The released $CO_2$ leavens (raises) bread and the ethanol is used to make alcoholic beverages.

16. The anaerobic catabolism of glucose can be 100 times faster than the catabolism of glucose in the presence of oxygen. However, fermentation produces 2 ATP per glucose, whereas oxidative metabolism (via the citric acid cycle and oxidative phosphorylation) generates 38 ATP for each glucose molecules that is converted to $CO_2$ and $H_2O$.

*Regulation of Glycolysis*

17. The reaction catalyzed by phosphofructokinase, which has a large negative $\Delta G$, is the first committed step in glycolysis and the primary regulatory point for the pathway. PFK is allosterically inhibited by ATP, which binds to an inhibitory site and stabilizes PFK's T (less active) state. This is an example of feedback inhibition, since ATP is a product of the pathway. AMP, ADP, and fructose-2,6-bisphosphate (F2,6P) relieve the inhibition of PFK by ATP by preferentially binding to the R (more active) state. PFK thereby senses the energy state of the cell and adjusts the flux through glycolysis accordingly.

18. Additional control of glycolytic flux is provided by a substrate cycle. PFK catalyzes the reaction F6P + ATP $\rightarrow$ FBP + ADP, whereas fructose-1,6-bisphosphatase (FBPase) catalyzes the opposing reaction FBP + $H_2O$ $\rightarrow$ F6P + $P_i$. The sum of these reactions is the hydrolysis of ATP. Both enzymes exist in the same cell and their relative activity is under hormonal and neuronal control. Substrate cycling, which produces heat, can provide a form of nonshivering thermogenesis.

*Metabolism of Hexoses Other than Glucose*

19. Three other sugars—fructose, galactose, and mannose—are major sources of cellular energy. In muscle, fructose can be directly phosphorylated by hexokinase to F6P. However, liver glucokinase cannot directly phosphorylate fructose. In the liver, fructokinase phosphorylates C1 to generate fructose-1-phosphate. Fructose-1-phosphate aldolase generates dihydroxyacetone phosphate and glyceraldehyde by an aldol cleavage. Glyceraldehyde kinase phosphorylates C3 of glyceraldehyde to produce the glycolytic intermediate GAP. Three other enzymes (alcohol dehydrogenase, glycerol kinase, and glycerol phosphate dehydrogenase) convert glyceraldehyde to dihydroxyacetone phosphate, which is then converted to GAP by triose phosphate isomerase. Individuals who have

defective fructose-1-phosphate aldolase have fructose intolerance and quickly develop a strong distaste for anything sweet.

20. Galactose, which differs from glucose in the configuration of the OH group at C4, requires four reactions to enter glycolysis. First, galactose is converted to galactose-1-phosphate by galactokinase. Next, galactose-1-phosphate uridylyl transferase transfers the UMP group of UDP–glucose to produce UDP–galactose and glucose-1-phosphate. An epimerase converts UDP–galactose to UDP–glucose. Finally, phosphoglucomutase converts glucose-1-phosphate to glucose-6-phosphate.

21. Mannose, the C2 epimer of glucose, is recognized by hexokinase. The resulting mannose-6-phosphate is then converted to the glycolytic intermediate fructose-6-phosphate by phosphomannose isomerase.

*The Pentose Phosphate Pathway*

22. Besides ATP, cells require the reducing power of NADPH for the biosynthesis of macromolecules (anabolism). NADPH is used in biosynthesis, whereas NADH is used in oxidative metabolism (catabolism). Cells keep the $[NAD^+]/[NADH]$ ration near 1000 (which favors metabolite oxidation) and the $[NADP^+]/[NADPH]$ ratio near 0.01 (which favors reductive biosynthesis). The oxidation of glucose by the pentose phosphate pathway generates NADPH.

23. The first stage of the pentose phosphate pathway consists of three steps, its oxidative reactions:
    (a) Glucose-6-phosphate is oxidized to 6-phosphoglucono-δ-lactone by glucose-6-phosphate dehydrogenase, producing the first NADPH.
    (b) 6-Phosphogluconolactonase hydrolyzes the lactone (cyclic ester) to yield 6-phosphogluconate.
    (c) 6-Phosphogluconate dehydrogenase then catalyzes the oxidative decarboxylation of 6-phosphogluconate by $NADP^+$ to yield ribulose-5-phosphate, $CO_2$, and the second NADPH.

24. Stage two of the pentose phosphate pathway is catalyzed by two enzymes that act on ribulose-5-phosphate (Ru5P): Ribulose-5-phosphate epimerase converts Ru5P to xylulose-5-phosphate (Xu5P), and ribulose-5-phosphate isomerase converts Ru5P to ribose-5-phosphate (R5P). The R5P can be used to produce nucleosides for RNA and DNA synthesis.

25. In the third stage of the pentose phosphate pathway, 3 five-carbon sugars are converted to 2 fructose-6-phosphate and 1 glyceraldehyde-3-phosphate. First, transketolase transfers a two-carbon unit from Xu5P to R5P yielding the seven-carbon sugar sedoheptulose-7-phosphate (S7P) and glyceraldehyde-3-phosphate (GAP). Next, transaldolase transfers a three-carbon unit from S7P to GAP to form fructose-6-phosphate (F6P) and the four-carbon sugar erythrose-4-phosphate (E4P). Finally, another transketolase reaction converts Xu5P and E4P to F6P and GAP.

26. The reversible nature of the second and third stages of the pentose phosphate pathway permits the cell to meet its need for R5P (a nucleic acid precursor) and NADPH. For example, if the need for NADPH is greater than that for R5P, the excess R5P is converted to F6P and GAP for consumption via glycolysis. Conversely, if the demand for R5P outstrips the need for NADPH, the glycolytic intermediates F6P and GAP can be diverted to the pentose phosphate pathway to synthesize R5P.

27. The flux of glucose-6-phosphate through the pathway is regulated by the rate of the glucose-6-phosphate dehydrogenase (G6PD) reaction. This enzyme is controlled by the availability of its substrate $NADP^+$ so that the pathway flux increases in response to increasing levels of $NADP^+$ (which indicates increased cellular demand for NADPH).

28. A deficiency of G6PD is the most common human enzyme deficiency. The resulting shortage of NADPH increases the sensitivity of red blood cells to oxidative stress, since NADPH is required to maintain the supply of reduced glutathione, which removes organic hydroperoxides that occasionally form. Certain compounds such as the antimalarial drug primaquine stimulate peroxide formation, which induces hemolytic anemia in G6PD-deficient individuals. However, mutations in G6PD confer resistance to malaria.

### Questions

*Overview of Glycolysis*

1. Describe the two-stage "chemical strategy" of glycolysis and write a balanced equation for each phase.

2. The two "high-energy" compounds produced in glycolysis are _____ and _____.

*The Reactions of Glycolysis*

3. Examine the following five glycolytic intermediates:

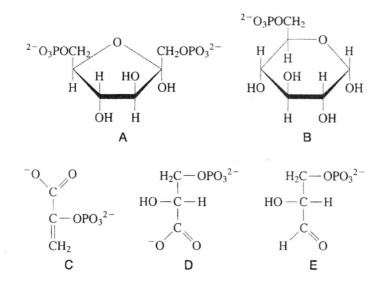

(a) Name each intermediate.
(b) Write the order in which they appear in glycolysis.
(c) Which intermediate is a reactant in substrate-level phosphorylation?
(d) List the glycolytic enzyme for which each intermediate is a substrate.
(e) Which phosphorylated intermediates of glycolysis are not shown above?

4. List the four kinases of glycolysis.

5. There are _____ isomerization reactions in glycolysis. The enzymes that catalyze them are _____.

6. For the reaction catalyzed by triose phosphate isomerase, what is the equilibrium ratio of reactants and products under standard biochemical conditions? How does this ratio differ from the ratio observed in the cell at 37°C?

7. How does the GAPDH-catalyzed exchange of $^{32}$P between $P_i$ and 1,3-bisphosphoglycerate corroborate the existence of an acyl–enzyme intermediate in the GAPDH reaction?

8. 2,3-BPG is an intermediate in the reaction catalyzed by phosphoglycerate mutase. Why does the cell require trace amounts of 2,3-BPG?

9. If the cytosolic [NAD$^+$]/[NADH] ratio is 100 and the [ATP]/[ADP][$P_i$] ratio is 10, what is the actual (not equilibrium) ratio of [GAP]/[3PG] at 37°C in the cell? Assume that [H$^+$] =1, its value in the biochemical standard state.

10. Fluoride ions specifically inhibit enolase in the presence of $P_i$ in cell extracts.
(a) Explain why both 2PG and 3PG accumulate in the presence of F$^-$ and $P_i$.
(b) Explain why 1,3-BPG does not accumulate.

11. Match each $^{14}$C-labeled glucose with the $^{14}$C-labeled pyruvate produced by glycolysis.

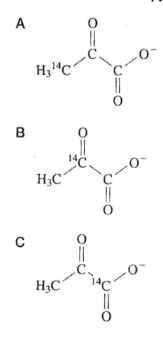

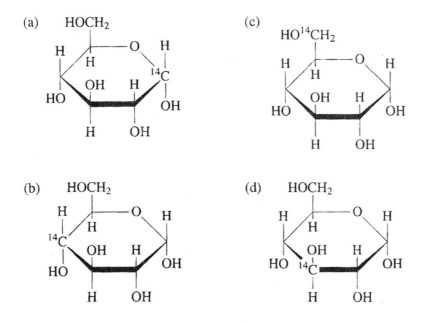

## Fermentation: The Anaerobic Fate of Pyruvate

12. How does a muscle cell maintain the $[NAD^+]/[NADH]$ ratio during the catabolic breakdown of glucose?

13. Which of the carbon(s) of glucose must be labeled with $^{14}C$ for the end products of alcoholic fermentation to be unlabeled ("cold") ethanol and $^{14}CO_2$?

14. Compare the rates of ATP production in fermentation versus oxidative phosphorylation. Which process is utilized in rapid bursts of muscular activity?

## Regulation of Glycolysis

15. In many metabolic pathways the first reaction is the rate-determining step of the pathway.
    (a) What is the rate-determining step of glycolysis?
    (b) What rationale can you offer for this "unusual" regulation?

16. What is meant by a futile cycle?

## Metabolism of Hexoses Other than Glucose

17. The catabolism in the liver of which sugar—mannose, fructose, or galactose—bears the closest similarity to the first two reactions of glycolysis?

18. Defend or refute the following statement: The association of $NAD^+$ with UDP–galactose-4-epimerase suggests the generation of an additional NADH in the oxidation of galactose to pyruvate.

19. Why is galactosemia especially dangerous to nursing infants?

20. Some organisms use glycerol as a carbon energy source, and it is also an intermediate in fructose metabolism.
    (a) Write equations for the reactions required to oxidize glycerol to pyruvate.
    (b) Compare the ATP yield of two molecules of glycerol versus one molecule of glucose in glycolysis.
    (c) Does the anaerobic fermentation of glycerol maintain the redox balance of the cell? Explain.

*The Pentose Phosphate Pathway*

21. The diagram below shows the interconversions of the nonoxidative reactions of the pentose phosphate pathway.

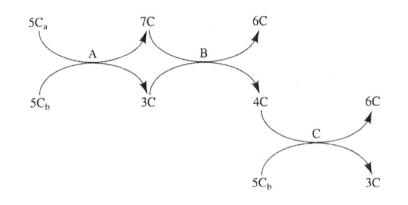

    (a) Which sugar phosphates correspond to the 5C compounds?
    (b) Which sugar phosphate corresponds to the 6C compound?
    (c) Which sugar phosphate corresponds to the 3C compound?
    (d) Which reaction(s) involves transketolase? Transaldolase?

22. The nonoxidative reactions of the pentose phosphate pathway convert pentose phosphates into hexose phosphates.
    (a) For every 3 glucose phosphates that enter the pentose phosphate pathway, how many fructose-6-phosphates are recovered?
    (b) For every 3 glucose-6-phosphates that enter the pentose phosphate pathway, how many glyceraldehyde-3-phosphates are recovered?

23. Ribulose-5-phosphate is converted to xylulose-5-phosphate and ribose-5-phosphate by an epimerase and an isomerase, respectively. What distinguishes these isomerizations?

24. You obtain a mutant transketolase from yeast that binds R5P and E4P poorly. What unique side product of the reaction catalyzed by this enzyme might you find in these cells?

25. Indicate the appropriate choices in parentheses: Transaldolase transfers a (1, 2, 3)-carbon unit from a(n) (ketose / aldose) to a(n) (ketose / aldose) to form a(n) (ketose / aldose).

26. Indicate the appropriate choices in parentheses: Transketolase transfers a (1, 2, 3)-carbon unit from a(n) (ketose / aldose) to a(n) (ketose / aldose) to form a(n) (ketose / aldose).

27.  Which step commits glucose to oxidation via the pentose phosphate pathway? How is this enzyme regulated?

28.  The conversion of G6P to R5P via the reactions of glycolysis and the pentose phosphate pathway, *without* the production of NADPH, can be summarized as

$$5 \text{ G6P} + \text{ATP} \rightleftharpoons 6 \text{ R5P} + \text{ADP} + \text{H}^+$$

What reaction requires ATP? How do you account for the stoichiometry?

# 16

# Glycogen Metabolism and Gluconeogenesis

This chapter discusses glycogen breakdown and synthesis, gluconeogenesis, and oligosaccharide synthesis, with an emphasis on the mechanisms that regulate these metabolic pathways. Six major enzymes are involved in the breakdown and synthesis of glycogen: glycogen phosphorylase, glycogen debranching enzyme, and phosphoglucomutase for glycogen breakdown, and UDP–glucose pyrophosphorylase, glycogen synthase, and glycogen branching enzyme for glycogen synthesis. Box 16-2 shows how inherited metabolic diseases contribute to our understanding of glycogen metabolism. The chapter explains how the synthesis of glycogen from glucose-1-phosphate requires the free energy of nucleotide hydrolysis in a reaction that is the opposite of the exergonic breakdown of glycogen. The mechanisms of glycogen phosphorylase and glycogen synthase are examined, including the role of the oxonium ion transition state in each case. Box 16-3 explores the strategies for optimizing the branched structure of glycogen. The chapter then discusses the regulation of glycogen phosphorylase and glycogen synthase by allosteric effectors and covalent modification. Regulation involves an enzyme cascade where extracellular signals (hormones), acting via the second messengers cyclic AMP and $Ca^{2+}$, initiate the activation of successive kinases, resulting in the reciprocal activation of glycogen phosphorylase and inactivation of glycogen synthase. . The role of phosphoprotein phosphatase-1 in modulating the ratio of phosphorylated to dephosphorylated enzymes is also explored as an additional layer of complexity in the regulation of the enzymes involved in glycogen breakdown and synthesis.

The chapter then moves on to gluconeogenesis, the process by which pyruvate and related metabolites can be converted to glucose. Gluconeogenesis uses many of the same enzymes as glycolysis but requires other enzymes to bypass the exergonic steps of glycolysis. The chapter discusses the interesting role of the allosteric effector fructose-2,6-bisphosphate in the regulation of gluconeogenesis and glycolysis in liver and heart muscle. The last section of this chapter discusses the roles of nucleotide sugars and dolichol pyrophosphate in oligosaccharide synthesis.

## *Essential Concepts*

1.  The mobilization of glucose from glycogen stores in the liver provides a constant supply of glucose to the central nervous system and red blood cells, which use glucose as their sole energy source. Under fasting conditions, amino acids (mainly from muscle protein degradation) serve as precursors for new glucose synthesis (gluconeogenesis).

2.  Glucose-6-phosphate (G6P) is a key branch point in glucose metabolism in the liver, as it can be polymerized to glycogen, degraded to pyruvate, converted to ribose-5-phosphate, or hydrolyzed to glucose.

## Glycogen Breakdown

3.  Glycogen breakdown (glycogenolysis) utilizes three enzymes:
    (a) Glycogen phosphorylase, which catalyzes the phosphorolysis of the glucose residues at the nonreducing ends of glycogen to yield glucose-1-phosphate (G1P).
    (b) Glycogen debranching enzyme, which transfers a tri- or tetrasaccharide and hydrolyzes the $\alpha(1{\to}6)$ linkage at branch points.
    (c) Phosphoglucomutase, which converts G1P to G6P.

4.  Phosphorylase utilizes the cofactor pyridoxal-5′-phosphate (PLP) in the general acid–base catalytic mechanism in glycogen phosphorolysis. Inorganic phosphate attacks the terminal glucose residue, which passes through an oxonium ion intermediate before being released as G1P. The activity of the dimeric enzyme is regulated by allosteric effectors and by phosphorylation/dephosphorylation of the protein at Ser 14.

5.  Glycogen debranching enzyme acts as an $\alpha(1{\to}4)$ transglycosylase by transferring a tri- or tetrasaccharide from a limit branch (a four- or five-glucose-residue segment that phosphorylase cannot further degrade) to the nonreducing end of another branch. The remaining glucosyl residue, which is attached to glycogen by an $\alpha(1{\to}6)$ linkage, is hydrolyzed at a separate active site on the enzyme to release free glucose.

6.  Phosphoglucomutase converts G1P to G6P by way of a G1,6P intermediate. A phosphate group from a Ser residue is transferred to C6 of G1P, followed by transfer of the C1 phosphate back to the Ser residue, in a manner similar to the phosphoglycerate mutase reaction.

7.  The liver expresses glucose-6-phosphatase, an enzyme that hydrolyzes G6P. The resulting free glucose equilibrates with glucose in the blood so that the breakdown of liver glycogen leads to an elevation of blood glucose levels.

## Glycogen Synthesis

8.  Glycogen synthesis requires three enzymes to covert G1P to glycogen. First, UDP–glucose pyrophosphorylase catalyzes the transfer of UMP from UTP to the phosphate group of G1P to form UDP–glucose and $PP_i$. $PP_i$ is eventually hydrolyzed to $P_i$ by inorganic pyrophosphatase, which provides the exergonic push for this reaction.

9.  Glycogen synthase catalyzes a transfer reaction in which the glucosyl residue of UDP–glucose is added to the nonreducing end of glycogen through an $\alpha(1{\to}4)$ bond. Glycogen synthase can only extend a pre-existing $\alpha(1{\to}4)$-linked chain. The glycogen molecule originates through the action of the protein glycogenin, which assembles a seven-residue glycogen "primer" for glycogen synthase to act on.

10. Glycogen branching enzyme transfers a seven-residue segment from the end of an $\alpha(1{\to}4)$-linked glucan chain to the C6-hydroxyl group of a glucosyl residue on the same chain or another chain, thereby forming an $\alpha(1{\to}6)$-linked branch.

*Control of Glycogen Metabolism*

11. The opposing processes of glycogen breakdown and synthesis are coordinately controlled by allosteric regulation and covalent modification of key enzymes. The major allosteric regulators are ATP, AMP, and G6P. Covalent modification occurs with the transfer of $P_i$ from ATP to certain enzymes by the action of specific kinases.

12. Glycogen phosphorylase exists in two forms: phosphorylase *a* (the phosphorylated, more active enzyme) and phosphorylase *b* (the dephosphorylated, less active enzyme). Each form also has two conformations: the T (relatively inactive) state and the R (relatively active) state. AMP promotes the T → R conformational change and thereby activates phosphorylase *b*, whereas ATP and G6P inhibit this conformational change. Phosphorylase *a* is less sensitive to allosteric effectors and is mainly in the R state. However, high concentrations of glucose promote the R → T transition.

13. Covalent modification regulates glycogen phosphorylase and glycogen synthase in a reciprocal fashion. Glycogen phosphorylase tends to be more active when it is phosphorylated, so the ratio of phosphorylase *a* to phosphorylase *b* largely determines the rate of glycogen phosphorolysis. This ratio is set by the activity of phosphorylase kinase, which is regulated by protein kinase A (PKA), and by phosphoprotein phosphatase-1 (PP1). In contrast, glycogen synthase is activated by dephosphorylation so that PKA promotes glycogen breakdown while inhibiting glycogen synthesis, and PP1 inhibits glycogen breakdown while promoting glycogen synthesis.

14. Phosphoprotein phosphatase-1 (PP1) can dephosphorylate glycogen phosphorylase, phosphorylase kinase, and glycogen synthase. In muscle, insulin-stimulated protein kinase activates PP1 by phosphorylating its $G_M$ subunit, which binds to glycogen in muscle . The PKA-mediated phosphorylation of another site on the $G_M$ subunit causes the catalytic subunit of PP1 to be released in the cytoplasm, where it cannot dephosphorylate glycogen-bound enzymes. In addition, phosphoprotein phosphatase inhibitor 1 (inhibitor-1) inhibits PP1. Inhibitor-1 activity is stimulated by phosphorylation by PKA, which helps preserve the phosphorylated (active) forms of phosphorylase kinase and phosphorylase *a*.

15. Glycogen synthase is inactivated by phosphorylation by the same enzyme system that phosphorylates glycogen phosphorylase. Hence, activation of phosphorylase kinase, which activates phosphorylase *a*, inactivates glycogen synthase. This regulatory mechanism provides a rapid and large-scale control of flux in the substrate cycle that links glycogen and G1P. Glycogen synthase activity is also controlled by other kinases.

16. Hormones, including glucagon, insulin, epinephrine, and norepinephrine, ultimately control glycogen metabolism. These hormones bind to transmembrane protein receptors and initiate a series of reactions that lead to the production of molecules called second messengers (e.g., cAMP and $Ca^{2+}$), which modulate the activities of numerous intracellular proteins. Glucagon binding to its receptor in the liver results in an elevation of cAMP, which favors glycogen breakdown. Epinephrine and norepinephrine bind to α- and β-adrenoreceptors in the liver and to β-adreno receptors in muscle. The binding of these

hormones to β-adrenoreceptors increases [cAMP], whereas their binding to α-adrenoreceptors increases cytosolic [$Ca^{2+}$]. Insulin binding to its receptor in tissues other than the liver decreases [cAMP] and promotes glycogen synthesis.

*Gluconeogenesis*

17. The principal noncarbohydrate precursors of glucose are lactate, pyruvate, and amino acids. In animals, these compounds (except for leucine and lysine) are converted, at least in part, into oxaloacetate, which is required for gluconeogenesis.

18. The conversion of pyruvate (or lactate) to glucose follows a pathway that is the reverse of glycolysis except where it bypasses the exergonic steps catalyzed by pyruvate kinase, phosphofructokinase, and hexokinase.

19. To bypass the pyruvate kinase reaction, pyruvate is first carboxylated by pyruvate carboxylase in a reaction that is driven by ATP hydrolysis. The enzyme's biotin prosthetic group is converted to a carboxybiotinyl group in order to transfer $CO_2$ to pyruvate. The product of this reaction is oxaloacetate, which is subsequently decarboxylated to phosphoenolpyruvate (PEP). The $HCO_3^-$ added to pyruvate therefore leaves as $CO_2$. This second reaction, catalyzed by PEP carboxykinase (PEPCK) is driven by the hydrolysis of GTP.

20. Oxaloacetate is produced in the mitochondria, while the reactions that convert PEP to glucose occur in the cytosol. In species with mitochondrial PEPCK, the PEP formed in the mitochondria is exported to the cytosol via a specific transporter. In species with cytosolic PEPCK, oxaloacetate is first converted to malate or aspartate, each of which has a transporter that allows mitochondrial–cytosolic exchange, and, in the cytosol, is reconverted to oxaloacetate.

21. PEP is converted to fructose-1,6-bisphosphate (FBP) by the enzymes of glycolysis operating in reverse. FBP is then hydrolyzed to fructose-6-phosphate and $P_i$ by the action of fructose bisphosphatase (FBPase-1). Similarly, G6P is hydrolyzed by glucose-6-phosphatase, yielding glucose and $P_i$. These reactions therefore bypass the exergonic hexokinase and phosphofructokinase reactions.

22. Gluconeogenesis and glycolysis are reciprocally regulated in the liver. The principal allosteric regulator is the metabolite fructose-2,6-bisphosphate (F2,6P), which is a potent activator of PFK-1 and inhibitor of FBPase-1. F2,6P is formed and degraded by a bifunctional enzyme referred to as PFK-2/FBPase-2. PKA phosphorylates this enzyme, thereby activating FBPase-2 and inactivating PFK-2, which results in a net decrease in F2,6P (and favors gluconeogenesis). In heart muscle, the situation is reversed, so that phosphorylation activates PFK-2, which facilitates the muscle's ability to extract energy from glucose via glycolysis.

23. Glucose metabolism is also regulated by other mechanisms:
    (a) Acetyl-CoA activates pyruvate carboxylase.
    (b) Alanine inhibits pyruvate kinase. The amino group of alanine is transferred to an α-keto acid by transamination to yield pyruvate and a new amino acid. The resulting pyruvate then serves as a substrate for gluconeogenesis.
    (c) Long-term regulation of gluconeogenesis occurs via changes in gene expression. Prolonged low concentrations of insulin or high concentrations of cAMP stimulate the transcription of the genes for PEPCK, FBPase, and glucose-6-phosphatase, and repress the transcription of the genes for glucokinase, PFK, and the PFK-2/FBPase-2 bifunctional enzyme.

### Other Carbohydrate Biosynthetic Pathways

24. The formation of glycosidic bonds in oligo- and polysaccharides is facilitated by nucleotide sugars, as in the polymerization of glucose by glycogen synthase. The principal nucleotides used are ADP and CDP. Nucleotide sugars are glycosyl donors in the synthesis of O-linked oligosaccharides and in the processing of N-linked oligosaccharides.

25. N-linked oligosaccharides are initially built on dolichol, an isoprenoid lipid carrier in the endoplasmic reticulum (ER). This process begins in the cytosol but finishes in the lumen of the ER, as the dolichol-oligosaccharide flips back and forth across the ER membrane. After the oligosaccharide reaches 14 residues, it is transferred to a protein, leaving dolichol pyrophosphate.

26. Lactose synthesis in mammals involves the mammary gland protein α-lactalbumin, which changes the substrate specificity of a galactosyltransferase so that it synthesizes lactate rather than N-acetyllactosamine.

### Questions

### Glycogen Breakdown

1. List the three enzymes required for the breakdown of glycogen.

2. What feature of the structure of glycogen phosphorylase is consistent with the observation that the enzyme cannot cleave a glycosidic bond within five residues of a branch point?

3. What role does pyridoxal phosphate (PLP) play in the mechanism of glycogen phosphorylase?

4. What would be the product of the nonenzymatic phosphorolysis of glycogen?

5.   In the diagram of glycogen shown below, circle the substrates for glycogen debranching enzyme.

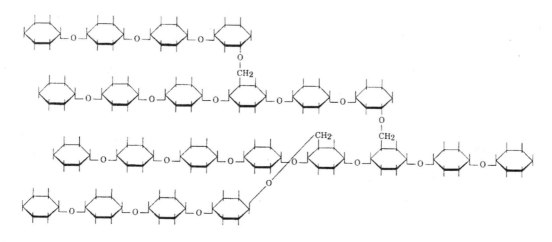

6.   The rate of debranching is much slower than that of phosphorolysis. Explain how highly branched glycogen molecules release glucose-1-phosphate at a greater rate than relatively unbranched ones.

7.   Write an equation for the equilibrium constant for the phosphorylase reaction. $\Delta G^{\circ\prime}$ for this reaction is $+3.1$ kJ·mol$^{-1}$, yet glycogen breakdown in the liver and muscle is thermodynamically favored at 37°C. What is the minimum ratio of $[P_i]/[G1P]$ required to make the phosphorylase reaction exergonic? Assume that the concentration of glycogen does not change significantly.

*Glycogen Synthesis*

8.   How is the thermodynamic barrier to glycogen synthesis overcome by cells?

9.   In the diagram below, each circle represents a glycosyl unit. By the action of branching enzyme, show the most branched structure this molecule can assume. Indicate the reducing and nonreducing ends of the molecule and the position where a glycogenin molecule would be found. There are 50 glycosyl residues.

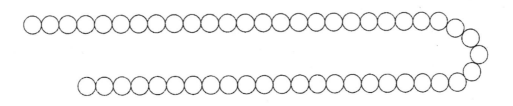

10. Below is a diagram showing the synthesis and breakdown of glycogen [(glucose)]. Fill in the blanks and identify the enzyme that catalyzes each numbered reaction.

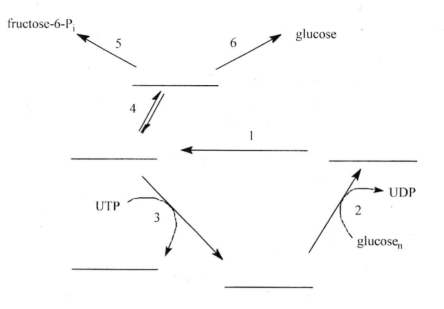

*Control of Glycogen Metabolism*

11. What structural features distinguish the T conformation of glycogen phosphorylase from the R conformation of the enzyme? (*Hint:* See Figure 12-15.)

12. Glycogen phosphorylase is activated in vigorously active muscle without significant changes in intracellular [cAMP]. Explain.

13. The presence of epinephrine results in the stimulation of PFK-2 in heart muscle. How does epinephrine affect glycolysis in this organ?

14. Which target enzyme in glycogen metabolism requires both $\alpha$- and $\beta$-adrenoreceptors to be activated for full enzyme activity?

15. Which of the hereditary glycogen storage diseases results in a pronounced decrease of stored glycogen?

*Gluconeogenesis*

16. What is the fate of $H^{14}CO_3^-$ added to a liver homogenate that is active in gluconeogenesis?

17. Write a balanced equation for the formation of PEP from pyruvate and compare it with the reverse reaction of glycolysis.

18. While acetyl-CoA, the end product of fatty acid oxidation, cannot be converted into glucose, another product of fat degradation, glycerol, can be converted to glucose. Where does glycerol enter the gluconeogenic pathway?

*Other Carbohydrate Biosynthetic Pathways*

19.  UDP–galactose can donate is sugar residue to glucose to form the disaccharide lactose. For the structures of UDP–galactose and glucose below, indicate the reactive electrophilic and nucleophilic centers and the products of the reaction.

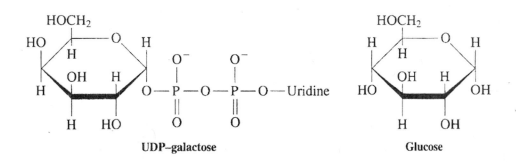

# 17

# Citric Acid Cycle

This chapter discusses the citric acid cycle as both a catabolic and an anabolic process. The citric acid cycle was elucidated by several investigators, but the key insights were provided by Hans Krebs, so the cycle is often referred to as the Krebs cycle. It is also called the tricarboxylic acid (TCA) cycle, which refers to its first intermediate, citrate. The chapter first provides an overview of the key features of the citric acid cycle. It then takes a closer look at each step of the cycle, beginning with a discussion of the pyruvate dehydrogenase complex that converts pyruvate to acetyl-CoA, the fuel molecule that enters the cycle. In this chapter, you will encounter several coenzymes that are critical to the functioning of the citric acid cycle, all derived from water-soluble vitamins. A discussion of the regulation of the citric acid cycle then ensues, in which you will encounter familiar regulatory mechanisms as well as new ones. The chapter then turns to the anabolic features of the citric acid cycle. The citric acid cycle is one of the key metabolic hubs in the cell: Its intermediates are generated by a variety of degradative reactions, and they provide precursors for several biosynthetic pathways. As citric acid cycle intermediates are diverted to anabolic pathways, they are replaced through anaplerotic reactions. The final section of the chapter introduces the glyoxylate cycle found in plants. This pathway is the only known pathway for the net conversion of acetyl-CoA to glucose. The chapter also discusses arsenic poisoning in the citric acid cycle (Box 17-2). Box 17-3 discusses the evolution of the citric acid cycle from a series of reductive reactions.

## *Essential Concepts*

### *Overview of the Citric Acid Cycle*

1.  The citric acid cycle is a central pathway for recovering energy from the three major metabolic fuels: carbohydrates, fatty acids, and amino acids. These fuels are broken down to yield acetyl-CoA, which enters the citric acid cycle by condensing with the $C_4$ compound oxaloacetate. The citric acid cycle is a series of reactions in which 2 $CO_2$ are released for every acetyl-CoA that enters the cycle, so that oxaloacetate is always reformed. Hence, the cyclical series of reactions acts catalytically to process acetyl-CoA continuously.

2.  The oxidation of the acetyl carbon skeleton in the citric acid cycle is coupled to the reduction of $NAD^+$ and FAD. Oxidation of the resulting NADH and $FADH_2$ by the electron-transport chain supplies free energy for ATP synthesis and regenerates $NAD^+$ and FAD for the oxidation of additional acetyl-CoA. In aerobic respiration, $O_2$ serves as the terminal acceptor of the acetyl group's electrons. In anaerobic respiration, this function is carried out by molecules such as $NO_3^-$, $SO_4^{2-}$, and $Fe^{3+}$.

3.  One complete round of the cycle yields 2 $CO_2$, 3 NADH, 1 $FADH_2$, and 1 GTP (which is the equivalent of 1 ATP). Hence, the net reaction of the citric acid cycle is

$$3\,NAD^+ + FAD + GDP + P_i + acetyl\text{-}CoA + 2\,H_2O \rightarrow$$
$$3\,NADH + FADH_2 + GTP + CoA + 2\,CO_2 + 3\,H^+$$

The carbon atoms of the acetyl group entering the cycle do not exit the cycle as $CO_2$ in the first round but do so in subsequent rounds.

### Synthesis of Acetyl-Coenzyme A

4.  In eukaryotes, all the enzymes of the citric acid cycle (and the pyruvate dehydrogenase complex) occur in the matrix (inner compartment) of the mitochondrion. The pyruvate produced by glycolysis in the cytosol is transported into the mitochondrion via a pyruvate–$H^+$ symport protein.

5.  The pyruvate dehydrogenase complex is a large multienzyme complex containing multiple copies of three enzymes ($E_1$, $E_2$, and $E_3$) organized in a polyhedral array. For example, the *E. coli* pyruvate dehydrogenase complex consists of 24 copies of $E_1$ and 24 copies of $E_2$ arranged about the corners of concentric cubes, and 12 copies of $E_3$. In mammals, the pyruvate dehydrogenase complex has dodecahedral symmetry.

6.  The pyruvate dehydrogenase complex catalyzes five reactions in the oxidative decarboxylation of pyruvate. Five different coenzymes are involved.
    (a) Pyruvate dehydrogenase ($E_1$) decarboxylates pyruvate in a reaction identical to that catalyzed by pyruvate decarboxylase in alcoholic fermentation (Figure 15-20). However, in the pyruvate dehydrogenase reaction, the hydroxyethyl group remains linked to the thiamine pyrophosphate (TPP) prosthetic group rather than being released as acetaldehyde.
    (b) Dihydrolipoyl transacetylase ($E_2$) transfers the hydroxyethyl group from TPP to a lipoic acid residue that is tethered to an enzyme Lys side chain via an amide linkage (lipoamide) to form acetyl-dihydrolipoamide.
    (c) $E_2$ then transfers the acetyl group to CoA to form acetyl-CoA and dihydrolipoamide.
    (d) Dihydrolipoyl dehydrogenase ($E_3$) oxidizes the dihydrolipoamide group of $E_2$ via the reduction of $E_3$'s reactive Cys—Cys disulfide bond.
    (e) $E_3$ is then reoxidized by $NAD^+$, in a reaction involving the transfer of electrons via the enzyme's FAD group, to regenerate the reactive disulfide bond, thereby preparing the pyruvate dehydrogenase complex for another round of hydroxyethyl transfer and oxidation.

### Enzymes of the Citric Acid Cycle

7.  Citrate synthase catalyzes the condensation of acetyl-CoA and oxaloacetate to form citrate in a highly exergonic reaction. The enzyme undergoes a large conformational change as part of an Ordered Sequential reaction mechanism.

8. Aconitase catalyzes the isomerization of citrate to isocitrate via stereospecific dehydration and rehydration.

9. Isocitrate dehydrogenase catalyzes the oxidative decarboxylation of isocitrate to produce $\alpha$-ketoglutarate and the first $CO_2$ and NADH of the citric acid cycle.

10. $\alpha$-Ketoglutarate dehydrogenase (a multienzyme complex similar to the pyruvate dehydrogenase complex) catalyzes the oxidative decarboxylation of $\alpha$-ketoglutarate by a mechanism identical to that of the pyruvate dehydrogenase complex. The reaction products are succinyl-CoA and the second $CO_2$ and second NADH of the cycle. Note that the two carbons released as $CO_2$ in this round of the citric acid cycle are not the carbons that entered the cycle as acetyl-CoA.

11. Succinyl-CoA synthetase catalyzes a coupled reaction in which thioester hydrolysis is coupled to the phosphorylation of GDP via succinyl-phosphate and phospho-His intermediates. This enzyme therefore catalyzes substrate-level phosphorylation. The GTP produced in this step is easily interconverted with ATP by nucleoside diphosphate kinase.

12. The remaining reactions of the citric acid cycle regenerate oxaloacetate.
    (a) Succinate dehydrogenase catalyzes the stereospecific dehydrogenation of succinate to fumarate. The enzyme's FAD prosthetic group is reduced in this redox reaction and is reoxidized when it gives up its electrons to the respiratory electron-transport chain.
    (b) Fumarase catalyzes the hydration of fumarate to malate.
    (c) Malate dehydrogenase catalyzes the oxidation of malate, which regenerates oxaloacetate and produces the third NADH of the cycle. This highly endergonic reaction is driven by the exergonic reaction that follows, namely the condensation of oxaloacetate with acetyl-CoA, which begins the next round of the citric acid cycle.

*Regulation of the Citric Acid Cycle*

13. The citric acid cycle generates ATP through substrate-level phosphorylation (GTP production) and by the subsequent reoxidation of its 3 NADH and $FADH_2$ products by the electron-transport chain. Each NADH generates ~2.5 ATP and $FADH_2$ generates ~1.5 ATP, so each round of the citric acid cycle yields ~10 ATP.

14. The pyruvate dehydrogenase complex is regulated by product inhibition by NADH and acetyl-CoA, and, in eukaryotes, by covalent modification via the phosphorylation/dephosphorylation of $E_1$. Pyruvate dehydrogenase kinase represses $E_1$ activity by phosphorylating it at a specific Ser residue, and pyruvate dehydrogenase phosphatase stimulates $E_1$ activity by dephosphorylating it.

15. The citric acid cycle is regulated principally at the steps catalyzed by citrate synthase, isocitrate dehydrogenase, and α-ketoglutarate dehydrogenase. Regulatory mechanisms include:
    (a) Substrate availability. The most critical regulators are acetyl-CoA, oxaloacetate, and the ratio of [NADH] to [NAD$^+$].
    (b) Product inhibition. Citrate competes with oxaloacetate in the citrate synthase reaction, and succinyl-CoA and NADH inhibit α-ketoglutarate dehydrogenase.
    (c) Competitive feedback inhibition. Succinyl-CoA competes with acetyl-CoA in the citrate synthase reaction.
    (d) Allosteric regulation. ADP activates isocitrate dehydrogenase, while ATP allosterically inhibits it. Ca$^{2+}$ activates pyruvate dehydrogenase, isocitrate dehydrogenase, and α-ketoglutarate dehydrogenase.

### Reactions Related to the Citric Acid Cycle

16. The citric acid cycle is amphibolic (both anabolic and catabolic). As an anabolic cycle, the citric acid cycle provides intermediates for gluconeogenesis (oxaloacetate, which must be first converted to malate for export from the mitochondrion), amino acid biosynthesis (oxaloacetate, α-ketoglutarate), porphyrin synthesis (succinyl-CoA), and lipid biosynthesis (citrate). The citric acid cycle also functions catabolically to complete the degradation of carbohydrates and fatty acids (which yield acetyl-CoA) and amino acids that are converted to fumarate, succinyl-CoA, α-ketoglutarate, and oxaloacetate.

17. Anaplerotic reactions replenish citric acid cycle intermediates that have been siphoned off into anabolic reactions. The most important of these reactions is catalyzed by pyruvate carboxylase:

$$\text{Pyruvate} + CO_2 + \text{ATP} + H_2O \rightarrow \text{oxaloacetate} + \text{ADP} + P_i$$

18. Acetyl-CoA cannot serve as a precursor for gluconeogenesis in animals, but it can do so in plants via the glyoxylate pathway. This pathway is a variation of the citric acid cycle that takes place in two organelles, the mitochondrion and the glyoxysome (a specialized peroxisome found in plants). The glyoxysome contains isocitrate lyase (which cleaves isocitrate to succinate and glyoxylate) and malate synthase (which condenses glyoxylate with acetyl-CoA to form malate). The net reaction for the glyoxylate pathway is

$$2 \text{ Acetyl-CoA} + 2 \text{ NAD}^+ + \text{FAD} + 3 H_2O \rightarrow$$
$$\text{oxaloacetate} + 2 \text{ CoA} + 2 \text{ NADH} + \text{FADH}_2 + 4 H^+$$

Note that in the glyoxylate pathway, no carbons are lost as $CO_2$.

## Questions

### Overview of the Citric Acid Cycle

1. The citric acid cycle can be divided into two phases with respect to the oxidation of acetyl-CoA. Describe each phase and write its balanced equation.

2.  Early experiments showed that malonate, which inhibits succinate dehydrogenase, blocks cellular respiration. This led to the idea that succinate participates in oxidative metabolism as an intermediate and not as just another metabolic fuel. Using isotopically-labeled reagents, what observation(s) would demonstrate the validity of this interpretation?

### Synthesis of Acetyl-Coenzyme A

3.  Name the three enzymes that form the pyruvate dehydrogenase complex.

4.  For each reaction listed below, indicate the appropriate enzyme(s) in the pyruvate dehydrogenase complex and the relevant cofactor(s), if applicable.

| Reaction | Enzyme | Cofactor |
|---|---|---|
| (a) Oxidative formation of an enzymatic disulfide bond | _____ | _____ |
| (b) Transfer of hydroxyethyl group bound to TPP | _____ | _____ |
| (c) Liberation of $CO_2$ | _____ | _____ |
| (d) Oxidation of dihydrolipoamide | _____ | _____ |
| (e) Formation of acetyl-CoA | _____ | _____ |

5.  What are the roles of FAD and $NAD^+$ in the pyruvate dehydrogenase catalytic mechanism?

6.  You are given two preparations of purified pyruvate dehydrogenase complex enzymes with all the required cofactors. You add pyruvate to each preparation and then measure the rate of production of acetyl-CoA and acetaldehyde under aerobic conditions.

| | | *Acetyl-CoA (molecules/s)* | *Free acetaldehyde (molecules/s)* |
|---|---|---|---|
| Preparation | A | $10^5$ | $10^{-6}$ |
| Preparation | B | $10^{-2}$ | $10^3$ |

How might the pyruvate dehydrogenase complex enzymes differ in each preparation?

7.  Which of the following labeled glucose molecules would yield $^{14}CO_2$ following glycolysis and the pyruvate dehydrogenase reaction?
    (a) 1-[$^{14}C$]-glucose
    (b) 3-[$^{14}C$]-glucose
    (c) 4-[$^{14}C$]-glucose
    (d) 6-[$^{14}C$]-glucose

### Enzymes of the Citric Acid Cycle

8.  What is the energetic function of the thioester bond of CoA in the citrate synthase reaction?

9.   For the half-reaction

$$FAD + 2\,e^- + 2\,H^+ \rightleftharpoons FADH_2$$

$\mathcal{E}^{\circ\prime} \approx 0.00$ V for FAD bound to succinate dehydrogenase.
(a)  Calculate $\Delta G^{\circ\prime}$ for the oxidation of succinate to fumarate by enzyme-bound FAD.
(b)  How does this compare to the $\Delta G^{\circ\prime}$ for the oxidation of succinate to fumarate by free NAD$^+$?
(c)  Based on these results, explain why nature chose FAD rather than NAD$^+$ as the oxidizing agent in the succinate dehydrogenase reaction.

10.  What is the fate of C4 (the carboxyl that is $\beta$ to the carbonyl) of oxaloacetate?

*Regulation of the Citric Acid Cycle*

11.  How would the rapid accumulation of succinyl-CoA affect the rate of glucose oxidation?

*Reactions Related to the Citric Acid Cycle*

12.  List four anabolic pathways that utilize citric acid cycle intermediates as starting material.

13.  How does acetyl-CoA affect the activity of pyruvate carboxylase? Why is this advantageous for the cell?

14.  Write a balanced equation for the synthesis of glucose from acetyl-CoA via the glyoxylate cycle.

15.  Which reactions of the glyoxylate cycle deviate from the citric acid cycle?

16.  Which pathway intermediates pass between the plant mitochondrion and the glyoxysome during the functioning of the glyoxylate cycle?

# 18

# Electron Transport and Oxidative Phosphorylation

This chapter introduces you to the remarkable process by which cells harness the free energy of oxidation and use it to synthesize ATP. (You have already seen how ATP can be synthesized by the phosphorylation of ADP by a metabolite with a higher phosphoryl group-transfer potential.) The reduced cofactors generated in metabolic reactions, NADH and FADH$_2$, are reoxidized in the mitochondrion by a set of reactions in which electrons flow through a series of redox carriers, finally reducing oxygen to water. During electron transport, an electrochemical potential is developed across the inner mitochondrial membrane by the vectorial transfer of protons. This proton gradient is stable because the inner mitochondrial membrane is impermeable to ions. The free energy of the electrochemical proton gradient is utilized by an ATP synthase to catalyze the endergonic reaction ADP + P$_i$ $\rightarrow$ ATP. The coupling of the electrochemical gradient to ATP synthesis is described by the chemiosmotic hypothesis, which is supported by considerable evidence.

## Essential Concepts

1. The complete oxidation of glucose carbons by glycolysis and the citric acid cycle can be written as

$$C_6H_{12}O_6 + 6\,H_2O \rightarrow 6\,CO_2 + 24\,H^+ + 24\,e^-$$

The reducing equivalents (electrons) are captured in the form of reduced coenzymes (NADH and FADH$_2$), which eventually transfer the electrons to molecular oxygen:

$$6\,O_2 + 24\,H^+ + 24\,e^- \rightarrow 12\,H_2O$$

This process regenerates NAD$^+$ and FAD and generates a proton concentration gradient across the inner mitochondrial membrane, whose dissipation provides the free energy for ATP synthesis. This process is known as oxidative phosphorylation.

## The Mitochondrion

2. The mitochondrion is surrounded by a relatively porous outer membrane. An inner membrane is folded to form cristae and encloses the gel-like matrix, which contains the enzymes of the citric acid cycle and fatty acid oxidation. The matrix also contains genetic machinery (DNA, RNA, and ribosomes), reflecting the bacterial origin of this organelle. The proteins involved in electron transport and oxidative phosphorylation are located in the inner mitochondrial membrane. The inner and outer membranes also contain proteins that mediate the transport of ions and metabolites.

3.  NADH produced in the cytosol as a result of glycolysis must enter the mitochondrion to be aerobically oxidized. There are two shuttles for NADH.
    (a) The malate–aspartate shuttle allows NADH to be indirectly transported into the mitochondrion by reducing oxaloacetate to malate in the cytosol and transporting it into the mitochondrion, where it is reoxidized to produce oxaloacetate and NADH. The oxaloacetate is converted by transamination to aspartate and transported out again.
    (b) The glycerophosphate shuttle in insects first reduces cytosolic dihydroxyacetone phosphate to 3-phosphoglycerate and $NAD^+$. The 3-phosphoglycerate is oxidized by an inner mitochondrial membrane enzyme, flavoprotein dehydrogenase, which introduces electrons directly into the electron-transport pathway.

4.  Most of the ATP generated in the mitochondria is used in the cytosol. The ADP–ATP translocator exports ATP out of the matrix while importing ADP. ATP has one more negative charge than ADP, so transport is electrogenic. Transport is driven by the electrochemical potential of the proton concentration gradient (positive outside). The proton gradient also favors the transport of $P_i$ into the matrix by a $P_i$–$H^+$ symport system.

*Electron Transport*

5.  In an electron transfer reaction, electrons flow from a substance with a lower reduction potential to a substance with a higher reduction potential. The standard reduction potential, $E^{\circ\prime}$, is a measure of a substance's affinity for electrons. For a redox reaction, $\Delta \mathscr{E}^{\circ\prime} = \mathscr{E}^{\circ\prime}_{(e-\ acceptor)} - \mathscr{E}^{\circ\prime}_{(e-\ donor)}$. When $\Delta \mathscr{E}^{\circ\prime}$ is positive, the reaction is spontaneous, since $\Delta G^{\circ\prime} = -n \mathscr{F} \Delta \mathscr{E}^{\circ\prime}$, where $n$ is the number of electrons transported and F is the faraday ($96{,}485$ $J \cdot V^{-1} \cdot mol^{-1}$). The transfer of electrons from NADH to $O_2$ ($\Delta \mathscr{E}^{\circ\prime} = 1.13$ V and $\Delta G^{\circ\prime} = -218$ $kJ \cdot mol^{-1}$) provides enough free energy to synthesize three ATP molecules.

6.  Four large protein complexes in the inner mitochondrial membrane are involved in transferring electrons from reduced coenzymes to $O_2$. Complexes I and II transfer electrons to the lipid-soluble electron carrier ubiquinone (coenzyme Q or CoQ), which transfers electrons to Complex III. From there, electrons pass to cytochrome $c$, a peripheral membrane protein with a heme prosthetic group, which transfers electrons to Complex IV. The reactions of Complexes I–IV are as follows:

    (I)  NADH + CoQ (*ox*) → $NAD^+$ + CoQ (*red*)
         $\Delta \mathscr{E}^{\circ\prime} = 0.360$ V and  $\Delta G^{\circ\prime} = -69.5$ $kJ \cdot mol^{-1}$
    (II) $FADH_2$ + CoQ (*ox*) → FAD + CoQ (*red*)
         $\Delta \mathscr{E}^{\circ\prime} = 0.085$ V and  $\Delta G^{\circ\prime} = -16.4$ $kJ \cdot mol^{-1}$
    (III) CoQ (*red*) + cytochrome $c$ (*ox*) → CoQ (*ox*) + cytochrome $c$ (*red*)
          $\Delta \mathscr{E}^{\circ\prime} = 0.190$ V and  $\Delta G^{\circ\prime} = -36.7$ $kJ \cdot mol^{-1}$
    (IV) Cytochrome $c$ (*red*) + ½ $O_2$ → cytochrome $c$ (*ox*) + $H_2O$
         $\Delta \mathscr{E}^{\circ\prime} = 0.580$ V and  $\Delta G^{\circ\prime} = -112$ $kJ \cdot mol^{-1}$

7.  Complex I is an enormous protein complex containing flavin mononucleotide (FMN, which is FAD minus its AMP group) and multiple iron–sulfur clusters (which are one-electron carriers). The two electrons donated by NADH are transferred through these

redox-active prosthetic groups and then to CoQ. As electrons are transferred, four protons are translocated from the matrix to the intermembrane space via a proton wire. Proton movement most likely occurs by protein conformational changes similar to those in bacteriorhodopsin.

8.  Complex II, which contains the citric acid cycle enzyme succinate dehydrogenase, transfers electrons from succinate to FAD and then to CoQ. No protons are translocated by Complex II, which serves mainly to feed electrons into the electron transport chain.

9.  Complex III (cytochrome $bc_1$ or cytochrome $c$ reductase) contains two $b$-type cytochromes, cytochrome $c_1$, and an iron–sulfur protein, which contains a [2Fe–2S] cluster. Electron flow from CoQ through Complex III follows a bifurcated cyclic pathway known as the Q cycle. In the first round of the Q cycle, fully reduced ubiquinone (ubiquinol; $QH_2$) donates one electron to the iron–sulfur protein, which then transfers it to cytochrome $c_1$ and then to cytochrome $c$. This one-electron donation yields the ubisemiquinone anion ($Q \cdot^-$), which donates its remaining electron to the low potential cytochrome $b$ ($b_L$), and then to the high potential cytochrome $b$ ($b_H$). The resulting ubiquinone diffuses to the other side of the membrane, where it accepts the electron from $b_L$ to reform $Q \cdot^-$. A second round of electron transfers completes the cycle: Another $QH_2$ donates its electrons, one to the iron–sulfur protein and one to cytochrome $b_L$. The net result is that two electrons are transferred, one at a time, to two cytochrome $c$ molecules, and four protons are transferred from the matrix to the intermembrane space, two from each $QH_2$ that participates in the Q cycle.

10. Cytochrome $c$ shuttles electrons between Complexes III and IV. Cytochrome $c$ is a small water-soluble protein whose heme group is largely buried in a crevice surrounded by a ring of Lys residues. Both cytochrome $c_1$ and cytochrome $c$ oxidase (Complex IV) have a corresponding patch of negatively charged amino acid residues to facilitate cytochrome $c$ binding and electron transfer.

11. Complex IV (cytochrome $c$ oxidase) has four redox centers [cytochrome $a$, cytochrome $a_3$, $Cu_A$ (which contains two Cu ions), and $Cu_B$], and it carries out the following reaction:

$$4 \text{ Cytochrome } c \text{ (Fe}^{2+}) + 4 \text{ H}^+ + O_2 \rightarrow 4 \text{ cytochrome } c \text{ (Fe}^{3+}) + 2 \text{ H}_2\text{O}$$

$O_2$ reduction takes place at the cytochrome $a_3$–$Cu_B$ binuclear complex, which mediates four one-electron transfer reactions. Four protons are consumed in the production of $H_2O$, and four additional proteins are pumped, most likely via a proton wire, from the matrix to the intermembrane space (two for each pair of electrons that enter the electron-transport chain).

*Oxidative Phosphorylation*

12. ATP synthase (Complex V) phosphorylates ADP by a mechanism driven by the free energy of electron transport, which is conserved in the formation of an electrochemical proton gradient across the inner mitochondrial membrane. The two processes are coupled as described by the chemiosmotic hypothesis. Four observations support this hypothesis:
    (a) Mitochondrial ATP formation requires an intact inner membrane.
    (b) The inner membrane is impermeable to ions, so an electrochemical gradient across the membrane can be sustained.
    (c) Electron transport pumps protons out of the mitochondrion to create a measurable electrochemical gradient.
    (d) Agents that increase the permeability of the inner mitochondrial membrane to protons inhibit ATP synthesis but not electron transport.

13. The protonmotive force results from the difference in concentration of protons (pH) in the matrix and the intermembrane space and from the difference in charge (membrane potential, $\Delta\Psi$) across the membrane. Thus, $\Delta G = 2.3\ RT\ [\text{pH}(side\ 1) - \text{pH}(side\ 2)] + Z\mathcal{F}\Delta\Psi$, where $Z$ is the charge of the proton. $\Delta\Psi$ is positive when a proton is transported from negative to positive, or against its potential. Thus, pumping protons out of the matrix (against the gradient) is an endergonic process, whereas transporting them back in (with the gradient) is an exergonic process. About three protons are needed to supply sufficient energy to synthesize one ATP from ADP + $P_i$.

14. ATP synthase, also called $F_1F_0$-ATPase, has two functional units. The $F_0$ component includes the transmembrane proton channel, which, depending on the species, is a ring of 9 to 14 $c$ subunits that rotates within the plane of the membrane such that a proton is picked up from the intermembrane space, completes about one rotation, and is released into the matrix. The $F_0$ component also includes a stator assembly that holds the $F_1$ component in place.

15. The $F_1$ component is a water-soluble protein of subunit composition $\alpha_3\beta_3\gamma\delta\epsilon$ that associates with the membrane via $F_0$ to form a lollipoplike structure. The 3 $\alpha$ and 3 $\beta$ subunits of $F_1$ form a structure of alternating $\alpha$ and $\beta$ subunits with pseudo-threefold rotational symmetry. The elongated $\gamma$ subunit, which also contacts $F_0$, rotates within the center of the $\alpha_3\beta_3$ ring. As the proton-translocating $c$-ring rotates, the $\gamma$ subunit also rotates.

16. The binding change mechanism describes ATP synthesis in terms of three processes:
    (a) Translocation of protons carried out by $F_0$.
    (b) Formation of the phosphoanhydride bond of ATP, catalyzed by $F_1$.
    (c) Interaction of $F_0$ and $F_1$ to couple the dissipation of the proton gradient to formation of the phosphoanhydride bond.

17. According to the binding change mechanism, ATP is synthesized as each $\alpha\beta$ protomer shifts though three conformations in sequence. The three possible conformations are called open (O), loose (L), and tight (T). ADP and $P_i$ bind to a protomer with the L conformation and are converted to ATP when the conformation shifts to the T state. The free energy of

the proton concentration gradient converts the T state to the O state (this is the rate-limiting step), thereby releasing ATP. This three-step mechanism is consistent with the pseudo-threefold axis of rotation. The γ subunit and *c*-ring rotate with respect to the $\alpha_3\beta_3$ assembly, and the geometric relationship of individual αβ protomers to the γ subunit dictates their conformational state. Protons equal in number to the number of *c* subunits in the *c*-ring are required for one complete rotation, promoting three conformational shifts, each of which synthesizes one ATP.

18. The ratio of the amount of ATP produced to the amount of substrate oxidized (measured as oxygen consumed) is called the P/O ratio. [The P/O ratio refers to atomic oxygen, O, rather than molecular oxygen, $O_2$, because each substrate (NADH or $FADH_2$) transfers two electrons, not four.] Depending on where a substrate's electrons enter the electron-transport chain, the P/O ratio is ~2.5, ~1.5, or ~1. For example, the two electrons transferred from NADH through Complexes I, III, and IV pump 10 protons, which would yield ~ 3 ATP, whereas the two electrons transferred from $FADH_2$ through Complexes II, III, and IV pump 6 protons, which would yield ~2 ATP. However, these numbers are lower due to leakage of protons and $P_i$ transport. The complete oxidation of glucose therefore yields ~32 ATP. The P/O ratio is not necessarily a whole number, because protons are contributed to the gradient by more than one process and some protons leak back into the matrix.

19. Electron transport and oxidative phosphorylation are normally strongly coupled processes; that is, neither process occurs in the absence of the other. This is because, if the rate of electron transport were to outpace the rate of ATP synthesis, the proton gradient would build up to the level that it would resist additional proton pumping by Complexes I, III, and IV and hence the rate of electron transport would be slowed. However, when uncoupling agents, which dissipate the proton gradient, are added to respiring mitochondria, electron transport proceeds unchecked while ATP synthesis stops. The free energy of electron transport is then redirected from ATP synthesis to generate heat. 2,4-Dinitrophenol is an uncoupling agent because it carries protons through the membrane from the intermembrane space to the matrix, thereby providing a route for dissipation of the gradient that bypasses $F_0$.

### Control of Oxidative Metabolism

20. Electron transfer from NADH to cytochrome *c* is nearly at equilibrium. In contrast, the cytochrome *c* oxidase reaction is irreversible and hence its rate depends on the concentration of its substrate, reduced cytochrome *c*. Increased NADH concentrations and decreased ATP concentrations lead to the production of more reduced cytochrome *c* and hence to increased electron transfer rates. Thus, the overall rate of oxidative phosphorylation depends on the ratios [NADH]/[NAD$^+$] and [ATP]/[ADP][$P_i$], which in turn may depend on the activities of the respective mitochondrial transporters.

21. The coordinated control of oxidative metabolism centers on several key enzymes: hexokinase (HK), phosphofructokinase (PFK), pyruvate kinase (PK), pyruvate dehydrogenase (PDH), citrate synthase (CS), isocitrate dehydrogenase (IDH) and α-ketoglutarate dehydrogenase (KDH). High levels of ATP inhibit PFK and PK while high

[NADH]/[NAD$^+$] ratios inhibit PDH, IDH, and KDH. Citrate inhibits both PFK and CS. The need for ATP, represented by high concentrations of either AMP or ADP, activates PFK, PK, PDH, and IDH, while Ca$^{2+}$ stimulates PDH, IDH, and KDH.

22. Anaerobic glycolysis produces 2 ATP per glucose consumed, whereas oxidative metabolism generates 32 ATP per glucose, a 16-fold increase. However, there are several drawbacks of oxygen-based metabolism. Many organisms depend on oxidative metabolism and would perish without a steady supply of oxygen. Reactive oxygen species generated by incomplete oxygen reduction are potentially dangerous.

23. Reactive oxygen species include the superoxide radical O$_2\cdot^-$ (produced by the reaction O$_2$ + $e^-$ → O$_2\cdot^-$), which is a precursor of even more powerful oxidizing species such as HO$_2\cdot$ and ·OH. These free radicals readily extract electrons from other substances, creating a chain reaction. Neurodegenerative conditions such as Parkinson's, Alzheimer's, and Huntington's diseases are associated with mitochondrial oxidative damage. Free radical reactions arising from normal oxidative metabolism appear to be partially responsible for the aging process.

24. Antioxidants limit oxidative damage by destroying free radicals. Superoxide dismutase (SOD) catalyzes the production of water and hydrogen peroxide from superoxide:

$$2\,O_2\cdot^- + 2\,H^+ \rightarrow H_2O_2 + O_2$$

This enzyme electrostatically guides its substrate to the active site to catalyze a reaction near the diffusion-controlled limit. Mutations in Cu,Zn SOD are associated with amyotrophic lateral sclerosis (Lou Gehrig's disease).

25. Catalase and glutathione peroxidase degrade hydroperoxides. Some types of glutathione peroxidase require selenium, so Se also appears to be an antioxidant.

## Questions

### The Mitochondrion

1. Draw a cross-section of a mitochondrion and label the following structural features:

   Outer membrane (OM)                Inner membrane (IM)
   Matrix (M)                         Intermembrane space (IMSP)
   ATP synthase complex (ASC)         Direction of proton flux
   ATP and P$_i$ transporters (T)     Cristae (CR)

2. Match the following enzyme or other molecule with its location:

   ___ Pyruvate dehydrogenase
   ___ 3-Phosphoglycerate dehydrogenase   A. Cytosol
   ___ Flavoprotein dehydrogenase

_____ Malate dehydrogenase          B. Mitochondrial outer membrane
_____ Cytochrome $c$
_____ Cytochrome $c_1$          C. Mitochondrial inner membrane
_____ Fatty acid oxidation enzymes
_____ Mitochondrial DNA          D. Mitochondrial intermembrane space
_____ ADP–ATP translocator
_____ Mitochondrial porin          E. Mitochondrial matrix

3.     About half the volume of the mitochondrial matrix is water, and the rest is protein. If a single protein of molecular mass 40,000 were as concentrated, what would be its molar concentration? Assume the protein's density is 1.37 $g \cdot mL^{-1}$.

4.     Oxidative phosphorylation requires the transfer of electrons donated by NADH. (a) Is NADH imported directly into the mitochondria? Explain. (b) Describe two import mechanisms that transfer cytosolic electrons from NADH into the mitochondrion. (c) Why is it important to maintain a relatively constant level of cytosolic $NAD^+$?

*Electron Transport*

5.     Which reactions of the citric acid cycle donate electron pairs to the mitochondrial electron-transport chain?

6.     The half-cell reduction potential is provided by the Nernst equation (Equation 14-8):

$$\mathscr{E}_A = \mathscr{E}°_A - \frac{RT}{n\mathscr{F}} \ln \left( \frac{[A_{red}]}{[A_{ox}^{n+}]} \right)$$

(a)   On the graph below, plot the reduction potentials for the $FADH_2/FAD$ half-cell ($E°' = -0.219$ V) when the $[FADH_2]/[FAD]$ ratios are 100, 10, 5, 2, 1, 0.5, 0.2, 0.1, and 0.01 at 25°C versus the percent reduction.

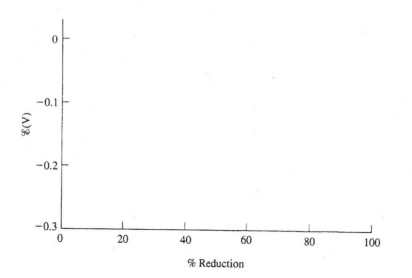

(b) Using the same [Reduced]/[Oxidized] ratios, plot the reduction potentials of cytochrome $c$ ($E°' = 0.235$ V).

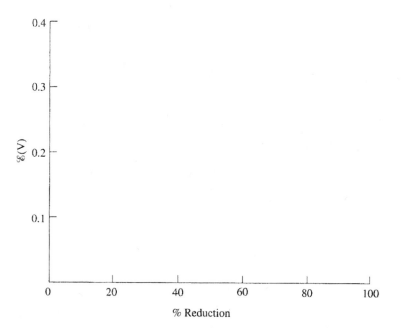

(c) What is $\Delta E$ for the oxidation of $FADH_2$ by cytochrome $c$ when the $[FADH_2]/[FAD]$ ratio is 10 and the [cytochrome $c$ ($Fe^{2+}$)]/[cytochrome $c$ ($Fe^{3+}$)] ratio is 0.1?

7.   Inhibitors of electron transport have been used to determine the order of electron carriers. What would be the expected redox states of cytochromes $a$, $b_L$, and $c$ when (a) stigmatellin, (b) antimycin A, or (c) rotenone is added to succinate-driven respiring mitochondria?

8.   What is the most abundant type of redox carrier in Complex I?

9.   Which mitochondrial electron carriers are potentially proton carriers? Which are more abundant—proton carriers or electron carriers? What does this suggest about the mechanism of transmembrane proton transport?

10.   What amino acid residues would you expect to find at the cytochrome $c$–binding sites of Complex III and Complex IV?

11.   Which redox group(s) in Complex IV accepts electrons from cytochrome $c$? Which redox group(s) binds oxygen during its four-electron reduction?

*Oxidative Phosphorylation*

12.   What key observations support the chemiosmotic hypothesis?

13.   Can a pH gradient exist without $\Delta\Psi$? Can $\Delta\Psi$ exist without a $\Delta$pH?

14. (a) What is a P/O ratio? (b) What happens to oxygen consumption when electron donors and inorganic phosphate are present in a suspension of mitochondria in the absence of ADP? (c) What happens when ADP is added?

15. (a) Valinomycin, an antibiotic ionophore, allows the free passage of only $K^+$ ions across a membrane. If $K^+$ and valinomycin are added to respiring, fully coupled ATP-synthesizing mitochondria, what happens to the pH gradient and the $\Delta\Psi$? (b) Nigericin, another ionophore, exchanges one $K^+$ for one $H^+$. How does this affect ATP synthesis and electron transport in mitochondria? (c) Gramicidin allows the free passage of many small molecules and ions across the membrane. What happens to ATP production and electron transport in the presence of gramicidin?

16. The free energy-requiring step in the synthesis of ATP is not the formation of ATP from ADP and $P_i$ ($\Delta G \approx 0$), but the release of tightly bound ATP. Explain why this is not inconsistent with the $+30.5$ kJ·mol$^{-1}$ free energy of formation of ATP in solution.

17. Dicyclohexylcarbodiimide (DCCD) reacts with Asp and Glu residues in the $c$ subunits of $F_0$ and blocks ATP synthase activity. What happens to the rate of electron transport when DCCD is added to actively respiring mitochondria?

18. What does an $H^+/P$ ratio measure? Why would it be impractical to determine an $H^+/P$ ratio?

## Control of Oxidative Metabolism

19. The conversion of glucose to 2 lactate has $\Delta G^{\circ\prime} = -196$ kJ·mol$^{-1}$. The complete oxidation of glucose to 6 $CO_2$ has $\Delta G^{\circ\prime} = -2823$ kJ·mol$^{-1}$. Compare the efficiencies of ATP synthesis by each of these processes under standard conditions.

20. On an average day, an adult dissipates about 7000 kJ of free energy. Assuming that this occurs under standard conditions, (a) how many moles of ATP must be hydrolyzed to provide this quantity of free energy? (b) What is the mass of this quantity of ATP? (c) If the amount of ATP in an adult is about 0.1 mole, how many times per day, on average, is a molecule of ADP recycled? The molecular mass of ATP is 507.

21. What is the irreversible step in electron transport and how is its rate controlled?

22. Cytochrome P450 (Section 12-4D) catalyzes a reaction in which two electrons supplied by NADPH reduce the heme Fe atom so that it can then reduce $O_2$ preparatory to the hydroxylation of a substrate molecule. Substrate binding to the enzyme displaces a water molecule that forms a ligand to the heme iron atom. This changes the reduction potential of the Fe from $-0.300$ V to $-0.170$ V. Why is this change necessary for efficient catalysis?

23. Rats that are fed a "cafeteria" diet (in which food is always available) tend to die sooner than rats whose dietary intake is limited. Propose an explanation for this observation.

# 19
# Photosynthesis

In the preceding chapters, you learned that free energy is derived from reduced foodstuffs such as glucose and that the energy produced from catabolic pathways is used to generate both ATP and NAD(P)H. Photosynthesis, an ancient and important process, allows energy to be harvested directly from the most abundant and renewable source, the sun. Photosynthesis is a light-driven process in which carbon dioxide is "fixed" to produce carbohydrates. This occurs in two phases: (1) The light reactions (requiring light) produce ATP and NADPH, and (2) the dark reactions (not requiring light) use ATP and NADPH to synthesize carbohydrates. This chapter describes how different pigments (e.g., chlorophylls in plants and bacteria) efficiently capture light energy and redistribute it to specific reaction centers. Purple photosynthetic bacteria contain one photosystem that recycles its electrons, whereas higher plants have two photosystems that use water as a source of electrons to reduce NADPH. The oxidation of water in higher plants generates $O_2$ as a by-product of photosynthesis. As in mitochondria, the topology of chloroplasts is central to the biochemistry of photosynthesis, starting with the light-driven reactions in the thylakoid membrane and finishing with the dark reactions in the stroma. The dark reactions occur via the Calvin cycle, a set of reactions that synthesize glyceraldehyde-3-phosphate from 3 $CO_2$. The chapter also discusses the control of the Calvin cycle along with a variant called the $C_4$ pathway.

## Essential Concepts

1. Photosynthesis is divided into two processes:
   (a) In the light reactions, organisms capture light energy to synthesize ATP and generate reducing equivalents in the form of NADPH.
   (b) In the dark reactions, carbon dioxide is converted to carbohydrates using the ATP and NADPH generated in the light reactions. Although the dark reactions are not light-driven, they occur only when it is light and hence are better described as light-independent.

### Chloroplasts

2. Plants differ from bacteria by providing a separate organelle for the photosynthetic machinery, the chloroplast. A chloroplast is enveloped by a highly permeable outer membrane and a nearly impermeable inner membrane. The inner membrane encloses the stroma, which contains the soluble enzymes of carbohydrate synthesis, and the thylakoid membrane, which is organized in stacks of pancake-like disks (grana) that enclose the thylakoid compartment and that are linked by unstacked stromal lamellae. The proteins that capture light energy and mediate electron-transport processes are embedded in the thylakoid membrane.

3.  Various pigment molecules absorb light of different wavelengths. The principal photosynthetic pigment is chlorophyll, a cyclic tetrapyrrole that ligands a central $Mg^{2+}$ ion. Photosynthetic organisms also contain other pigments, such as carotenoids, phycoerythrin, and phycocyanin, which together with chlorophyll absorb most of the visible light in the solar spectrum.

4.  Multiple pigment molecules are arranged in light-harvesting complexes (LHCs), which are proteins that act as antennae to gather light energy and redirect it to photosynthetic reaction centers, where the light energy is converted to chemical energy in the form of ATP and NADPH. The accessory pigments in the LHCs boost light absorption at wavelengths at which chlorophyll does not absorb strongly.

*The Light Reactions*

5.  Photons propagate as discrete energy packets called quanta, whose energy, $E$, is given by Planck's law: $E = h\nu = hc/\lambda$, where $h$ is Planck's constant ($6.626 \times 10^{-34}$ J·s) $c$ is the speed of light ($2.998 \times 10^8$ m·s$^{-1}$), $\lambda$ is the wavelength of light, and $\nu$ is the frequency of the radiation.

6.  When a molecule absorbs a photon, one of its electrons is promoted to a higher energy orbital. The excited electron can return to the ground state in several ways:
    (a) In internal conversion, electronic energy is converted to kinetic energy (heat).
    (b) Fluorescence results when the molecule emits a photon at a lower wavelength.
    (c) The excitation energy can be transferred to another molecule by exciton transfer (resonance energy transfer). This occurs in the transfer of light energy from LHCs to the photosynthetic reaction center.
    (d) The molecule may undergo photooxidation by transfer of an electron to another molecule. The excited chlorophyll at the reaction center transfers electrons in this manner.

7.  The bacterial photosynthetic reaction center of *Rps. viridis* consists of a series of prosthetic groups arranged with nearly twofold symmetry: 2 closely associated bacteriochlorophyll *a* (BChl *a*) molecules known as the special pair, 2 bacteriopheophytin *a* (BPheo *a*; BChl *a* that lacks an $Mg^{2+}$ ion), 2 additional BChl *a* molecules, a menaquinone, a ubiquinone, and an Fe(II) ion.

8.  In purple photosynthetic bacteria, the special pair undergoes photooxidation virtually every time it absorbs a photon. The transferred electron is first passed to the BPheo *a* on the "right" side of the photosynthetic reaction center and then to the menaquinone and then the ubiquinone to yield a semiquinone radical anion ($Q_B^-$). A second photon absorption then transfers a second electron to yield $Q_B^{2-}$, which picks up two protons from the cytoplasm to form $QH_2$ and then exchanges with the membrane-bound pool of ubiquinone.

9.  The electrons ejected from the special pair return to the photosynthetic reaction center via an electron-transport chain consisting of a cytochrome $bc_1$ complex and cytochrome $c_2$. Electrons flow follows a Q cycle in cytochrome $bc_1$, which translocates four protons to the

periplasmic space for every two electrons transferred. The free energy of the resulting transmembrane proton gradient drives ATP synthesis.

10. In plants and cyanobacteria, photooxidation occurs at two reaction centers, and electron transport is noncyclical. The path of electrons from water to NADPH is described by the Z-scheme, in which photosystem II (PSII) passes its electrons to the cytochrome $b_6f$ complex via the mobile electron carrier ubiquinol ($QH_2$), and cytochrome $b_6f$ then transfers these electrons to photosystem I (PSI) via the mobile Cu-containing protein plastocyanin. Since PSII and PSI are thereby "connected in series," the energy of each electron is boosted by two photon absorptions.

11. PSII includes the Mn-containing oxygen-evolving center (OEC), which cycles through five electronic states ($S_0$–$S_4$) in the conversion of $H_2O$ to $O_2$ and is driven by four consecutive excitations of the PSII reaction center (called P680). The four electrons released from $H_2O$ follow a path similar to that of the bacterial photosynthetic reaction center, eventually reaching the membrane plastoquinone pool.

12. Electron transport through the cytochrome $b_6f$ complex (which resembles the mitochondrial Complex III) generates a transmembrane proton gradient via a Q cycle. Eight protons are translocated to the thylakoid lumen for the four electrons released from each $H_2O$. Plastocyanin, a peripheral membrane protein, has a Cu redox center than ferries one electron at a time from cytochrome $b_6f$ to PSI.

13. PSI contains multiple pigments and redox groups, including chlorophylls, carotenoids, [4Fe–4S] clusters, and phylloquinone. Photooxidation of the PSI special pair (called P700) allows the electron received from plastocyanin to pass through a series a electron carriers in one of two routes:
    (a) In the noncyclic pathway, electrons flow through PSI to the [2Fe–2S]-containing one-electron carrier ferredoxin (Fd), which is a soluble stromal protein. Two Fd's deliver their electrons to ferredoxin–NADP$^+$ reductase, which thereupon carries out the two-electron reduction of NADP$^+$ to NADPH.
    (b) In the cyclic pathway, electrons return from PSI through cytochrome $b_6f$ to the plastoquinone pool and thereby participate in the Q cycle. This pathway augments the proton gradient across the thylakoid membrane and hence contributes additional free energy for the synthesis of ATP but does not yield NADPH.

14. The free energy of the proton gradient is tapped by chloroplast $CF_1CF_0$-ATP synthase, which closely resembles the mitochondrial $F_1F_0$-ATPase. Approximately 12 protons enter the thylakoid lumen for each $O_2$ generated in noncyclic electron transport (4 H$^+$ from the OEC reaction and 8 H$^+$ from the Q cycle). The synthesis of ATP requires the transport of ~3 protons from the thylakoid lumen to the stroma.

*The Dark Reactions*

15. $CO_2$ is incorporated into carbohydrates by carboxylation of a 5-carbon sugar, ribulose-5-phosphate (R5P). The resulting 6-carbon compound is split into two molecules of 3-phosphoglycerate (3PG), which is then converted to glyceraldehyde-3-phosphate (GAP). Some of the GAP is diverted to carbohydrate synthesis, and the rest is converted back to Ru5P. This set of 13 reactions, called the Calvin cycle, has two stages:
    (a) In the production phase, 3 Ru5P react with 3 $CO_2$ to yield 6 GAP (for a net yield of 1 GAP from 3 $CO_2$), at a cost of 9 ATP and 6 NADPH.
    (b) In the recovery phase, the carbons of 5 GAP are shuffled via aldolase- and transketolase-catalyzed reactions to reform 3 Ru5P, without consuming ATP or NADPH.
    The GAP product of the cycle is then converted to glucose-1-phosphate, a precursor of sucrose and starch.

16. Ribulose bisphosphate carboxylase, which accounts for up to 50% of leaf proteins, catalyzes the carboxylation of ribulose-1,5-bisphosphate (RuBP). Enzymatic abstraction of a proton from RuBP generates an enediolate that attacks $CO_2$. $H_2O$ then attacks the resulting β-keto acid to yield two 3PG.

17. The activity of RuBP carboxylase is controlled in several ways so that the dark reactions proceed only when the light reactions are able to provide the ATP and NADPH necessary to drive them:
    (a) RuBP carboxylase is most active at pH 8.0, which occurs when protons are pumped out of the stroma during the light reactions.
    (b) The $Mg^{2+}$ that enters the stroma to compensate for the efflux of $H^+$ stimulates RuBP carboxylase.
    (c) Plants synthesize 2-carboxyarabinitol-1-phosphate, an inhibitor of RuBP carboxylase, only in the dark.

18. The Calvin cycle enzymes fructose bisphosphatase and sedoheptulose bisphosphatase are also activated by increases in pH, $[Mg^{2+}]$, and [NADPH]. The redox state of ferredoxin is sensed by a thiol-exchange cascade involving ferredoxin–thioredoxin reductase, thioredoxin, and disulfide groups on the bisphosphatases, so that the Calvin cycle enzymes are stimulated when Fd is reduced (i.e., when the light reactions are occurring).

19. RuBP carboxylase can also react with oxygen, which competes with $CO_2$ at the carboxylase active site. This process, called photorespiration, converts RuBP into 3PG and 2-phosphoglycolate, a two-carbon compound. A series of reactions in the chloroplast and peroxisome convert two 2-phosphoglycolate to two glycine, which are converted in the mitochondria to serine + $CO_2$. The serine is converted back to the Calvin cycle intermediate 3PG by reactions that require NADH and ATP. Thus, photorespiration consumes $O_2$ and produces $CO_2$, at the expense of ATP and NADH, thereby reversing the results of photosynthesis. All known RuBP carboxylases have this activity, which may protect chloroplasts from photoinactivation when high light intensity has greatly reduced the local $CO_2$ concentration.

20.  The rate of photorespiration becomes significant on hot bright days when photosynthesis has depleted the level of $CO_2$ at the chloroplast and raised the concentration of $O_2$. However, $C_4$ plants such as corn and sugar cane prevent photorespiration by concentrating $CO_2$ at the chloroplast. They do so by using phosphoenolpyruvate (PEP) carboxylase to make oxaloacetate. Oxaloacetate is converted to malate in the mesophyll cells (which lack RuBP carboxylase) and transported to the bundle-sheath cells, where the Calvin cycle operates. There, malic enzyme cleaves malate into pyruvate and $CO_2$. The $CO_2$ is thereby delivered to RuBP carboxylase at a high enough concentration to essentially eliminate photorespiration. The pyruvate is transported back to the mesophyll cells and converted to PEP at the expense of two "high-energy" bonds. In the tropics, $C_4$ plants grow faster than so-called $C_3$ plants. In more temperate climates, where the rate of photorespiration is reduced, $C_3$ plants have an advantage because they require less energy to fix $CO_2$.

21.  Many desert succulent plants conserve water by opening their stomata only at night to acquire $CO_2$. The $CO_2$ is converted to PEP at night and is released as $CO_2$ in the day, to be fixed by RuBP carboxylase. This process is called Crassulacean acid metabolism (CAM) because it was discovered in the family Crassulaceae.

## Key Equations

$$E = h\nu = \frac{hc}{\lambda}$$

## Questions

1.  Define the terms light reactions and dark reactions. Do the dark reactions occur in the dark? Explain.

### Chloroplasts

2.  Draw a cross-section of a chloroplast and indicate the locations of the following proteins and other structural features:

| | |
|---|---|
| Outer membrane | Thylakoid lumen |
| Stromal lamella | Photosystem I |
| Inner membrane | Photosystem II |
| Intermembrane space | Cytochrome $b_6f$ |
| Grana | $CF_1CF_0$-ATP synthase |
| Stroma | Direction of proton pumping |

### The Light Reactions

3.  Calculate the energy of a mole of photons with wavelengths of (a) 400 nm, (b) 500 nm, (c) 600 nm, and (d) 700 nm.

4. Purple photosynthetic bacteria have different pigments than higher plants. Why is this an advantage for these bacteria?

5. What distinguishes the chlorophyll in a reaction center from the antennae chlorophyll?

6. The initial electron transfers in the bacterial photosynthetic reaction center are extremely rapid, but the lifetime of the terminal semiquinone is relatively long. (a) Why is it essential for the electrons to quickly leave the vicinity of the special pair? (b) Why does the terminal semiquinone persist?

7. Compare the electron flow in purple photosynthetic bacteria to that in higher plant chloroplasts. What is the origin of the electrons and their eventual fates? How many protons are translocated?

8. What is the standard reduction potential for the oxidation of $H_2O$ (see Table 14-5)? Can this value be obtained from purple bacterial photosynthesis? Compare this to two-center photosynthesis.

9. The number of $O_2$ molecules released per photon absorbed by a suspension of algae can be measured at different wavelengths. When algae are illuminated by 700 nm light, very little $O_2$ is produced. However, when they are also illuminated by 500 nm light, the $O_2$ production is well in excess of that produced with only the 500 nm light. Explain.

10. What is the fate of water-derived electrons in chloroplasts treated with DCMU? What simple screening method could be used to identify plants that had been exposed to DCMU?

11. What do the $S$ states represent in the oxygen-evolving center (OEC)? What $S$ state predominates in dark-adapted chloroplasts?

12. What chloroplast component generates the majority of the proton gradient used for ATP production? Does it have a mitochondrial counterpart?

13. Why does the Cu atom of plastocyanin have an unusually high standard reduction potential?

14. What are the similarities and differences between photosystem I and photosystem II?

15. Estimate the minimum reduction potential for P680. Estimate the maximum reduction potential for ferredoxin. Explain your answers.

16. Describe the distribution of LHCs between grana (stacked membranes) and stromal lamellae (unstacked membranes) in chloroplasts under a bright sun and in shady light.

*The Dark Reactions*

17. What is the first stable radioactive sugar intermediate seen when $^{14}CO_2$ is added to algae such as *Chlorella*? When the supply of $^{14}CO_2$ is cut off, what compound accumulates? What do these results suggest about the pathway of $CO_2$ incorporation into carbohydrates?

18. Using Figure 19-26, write out the 13 reactions of the Calvin cycle. What is the main difference between Stage I and Stage II reactions? Write a net equation for each stage.

19. Glycolysis and the Calvin cycle are opposing pathways. Which reactions form a potential futile cycle? How are these reactions controlled in plant chloroplasts?

20. What is photorespiration? What is the ultimate result of this process?

21. The concentration of atmospheric $CO_2$ has been increasing for many decades. If this trend continues, how might it affect the relative abundance of $C_3$ and $C_4$ plants?

# 20

# Lipid Metabolism

As we saw in Chapter 9, cells contain a diverse array of lipids. The major functions of lipids include forming a barrier to the extracellular environment, maintaining membrane fluidity, and serving as an important energy store, principally in the form of triacylglycerols. This chapter focuses on the diverse metabolic pathways of cellular lipids. First, the text considers the means by which ingested dietary lipids are degraded, absorbed, and packaged in lipoproteins for transport between tissues. The chapter then presents the oxidative pathways by which long acyl chains are converted to successive two-carbon acetyl-CoA fragments, which are either further oxidized via the citric acid cycle and the electron-transport system, converted to ketone bodies, or used in biosynthesis. Next, mechanisms of fatty acid biosynthesis by successive condensations of two-carbon fragments are presented. The steps leading to the biosynthesis of membrane phospholipids and sphingoglycolipids are outlined, followed by the pathway of cholesterol biosynthesis and its regulation. The reader should contemplate the varied metabolic reactions that result in the formation or degradation of lipids in the context of the nutritive, structural, and regulatory roles that lipids fulfill in the cell.

## Essential Concepts

### Lipid Digestion, Absorption, and Transport

1. The enzymatic digestion of triacylglycerols occurs at the lipid–water interface and is aided by the presence of bile acids, which help solubilize the lipids. Bile acids are cholesterol derivatives that are synthesized in the liver as taurine or glycine derivatives, stored in the gall bladder, and released into the small intestine.

2. Pancreatic lipase hydrolyzes triacylglycerols first to diacylglycerols and then to 2-monoacylglycerols plus free fatty acids. Binding of the enzyme to triacylglycerol at the lipid–water interface requires colipase. The interaction of this protein with lipase produces a hydrophobic surface which promotes binding of the protein complex to the lipid. Phospholipase $A_2$ also degrades phospholipids to lysophospholipids and fatty acids at a lipid–water interface.

3. Micelles containing bile acids promote the absorption of triacylglycerol and phospholipid hydrolysis products by the intestinal mucosa. Inside intestinal cells, fatty acids are complexed to a fatty acid–binding protein that, in effect, increases the aqueous solubility of these hydrophobic compounds.

4. Many lipids are transported between tissues through the circulation as components of lipoproteins. Lipoproteins are globular structures that are composed of a hydrophobic interior containing triacylglycerols and cholesteryl esters encased in an amphiphilic outer layer of protein, phospholipid, and cholesterol. There are five classes of lipoproteins: (a)

chylomicrons; (b) very low density lipoproteins (VLDL); (c) intermediate density lipoproteins (IDL); (d) low density lipoproteins (LDL); and (e) high density lipoproteins (HDL). The densities of lipoprotein classes increase as the quantity of lipid contained in the core decreases.

5.  At least nine apolipoproteins comprise the protein components of human lipoproteins. Most of these have a high content of α helices whose nonpolar and polar residues are on opposite sides of the α helix. The nonpolar faces of the apolipoproteins interact with the phospholipid hydrophobic tails, and the polar faces interact with the phospholipid polar head groups.

6.  Absorbed fatty acids are used for triacylglycerol synthesis in intestinal cells. These triacylglycerols are then incorporated into chylomicrons, which enter the bloodstream via the lymphatic system and eventually provide triacylglycerols to peripheral tissues, chiefly skeletal muscle and adipose tissue. The delivery process converts chylomicrons to much smaller chylomicron remnants, which are taken up by the liver.

7.  The liver synthesizes other lipoproteins, including very low density lipoproteins (VLDL), which transport triacylglycerols and cholesterol from the liver to skeletal muscle and adipose tissue. In the capillaries, triacylglycerols are degraded by lipoprotein lipase to yield fatty acids (which the cells can either oxidize or reincorporate into triacylglycerols) and glycerol (which can be transformed to the glycolytic intermediate dihydroxyacetone phosphate). As they lose their triacylglycerol component, VLDL are converted first to intermediate density lipoproteins (IDL) and then to low density lipoproteins (LDL). Cells take up the cholesterol-rich LDL by endocytosis mediated by the LDL receptor.

8.  Cholesterol is removed from cell surface membranes by high density lipoprotein (HDL), esterifed by HDL-associated lecithin-cholesterol acyltransferase (LCAT) and carried as cholesteryl esters through the bloodstream to the liver. There, HDL unloads its lipid cargo via interaction with a cell surface scavenger receptor.

*Fatty Acid Oxidation*

9.  When energy needs dictate, triacylglycerols stored in adipose tissue are broken down (mobilized) by hormone-sensitive lipase. Released free fatty acids are transported in complex with serum albumin to the liver and other tissues.

10. Before being degraded by oxidation, fatty acids are first activated by the formation of an acyl-CoA in an ATP-dependent reaction catalyzed by thiokinase.

11. Since β oxidation takes place in the mitochondrial matrix, the acyl groups must cross the inner mitochondrial membrane, which is impermeable to fatty acyl-CoA derivatives. Therefore, the acyl group is transferred to carnitine by carnitine palmitoyl transferase I. The resulting acyl-carnitine readily crosses the membrane via a carrier protein. Once in the mitochondrial matrix, the acyl group is transferred back to a CoA molecule by carnitine palmitoyl transferase II, and the liberated carnitine crosses the membrane back to the cytosol.

12. Fatty acyl groups are degraded by a process called β oxidation, in which successive two-carbon fragments are removed as acetyl-CoA units.
    (a) Acyl-CoA dehydrogenase catalyzes formation of a *trans*-2,3 (α,β) double bond. The enzyme's bound FAD is thereby reduced to $FADH_2$.
    (b) Enoyl-CoA hydratase catalyzes the hydration of the double bond to produce a 3-L-hydroxyacyl-CoA.
    (c) 3- L-Hydroxyacyl-CoA dehydrogenase catalyzes the formation of a β-ketoacyl-CoA with the reduction of $NAD^+$ to NADH.
    (d) Thiolase catalyzes the thiolysis of the C2—C3 bond, releasing acetyl-CoA and forming a new acyl-CoA which is two carbons shorter than the starting substrate.
    This sequence of reactions is repeated until the acyl-CoA has been converted entirely to acetyl-CoA. For oxidation of palmitoyl-CoA, the sequence occurs 7 times to yield 8 acetyl-CoA.

13. The oxidation of fatty acids is highly exergonic. For example, palmitate's 8 acetyl-CoA can enter the citric acid cycle, and the $FADH_2$ and NADH generated by β oxidation and the citric acid cycle can be reoxidized by the electron-transport chain, which yields a total of 106 ATP.

14. The oxidation of unsaturated fatty acids requires additional enzymatic reactions to accommodate the double bond at C9 in monounsaturated fatty acids (e.g., oleic acid) and at three-carbon intervals in polyunsaturated fatty acids (e.g., linoleic acid). When a *cis*-3,4 (β,γ) double bond is encountered after several rounds of β oxidation, enoyl-CoA isomerase converts it to a *trans*-2,3 double bond. Oxidation of a polyunsaturated fatty acid also requires a reaction catalyzed by 2,4-dienoyl-CoA reductase, which removes a double bond at the expense of an NADPH. In mammals, the resulting product has a *trans*-3,4 double bond, which must be converted to a *trans*-2,3-double bond by 3,2-enoyl CoA reductase before β oxidation can continue.

15. The final round of β oxidation of odd-chain fatty acids yields propionyl-CoA. This three-carbon compound is converted to succinyl-CoA, a citric acid cycle intermediate, by three reactions:
    (a) Propionyl-CoA carboxylase catalyzes an ATP-dependent carboxylation reaction that requires the coenzyme biotin and produces (*S*)-methylmalonyl-CoA.
    (b) Methylmalonyl-CoA racemase converts (*S*)-methylmalonyl-CoA to (*R*)-methylmalonyl-CoA.
    (c) Methylmalonyl-CoA mutase transforms (*R*)-methylmalonyl-CoA into succinyl-CoA in a reaction that requires the coenzyme 5′-deoxyadenosylcobalamin, which is derived from cobalamin (vitamin $B_{12}$).

16. The methylmalonyl-CoA mutase reaction rearranges the substrate's carbon skeleton. The reaction mechanism features homolytic cleavage of the C—Co bond in the coenzyme so that the C and Co atoms each retain one electron. Such homolytic cleavage is rare in biological systems; in biochemical reactions, bonds are usually broken by heterolytic cleavage in which one of the atoms acquires both electrons. The Co ion therefore functions as a generator of free radicals, which are essential for the reaction.

17.  The peroxisome also carries out β oxidation. In animals, peroxisomes oxidize very long acyl chains (>22 carbons). These are shortened in the peroxisome, and β oxidation is then completed in the mitochondrion. Peroxisomal oxidation differs from oxidation in mitochondria in two ways:
     (a)  No carnitine is required.
     (b)  The first step of acyl-CoA oxidation by acyl-CoA oxidase involves direct transfer of electrons to $O_2$ with formation of hydrogen peroxide ($H_2O_2$) and a *trans*-2,3-enoyl-CoA. Since these electrons do not pass through the electron transport chain, peroxisomal oxidation of fatty acids yields less energy than mitochondrial oxidation.

### Ketone Bodies

18.  Acetyl-CoA, in addition to undergoing oxidation via the citric acid cycle, can also undergo ketogenesis to form acetoacetate, D-β-hydroxybutyrate, and acetone. These water-soluble compounds are collectively called ketone bodies. Acetoacetate and D-β-hydroxybutyrate are important sources of metabolic energy under certain circumstances, such as starvation

19.  Ketogenesis, the formation of ketone bodies from acetyl-CoA, occurs as follows:
     (a)  Two acetyl-CoA units condense to form acetoacetyl-CoA in a reversal of the thiolase reaction.
     (b)  Hydroxymethylglutaryl-CoA synthase catalyzes the condensation of acetoacetyl-CoA with a third acetyl-CoA unit to form β-hydroxy-β-methylglutaryl-CoA (HMG-CoA).
     (c)  HMG-CoA lyase cleaves HMG-CoA to form acetyl-CoA and acetoacetate.
     (d)  β-Hydroxybutyrate dehydrogenase can reduce acetoacetate to β-hydroxybutyrate in an NADH-dependent reaction.
     (e)  Acetoacetate also undergoes spontaneous decarboxylation to form acetone.

20.  Acetoacetate and β-hydroxybutyrate formed in the liver are transported in the bloodstream to peripheral tissues and there are converted into two acetyl-CoA units. Succinyl-CoA supplies the CoA for this process and, when this happens, cannot therefore be utilized in the citric acid cycle to generate GTP and succinate.

### Fatty Acid Biosynthesis

21.  Although fatty acid biosynthesis involves the condensation of successive acetyl-CoA units, this metabolic pathway is distinct from β oxidation in several respects:
     (a)  It is a reductive process.
     (b)  It takes place in the cytosol.
     (c)  It utilizes NADPH as hydrogen donor.
     (d)  It uses the $C_3$ dicarboxylic acid derivative, malonyl-CoA, as its $C_2$ donor.
     (e)  The growing acyl chain is attached to acyl-carrier protein (ACP) rather than to CoA.
     (f)  It employs entirely different enzymes and is independently regulated.

22.  In order for fatty acid biosynthesis to proceed, sufficient amounts of both acetyl-CoA and NADPH must be available in the cytosol. Acetyl-CoA, which is generated by pyruvate dehydrogenase in the mitochondrion, cannot cross the inner mitochondrial membrane to reach the cytosol. Instead, transport occurs by means of the tricarboxylate transport system

in which acetyl-CoA reacts with oxaloacetate to form citrate, which readily crosses the membrane via a transporter. Once in the cytosol, citrate is converted to pyruvate in a series of reactions that liberates acetyl-CoA and also generates NADPH in a 1:1 ratio. Pyruvate then enters the mitochondrion and is converted to oxaloacetate.

23.  The first committed step in fatty acid biosynthesis is catalyzed by acetyl-CoA carboxylase, a biotin-dependent enzyme, which converts acetyl-CoA to malonyl-CoA. In the first reaction step, bicarbonate becomes covalently attached to the biotin prosthetic group. This "activated" $CO_2$ is then transferred from biotin to acetyl-CoA, forming malonyl-CoA.

24.  In eukaryotic cells, acetyl-CoA carboxylase is regulated both allosterically and by covalent modification. Citrate allosterically increase the $V_{max}$ of the enzyme, whereas long-chain acyl-CoAs inhibit the reaction. AMP-dependent kinase phosphorylates Ser 79, thereby inactivating the enzyme. Glucagon acts through a cAMP-dependent pathway to inhibit the enzyme, possibly by preventing its dephosphorylation. In contrast, insulin enhances enzyme activity by promoting its dephosphorylation.

25.  Fatty acid synthesis from acetyl-CoA and malonyl-CoA takes place by successive cycles of six enzymatic reactions. In *E. coli,* these reactions are catalyzed by separate enzymes. In eukaryotic cells, the process occurs within fatty acid synthase, a protein whose sequence contains all the required enzyme activities and an ACP domain. Fatty acid synthase catalyzes the following reactions:
    (a) In the first two "priming" reactions, the synthase is primed by transferring both an acetyl group from acetyl-CoA and a malonyl group from malonyl-CoA to the terminal SH of the phosphopantetheine prosthetic group of ACP.
    (b) In the third reaction, the acetyl group is transferred to a Cys –SH group of the enzyme. Malonyl-ACP is decarboxylated and condenses with the acetyl group to form a β-ketoacyl-ACP.
    (c) In reactions 4 to 6, the β-keto group is reduced by NADPH to form a hydroxyl group, the hydroxyl group is eliminated in a dehydration reaction to form a double bond, and a second reduction by NADPH reduces the double bond to produce butyryl-ACP.
    (d) This four-carbon acyl chain is then transferred from the ACP to the enzyme Cys SH. The cycle starts anew by the transfer of another malonyl group to the now vacant ACP site in the synthase.
    Seven cycles are required to synthesize a $C_{16}$ acyl chain (palmitate), a process that consumes 8 acetyl-CoA and 14 NADPH (which are provided both by glucose oxidation via the pentose phosphate pathway and by operation of the tricarboxylate transport system). The completed saturated acyl chain is released by thioester cleavage catalyzed by the seventh enzyme activity associated with the synthase.

26.  Mammalian fatty acid synthase consists of two identical monomers arranged such that the Cys-linked acyl group in one polypeptide is located close to the phosphopantetheine moiety in the other. This allows the synthase dimer to simultaneously synthesize two acyl chains.

27.  Palmitate may be lengthened and desaturated by elongases and desaturases, respectively. However, because animals, unlike plants, cannot introduce a double bond beyond C9 in an

acyl chain, linoleic acid (9,12-*cis*-octadecadienoic acid) must be obtained from the diet. It is therefore called an essential fatty acid.

28. The synthesis of triacylglycerols begins with successive acylations of glycerol-3-phosphate to yield lysophosphatidic acid and then phosphatidic acid. Dephosphorylation produces 1,2-diacylglycerol, which then accepts another fatty acyl group. An alternative pathway involves acylation of dihydroxyacetone phosphate, followed by an NADPH-dependent reduction to produce lysophosphatidic acid.

*Regulation of Fatty Acid Metabolism*

29. Triacylglycerol metabolism, like that of glycogen, is important for the well-being of the whole organism and is regulated by hormones. Fatty acid oxidation is controlled by the rate of triacylglycerol hydrolysis in adipose tissue by hormone-sensitive triacylglycerol lipase. The lipase is stimulated by glucagon through cAMP-dependent phosphorylation, which also inhibits acetyl-CoA carboxylase, so fatty acid oxidation is enhanced and fatty acid synthesis is inhibited.

30. Insulin opposes the effects of glucagon by reducing cAMP levels, thereby inactivating triacylglycerol lipase and stimulating fatty acid synthesis. The ratio of glucagon to insulin therefore controls the status of fatty acid metabolism.

31. A rise in the level of malonyl-CoA also enhances fatty acid synthesis while depressing fatty acid oxidation, because this compound inhibits carnitine palmitoyl transferase I, thereby preventing fatty acyl-CoA from entering the mitochondrion.

32. These mechanisms of short-term regulation are complemented by long-term hormonal regulation of fatty acid metabolism, which alters the levels of key enzymes such as acetyl-CoA carboxylase and triacylglycerol lipase.

*Synthesis of Other Lipids*

33. The formation of choline- and ethanolamine-containing glycerophospholipids involves three enzymatic steps:
    (a) The phosphorylation of the nitrogen-containing base.
    (b) Activation of the phosphocholine or phosphoethanolamine by CTP to form CDP–choline or CDP–ethanolamine.
    (c) Transfer of the activated base to 1,2-diacylglycerol to form the phospholipid.

34. The synthesis of phosphatidylglycerol and phosphatidylinositol involves activation of phosphatidic acid by reaction with CTP to form CDP–diacylglycerol. The activated lipid is then transferred to glycerol-3-phosphate or inositol. Cardiolipin is synthesized from two phosphatidylglycerol molecules

35. Enzymes that acylate glycerol-3-phosphate show a preference for introducing a saturated fatty acid in position 1 and an unsaturated fatty acid in position 2. However, there must be

additional remodeling reactions, catalyzed by phospholipases and acyltransferases and which result in the exchange of acyl groups, to account for the fatty acid compositions of all membrane phospholipids.

36. In sphingolipid synthesis, ceramide (*N*-acylsphingosine) is formed in four reactions from palmitoyl-CoA and serine. Phosphatidylcholine donates its phosphocholine group to ceramide to produce sphingomyelin. Ceramide can also be glycosylated, with UDP–glucose or UDP–galactose serving as the sugar donor, to form cerebrosides. Additional glycosylation of cerebrosides generates more complex sphingoglycolipids such as gangliosides.

37. Prostaglandin synthesis from arachidonic acid begins with the formation of a cyclopentane ring within the linear fatty acid. This reaction is catalyzed by prostaglandin $H_2$ synthase, which contains two catalytic activities, a cyclooxygenase and a peroxidase. Prostaglandin $H_2$ synthase has three isoforms: COX-1, COX-2, and the recently discovered COX-3.

38. Some common anti-inflammatory drugs interact with the various isoforms of prostaglandin $H_2$ synthase:
    (a) Aspirin irreversibly acetylates a specific serine residue of the enzyme, blocking access of arachidonic acid to the active site;
    (b) Ibuprofen and acetaminophen interact noncovalently and block the active site of prostaglandin $H_2$ synthase;
    (c) Vioxx and Celebrex specifically inhibit COX-2 but not COX-1.
    (d) Aspirin and ibuprofen are relatively nonspecific, but acetaminophen inhibits only COX-3.

*Cholesterol Metabolism*

39. All 27 carbon atoms of cholesterol are derived from acetate. The major stages in cholesterol formation are:
    (a) Acetate is converted to hydroxymethylglutaryl-CoA (HMG-CoA) and then via mevalonate to an isoprene unit, isopentenyl pyrophosphate.
    (b) Condensation of six isoprene units forms squalene, a linear 30-carbon compound.
    (c) Squalene is oxidized and cyclized to form lanosterol.
    (d) Further modification and removal of three carbons yields cholesterol.

40. After its synthesis in the liver, cholesterol may be transformed into bile acids or converted to cholesteryl esters for transport via lipoproteins. Cholesterol is also the precursor of steroid hormones.

41. Cholesterol is essential for cell membrane integrity, yet excess cholesterol may be harmful to the organism. Thus, its biosynthesis, utilization, and cellular distribution are carefully controlled. Cholesterol metabolism is regulated in several ways.

42. Cholesterol biosynthesis is controlled by HMG-CoA reductase, which catalyzes the rate-limiting conversion of HMG-CoA to mevalonate, in both a short- and long-term manner. In the short term, this enzyme is inactivated by phosphorylation (catalyzed by AMP-

dependent kinase, the same enzyme that inactivates acetyl-CoA carboxylase) and activated by dephosphorylation.

43. The primary regulatory mechanism for HMG-CoA reductase activity is long-term regulation of cellular enzyme concentration such that the level of this enzyme is in inverse proportion to the concentration of cholesterol.

   The control occurs at the level of transcription of the HMG-CoA reductase gene. The expression of this and 20 other genes participating in cholesterol biosynthesis and transport involves the binding of a specific regulatory protein, or transcription factor (see Chapter 28), at a DNA sequence called the sterol regulatory element (SRE). A SRE binding protein (SREBP) binds to the SRE to stimulate transcription.

44. Inactive SREBP, a transmembrane protein of the ER, is noncovalently bound to SREBP cleavage-activating protein (SCAP), another transmembrane protein, which remains localized to the ER when bound to cholesterol (as occurs when the cellular cholesterol concentration is high). As the cholesterol concentration falls, cholesterol dissociation from SCAP results in a conformational change in SCAP that promotes the migration of the SCAP–SREBP complex into the Golgi apparatus. There, two proteases cleave SREBP, releasing it into the cytosol and allowing its N-terminal fragment to migrate to the nucleus, where it interacts with available SRE sites on DNA.

45. The statins are a group of drugs that inhibit HMG-CoA reductase via complementary interactions at the active site of the enzyme.

46. Cholesterol transport and removal from blood is governed largely by the activity of LDL receptors on the liver cell surface, which in turn depend on the number of receptors and hence on the rate of LDL receptor synthesis.

47. Elevated levels of blood cholesterol, as part of LDL, are strongly correlated with atherosclerosis, a risk factor for cardiovascular disease. High blood cholesterol results from the genetic disease familial hypercholesterolemia (which is characterized by an absence of LDL receptors) or from high dietary cholesterol intake (which tends to repress LDL receptor synthesis). In Tangier disease, a defective transport protein prevents cholesterol efflux from cells, which also contributes to atherosclerosis.

## Questions

1. From a chemical perspective, why is the energy content of fats so much greater than that of carbohydrates or proteins?

### Lipid Digestion, Absorption, and Transport

2. Why do individuals who have their gall bladders removed encounter difficulties in digesting large amounts of fats?

3. How do pancreatic lipase and phospholipase $A_2$ differ in the mechanism by which they promote hydrolysis of glycerolipids at interfaces?

4. Match each term on the left with its description on the right.

    ____ Bile acid            A.   Helps bind lipase to the lipid–water interface

    ____ Intestinal fatty acid      B.   Hydrolyzes phospholipids to yield binding protein lysophospholipids and free fatty acids

    ____ Albumin              C.   Forms micelles that take up nonpolar lipid degradation products and helps transport them through the intestinal wall

    ____ Phospholipase $A_2$       D.   Transports lipid digestion products through the lymphatic system and then the bloodstream to the tissues

    ____ Colipase             E.   Transports through the bloodstream free fatty acids released from adipose tissue stores

    ____ Chylomicrons        F.   Forms complexes with free fatty acids to shield intestinal cells from their detergent-like effects and increase their effective solubility in the blood.

5. Determine the order of the following events involving lipoprotein-mediated transport of dietary triacylglycerols and cholesterol.

    ____ Triacylglycerols are removed from circulating VLDL by lipoprotein lipase.
    ____ Chylomicrons are transported through the lymphatic system and enter the bloodstream.
    ____ Chylomicrons are formed in the intestinal mucosa.
    ____ Cells take up cholesterol via receptor-mediated LDL endocytosis.
    ____ Chylomicrons are degraded by lipoprotein lipase to chylomicron remnants.
    ____ LDL components are rapidly degraded by lysosomal enzymes.
    ____ VLDL are synthesized in liver.

6. Why are cholesteryl esters located in the interior of a lipoprotein while cholesterol is located on the exterior?

7. If the LDL receptor underwent a mutation that increased its affinity for apoB-100, what would be the effect on serum cholesterol levels?

*Fatty Acid Oxidation*

8.   What products would be isolated from urine when dogs are fed (a) phenylheptanoic acid and (b) phenyloctanoic acid? How does this experiment, originally performed by Knoop, shed light on the process of fatty acid oxidation?

9.   A patient develops an enlarged fatty liver and low blood glucose. These symptoms can be partially overcome only when massive amounts of carnitine are included in the diet. What enzyme defect might be responsible for this problem?

10.   Which of the following statements is (are) correct? In the β oxidation of fatty acids,
   (a)   The activation of fatty acids by acyl-CoA synthetase is driven by the hydrolysis of pyrophosphate.
   (b)   The reaction catalyzed by acyl-CoA dehydrogenase uses FAD as electron acceptor.
   (c)   The reaction catalyzed by enoyl-CoA hydratase produces a 3-D-hydroxyacyl-CoA.

11.   Calculate the yield of ATP when one mole of stearic acid is completely oxidized to $CO_2$ and $H_2O$.

12.   In the β oxidation of oleic acid, the presence of the double bond at the _____ position necessitates modification of the β oxidation cycle. The modification occurs after the _____ round of β oxidation because the _____ enoyl-CoA is not a substrate for _____. The problem is overcome by the action of the enzyme _____ which converts the _____ bond to a _____ bond so that β oxidation can continue.

13.   In the β oxidation of linoleic acid, a second difficulty presents itself because of the additional double bond in the _____ position. In this case, after the fifth round of oxidation, a_____ CoA is formed, which is a poor substrate for _____. In mammals, two reactions are necessary for β oxidation to resume. In the first, NADPH-dependent _____ reductase reduces one double bond to form a _____ , and then the action of a _____ isomerase yields a _____ CoA that can participate in β oxidation.

14.   Describe the role of cobalamin (vitamin $B_{12}$) in the methylmalonyl-CoA mutase reaction.

15.   Explain why succinyl-CoA arising from the oxidation of odd-chain fatty acids cannot be directly oxidized by the citric acid cycle. How can it be further degraded?

16.   Why does the β oxidation of fatty acids in peroxisomes yield less ATP than the corresponding process in mitochondria?

*Ketone Bodies*

17.   Infants have high levels of ketone bodies in their blood and abundant 3-ketoacyl-CoA transferase in their tissues (except in liver) prior to weaning. What nutritional advantage does this confer?

18. Calculate the ATP that is produced when linoleic acid (9,12-octadecadienoic acid; 18:2) is (a) oxidized to $CO_2$ and $H_2O$ or (b) converted to the ketone body acetoacetate in the liver and then oxidized to $CO_2$ and $H_2O$ in the peripheral tissues.

19. Write equations for ketone body synthesis and degradation. What is the net result of combined synthesis and degradation?

*Fatty Acid Biosynthesis*

20. A rat liver cytosol preparation is incubated with acetate and all cofactors necessary for the biosynthesis of palmitate. Which carbon atoms of palmitate will be isotopically labeled when the preparation contains (a) $H^{14}CO_3^-$ or (b) $^{14}CH_3COO^-$?

21. What are the advantages of the multifunctional dimeric structure of animal fatty acid synthase in the formation of fatty acids?

22. Which of the following statements is(are) correct? The transport of acetyl-CoA from the mitochondrial matrix to the cytosol for fatty acid biosynthesis:
    (a) is necessary because the inner mitochondrial membrane is impermeable to acetyl-CoA.
    (b) uses the tricarboxylate transport system, which operates in both directions.
    (c) can generate equal numbers of acetyl-CoA and NADPH molecules in the cytosol .

*Regulation of Fatty Acid Metabolism*

23. Match each term on the left with its function in the short-term regulation of fatty acid metabolism.

    ____ Citrate                A. Activates acetyl-CoA carboxylase
    ____ cAMP-dependent         B. Inhibits carnitine palmitoyl transferase phosphorylation
    ____ Palmitate              C. Activates hormone-sensitive lipase
    ____ Malonyl-CoA            D. Inhibits acetyl-CoA carboxylase

24. Explain the role of AMP-dependent protein kinase in regulating fatty acid metabolism.

*Synthesis of Other Lipids*

25. Match each term on the left with its metabolic role on the right.

    ____ CDP–diacylglycerol        A. Precursor of cardiolipin
    ____ Ceramide                  B. Precursor of phosphatidylcholine
    ____ Phosphatidylglycerol      C. Contains a vinyl ether linkage
    ____ Plasmalogen               D. Precursor of sphingomyelin
    ____ CDP–Choline               E. Precursor of phosphatidylinositol

26. When tissues are incubated *in vitro* with $^{32}$P-labeled $P_i$, the incorporation of radioactivity into phospholipids can be readily demonstrated. High concentrations of the drug propranolol dramatically alter the pattern of phospholipid labeling, such that radioactivity in phosphatidylcholine and phosphatidylethanolamine markedly declines while that in phosphatidylinositol and phosphatidylglycerol is greatly elevated. From this information and your knowledge of phospholipid biosynthetic pathways, determine which enzymatic reaction is principally affected by propranolol and explain how the drug alters phospholipid biosynthesis.

27. Fumonisin B is a mycotoxin that inhibits the synthesis of dihydroceramide from sphinganine. When this inhibitor is added to cells in culture that actively synthesize spingoglycolipids, predict how this inhibitor would affect the synthesis of cerebrosides.

28. A metabolic disease may result from an enzyme deficiency that prevents the synthesis of an essential metabolite or that prevents the normal breakdown of a metabolite. Which type of defect is exhibited in Tay–Sachs disease and what enzyme is affected?

29. In lipid metabolism, the production of _____ is blocked by aspirin, which inhibits the reaction catalyzed by _____.

## Cholesterol Metabolism

30. What are the two principal ways in which the cholesterol needs of many tissues are met?

31. Place the following intermediates in cholesterol biosynthesis in the correct order: squalene, farnesyl pyrophosphate, 2,3-oxidosqualene, hydroxymethylglutaryl-CoA (HMG-CoA), lanosterol, mevalonate, geranyl pyrophosphate.

32. Explain how the regulation of HMG-CoA reductase, the principal control site for cholesterol synthesis, can conserve cellular ATP.

33. Which of the following statements is(are) correct? Inhibitors of HMG-CoA reductase are used to decrease serum cholesterol levels. Such inhibitors would also:
    (a) reduce the intracellular level of mevalonate.
    (b) reduce the synthesis of LDL receptors.
    (c) reduce the synthesis of ubiquinone.

34. Match the following biochemical role or event with the substance listed on the right.

| | |
|---|---|
| ____ Cholesterol | A. Exposes proteolytic site in the cytosolic domain of SREBP |
| ____ Site-1 protease | B. Interacts with SRE sequences |
| ____ Site-2 protease | C. Binds to SCAP |
| ____ WD repeat | D. Required for transport to the nucleus |
| ____ bHLH | E. Domain of SCAP that binds to SREBP |

# 21

# Amino Acid Metabolism

This chapter surveys nitrogen metabolism, with its key focus on amino acid metabolism. The chapter begins with a brief discussion of protein turnover and pathways of protein degradation, including the lysosomal and ubiquitin-dependent pathways. The following sections describe the mechanisms by which amino acids are catabolized, beginning with amino acid deamination by transamination and oxidative deamination. Next is the urea cycle, a pathway that transfers an ammonium ion and the amino group of aspartate through a series of intermediates to arginine, which is cleaved to generate urea and regenerate the amino acid ornithine. The chapter then discusses the degradation of amino acids, which yield common intermediates that can be used for gluconeogenesis (from glucogenic amino acids) or fatty acid synthesis (from ketogenic amino acids). Many of these degradative pathways involve enzymatic reactions mediated by the coenzyme pyridoxal-5′-phosphate (PLP), which catalyzes transamination reactions, decarboxylation reactions, and reactions that break the $C_\alpha$—$C_\beta$ covalent bond of amino acids. This coenzyme, like thiamine pyrophosphate, is a resonance-stabilized electron sink that stabilizes carbanions. Some defects in amino acid metabolism are discussed in Boxes 21-1 and 21-2.

The sections in the final part of the chapter include the anabolic reactions of amino acid metabolism. Amino acid biosynthetic reactions are divided into two groups, those synthesizing the nonessential (for mammals) and essential amino acids. Within each group, the derivation of amino acids from common precursors is emphasized. The chapter then examines the synthesis of heme, which is assembled in a modular fashion beginning with glycine and succinyl-CoA. Disorders of heme metabolism are discussed in the text and in Box 21-3. The following section describes the conversion of select amino acids to hormones and neurotransmitters. This chapter concludes with a discussion of how microorganisms fix $N_2$ into $NH_3$ by an energetically costly process.

## Essential Concepts

### Protein Degradation

1. Proteins are continuously synthesized and degraded into amino acids. This serves three functions:
   (a) Proteins serve as a form of long-term energy storage that is utilized to provide gluconeogenic precursors and ketone bodies.
   (b) Degradation eliminates abnormal and damaged proteins, whose accumulation is potentially hazardous to cell function.
   (c) The regulation of metabolic activity under changing physiological conditions requires the degradation of one set of regulatory proteins and its replacement by a new set.

2.  Proteins have highly variable half-lives, ranging from a few minutes to several days. The half-life of a protein varies with the identity of its N-terminal residue, a phenomenon which is referred to as the N-end rule. Other signals are contained in the protein's amino acid sequence. For example, proteins with segments rich in Pro (P), Glu (E), Ser (S), and Thr (T), which are called PEST sequences, have very short half-lives.

3.  Intracellular proteins are degraded either in lysosomes or, after ubiquitination, in proteasomes.
    (a) Lysosomes degrade proteins (and other biological molecules) that are taken up by endocytosis, and they recycle cellular proteins that are enclosed in vacuoles. In well-nourished cells, lysosomal protein degradation is nonselective, but in starved cells, this pathway is more selective.
    (b) Intracellular proteins may be tagged for degradation by covalently linking them to a small protein called ubiquitin. Ubiquitination requires three enzymes: ubiquitin-activating enzyme (E1), ubiquitin-conjugating enzyme (E2), and ubiquitin-protein ligase (E3). E1 forms a thioester bond with ubiquitin in a reaction that consumes one ATP. E2 transfers this activated ubiquitin from E1 to itself, forming another thioester bond. E3 transfers activated ubiquitin to the Lys ε-amino group of a previously selected target protein.

4.  Ubiquitinated proteins are degraded in an ATP-dependent manner inside a large multiprotein complex called the 26S proteasome. The complex has a barrel-like core (20S subunit) that degrades proteins into ~8 residue fragments, which diffuse out of the proteasome and are subsequently hydrolyzed by cytosolic peptidases. The 19S cap recognizes ubiquitinated proteins and unfolds them for degradation.

*Amino Acid Deamination*

5.  The degradation of an amino acid usually begins with its transamination by a pyridoxal-5′-phosphate (PLP)-containing enzyme, in which the amino group is transferred to an α-keto acid (usually α-ketoglutarate). Transamination utilizing α-ketoglutarate generates glutamate, which can be subsequently oxidatively deaminated by glutamate dehydrogenase to regenerate α-ketoglutarate and release a free ammonium ion. High concentrations of ammonia are toxic, so the steady-state level of ammonia is kept low.

6.  PLP functions as an electron sink in the resonance stabilization of a $C_\alpha$ carbanion derived from an amino acid. This carbanion is an intermediate in amino acid transamination, decarboxylation, and the breakage of the $C_\alpha$—$C_\beta$ bond.

7.  Glutamate dehydrogenase catalyzes the oxidative deamination of glutamate to yield α-ketoglutarate and ammonium ion.

*The Urea Cycle*

8.  Living organisms excrete excess nitrogen either as free ammonia (aquatic organisms only), urea (most terrestrial vertebrates), or uric acid (birds and terrestrial reptiles). The overall reaction of the urea cycle is

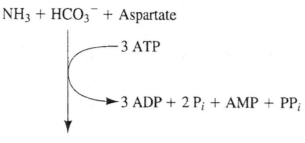

NH$_3$ + HCO$_3^-$ + Aspartate

― 3 ATP

► 3 ADP + 2 P$_i$ + AMP + PP$_i$

Urea + Fumarate

9. The urea cycle takes place in the liver of adult vertebrates. Of the five reactions in the urea cycle, two are mitochondrial and three are cytosolic.
   (a) Ammonia (a product of the glutamate dehydrogenase reaction) reacts with HCO$_3^-$ and two ATP to form the "activated" product carbamoyl phosphate in a reaction catalyzed by mitochondrial carbamoyl phosphate synthetase.
   (b) Transfer of the carbamoyl group to ornithine results in the formation of citrulline, which is transported out of the mitochondrion to the cytosol.
   (c) Condensation of aspartate with citrulline in an ATP-dependent reaction forms argininosuccinate.
   (d) Argininosuccinate is cleaved into fumarate and arginine.
   (e) In the final reaction, hydrolytic cleavage of arginine yields urea and regenerates ornithine, which is transported back into the mitochondrion.

10. The urea cycle is regulated by the activity of carbamoyl phosphate synthetase I, which is allosterically activated by N-acetylglutamate. Increasing concentrations of glutamate, resulting from transamination during protein degradation, stimulate the synthesis of N-acetylglutamate, thereby boosting urea cycle activity to increase the excretion of amino groups as urea.

*Breakdown of Amino Acids*

11. The carbon skeletons of the 20 "standard" amino acids can be completely oxidized to CO$_2$ and H$_2$O. They can also be used to synthesize glucose or fatty acids:
    (a) Glucogenic amino acids are broken down to pyruvate, α-ketoglutarate, succinyl-CoA, fumarate, and oxaloacetate, which can be used as glucose precursors.
    (b) Ketogenic amino acids can be broken down to the ketone body acetoacetate or to acetyl-CoA, which can be used for fatty acid synthesis.

12. The amino acids can be grouped according to their common degradative products.
    (a) Alanine, cysteine, glycine, serine, and threonine are degraded to pyruvate.
    (b) Asparagine and aspartate are degraded to oxaloacetate.
    (c) Arginine, glutamate, glutamine, histidine, and proline are degraded to α-ketoglutarate.
    (d) Isoleucine, methionine, and valine are degraded to succinyl-CoA
    (e) Leucine and lysine are degraded to acetoacetate and acetyl-CoA.
    (f) Tryptophan is degraded to alanine and acetoacetate.
    (g) Phenylalanine and tyrosine are degraded to fumarate and acetoacetate.

13. Important cofactors involved in the degradation of amino acids are PLP, tetrahydrofolate (THF), biopterin, and *S*-adenosylmethionine (SAM).
    (a) PLP mediates the catalytic breakdown of the $C_\alpha$—$C_\beta$ covalent bond in threonine to generate acetaldehyde and glycine.
    (b) THF is composed of the following functional groups: 2-amino-4-oxo-6-methylpterin, *p*-aminobenzoic acid, and several glutamates. THF is derived from the vitamin folic acid, which must be reduced to form THF. THF serves as a one-carbon carrier in several reactions involved in amino acid catabolism. The carbon unit may exist in one of several oxidation states ranging from a relatively oxidized formyl group to a methyl group.
    (c) Biopterin is a redox cofactor involved in the oxidation of phenylalanine to tyrosine. Biopterin contains a pteridine ring that is similar to the isoalloxazine ring of the flavin coenzymes.
    (d) The reaction of ATP with methionine yields SAM, a potent methylating reagent in several reactions in which the transfer of a methyl group from THF is not favored.

14. The catabolism of the branched-chain amino acids (isoleucine, leucine, and valine), lysine, and tryptophan employs enzymatic conversions resembling those in the $\beta$ oxidation of fatty acids and in the oxidative decarboxylation of $\alpha$-keto acids such as pyruvate and $\alpha$-ketoglutarate.

## *Amino Acid Biosynthesis*

15. Amino acids that cannot be synthesized by mammals and that must therefore be obtained from their diet are called essential amino acids. Nonessential amino acids can be synthesized from common intermediates:
    (a) Alanine, aspartate, and glutamate are formed by one-step transamination reactions. Asparagine and glutamine are formed by amidation of aspartate and glutamate. The activation of glutamate by glutamine synthetase, which occurs prior to its amidation, is a key regulatory point in bacterial nitrogen metabolism.
    (b) Glutamate gives rise to proline and arginine (which is also considered to be an essential amino acid because, when synthesized, it is largely degraded to urea).
    (c) 3-Phosphoglycerate is the precursor of serine, which can be converted to cysteine and glycine. Glycine can also be produced from $CO_2$, $NH_4^+$, and $N^5,N^{10}$-methylene-THF.

16. The essential amino acids can be categorized into four groups based on their synthetic pathways:
    (a) The aspartate family. Aspartate serves as the precursor for the synthesis of lysine, threonine, and methionine. This pathway also produces homoserine and homocysteine.
    (b) The pyruvate family. Pyruvate serves as a precursor for the synthesis of valine, leucine, and isoleucine.
    (c) Aromatic amino acids. Phosphoenolpyruvate and erythrose-4-phosphate serve as the precursors for tyrosine, phenylalanine, and tryptophan. Serine is also required for the synthesis of tryptophan.
    (d) Histidine. 5-Phosphoribosyl-$\alpha$-pyrophosphate is a precursor for the synthesis of histidine.

## Other Products of Amino Acid Metabolism

17. Heme synthesis occurs in a modular fashion similar to the synthesis of cholesterol. Heme synthesis begins in the mitochondrion with the condensation of glycine and succinyl-CoA. The product, δ-aminolevulinic acid (ALA), diffuses into the cytosol, where two ALA condense to form porphobilinogen (PBG). Four PBG then condense to form the porphyrin uroporphyrinogen III. Decarboxylation reactions form coproporphyrinogen III, which diffuses into the mitochondrion and is converted to heme.

18. Heme synthesis occurs in erythroid cells and in the liver. In the liver, heme acts by feedback inhibition to regulate ALA synthase activity. In reticulocytes (immature red blood cells), heme stimulates the synthesis of globin and possibly itself. Once heme synthesis is "switched on," it appears to occur at its maximal rate.

19. Red blood cells survive in the bloodstream for about 120 days and then are removed by the spleen and destroyed. Heme catabolism results in the formation of bilirubin, which is transported to the gall bladder for secretion into the intestinal tract. Bilirubin is converted by microbes in the colon into urobilinogen then into stercobilin, which gives the feces their red-brown pigment. Some urobilinogen is reabsorbed and converted in the kidneys to urobilin, which gives urine its characteristic yellow color.

20. Several hormones and neurotransmitters are derived from amino acids. PLP-mediated decarboxylation forms histamine from histidine and γ-aminobutyric acid (GABA) from glutamate. Decarboxylation and hydroxylation of tryptophan yield serotonin. Hydroxylated tyrosine is the precursor of L-DOPA and melanin. PLP-dependent decarboxylation of L-DOPA yields dopamine, which is hydroxylated to form norepinephrine. Donation of a methyl group from SAM to norepinephrine yields epinephrine.

21. Nitric oxide (NO) is produced by a five-electron oxidation of arginine. NO, a relatively stable gaseous free radical, is synthesized by vascular endothelial cells and induces vasodilation. A variety of cells, including neuronal cells and leukocytes, synthesize NO, which acts as a signaling molecule and is involved in the generation of antimicrobial hydroxyl radicals. Sustained NO release is implicated in endotoxic shock and neuronal damage.

## Nitrogen Fixation

22. $N_2$, an extremely stable gaseous molecule, can be reduced, or fixed, by species of bacteria called diazotrophs, such as those of the genus *Rhizobium,* which live symbiotically in the root nodules of leguminous plants. Diazotrophs convert $N_2$ to $NH_3$ by the following net reaction

$$N_2 + 8\,H^+ + 8\,e^- + 16\,ATP + 16\,H_2O \rightarrow 2\,NH_3 + H_2 + 16\,ADP + 16\,P_i$$

The ammonia formed is added to α-ketoglutarate or glutamate by glutamate dehydrogenase or glutamine synthetase, respectively.

23.   Nitrogenase, which catalyzes the reduction of $N_2$ to $NH_3$, is a two-subunit enzyme. Its Fe-protein contains one [4Fe–4S] cluster and two ATP-binding sites, and its MoFe-protein contains a P-cluster (a [4Fe–4S] cluster linked to a [4Fe–3S] cluster) and a FeMo cofactor (a [4Fe–3S] cluster and a [Mo–3Fe–3S] cluster bridged by 3 sulfide ions). The electrons to reduce $N_2$ come from oxidative or photosynthetic reactions, depending on the species. Although electron transfer must occur at least 8 times per $N_2$ molecule fixed (requiring a total of 16 ATP), the actual physiological cost of $N_2$ fixation is as high as 20–30 ATP.

24.   Several of the enzymes involved in amino acid metabolism, including *E. coli* carbamoyl phosphate synthetase I, tryptophan synthase, and glutamate synthetase, exhibit channeling, in which reaction intermediates travel some distance from one active site to the next without dissociating from the enzyme. Channeling increases the rate of a metabolic pathway by generating high local concentrations of intermediates and by preventing their loss or degradation.

## Questions

### Protein Degradation

1.   Indicate whether the following statements are true:
   (a)   Proteins with the sequence Lys-Phe-Glu-Arg-Gln are selectively degraded by proteasomes.
   (b)   Proteins containing sequences rich in Pro, Glu, Ser, and Thr often have short half-lives.
   (c)   The addition of ubiquitin protects segments of a protein from proteolysis.
   (d)   Lysosomal proteases degrade only extracellular proteins that enter the cell by endocytosis.
   (e)   The ubiquitin-transfer reactions catalyzed by E2 and E3 do not require the input of free energy in the form of ATP.

### Amino Acid Deamination

2.   Draw the chemical structures for the products of transamination reactions involving α-ketoglutarate and (a) threonine, (b) isoleucine, and (c) glycine.

3.   In what form is PLP found in aminotransferases prior to reacting with an amino acid? In what form is this coenzyme found after the release of the α-keto acid?

4.   Aminotransferase reactions occur via a Ping Pong mechanism (see p. 367.) Use Cleland notation to describe the reaction shown in Figure 21-8, including substrates, intermediates, and products.

5.   In which direction would you expect flux through the glutamate dehydrogenase reaction in starved individuals? Why?

*The Urea Cycle*

6.  Write the overall equation for the formation of urea. What is the total free energy cost (in terms of "high-energy" phosphoanhydride bonds) per mole of urea synthesized?

7.  Which reactions of the urea cycle take place in the mitochondrion? Which step is the first committed step?

8.  Complete the diagram below, which shows the flux of alanine's amino group from its entry into the liver to its exit as urea.

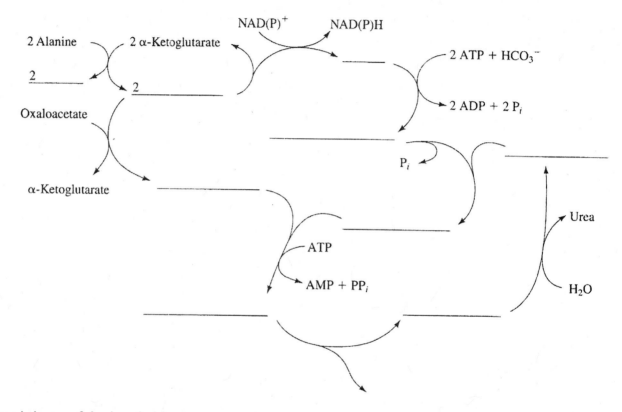

*Breakdown of Amino Acids*

9.  Which amino acids yield citric acid cycle intermediates upon transamination?

10. Reactions that involve PLP can be classified according to which bond to the alpha carbon is broken. For the amino acid–PLP adduct shown below, indicate what kind of reaction involves cleavage of bonds *a*, *b*, and *c* and provide an example of each.

Amino acid–PLP Schiff base
(aldimine)

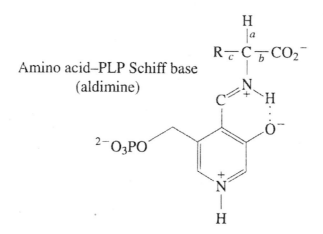

11. Many glucogenic amino acids are broken down to a citric acid cycle intermediate, which merely increases the catalytic activity of the cycle. How does the cell obtain a net yield of glucose from such amino acids?

12. Examine Figure 21-21 for the degradation of isoleucine, valine, and leucine.
    (a) Which reaction is analogous to the reaction catalyzed by the pyruvate dehydrogenase complex?
    (b) Which reaction is analogous to the reaction catalyzed by acyl-CoA dehydrogenase in fatty acid oxidation?

13. Some forms of maple syrup urine disease are amenable to treatment with large doses of dietary thiamine. What metabolic defect is being treated in these cases?

14. Which coenzymes mediate one-carbon transfer reactions?

15. How do sulfonamides selectively inhibit bacterial growth? Why aren't animals also adversely affected by these antibiotics?

16. *S*-adenosylmethionine (SAM) is a key methylating agent in several physiological processes, but its regeneration depends on the presence of another methylating agent. How is SAM regenerated?

17. Calculate the free energy yield for the complete oxidation of the carbon skeletons of (a) aspartate, (b) glutamine, and (c) lysine. Express your answer in ATP equivalents and assume that NADPH is equivalent to NADH.

Amino Acid Biosynthesis

18. Rationalize the significance of $\alpha$-ketoglutarate as an activator of glutamine synthetase.
19. The four intermediates of the urea cycle are $\alpha$-amino acids.
    (a) Which one of these is considered to be an essential amino acid in children?
    (b) Outline a pathway by which adults can synthesize this amino acid from glucose.

20. Chorismate is the seventh intermediate in the synthesis of tryptophan from phosphoenolpyruvate and erythrose-4-phosphate (Figure 21-34). Why is it considered a branch point in amino acid biosynthesis?

21. What is the advantage of channeling? Why is it especially important for indole in the tryptophan synthase reaction?

*Other Products of Amino Acid Metabolism*

22. Porphyrin is synthesized in a modular fashion, much like cholesterol. How many molecules of succinyl-CoA are required?

23. How many carbons of heme would be labeled with $^{14}C$ if the starting material included (a) glycine with a $^{14}C$-labeled carboxyl group and (b) succinyl-CoA with a $^{14}C$-labeled carboxyl group?

24. Heme oxygenase converts heme to the linear _____ and releases _____ and _____.

25. Bilirubin serves as an antioxidant (i.e., it undergoes oxidation to protect other cell components from oxidation). How is bilirubin regenerated after its oxidation?

*Nitrogen Fixation*

26. In mitochondrial respiration, electrons transferred from NADH and $FADH_2$ reduce $O_2$ to $H_2O$. Why can't these same coenzymes participate in the reduction of $N_2$ to $NH_3$?

# 22

# Mammalian Fuel Metabolism: Integration and Regulation

This chapter summarizes and integrates the major pathways involved in fuel metabolism. It first examines the variations of fuel metabolism in the brain, muscle, adipose tissue, liver, and kidney. Each of these organs exhibits specialized needs and functions, which influence the regulation of fuel metabolism in that organ. The discussion in this chapter then turns to the ways in which fuel metabolism is interconnected between the liver and skeletal muscle via the Cori cycle and the glucose–alanine cycle. Fuel metabolism is regulated principally by the action of hormones, which bind to specific receptor proteins. Three signal transduction pathways regulate fuel metabolism: the adenylate cyclase signaling system and the phosphoinositide pathway activated by G protein–coupled receptors, and signaling pathways activated by receptor tyrosine kinases. This chapter concludes with a discussion of disturbances in fuel metabolism, in particular, starvation, diabetes, and obesity, with a brief diversion into the discovery of insulin (Box 22-2).

## *Essential Concepts*

### *Organ Specialization*

1.  The biochemical pathways for the synthesis and breakdown of the three major metabolic fuels (carbohydrates, fatty acids, and protein) converge on a small number of metabolites: pyruvate, acetyl-CoA, and citric acid cycle intermediates. In animals, flux through these pathways is tissue specific. In mammals, the liver is one of the few organs that can carry out most of the biochemical pathways involved in energy metabolism (see Figure 22-1).

2.  Some organs do not synthesize fuel molecules but only catabolize these molecules. Muscle and brain are two such organs. The brain's exclusive fuel source is normally glucose, and because the brain stores no glycogen, it relies on glucose present in the bloodstream. Prolonged fasting results in a gradual switch from glucose catabolism to ketone body catabolism in the brain. Skeletal and heart muscles rely on fatty acids when resting and use glucose from stored glycogen or from the blood upon exertion.

3.  Adipose tissue is largely involved in the long-term storage of fatty acids (in the form of triacylglycerols). It releases fatty acids into the bloodstream in response to hormones (e.g., epinephrine). The synthesis of triacylglycerols requires the catabolism of glucose to glycerol-3-phosphate, to which fatty acids are esterified.

4.  The liver maintains homeostatic levels of circulating fuel molecules for use by all tissues. The liver metabolizes excess glucose by converting it to glucose-6-phosphate (G6P). This reaction is catalyzed by glucokinase (an isozyme of hexokinase) whose relatively high $K_M$ (~5 mM) makes it sensitive to changes in glucose concentration over the physiological

range (~5 mM). In contrast, hexokinase ($K_M$ < 0.1 mM) is saturated with glucose at these glucose levels.

5.   In the liver, G6P has several fates:
     (a) Polymerization into glycogen.
     (b) Conversion to glucose, which then equilibrates with the glucose in the bloodstream.
     (c) Catabolism into acetyl-CoA, which can be further catabolized for the production of ATP, or diverted to anabolic routes for the synthesis of fatty acids.
     (d) Conversion via the pentose phosphate pathway to ribose-5-phosphate, an important precursor for nucleotide synthesis, and NADPH for use in biosynthetic reactions.

6.   The liver can synthesize and degrade triacylglycerols based on the body's metabolic needs. It also degrades amino acids to a variety of intermediates, which are then used for fatty acid synthesis (from the ketogenic amino acids) and gluconeogenesis (from the glucogenic amino acids). The kidney also carries out gluconeogenesis from the α-ketoglutarate generated when ammonia is released from glutamine and glutamate.

7.   Some metabolic pathways, called interorgan pathways, are segregated between tissues that exchange metabolites (if such a pathway occurred entirely within a single organ, it would constitute a futile cycle). The Cori cycle is defined by the exchange of glucose and lactate between muscle and liver. The lactate generated by anaerobic glycolysis in muscle permits the muscle to anaerobically generate ATP required for contraction, while postponing the aerobic production of ATP in the liver necessary to resynthesize glucose.

8.   Another interorgan pathway is the glucose–alanine cycle. Glycolytically produced pyruvate serves as an amino-group acceptor in transamination reactions in the muscle during protein degradation. The resulting alanine is transported to the liver, where it is reconverted to pyruvate for gluconeogenesis with the released amino group appearing in urea.

## Hormonal Control of Fuel Metabolism

9.   Hormones are chemical signals that reach their target cells via the bloodstream. Each hormone binds to a specific receptor, and its binding results in the transduction of the hormone signal into a biochemical response inside the cell. Many factors may modulate the response of a cell to a specific hormone, such as the number of receptors and the activities of other receptors.

10.  Energy metabolism is regulated largely by pancreatic and adrenal hormones. The pancreas secretes the polypeptide hormones glucagon and insulin, which act antagonistically to each other to regulate blood glucose concentration. The adrenal medulla secretes the catecholamines epinephrine and norepinephrine, which bind to α- and β-adrenoreceptors. These receptors often respond to the same ligand in opposite ways in different tissues.

11.  A variety of signal transduction pathways regulate metabolism (see Figure 22-10). Glucagon and the aforementioned catecholamines exert their regulation via adenylate cyclase and/or the phosphoinositide pathways, while regulation by insulin occurs via the

receptor tyrosine kinase–Ras signaling cascade (though it can activate the phosphoinositide pathway as well (see Figure 13-31).

*Metabolic Homeostasis: Regulation of Energy Metabolism, Appetite, and Body Weight*

12. AMP-dependent protein kinase (AMPK) serves as a major sensor for the regulation of metabolic homeostasis. Allosteric control of the phosphorylated form of AMPK by AMP and ATP results in AMPK-mediated activation of metabolic breakdown pathways and inhibition of biosynthetic pathways in order to conserve ATP during metabolic stress. Adiponection is a key polypeptide hormone secreted by adipocytes that leads to the activation of AMPK.

13. Five peptide hormones that regulate appetite include leptin, neuropeptide Y, insulin, ghrelin, and $PYY_{3-36}$. Insulin, leptin, and $PYY_{3-36}$ suppress appetite, while ghrelin and neuropeptide Y stimulate appetite. Diminished response to leptin or the secretion of ghrelin by the stomach results in the secretion of neuropeptide Y by the hypothalamus. Conversely, both leptin and insulin inhibit the secretion of neuropeptide Y. While ghrelin appears largely to control short-term appetite, leptin may control both long-term and short-term appetite. The gastrointestinal tract secretes the appetite-suppressing peptide hormone $PYY_{3-36}$.

*Disturbances in Fuel Metabolism*

14. Starvation requires the mobilization of stored fuel molecules. Early in a fast, the liver accelerates glycogenolysis and increases gluconeogenesis from available materials. This makes glucose available continuously to the central nervous system. Skeletal muscle protein becomes a source of glucose via glucogenic amino acids released upon proteolysis. After several days, the liver shifts its metabolic machinery toward the breakdown of fatty acids in order to produce ketone bodies. The brain eventually adapts to use ketone bodies rather than glucose as a primary metabolic fuel.

15. Diabetes mellitus is a pathological condition that results from insufficient secretion of insulin or inefficient stimulation of its target cells. There are two major forms of diabetes:
    (a) Insulin-dependent or juvenile-onset diabetes mellitus (type I) results from an autoimmune response that destroys pancreatic β cells, which secrete insulin.
    (b) Non-insulin-dependent or maturity-onset diabetes mellitus (type II) may be caused by a decrease in the number of insulin receptors or their ability to bind insulin. This leads to the clinical condition called insulin resistance, in which insulin levels are far higher than normal but target tissues exhibit little or no response to the hormone.

16. While obesity can result from the loss of proper leptin production or responsiveness, most cases of obesity are caused by high-calorie diets that result in a new set point for metabolic control. Obesity and long-term high-calorie diets appear to be major contributors to a phenomenon referred to as metabolic syndrome. This syndrome is characterized by insulin resistance, systemic inflammation, and an increased likelihood of developing atherosclerosis, hypertension, and type 2 diabetes, which can lead to coronary heart disease.

## Questions

### Organ Specialization

1.  List the possible products of acetyl-CoA metabolism in animals. Include the pathway involved and, where appropriate, the key regulatory enzyme that commits acetyl-CoA to that pathway.

| Fate | Pathway | Key Enzyme |
| --- | --- | --- |
| 1. | | |
| 2. | | |
| 3. | | |
| 4. | | |
| 5. | | |

2.  List the metabolic products of pyruvate metabolism in animals. Include the pathway involved and, where appropriate, the key regulatory enzyme that commits pyruvate to that pathway.

| Fate | Pathway | Key Enzyme |
| --- | --- | --- |
| 1. | | |
| 2. | | |
| 3. | | |
| 4. | | |

3.  List the metabolic products of glucose-6-phosphate metabolism. Include the pathway involved and the key regulatory enzymes.

| Fate | Pathway | Key Enzyme |
| --- | --- | --- |
| 1. | | |
| 2. | | |
| 3. | | |
| 4. | | |

4.  Match the metabolic pathways below with the cellular compartments in which they occur.
    A.  Cytosol
    B.  Inner mitochondrial membrane
    C.  Mitochondrial matrix
    D.  Mitochondrial intermembrane space
    E.  Peroxisome
    F.  Endoplasmic reticulum

    ____ Glycolysis
    ____ Lactic fermentation
    ____ Pyruvate dehydrogenation
    ____ Citric acid cycle
    ____ Oxidative phosphorylation of ADP
    ____ Pentose phosphate pathway
    ____ Fatty acid elongation and desaturation

    ____ Palmitoyl-CoA oxidation
    ____ Palmitate synthesis
    ____ Lignoceric (24:0) acid synthesis
    ____ Amino acid degradation
    ____ Urea cycle
    ____ Gluconeogenesis
    ____ Cholesterol synthesis

5.    Match the metabolic function below with the appropriate pathway(s).

A.   Glycolysis                          D.   Fatty acid oxidation
B.   Pentose phosphate pathway          E.   Glycogen synthesis
C.   Gluconeogenesis                     F.   Amino acid degradation

(a)   Provides reducing equivalents for biosynthesis.
(b)   Provides glucose during the early phase of a fast, after glycogen stores are depleted.
(c)   Provides energy for heart muscle.
(d)   Stores metabolic fuel in the liver.
(e)   Provides ATP during a rapid burst of skeletal muscle activity.
(f)   Generates the nucleotide precursor ribose-5-phosphate.
(g)   Provides energy for gluconeogenesis during a fast
(h)   Provides energy immediately after a meal

6.    What is the significance of high levels of hexokinase activity in the brain?

7.    What similarities are shared by the Cori cycle and the glucose–alanine cycle? What distinguishes them?

8.    Muscle phosphofructokinase is allosterically stimulated by $NH_4^+$. What is the physiological function of this stimulation?

9.    How does an increase in blood glucose affect triacylglycerol metabolism in adipocytes?

*Hormonal Control of Fuel Metabolism*

10.   Without referring to the text, fill in the following table by writing increase, decrease or no effect:

| Tissue | Glucagon | Insulin | Epinephrine |
|---|---|---|---|
| Muscle | | | |
| Glucose uptake | | | |
| Glycogen synthesis | | | |
| Adipose | | | |
| Glucose uptake | | | |
| Lipolysis | | | |
| Lipogenesis | | | |
| Liver | | | |
| Glycogen synthesis | | | |
| Lipogenesis | | | |
| Gluconeogenesis | | | |

11.  Which glucose transporter (GLUT2 or GLUT4) would you expect to find in the insulin-secreting β-islet cells of the pancreas? Provide a rationale.

*Metabolic Homeostasis: Regulation of Energy Metabolism, Appetite, and Body Weight*

12.  Why does AMPK respond to AMP and ATP instead of ADP and ATP?

13.  Fill in the following table with the effects of AMPK in cardiac muscle, skeletal muscle, liver, and adipose tissue. What general metabolic flux is favored by the activities of AMPK in these tissues?

| *Tissue* | *AMPK activity* |
| --- | --- |
| Cardiac Muscle | |
| Skeletal Muscle | |
| Liver | |
| Adipose Tissue | |

14.  Which hormones regulate the production of the appetite-stimulating hormone neuropeptide Y? Which acts as a long-term regulator? Which is a short-term regulator?

*Disturbances in Fuel Metabolism*

15.  List the order in which the following energy sources are used by skeletal muscle to produce ATP: protein, phosphocreatine, fatty acids, and glycogen.

16.  List the order in which the liver uses the following substances to provide the body with metabolic fuel during starvation: glycogen, fatty acids, muscle protein, and nonmuscle protein. Explain your answer.

17.  You obtain a liver homogenate from a fasting mouse and "spike" it with a high concentration of glucose. In what form would you expect glycogen phosphorylase?

18.  During a fast to lose weight, is it important to be physically active or might it be better to remain sedentary? Explain.

# 23

# Nucleotide Metabolism

This chapter presents nucleotide metabolism, including the synthesis of purine and pyrimidine ribonucleotides, the conversion of ribonucleotides to deoxyribonucleotides, and the degradation of nucleotides. As you saw in Chapter 3, nucleotides are composed of a purine or pyrimidine which is linked to C1′ of either ribose or deoxyribose. The pentose in turn is esterified through C5′ to one or more phosphate groups. Nucleoside triphosphates are precursors of nucleic acids and also, through the hydrolysis of one or both of their phosphoanhydride bonds, provide the free energy that drives many biochemical reactions. This chapter begins with a detailed discussion of the synthesis of purine nucleotides, describing the source of each of the carbons and nitrogens of the purine ring. The chapter then discusses the intricate system of negative and positive feedback regulation that balances the synthesis of ATP and GTP. Cells recycle the purines resulting from the turnover of nucleic acids via salvage pathways. The following section describes the synthesis of pyrimidines, in which negative feedback mechanisms balance the amounts of UTP and CTP in the cell. Next, the chapter discusses the formation of deoxyribonucleotides by the action of ribonucleotide reductase. This enzyme has several unique features, including its mechanism of action and its intricate allosteric regulatory mechanisms. The formation of deoxyUTP provides the precursor for deoxythymidylate. The key enzyme involved here is thymidylate synthase, which is a target of cancer chemotherapy, as described in Box 23-1. Finally, the chapter considers the degradation of nucleotides, which is in some respects a continuation of nitrogen metabolism discussed in Chapter 21. The catabolism of purines leads to the formation of uric acid, and the catabolism of pyrimidines leads to the formation of malonyl-CoA and methylmalonyl-CoA. The fate of uric acid varies among animal groups depending on their strategies for excreting excess nitrogen.

## *Essential Concepts*

### *Synthesis of Purine Ribonucleotides*

1.  Early investigations of the nucleotide biosynthetic pathway identified the precursors that supply the carbon and nitrogen atoms of the purine ring: N1 from aspartate; C2 and C8 from formate; N3 and N9 from glutamine; C4, C5, and N7 from glycine; and C6 from bicarbonate.

2.  The purines are synthesized as ribonucleotides rather than as free bases. The purine nucleotide biosynthetic pathway produces inosine 5′-monophosphate (IMP) in 11 steps:
    (a) Ribose is activated by reaction with ATP to form 5-phosphoribosyl-α-pyrophosphate (PRPP).
    (b) The amide nitrogen of glutamine (which becomes N9) displaces the pyrophosphate group of PRPP, inverting the configuration at C1′ to form β-5-phosphoribosylamine.

(c) A molecule of glycine is added, providing C4, C5, and N7 and forming glycinamide ribotide (GAR).

(d) The free amino group of GAR reacts with the one-carbon carrier N10-formyltetrahydrofolate to add C8 and give formylglycinamide ribotide (FGAR).

(e) The amide nitrogen of a second glutamine is added to FGAR in an ATP-dependent reaction, which provides N3 and yields formylglycinamidine ribotide (FGAM).

(f) The purine imidazole ring is closed in a reaction requiring ATP and producing 5-aminoimidazole ribotide (AIR).

(g) Purine C6 is incorporated into AIR in a carboxylation reaction to give carboxyaminoimidazole ribotide (CAIR).

(h) Aspartate contributes N1 by forming an amide bond with C6 to yield 5-aminoimidazole-4-(N-succinylocarboxamide) ribotide (SACAIR).

(i) Fumarate is cleaved from SACAIR to produce 5-aminoimidazole-4-carboxamide ribotide (AICAR). Note that reactions (h) and (i) resemble reactions of the urea cycle.

(j) The last purine ring atom, C2, is added through formylation by N10-formyltetrahydrofolate to form 5-formaminoimidazole-4-carboxamide ribotide (FAICAR).

(k) In the final step, the larger ring is closed to form IMP.

3.  IMP is rapidly transformed into AMP and GMP. To form AMP, the amino group of aspartate displaces the carbonyl oxygen at C6 of the purine base, followed by removal of fumarate. To form GMP, IMP is oxidized in an $NAD^+$-dependent reaction to yield xanthosine monophosphate (XMP), which then accepts a glutamine amide nitrogen to produce the guanine nucleotide. Note that AMP synthesis from IMP requires cleavage of GTP to GDP + $P_i$, whereas GMP formation requires ATP cleavage to AMP + $PP_i$.

4.  Nucleoside monophosphates are converted to di- and triphosphate derivatives in two steps. First, a specific nucleoside monophosphate kinase catalyzes phosphorylation of a base-specific nucleoside monophosphate by a nucleoside triphosphate to yield two nucleoside diphosphates. Second, a nucleoside diphosphate is phosphorylated to the corresponding triphosphate by the action of nucleoside diphosphate kinase. This enzyme can utilize nucleotides containing any base and either ribose or deoxyribose as substrates.

5.  Purine nucleotide biosynthesis is carefully regulated. IMP production is controlled at its first two steps. Ribose phosphate pyrophosphokinase, which generates PRPP, is inhibited through negative feedback by ADP and GDP. Amidophosphoribosyl transferase, which produces phosphoribosylamine, is inhibited by adenine nucleotides and by guanine nucleotides, which bind at separate inhibitory sites. PRPP also allosterically stimulates amidophosphoribosyl transferase in a feed-forward mechanism.

6.  Additional regulation occurs immediately beyond the IMP branch point, because AMP and GMP inhibit their own synthesis. Moreover, a balance in the synthesis of the two purine nucleotides is achieved because GTP is necessary for AMP synthesis and ATP is required for GMP formation (see Figure 23-4).

7.  Purines released from the degradation of nucleic acids are re-utilized via metabolic salvage pathways. In mammals, distinct enzymes catalyze the reactions of adenine and guanine with PRPP to form AMP and GMP, respectively. In an analogous reaction, hypoxanthine–guanine phosphoribosyltransferase (HGPRT) uses guanine as a substrate but also transforms the purine base hypoxanthine into IMP. An inherited deficiency in HGPRT results in Lesch–Nyhan syndrome, a disease characterized by bizarre behavioral abnormalities.

## Synthesis of Pyrimidine Ribonucleotides

8.  Pyrimidine biosynthesis utilizes only three precursors. Aspartate contributes N1, C4, C5, and C6 of the pyrimidine ring, whereas C2 is derived from bicarbonate and N3 is donated by glutamine.

9.  In pyrimidine nucleotide formation, the ring is completed before being linked to ribose-5-phosphate. The six steps of the UMP biosynthetic pathway are as follows:
    (a) Bicarbonate and the amide nitrogen of glutamine combine to form carbamoyl phosphate in a reaction that requires 2 ATP and is catalyzed by cytosolic carbamoyl synthetase II. This enzyme is distinct from mitochondrial carbamoyl synthetase I, which forms carbamoyl phosphate during the urea cycle, a process in which ammonia is the nitrogen donor.
    (b) Carbamoyl phosphate reacts with aspartate to yield carbamoyl aspartate. The reaction is catalyzed by aspartate transcarbamoylase, a key enzyme for pyrimidine biosynthesis regulation in *E. coli* but not in animals.
    (c) Carbamoyl aspartate is converted to the ring compound dihydroorotate by elimination of water.
    (d) Dihydroorotate is dehydrogenated to form orotate, with a mitochondrial quinone serving as the ultimate electron acceptor.
    (e) Orotate reacts with PRPP to produce orotidine-5′-monophosphate (OMP). The same enzyme can also salvage uracil and cytosine by catalyzing their conversion to UMP and CMP, respectively.
    (f) OMP undergoes decarboxylation to form UMP.

10. In microorganisms, the six reactions of UMP biosynthesis are carried out by separate enzymes. However, in animals, several of the reactions are catalyzed by polypeptides that possess more than one enzyme activity. This is also true for certain steps in purine formation. In these instances, pathway intermediates are not liberated from the multifunctional proteins but are channeled directly to the next active site.

11. As in purine nucleotide biosynthesis, UTP is formed from UMP by the successive actions of a nucleoside monophosphate kinase and nucleoside diphosphate kinase.

12. CTP is synthesized through amination of UTP by glutamine (in animals) or ammonia (in bacteria) in an ATP-dependent reaction.

13. The regulation of pyrimidine biosynthesis in bacteria is accomplished by the allosteric modulation of aspartate transcarbamoylase, such that the binding of ATP to the enzyme stimulates activity and the binding of CTP or UTP inhibits it. Pyrimidine formation in animals is regulated by changes in carbamoyl synthetase II activity. This enzyme is inhibited by UDP and UTP and stimulated by ATP and PRPP. In addition, decarboxylation of OMP is diminished by elevated UMP levels.

## Formation of Deoxyribonucleotides

14. Deoxyribonucleotides are synthesized by the reduction of the hydroxyl group at C2′ of the corresponding ribonucleotide. Enzymes that catalyze these reactions are called ribonucleotide reductases.

15. Ribonucleotide reductase in eukaryotes and some prokaryotes converts a ribonucleoside diphosphate (NDP) to a deoxyribonucleoside diphosphate (dNDP). The enzyme is a heterotetramer that can be dissociated into inactive homodimers, R1 and R2. R1 contains a substrate-binding site and three independent allosteric sites that control enzymatic activity and substrate specificity. Each subunit of R2 has an Fe(III) prosthetic group that interacts with Tyr 122 to form a tyrosyl free radical, which plays a critical role in the reaction mechanism that leads to reduction at the C2′ of ribose. Five cysteine residues in each subunit of R1 also participate, some acting as electron donors to reduce the substrate and generate a disulfide bond.

16. Following NDP reduction, the ribonucleotide reductase disulfide bond is reduced via disulfide interchange with the small protein thioredoxin, which regenerates the active enzyme. Oxidized thioredoxin is subsequently reduced by thioredoxin reductase, which obtains electrons from NADPH. This coenzyme is thus the ultimate reducing agent in the conversion of an NDP to a dNDP.

17. The production of balanced amounts of deoxyribonucleotides needed for normal DNA synthesis is achieved by elaborate allosteric control of ribonucleotide reductase. Binding of nucleotides to the specificity site, activity site, and hexamerization site govern the enzyme's substrate specificity, catalytic activity, and oligomerization (which also affects catalytic activity).

18. Nucleoside diphosphate kinase catalyzes the ATP-dependent phosphorylation of dNDPs to produce the dNTPs needed for DNA synthesis.

19. In the synthesis of thymine deoxyribonucleotides, dUTP is first hydrolyzed to dUMP and then methylated by thymidylate synthase to form dTMP. The methyl donor in this reaction is $N^5,N^{10}$-methylenetetrahydrofolate, which is simultaneously oxidized to dihydrofolate. Dihydrofolate is reduced to tetrahydrofolate with NADPH by dihydrofolate reductase. Tetrahydrofolate then accepts a hydroxymethyl group from serine to regenerate the methyl donor. Finally, dTMP is phosphorylated to dTTP. Drug-induced inhibition of dTMP synthesis is an important tool in cancer therapy (see Box 23-1).

*Nucleotide Degradation*

20. The breakdown of purine nucleotides leads to production of uric acid. Degradation involves the dephosphorylation of the nucleotides to nucleosides, followed by the hydrolysis or phosphorolysis of the nucleosides to yield the free purine base and ribose or ribose-1-phosphate. The bases may be reincorporated into purine nucleotides via the salvage reactions. Adenosine is deaminated to form inosine prior to further degradation.

21. Both hypoxanthine, which is derived by phosphorolysis of inosine, and guanine are oxidized sequentially to xanthine and then uric acid. Xanthine oxidase catalyzes the two reactions that convert hypoxanthine to uric acid, via a complex series of electron-transfer events.

22. Uric acid is the final product of purine degradation for primates, birds, terrestrial reptiles, and some insects. Excretion of uric acid is advantageous in that it requires little accompanying water. Other organisms oxidize uric acid to allantoin, allantoic acid, urea, or ammonia prior to excretion.

23. A variety of factors may lead to abnormally high levels of uric acid, which can result in gout.

24. Pyrimidine nucleotide catabolism occurs through reactions similar to those for purine nucleotides to yield the free bases uracil and thymine. These are further broken down to the amino acids β-alanine and β-aminoisobutyrate, which can be further metabolized to furnish malonyl-CoA and methylmalonyl-CoA.

## Questions

*Synthesis of Purine Ribonucleotides*

1. Which of the following statements is (are) true? The enzyme ribose phosphate pyrophosphokinase:
   (a) Catalyzes the transfer of pyrophosphate from 5-phosphoribosyl-α-pyrophosphate (PRPP) to activate 5-phosphoribosylamine.
   (b) Catalyzes the formation PRPP from ribose-5-phosphate and ATP.
   (c) Transfers a pyrophosphoryl group from ATP to the C1 of ribose.
   (d) Produces pyrophosphate as a reaction product.

2. Which of the following compounds contribute atoms in the synthesis of the purine ring?

   | | |
   |---|---|
   | $NH_4^+$ | $N^5,N^{10}$-methylenetetrahydrofolate |
   | *S*-adenosylmethionine | Aspartate |
   | Glutamine | Fumarate |
   | $N^{10}$-formyltetrahydrofolate | $CO_2$ |
   | Serine | Glycine |

3. In the synthesis of AMP from inosine monophosphate, the amino group that replaces the 6-keto group is obtained from_____ with release of _____, whereas GMP formation involves an oxidation reaction in which the electron acceptor is_____, followed by transfer of an amino group from _____.

4. Which of the following statements is (are) true? Purine nucleotide biosynthesis is regulated in part by:
   (a) Feedback inhibition of amidophosphoribosyl transferase by CTP and UMP.
   (b) Feedback inhibition of amidophosphoribosyl transferase by PRPP.
   (c) Feedback inhibition of amidophosphoribosyl transferase by ADP and GDP.
   (d) Feedback inhibition of ribose phosphate pyrophosphokinase by ADP and GDP.

5. In most tissues, most purine bases have been salvaged rather than synthesized *de novo*. What is the advantage of this metabolic strategy?

**Synthesis of Pyrimidine Ribonucleotides**

6. Identify the compound shown below.

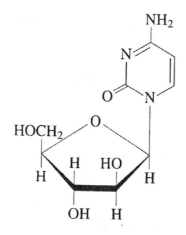

7. Determine which of the following compounds participate in CTP biosynthesis and place them in the correct order:

   Orotic acid                Aspartic acid
   Glutamine                  $N^{10}$-formyltetrahydrofolate
   Uracil                     $HCO_3^-$
   Cytidine                   Uridylic acid
   Dihydroorotate             PRPP

8. In what way is UMP synthesis similar to fatty acid synthesis in mammals?

9. Why is the administration of uridine an effective treatment for human orotic aciduria? Why is it preferable to use uridine instead of UMP?

*Formation of Deoxyribonucleotides*

10. The pentose phosphate pathway converts glucose to the ribose-5-phosphate needed for DNA synthesis. What other pentose phosphate pathway product is required for DNA synthesis?

11. Match each component of the ribonucleotide reductase reaction on the left with its function on the right.

    _____ dATP          A.   Location of tyrosyl radical
    _____ R1 subunit    B.   Stimulates GDP reduction and inhibits CDP and UDP
                             reduction
    _____ R2 subunit    C.   Physiological reducing agent of the enzyme
    _____ dTTP          D.   Inhibits reduction of all NDPs
    _____ ATP           E.   Location of substrate specificity site
    _____ thioredoxin   F.   Stimulates CDP and UDP reduction

12. Why is 5-fluorouracil a powerful antitumor agent? Why is it considered a suicide substrate?

13. What is the rationale for administering a high dose of methotrexate followed by a massive amount of $N^5$-formyltetrahydrofolate and thymidine to a cancer patient?

*Nucleotide Degradation*

14. Adenosine deaminase (ADA) catalyzes the first step in the catabolism of adenine nucleotides. Why does an absence of ADA seriously compromise the immune system?

15. Summarize the metabolic pathway by which $^{14}C$ initially present at position 5 of uracil could appear in long-chain fatty acids.

16. Some terrestrial organisms secrete excess nitrogen as uric acid rather than urea. Why is this an advantage?

# 24
# Nucleic Acid Structure

This chapter takes a closer look at the structures of nucleic acids, which were first introduced in Chapter 3. This chapter begins by examining the structure of DNA in some detail, focusing on the different helical conformations and the limited flexibility of DNA that results from restrictions on rotation of various covalent bonds. Next, you are introduced to supercoiling, a structural feature of DNA that has important consequences for the biological activity of DNA, and topoisomerases, the enzymes that alter supercoiling. The forces that stabilize double helical structures are considered next. Knowledge of these forces is central to understanding the techniques used to isolate, analyze, and manipulate nucleic acids.

The chapter then turns to the interactions of DNA with proteins, giving as examples the restriction endonucleases and well-characterized prokaryotic and eukaryotic sequence-specific regulators of transcription. These transcription factors include the repressor from bacteriophage 434, the E. coli trp repressor, and the E. coli met repressor. The eukaryotic transcription factors that are discussed are proteins that contain zinc finger DNA-binding motifs and the leucine zipper dimerization motif (including those containing the basic helix–loop–helix motif near the DNA-binding region). In all cases, your attention is drawn to the common molecular interactions found in all of these proteins.

Finally, the structure of eukaryotic chromosomes is considered. This section explores how the histone proteins interact with DNA to generate higher-order structures that condense DNA over 50,000-fold. First, the structure of the nucleosome is discussed. This is followed by descriptions of the 30-nm filament of chromatin and the organization of highly condensed metaphase chromosomes.

## Essential Concepts

1.  DNA and RNA are the two kinds of nucleic acids that store genetic information and make it available to the cell. These molecules must therefore have the following properties:
    (a) Genetic information must be stored in a form that is stable and manageable in size.
    (b) Genetic information must be decoded by transcription and translation, which together convert nucleic acid sequences into protein sequences.
    (c) The information in DNA or RNA must be accessible to proteins and other nucleic acids.
    (d) Replication, in the case of DNA, must be template-driven so that each daughter cell receives the same genetic information.

## The DNA Helix

2.  DNA occurs in three major structural forms called A-DNA, B-DNA, and Z-DNA. B-DNA is the most common form and has the structural form first described by Watson and Crick.

3.  The key structural features of B-DNA include:
    (a) The two antiparallel strands wind in a right-handed manner around a common axis, producing a double helix that is about 20 Å in diameter.
    (b) Pairs of nucleotide bases are nearly perpendicular to the axis of the helix. The base pairs are in the interior of the double helix, while the sugar–phosphate backbone is on the outside, thus giving the appearance of a spiral staircase.
    (c) The "ideal" B-DNA double helix has ten base pairs (bp) per turn with a pitch of about 34 Å.
    (d) The double helix contains a wide and deep major groove and a narrow and deep minor groove.

4.  A-DNA is a wider and flatter right-handed helix than B-DNA. The planes of its base pairs are tilted about 20° to the helix axis. Its major groove is deep and narrow, while its minor groove is shallow and wide. A-DNA forms under dehydrating conditions.

5.  Z-DNA is a left-handed DNA helix that contains a deep minor groove but no discernible major groove. It forms *in vitro* in nucleic acids containing alternating purines and pyrimidines. The existence of Z-DNA-binding proteins suggests that Z-DNA forms *in vivo*.

6.  RNA is single-stranded but can form regions of double helix by folding back on itself. 2′-OH groups preclude B-form structures; instead, double helical regions assume conformations resembling A-DNA. DNA–RNA hybrids also show A-form conformations.

7.  Segments of B-DNA deviate from the ideal conformation, often in a sequence-dependent manner, which may be important for sequence-specific recognition of DNA by proteins involved in regulating specific gene activity.

8.  The conformational flexibility of DNA is limited by steric hindrance at the glycosidic bond between the nitrogenous base and the ribose moiety. Steric hindrance also induces a specific pucker in the ribose and restricts rotation in the bonds of the sugar–phosphate backbone.

9.  The twisting of a double helix that is often observed in covalently closed, circular, double-stranded DNA is called supercoiling or superhelicity. A key topological property of a closed circular double helix is that the number of supercoils cannot be altered without first cutting at least one strand of DNA. In the mathematical relationship

$$L = T + W$$

$L$ is the linking number (the number of times that one strand of the double helix winds around the other), $T$ is the twist (the number of complete revolutions that one strand makes around the axis of the double helix), and $W$ is the writhing number (the number of turns the duplex axis makes around the superhelix axis, a measure of the supercoiling of the circular DNA). When $W < 0$, the DNA is negatively supercoiled; when $W > 0$, the DNA is positively supercoiled. Naturally occurring DNA is negatively supercoiled, which promotes strand separation for processes such as DNA replication and transcription.

10. Topoisomerases alter DNA supercoiling by making transient single-strand breaks (Type I) or double-strand breaks (Type II).

    (a) Type IA topoisomerases relax only negatively supercoiled DNA by changing the linking number in increments of one; while type IB topoisomerases can relax negative and positive supercoils by a controlled rotation mechanism. In both cases, the energy of the broken phosphodiester bond is conserved in the form of a phospho-Tyr diester linkage.

    (b) Type II topoisomerases of both prokaryotes and eukaryotes relax negative and positive supercoils in an ATP-dependent manner. Only the prokaryotic version (also called DNA gyrase) can introduce negative supercoils. Negative supercoils in eukaryotes result primarily from packaging DNA into nucleosomes.

*Forces Stabilizing Nucleic Acid Structures*

11. Like the denaturation of proteins, the denaturation of DNA is cooperative; i.e., the unwinding of one part of DNA destabilizes the remaining double helix. This can be measured by observing the increase in ultraviolet light absorbance in going from double-stranded to single-stranded DNA. The midpoint of this "melting" curve is called the melting temperature, $T_m$.

12. The $T_m$ of double-stranded DNA depends on the solvent, the kind and concentration of ions in solution, the pH, and the mole fraction of G · C base pairs.

13. While base pairing provides specificity to the structure of DNA, it contributes little to the stability of DNA. Hydrophobic interactions (which tend to cause free base pairs to aggregate) and van der Waals interactions between base pairs (called stacking interactions) contribute the most to the stability of double-stranded nucleic acids. Stacking interactions between G · C base pairs are stronger than those between A · T base pairs. Hence, the greater stability of GC-rich DNA is a reflection not of the greater number of hydrogen bonds but of greater stacking energies.

14. Most RNA is single-stranded and adopts a much wider variety of shapes than does DNA. These varied shapes are due to various sections of RNA forming double-stranded regions via specific base pairing. For example, 5S ribosomal RNA (rRNA) contains helices, hairpin loops, internal loops, and bulges. The compact shape of transfer RNA (tRNA) is a result of base pairing and extensive stacking interactions.

15. RNA also has catalytic activity. Catalytic RNAs are similar to protein enzymes in their rate enhancement, substrate specificity, ability to use metal cofactors or imidazole groups, and regulation by small allosteric effectors. An RNA found in certain plant viruses, called the hammerhead ribozyme, catalyzes the cleavage of a specific RNA during posttranscriptional processing. Naturally occurring and synthetic versions of ribozymes can catalyze many other reactions, including some required for RNA synthesis.

*Fractionation of Nucleic Acids*

16.  DNA and RNA in cells are invariably associated with proteins, so nucleic acid preparations must be deproteinized as part of their purification. Deproteinization can be accomplished by treating cell lysates with detergents, chaotropic agents, or proteases. The nucleic acid can then be recovered by precipitation with ethanol.

17.  Double-stranded DNA can be separated from single-stranded nucleic acid, nucleotides, and other soluble molecules by chromatography using a column of hydroxyapatite, a form of calcium phosphate.

18.  Eukaryotic messenger RNA (mRNA) can be separated from total RNA (of which it constitutes no more than 5% by mass), DNA, nucleotides, and other soluble molecules by affinity chromatography. Most eukaryotic mRNAs contain a poly(A) sequence at their 3′ ends, which is not found in rRNA or tRNA. Poly(U) or poly(T) covalently attached to a cellulose or agarose matrix can bind eukaryotic mRNA by complementary base pairing under high salt conditions; the mRNA is subsequently recovered by exposing the matrix to a low salt buffer.

19.  Small nucleic acids (< 1000 nucleotides) can be easily resolved by polyacrylamide gel electrophoresis. Larger nucleic acids (up to 100,000 bp) can be separated using less cross-linked agarose gels. For extremely large DNAs (up to $10^7$ bp), pulsed-field gel electrophoresis is used. Nucleic acids can be visualized in these gels by adding planar aromatic cations such as ethidium bromide, acridine orange, or proflavin. These dyes are called intercalation agents because they intercalate between the stacked bases, where they exhibit much greater fluorescence than do the free dyes. Hybridization with a labeled oligonucleotide probe is the basis for Southern blotting (to detect DNA) and Northern blotting (to detect RNA).

*DNA–Protein Interactions*

20.  Many proteins bind DNA nonspecifically, primarily through interactions between the protein's functional groups and the sugar–phosphate backbone of DNA. Histones and certain DNA replication proteins are examples of such proteins.

21.  Sequence-specific DNA-binding proteins usually interact with base pairs in the major groove of DNA by hydrogen bonding, either directly or indirectly via intervening water molecules. Ionic interactions with the sugar–phosphate backbone also occur, probably to facilitate contact between these proteins and DNA, after which the DNA can be "scanned" for a specific binding site. Dissociation constants for sequence-specific DNA-binding proteins are $10^3$ to $10^7$ times lower than those for nonspecific DNA-binding proteins.

22.  Prokaryotic sequence-specific DNA-binding proteins recognize base pairs directly or through indirect readout, in a manner that depends on the sequence-dependent conformation and/or flexibility of the DNA backbone. Many DNA-binding proteins recognize palindromic (or nearly palindromic) sequences.

23. A large class of eukaryotic sequence-specific DNA-binding proteins are transcription factors. Many of these proteins form dimers and promote transcription at specific sites.

24. Many eukaryotic transcription factors contain structural motifs called zinc fingers. Zinc fingers are compact ~30 residue modules containing one or two $Zn^{2+}$ ions liganded by His and/or Cys residues.

25. Another type of eukaryotic transcription factor contains a structural motif called the leucine zipper. The leucine zipper is a seven-residue pseudorepeating unit $(a\text{-}b\text{-}c\text{-}d\text{-}e\text{-}f\text{-}g)_n$ in which a hydrophobic strip along one side of the $\alpha$ helix formed by these residues promotes dimerization. The DNA-binding domain of such proteins may consist of extensions of the zipper helices or may be separated from the zipper by a basic helix–loop–helix motif.

## Eukaryotic Chromosome Structure

26. Prokaryotic genomes typically consist of a single circular DNA molecule ranging from several hundred thousand bp to several million bp. The 23 chromosomes of the 3-billion-bp human genome have a total extended length of nearly one meter. Individual human chromosomes, with extended lengths between 1.6 and 8.2 cm, are condensed to varying degrees in different cells at different times, down to 1.3 to 10 μm long during mitosis.

27. Eukaryotic DNA is packaged into units called nucleosomes. Each nucleosome contains an octamer of four different basic proteins called histones $[(H2A)_2(H2B)_2(H3)_2(H4)_2]$ and 146 bp of DNA. Approximately 55 bp of DNA (called linker DNA) links two nucleosomes and is associated with a fifth histone, H1. Histones are basic proteins with a large proportion of Arg and Lys residues, which interact with the phosphate oxygens of the sugar–phosphate backbone via salt bridges, hydrogen bonds, and helix dipoles.

28. The winding of DNA in nucleosomes to form the beaded chain seen in electron micrographs reduces its length by sevenfold. Further reduction of length is achieved by coiling the beaded chain into a ~30-nm-diameter filament. Higher-order structures, requiring nonhistone proteins are less well characterized. The most condensed structure is the metaphase chromosome, which has a central scaffold of fibrous protein that anchors loops of DNA and has an overall diameter of ~1.0 μm.

29. Prokaryotic DNA is also packaged into nucleosome-like structures with highly basic proteins that functionally resemble histones.

## Key Equation

$$L = T + W$$

### Behind the Equations: Topology and Eukaryotic DNA structure

The relationship between linking number ($L$), twist ($T$), and writing number ($W$),

$$L = T + W$$

has significance beyond the examination of simple circular DNA.  Relatively precise experimental work has been developed to test this mathematical model for superhelical conformations of DNA using naked, circular DNA in solution.  However, the DNA of eukaryotes is linear. So how does this equation bear relevance to longer linear DNA?

As our textbook points out, the linking number is invariant for circular double-stranded DNA since the number of times one polynucleotide winds around its complementary polynucleotide cannot be changed without breaking at least one strand.  In a linear DNA molecule, the linking number can change, as well as the twist and writing number. Eukaryotic DNA molecules are linear, however *in vivo*, they are tethered along their length. As such one can view the topological structure of DNA to a good approximation as bounded, similar to circular DNA. Hence, the topological invariance of linking number can still apply under many conditions.

Observations show that DNA inside of cells is underwound, which results in tension leading to negative supercoiling. If the DNA were naked, it would be compactly coiled, as shown in Figure 24-11. Proteins that bind to DNA, such as histones, can affect the topology of DNA. In fact, when DNA is wrapped around histones to form nucleosomes, it appears to make two negative superhelical turns per nucleosome. The DNA appears to have just one negative superhelical tern when the nucleosomes are removed. However, proteins that bind DNA electrostatically such as histones can impact twist and observations suggest that they do. Although the equilibrium values for $L$, $T$, and $W$ are not well described for nucleosomal structures, determining their values can help in understanding the energetics and mechanisms of DNA replication and transcription.

A long standing observation shows that DNA bound to nucleosomes appear to have two negative superhelical turns while only one negative superhelical turn appears to be present when nucleosomes are removed.

*Question*: If $L$ is constant, what parameter appears to change as nucleosomes are removed? What structural change is occurring?

*Answer*: With $L$ constant, any change in $W$ (supercoiling) must be exactly compensated by $T$. Hence if $W$ increases by +1, $T$ must decrease by -1.  If the twist decreases by one, then, the angle of base pairs to the central axis decreases, closer to what is observed in relaxed B-form DNA.

DNA replication and transcription require melting of the DNA helix. As the DNA helix melts, helical coiling is reduced where the strands are separated but increases ahead of the point of helix separation, resulting in overcoiling. If the linking number does not change, then twist and writhe must. Consider what happens to the structure of DNA: (1) the angles of base pairs change and (2) tension is relieved by increased supercoiling. While the impact of changes in twist may have minimal impact on further DNA melting, supercoiling will certainly increase the energy barrier for helicases. Helicases are enzymes that use ATP to catalyze DNA strand separation during replication. But even the free energy of ATP is insufficient to melt DNA in supercoiled DNA. For this reason, topoisomerases are needed to relieve the tension due to supercoiling and allow the helicases to unwind the DNA. As you have read, this is where topoiosomerases provide a mechanism by which to relieve tension. As technology to produce nucleosomal structures *in vivo* improve, the structural features of DNA replication and transcription will be more precisely understood using this topological equation.

### Questions

*The DNA Helix*

1. What form of DNA might you expect to see in desiccated (but viable!) brine shrimp eggs? Why?

2. Double-stranded DNA is relatively stiff, whereas single-stranded DNA is a flexible coil. What factors influence the structure of single-stranded versus double-stranded DNA?

3. How does the ribose pucker affect DNA structure?

4. A sample of a circular plasmid is digested with Type IA topoisomerase and analyzed by agarose gel electrophoresis followed by staining with ethidium bromide. Fifteen bands of DNA are visible. What do these bands represent and how do their linking numbers differ?

*Forces Stabilizing Nucleic Acid Structures*

5. Why is DNA not susceptible to hydrolysis by NaOH?

6. High concentrations of denaturing agents such as urea or formamide ($HCONH_2$) tend to favor a rodlike conformation for single-stranded DNA, rather than a flexible coil. What molecular interactions promote this behavior?

7. Explain how the following affect the $T_m$ of double-stranded DNA:
   (a) Increasing the monovalent salt concentration.
   (b) Decreasing the pH.
   (c) Increasing the pH.
   (d) Increasing the concentration of formamide.

8.   Would the UV absorbance of a solution containing partially stacked poly(A) increase or decrease when [Na$^+$] increases?

9.   What forces stabilize the tertiary structure of tRNA?

*Fractionation of Nucleic Acids*

10.  Ice-cold ethanol is used to precipitate DNA. What is the significance of the temperature of the ethanol? Is this more important for shorter or longer DNA strands?

11.  Samples of a circular plasmid are incubated with increasing concentrations of ethidium bromide and then analyzed on an agarose gel. Which gel below shows the results of this procedure?

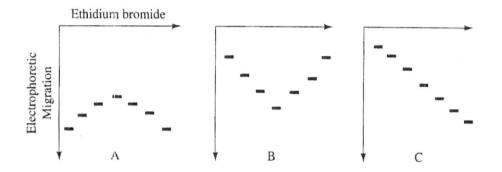

*DNA–Protein Interactions*

12.  Specific DNA-binding proteins mainly contact DNA through hydrogen bonds in the major groove of B-DNA.
     (a)  Why might sequence-specific binding be more common in the major groove than in the minor groove?
     (b)  What hydrogen-bond contacts can proteins make with the bases in the major groove? Are these different from those in the minor groove?

13.  In what ways do HTH proteins and the *met* repressor represent two general modes of DNA–protein interaction?

*Eukaryotic Chromosome Structure*

14.  What insight into the structure of the eukaryotic chromosome was obtained from nuclease digestion studies?

15.  Is histone H1 present in the electron micrograph shown in Figure 24-41? What is the relationship of H1 to the nucleosome structure?

16. Estimate the packing ratio of the DNA in human metaphase chromosomes, which have an aggregate length of 200 μm and contain $6 \times 10^9$ bp.

17. Limited digestion of chromatin with a bacterial nuclease yields a 166-bp DNA fragment, whereas limited digestion with pancreatic DNase I yields 10-bp fragments. Explain the difference in these results.

# 25

# DNA Replication, Repair, and Recombination

This chapter introduces you to the remarkably complex biochemistry of DNA replication, repair, and recombination. The chapter begins with a review of the classic Meselson and Stahl experiment showing that DNA is replicated semiconservatively; that is, each daughter DNA contains one new strand hydrogen bonded to one old, or parental, strand. The chapter reviews the essential steps in the replication of DNA and then takes a more detailed look at the proteins and enzymes involved in this remarkable process. Our best information comes from the enzymology of prokaryotic DNA replication, so this is the focal point of the discussion. Unique features of eukaryotic DNA replication are covered, particularly the problems presented by replicating a linear DNA molecule. The chapter then looks at the genesis of point mutations and small insertions and deletions of nucleotides during DNA replication. This naturally leads to a discussion of the mechanisms cells use to repair damaged DNA or mistakes incurred during replication. Again, prokaryotic mechanisms are understood best, and hence are the focus of the discussion. The chapter then considers the mechanisms by which prokaryotes engage in DNA recombination. This chapter concludes with an introduction to mobile DNA elements called transposons (and retrotransposons in eukaryotes), which can move to new locations in the genome by recombination during their own replication.

## Essential Concepts

### Overview of DNA Replication

1. The structure of DNA provides a template-driven mechanism for its replication. Experiments by Meselson and Stahl showed that each polynucleotide strand serves as a template for a new daughter strand. On completion of replication, each daughter strand, which is hydrogen bonded to its template, or parental strand, segregates to one of the daughter cells. This mode of DNA replication is called semiconservative DNA replication.

2. DNA polymerase requires a template, all four deoxynucleoside triphosphates (NTPs), and a primer from which to extend the chain. The polymerization reaction involves the nucleophilic attack of the growing DNA chain's (or a primer's) 3′-OH group on the $\alpha$-phosphoryl group of a free NTP that is hydrogen bonded to the template. The liberated $PP_i$ is subsequently hydrolyzed, making the polymerization irreversible. Because the 3′ end of the chain grows, polymerization is said to proceed from 5′ to 3′.

3. DNA replication is bidirectional, meaning that DNA synthesis proceeds in both directions from the site of initiation of replication. Circular DNA that contains a bubble where the parental DNA has unwound to allow replication is said to be undergoing $\theta$ replication.

4.  Since DNA replication can proceed only in the 5'→3' direction, the daughter strand that extends 3'→5' in the direction that the replication fork travels, the lagging strand, must be synthesized as a series of fragments,that is, semidiscontinuously. The fragments, which are are called Okazaki fragments (named after the scientist who first observed their synthesis), are eventually joined together by DNA ligase. In the daughter strand that extends 5'→ 3' in the direction that the replication fork travels, the leading strand, DNA synthesis is continuous from the site of initiation.

5.  DNA synthesis *in vivo* begins as the extension of an RNA primer synthesized by primase (or RNA polymerase) at the site of initiation of DNA replication. Primers are subsequently removed by the 5' → 3' exonuclease activity of a DNA polymerase, and the polymerase fills in the gap, which is then sealed by DNA ligase. Ligation is endergonic and requires the free energy of ATP or $NAD^+$ hydrolysis.

*Prokaryotic DNA Replication*

6.  There are three DNA polymerases in prokaryotes:
    (a) DNA polymerase I, the enzyme discovered by Arthur Kornberg, removes RNA primers (via its 5'→3' exonuclease activity) and subsequently fills in the gaps (via its 5'→3' polymerase). These two reactions together are referred to as nick translation. Pol I also has 3'→5' exonuclease activity to eliminate misincorporated nucleotides.
    (b) DNA polymerase II is involved in DNA repair.
    (c) DNA polymerase III is the primary DNA replicating enzyme. It is the largest of the three with at least 10 subunits. Pol III has 5'→3' polymerase activity and a 3'→5' exonuclease activity that eliminates misincorporated nucleotides.
    The polymerase active site forms sequence-independent hydrogen bonds with double-stranded DNA and can therefore detect mispairings. Catalysis requires two metal ions, usually $Mg^{2+}$.

7.  The initiation of DNA replication in *E. coli* occurs at a specific sequence called the *oriC* locus, where the DNA is melted apart, unwound, and the leading strand primer is synthesized.
    (a) DNA melting, or local denaturation, requires ATP and is accomplished by a complex of DnaA protein.
    (b) The hexameric protein DnaB (a helicase) further unwinds DNA in an ATP-dependent reaction.
    (c) The tetrameric single-strand binding protein (SSB) prevents reannealing of the DNA. Numerous molecules of SSB coat single-stranded DNA.
    (d) A complex of proteins called the primosome, which contains DnaB, primase (DnaG protein), and five other proteins, carries out primer synthesis. The primosome displaces SSB in an ATP-dependent process.

8.  DNA synthesis of both leading and lagging strands is carried out by the replisome, a complex containing two Pol III enzymes. In order for the replisome to move as a single unit in the 5'→3' direction, the lagging strand template must loop around once so that the Okazaki fragment and the leading strand can be synthesized in the same direction.

(a)   The β subunit of Pol III forms a sliding clamp that moves along DNA, allowing Pol III to replicate the DNA with a processivity of >5000 residues.

(b)   On the lagging strand, the γ complex (or clamp loader) of Pol III reloads the enzyme back onto the DNA template every ~1000 bp (the size of an Okazaki fragment).

9.   Specific DNA sequences (*Ter* sites) signal the termination of DNA synthesis. However, arrest of DNA replication requires the activity of Tus protein, which binds to a *Ter* site and prevents strand displacement by DnaB.

10.   The fidelity of DNA replication in *E. coli* is very high: ~1 error per 1000 bacteria per generation! Four factors contribute to this high fidelity:

(a)   Cells maintain balanced levels of all four deoxynucleotides.

(b)   Polymerization by Pol I, II, or III occurs only after an incoming nucleotide is properly positioned in the catalytic site and on the template.

(c)   The $3' \rightarrow 5'$ exonuclease activity of DNA polymerases detects and excises mismatches in the new DNA strand.

(d)   DNA repair systems detect and correct residual errors after DNA synthesis is complete.

## Eukaryotic DNA Replication

11.   Eukaryotes contain at least 13 DNA polymerases that are grouped, along with prokaryotic polymerases, in six families according to sequence homology.

12.   The major polymerases of DNA replication in eukaryotes are polymerases α and δ.

(a)   DNA polymerase α, which lacks exonuclease activity and is only moderately processive, associates with a primase and is responsible for initiating DNA replication.

(b)   DNA polymerase δ is the main polymerase for the leading and lagging strands and has $3' \rightarrow 5'$ exonuclease activity. It has unlimited processivity when it is in complex with a protein called proliferating cell nuclear antigen (PCNA), which functions as a sliding clamp.

13.   DNA polymerase ε appears to have a regulatory function. DNA polymerase γ is found in mitochondria (chloroplasts contain a similar enzyme), where it presumably replicates mitochondrial DNA.

14.   A viral RNA-directed DNA polymerase is a reverse transcriptase. It has no primase or $3' \rightarrow 5'$ exonuclease activity but does contain ribonuclease (RNase H) activity. Reverse transcriptase uses a tRNA primer to replicate the viral genome. Reverse transcriptase is an essential tool for cloning mRNA sequences, because it can be used to synthesize a complementary DNA molecule (cDNA).

15. In yeast, DNA replication begins at DNA sequences called autonomously replicating sequences (ARS). In higher eukaryotes, no analogous sequences have been identified. Many initiation sites appear simultaneously in higher eukaryotes and presumably require some alteration of nucleosome structure. As replication of the genome ensues, the RNA primers are removed by RNase H1 and flap endonuclease-1 (FEN1) in a two-step process. There are no known termination sequences in eukaryotic DNA.

16. Eukaryotic chromosomes are linear, so the RNA primers at the 3′ ends of a chromosome cannot be replaced with DNA (DNA polymerase must extend a DNA strand from 5′ to 3′). This would lead to progressively shorter chromosomes, with the concomitant loss of genetic information, with each cell generation. However, eukaryotic chromosomes contain telomeres, repetitive sequences at the ends of chromosomes. These sequences are added by an enzyme called telomerase. Telomerase, a ribonucleoprotein, contains an RNA molecule that serves as a template for a reverse transcriptase activity that extends the telomere. Telomeric sequences tend to fold back on themselves and also serve as binding sites for certain proteins.

17. Telomerase activity is absent in somatic tissues of animals, which exhibit a finite replicative lifespan (about 60–80 divisions after birth in humans). Telomeres appear to shorten as cells age in culture. Hence, progressive loss of telomeres may have an impact on aging in at least some highly proliferative tissues. Many immortal cells (such as cancer cells) contain telomerase activity. Hence, telomerase is an attractive target for antitumor drug development.

## DNA Damage

18. A mutation is a random, heritable alteration of genetic information and can arise in several ways:
    (a) Misincorporation of the wrong base during DNA relication.
    (b) Addition or deletion of one or more nucleotides.
    (c) Chemical modification of a base so that it pairs with a different base during DNA replication.
    (d) Translocation of a segment of DNA from one region of the genome to another.

19. Chemicals that induce mutations, called chemical mutagens, produce two classes of DNA damage:
    (a) Point mutations, in which one base pair replaces another. In a transition, a purine (or pyrimidine) is replaced by another. In a transversion, a purine is replaced by a pyrimidine, or vice versa.
    (b) Insertion/deletion mutations, in which one or more nucleotide pairs are inserted or deleted.

20. The bases in DNA may be chemically altered by deamination, oxidation, and alkylation, and the base itself may be lost by hydrolysis of the glycosidic bond. Intercalating agents can generate insertion/deletion mutations. Adenine and cytosine in cellular DNA are modified by methylation in a sequence-specific manner that regulates gene expression.

21. Many mutagens are also carcinogens (chemicals that cause cancer). This property is used in the Ames test to assess the carcinogenic potential of chemicals. In the Ames test, a *his⁻* strain of bacteria is exposed to a potential mutagen on an agar plate containing no His. The mutagenicity of a substance is scored as the number of *his⁺* colonies that appear, minus the few spontaneous ones that appear in the absence of the putative mutagen.

*DNA Repair*

22. DNA repair mechanisms range from simple one-enzyme systems (such as recognition of alkylated bases or pyrimidine dimers) to more elaborate multienzyme systems (such as nucleotide excision repair, recombination repair, and the SOS response).

23. Simple single-enzyme systems correct specific alkylated bases: e.g., $O^6$-methylguanine residues are recognized by $O^6$-alkylguanine–DNA alkyltransferase, which transfers the aberrant methyl group to a Cys residue on the enzyme. Pyrimidine dimers, which form from the absorption of UV light by two stacked pyrimidine residues, are repaired by light-activated DNA photolyases.

24. In base excision repair, the damaged base is removed by a DNA glycosylase, an endonuclease cleaves the backbone at the resulting abasic site, and an exonuclease removes several additional residues. The gap is then filled in by a DNA polymerase and the nick is sealed by DNA ligase.

25. Nucleotide excision repair is a more sophisticated method to correct pyrimidine dimers or other DNA lesions involving several base pairs. Nucleotide excision repair in *E. coli* is carried out by several enzymes:
    (a) UvrABC endonuclease excises an oligonucleotide containing the damaged bases.
    (b) Pol I fills in the single-stranded region.
    (c) DNA ligase catalyses the formation of a phosphodiester linkage to restore the intact DNA molecule.

26. The mismatch repair (MMR) system corrects replication mispairings overlooked by the DNA polymerase proofreading activity and corrects insertions and deletions up to 4 base pairs in length. Defects in enzymes in this system have been implicated in high incidence of cancer, including hereditary nonpolyposis colorectal cancer syndrome. Three proteins carry out MMR in *E. coli*:
    (a) the MutS homodimer recognizes and binds to mismatched base pairs;
    (b) the MutL homodimer forms a tetramer with MutS to loop out the region of DNA containing the mismatched base pairs in a ATP-driven reaction;
    (c) the MutS/MutL tetramer recruits MutH and activates MutH endonuclease activity. MutH makes a single-strand cut (nick) adjacent to an unmethylated GATC sequence. The UvrD helicase and an exonuclease remove the DNA, and the gap is filled in by DNA pol III.

27.  Eukaryotes contain at least 5 DNA polymerases, called error-prone polymerases, that can replicate past various DNA lesions. These polymerases have no proofreading exonuclease activity and, in the case of DNA polymerase η, insert incorrect bases once every ~30 nucleotides.

28.  Double-stranded DNA breaks are repaired by two mechanisms: recombination repair and nonhomologus end-joining (NHEJ) repair.

29.  A complex DNA repair system in *E. coli* is the SOS response, which is activated when DNA damage prevents replication. SOS repair is error prone and therefore mutagenic. The SOS response involves RecA, which also participates in recombination and error-prone DNA polymerases.

*Recombination*

30.  Homologous (general) recombination occurs between two sections of DNA with extensive homology. The aligned strands of DNA are cleaved, and the strands cross over to form a four-branched structure called a Holliday junction. The junction can dissociate into two new duplexes in two ways, both yielding new DNA molecules.

31.  Recombination in *E. coli* is mediated by RecA, which polymerizes on single-stranded gaps in DNA duplexes. RecA partially unwinds the duplexes and exchanges two strands in an ATP-dependent reaction. All recombination events require additional proteins to further unwind DNA, produce nicks, maintain proper supercoiling, drive branch migration, and seal nicks.

32.  In recombination repair, DNA that contains a nick or gap (and which therefore cannot be further replicated) undergoes recombination so that an intact homologous daughter strand can serve as a template in place of the damaged parental strand, allowing replication to proceed. Double-strand breaks can also be repaired by recombination with an intact homologous chromosome.

33.  Transposons are DNA segments in prokaryotes and eukaryotes that can move from one site in the genome to another by a mechanism that involves their replication. The simplest transposons are insertion elements (IS), which contain a transposase gene that mediates the recombination reactions involved in transposition. This recombination occurs between short target sequences located at the ends of the transposon. More complex transposons contain genes required for transposition as well other genes, such as antibiotic resistance genes.

34.  Transposons promote inversions, deletions, and rearrangements of host DNA and thereby affect gene expression and permit chromosome evolution. Most eukaryotic transposons resemble the DNA from retroviruses rather than bacterial transposons; hence they are called retrotransposons. Transposition of a retrotransposon begins with its transcription to RNA, followed by synthesis of DNA from the RNA template by reverse transcriptase; the DNA then inserts randomly into the host genome.

### Behind the Equations: Mutational load in humans and human evolution

Mutations in human germ cells can result in inherited disorders while mutations in somatic cells can lead to cancer. A look at mutation rates and cell divisions during development reveals that the average mutational load a person accumulates during her lifetime is modest but provides the potential for lesions that can compromise the functioning of tissues or organs, or possibly lead to cancer.

Current estimates of mutation rates (post-DNA-repair mechanisms) range from $5 \times 10^{-8}$ to $10^{-9}$. (A mutation rate of $10^{-9}$ means that an error occurs with a rate of $1/10^{9}$, or once per billion nucleotides each cell generation.) Human somatic cells contain $6.6 \times 10^{9}$ base pairs, such that it is easy to see that after DNA replication one could expect each daughter cell to contain 6.6 to 13.2 mutations.

For example, $5 \times 10^{-8}$ mutations per base pair per cell division x $6.6 \times 10^{9}$ base pairs = 13.2 mutations per cell division

If one makes some simple assumptions about early human development, we can take a look at the mutational load a child is born with. The human body is estimated to contain $10^{13}$ cells. In order to explore mutational load, we need to make a couple of simplifying assumptions. First, assume that $10^{13}$ cells is a reasonable estimate for a human neonate. Second, assume that the fertilized ovum undergoes binary divisions with no cell loss during development (certainly not true but we will come back to that). How many cell divisions would be required to get to this number? With these assumptions, we have

$2^{x} = 10^{13}$, where $x$ represents the number of cell divisions; taking the log of each side of the equation and re-arranging, we have

$x = 13/\log 2 = 43$ cell divisions

43 cell divisions x 6.6 mutations per cell division suggest that cells at the time of birth carry minimally 283 mutations in each cell. Binary cell division does not occur during human development, as cells divide asynchronously, exit the cell cycle and differentiate into tissue-specific cells, and die or undergo programmed cell death. Hence, the model for cell replication leading to the total cell number in humans is more complex such that many cells undergoing division throughout human development probably undergo more than 43 cell divisions. Since only 3% of the genome encodes polypeptides, the mutation load at birth appears minimal, about 8.5 mutations in polypeptide-encoding genes per cell. With 22,000 genes there are at most mutations in 1.3% of the genes.

Human neonatal cells have the potential to undergo another 60-70 cell divisions: 60 cell divisions x 6.6 mutations per cell yield another 396 mutations. Hence, even across a lifetime, somatic cells accumulate 20.3 mutations per cell in polypeptide-encoding genes. At the time of death an individual may harbor mutations in up to 1.8% of their genes.

Given the redundancy observed in genetic regulatory systems, this mutational load appears relatively small. Interestingly, a recently published estimate of mutation rate in human oocytes suggests that mammalian germ cells may experience a mutation rate 10-fold lower than dividing somatic cells.

Another measure of mutation rate in protein-encoding genes is made by estimating the incidence of new cases of an autosomal dominant human disorder. Here, the mutation rate ranges from 4-10 x $10^{-5}$ (mutations per locus per generation) for neurofibromatosis to 3-5 x $10^{-6}$ for aniridia. Interestingly, the neurofibramatosis gene is very large, approaching 1 million base pairs, so this number is about one order of magnitude off from our simple calculation above, suggesting one mutation rate per 8.8 million base pairs. However, the *Pax6* gene, a mutation in which causes aniridia, is smaller, suggesting perhaps a hotspot for mutation. Indeed, human disorders like achrondroplasia show a high rate of mutation at one specific base pair in the growth factor receptor gene (fibroblast growth factor-3). Hence, our simple calculations above, while useful for appreciating the total mutational load, can hide the occurence of high mutation rates at specific genetic loci.

## Questions

*Overview of DNA Replication*

1. What feature of DNA replication, as shown in Figure 25-4, is inconsistent with the known enzymatic properties of DNA polymerases? What observation reconciled this inconsistency?

2. What prevents the reverse reaction of DNA polymerization *in vivo*?

*Prokaryotic DNA Replication*

3. Describe the roles of the exonuclease activities of Pol I.

4. For each of the enzymes and proteins listed below, match the function or feature related to DNA replication in *E. coli*.

   A. DNA polymerase I  B. DNA polymerase II  C. DNA polymerase III
   D. DNA ligase         E. DnaB                F. Single-strand binding protein
   G. Primase          H. Tus protein      I. DNA gyrase

   _____ Required for the initiation of DNA synthesis
   _____ Essential for the condensation of Okazaki fragments
   _____ A helicase required for unwinding the DNA duplex
   _____ Involved in the termination of DNA strand replication
   _____ The enzyme that is strictly involved in DNA replication
   _____ The enzyme critical for relieving buildup of positive supercoils
   _____ The enzyme that participates in DNA repair
   _____ Prevents the reannealing of DNA at the replication fork

_____ Excision and replacement of RNA primers
_____ Incapable of nick translation
_____ Requires NAD$^+$ in _E. coli_
_____ The most abundant DNA polymerase

5.  Radioactive DNA probes can be made by nick translation of linear pieces of DNA. Why is a trace amount of DNase I necessary to do so?

6.  From the gut of a slug in your garden you have identified a new circular, double-stranded DNA–containing bacteriophage. Electron micrographs of infected bacteria during production of phage progeny reveal the structures shown below (thick lines represent double-stranded DNA and thin lines represent single-stranded nucleic acids). Alkali treatment of these replicative DNA forms destroys their single-stranded "tails." Bacteria incubated with $^{15}$N-nucleosides were infected with this bacteriophage. CsCl equilibrium density centrifugation of DNA (which separates DNA roughly by mass) reveals two bands of phage DNA. The heavier band contains 100 times more DNA than the lighter band.

    (a) What is the composition of the single-stranded nucleic acid in the structures shown below?
    (b) Suggest a mechanism for DNA replication in this bacteriophage.

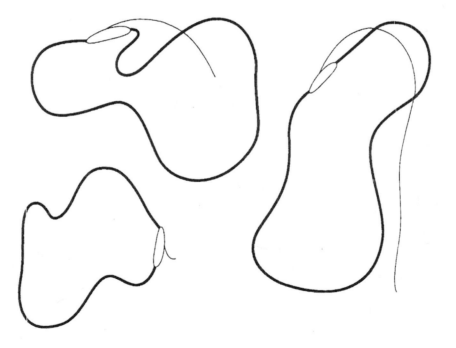

_Eukaryotic DNA Replication_

7.  Which eukaryotic DNA polymerase is most likely responsible for the following activities: (a) synthesizing the leading strand, (b) synthesizing the lagging strand, (c) synthesizing mitochondrial DNA, (n) initiating DNA replication?

8.   Why does reverse transcription of retroviral DNA require a specific host tRNA for DNA synthesis? What section of the tRNA is most likely to be involved?

9.   When reverse transcriptase is used in the oligo(T)-primed synthesis of double-stranded DNA from a specific cellular mRNA, the aggregate length of the resulting double-stranded cDNA is sometimes twice as long as that of the mRNA template. Explain this phenomenon.

10.   What mammalian cell type must retain its telomerase activity?

*DNA Damage*

11.   In the pairs of DNA sequences below, the lower duplex represents a mutagenized daughter duplex. Identify the mutation as a transition, a transversion, an insertion, or a deletion.

(a)   5′  GCCTAGAACCAGTAC  3′
      3′  CGGATCTTGGTCATG  5′

      5′  GCACTAGAACCAGTA  3′
      3′  CGTGATCTTGGTCAT  5′

(b)   5′  CGCATAGCTACTGGAA  3′
      3′  GCGTATCGATGACCTT  5′

      5′  CGCAGAGCTACTGGAA  3′
      3′  GCGTCTCGATGACCTT  5′

12.   What kinds of mutations does 5-bromouracil generate?

13.   Match the compound on the left with the kind of mutation on the right.

_____ Acridine orange            A. Transition
_____ Nitrous acid               B. Transversion
_____ Ethylnitrosourea           C. Insertion or deletion
_____ Dimethyl sulfate
_____ Ethidium bromide

*DNA Repair*

14.   Match the proteins below with their roles in DNA repair.

A. $O^6$-alkylguanine–DNA alkyltransferase          B. UvrABC endonuclease
C. Uracil–DNA glycosylase                                      D. RecA

_____ Removes 7-methyladenine residues from damaged DNA
_____ Serves as a sink for methyl residues abstracted from $O^6$-methylguanine residues
_____ Removes deaminated cytosine residues
_____ Removes damaged DNA via recombination repair
_____ Removes thymine dimers
_____ Initiates the SOS response

*Recombination*

15.   Why is transposition a misnomer in describing the operation of bacterial transposons?

16.   You transformed a strain of *E. coli* with a plasmid containing a transposon that includes an ampicillin-resistance gene. To recover the plasmid, you pick a colony from a culture plate containing ampicillin. Much to your surprise, the plasmid you recover is twice the size you expected. You then pick 100 more colonies and discover that only one of them has plasmid of the expected size.
(a)   Why is the recovery of plasmid of expected size so low?
(b)   Why can you recover any plasmid of expected size at all?

# 26

# Transcription and RNA Processing

This chapter focuses on the biochemistry of RNA synthesis, from the initiation of transcription through the subsequent modification of the transcription products into biologically functional RNAs. The chapter begins by discussing the structure and biochemical properties of prokaryotic RNA polymerase, comparing and contrasting them with those of DNA polymerase. The discussion then progresses to the three main activities of transcription: template binding and RNA chain initiation, chain elongation, and chain termination. The chapter then examines the remarkably more complex biochemistry of transcription in eukaryotic cells, which involves multiple DNA sequence elements (promoters, enhancers, and silencers) and many proteins (transcription factors). In addition, eukaryotic cells use three different isoforms of RNA polymerase to synthesize different classes of RNA. The chapter concludes with a discussion of posttranscriptional processing, focusing first on the extensive modifications of eukaryotic mRNAs. Here you are introduced to some of the minor forms of RNA involved in RNA processing (small nucleolar RNAs and guide RNAs). Finally, rRNA and tRNA processing in prokaryotes and eukaryotes is considered.

## Essential Concepts

1.  Cells contain several types of RNA, mainly ribosomal RNA (rRNA), transfer RNA (tRNA), and messenger RNA (mRNA). rRNA constitutes better than 80% of the total mass of cellular RNA and accounts for two-thirds of the mass of ribosomes. tRNAs are small RNAs that deliver amino acids to the ribosome. mRNAs contain the nucleotide sequences that encode the amino acid sequences of polypeptides. In addition to these three large classes, which account for over 95% of the mass of RNA in eukaryotic cells, there are small nuclear RNAs (snRNAs) involved in mRNA splicing, guide RNAs involved in RNA editing, and other small RNAs.

### Prokaryotic RNA Transcription

2.  The transcription of DNA into RNA is carried out by RNA polymerases, which contain multiple different subunits. Bacteria contain one RNA polymerase core enzyme, whereas eukaryotes have four or five isoforms. All RNA polymerases catalyze the reaction

$$(RNA)_{n\ residues} + NTP \rightleftharpoons (RNA)_{n+1\ residues} + PP_i$$

The enzyme uses the substrates ATP, CTP, GTP, and UTP, synthesizing an RNA chain in the $5' \rightarrow 3'$ direction. Hydrolysis of the released $PP_i$ makes polymerization irreversible.

3.  The RNA polymerase holoenzyme in *E. coli* contains five types of subunits: $\alpha_2\beta\beta'\omega\sigma$. The σ subunit dissociates from the complex on the initiation of transcription, leaving the core enzyme to continue transcription.

4.  Many prokaryotic genes exist in contiguous sets of functionally related genes called operons and are transcribed as polycistronic mRNAs. Most protein-encoding genes (structural genes) in eukaryotes, however, are transcribed individually as monocistronic mRNAs.

5.  The prokaryotic RNA holoenzyme binds to DNA sequences called promoters, which are "upstream" from the coding sequence of a gene or operon. RNA polymerase initiates transcription about 10 nucleotides "downstream" from a consensus DNA sequence called the Pribnow box.

6.  Different genes or operons have different promoter sequences, which allows for differential gene expression in response to environmental needs or developmental changes in the organism. Promoter selectivity is regulated by the σ subunit of RNA polymerase. The rate of transcription of a gene is directly proportional to the rate at which the holoenzyme forms a stable initiation complex.

7.  A promoter region can be identified by a technique called footprinting. RNA polymerase and DNA containing the putative promoter are incubated with an alkylating reagent at DNA bases that are not protected by contact with the RNA polymerase. This results in cleavage of the DNA backbone at the alkylated residues. The products of this reaction can be analyzed by electrophoresis to determine the region of DNA that binds to RNA polymerase.

8.  The RNA polymerase core enzyme binds DNA tightly ($K \approx 5 \times 10^{-12}$ M, with a half-life of ~60 min). The RNA polymerase holoenzyme, however, binds DNA relatively weakly ($K \approx 10^{-7}$ M) except at specific promoter sequences recognized by the σ subunit. Hence, the σ subunit allows the holoenzyme to rapidly scan many sites until it finds its promoter. RNA polymerase then melts ~11 bp of DNA to form an open complex. Once initiation of an RNA chain has occurred, the σ subunit is jettisoned, and the tight-binding core enzyme elongates the RNA chain with high processivity until a termination site is reached.

9.  Transcriptional initiation requires the formation of unwound DNA, which creates a mechanical strain as RNAP maintains a firm grip on the promoter. This frequently results in the release of short transcripts (abortive initiation). The strain eventually provides the energy for RNAP to pull or "clear" the promoter past the enzyme, allowing for successful initiation. Subsequent elongation of the initiated transcript is highly processive and rapid.

10. Transcription termination is relatively imprecise. In prokaryotes, termination occurs in one of two major ways:
    (a) RNA polymerase halts when it encounters a hairpin-forming GC-rich region that is followed by a series of U residues on the DNA sense strand.

(b) Termination is mediated by Rho factor, a hexameric protein that recognizes a nascent RNA and unwinds the DNA–RNA duplex in the open complex, thereby releasing the RNA transcript at a particular site.

## Transcription in Eukaryotes

11. Eukaryotes have three distinct isoforms of RNA polymerase (RNAP):
    (a) RNA polymerase I, located in nucleoli, synthesizes the precursors of most rRNAs.
    (b) RNA polymerase II, located in the nucleoplasm, synthesizes mRNA precursors.
    (c) RNA polymerase III, also located in the nucleoplasm, synthesizes the precursors of 5S rRNA, tRNA, and a variety of small nuclear and cytosolic RNAs.

12. Each RNA polymerase contains up to 15 subunits, several of which are homologous to the subunits of prokaryotic RNA polymerase. RNAP II has a positively charged DNA-binding cleft, two $Mg^{2+}$ ions at the active site, and a clamp structure that enhances processivity. It accommodates the growing RNA chain as a hybrid DNA–RNA helix for about one turn.

13. In eukaryotic promoters, which are much more complex and diverse than prokaryotic promoters, the position and orientation may vary considerably relative to the transcription start site.

14. RNA polymerase I recognizes only one promoter sequence, which is species specific. In contrast, RNA polymerases II and III recognize many different promoters. Some RNA polymerase III promoters are located within the gene's transcribed sequence.

15. There are two classes of RNA polymerase II promoters, both analogous to prokaryotic promoter sequences:
    (a) Most constitutively expressed genes (housekeeping genes) contain CpG islands (see Box 25-4) near the transcription start site.
    (b) Many genes that are selectively expressed in specific cells at specific times contain an AT-rich region known as the TATA box or Inr (initiator) element (see Figure 26-14).
    In addition, there are other upstream sequences that are critical for transcriptional regulation.

16. DNA sequences upstream of the gene play crucial roles in the regulation of transcription. These sequences include enhancers (which stimulate transcription initiation rates) and silencers (which repress transcription initiation rates). Consequently, the regulatory proteins that bind to these sequences are called activators or repressors, respectively. All these regulatory proteins are considered to be transcription factors.

17. In eukaryotes, at least six proteins (called general transcription factors or GTFs) are required to recruit RNA polymerase to a promoter and initiate transcription.

18. The multiprotein complex that initiates transcription is called the preinitiation complex (PIC). The formation of the PIC begins with the binding of the TATA-binding protein (TBP) to the TATA box of a promoter. Several other GTFs bind in succession, followed by RNA polymerase II and the remaining GTFs. One of the PIC proteins (TFIIH) is an ATP-dependent helicase, which forms the open complex.

19. Shortly after transcription commences, RNA polymerase undergoes a conformational change that involves phosphorylation of its C-terminal domain (CTD). Some GTFs dissociate from the transcribing polymerase, and an Elongator complex binds. In eukaryotes, transcription termination is linked to RNA processing.

20. RNA polymerases I and III also use TBP but different sets of GTFs at their TATA-less promoters.

*Posttranscriptional Processing*

21. In prokaryotes, primary transcripts are translated without further modification; however, in eukaryotes most primary transcripts must be specifically modified by:
    (a) The addition of nucleotides to the 5′ and 3′ ends and, in rare cases, in the interior of the transcript.
    (b) The exo- and endonucleolytic removal of polynucleotide sequences.
    (c) Modification of specific nucleotide residues.

22. Eukaryotic mRNAs have a cap structure consisting of a 7-methylguanosine residue joined to the 5′ end of the mRNA via a 5′– 5′ triphosphate linkage. Cap formation requires a triphosphatase, capping enzyme, and one or more methyltransferases.

23. Most eukaryotic mRNAs also have a string of adenosine residues, the poly(A) tail, at their 3′ end. The primary transcript is cleaved past an AAUAAA sequence and the poly(A) tail is added by poly(A) polymerase. The poly(A) tail, which ranges from 20 to 250 nt, protects the mature mRNA from ribonuclease activity in cells.

24. Unlike prokaryotic genes, most eukaryotic structural genes give rise to primary transcripts, called heterogeneous RNA (hnRNA), that are much longer than predicted from the size of the polypeptides they encode. These transcripts are processed by splicing, which is the excision of intervening sequences (introns) and the ordered joining of the flanking expressed sequences (exons) to form a mature mRNA.

25. Each splicing reaction involves two transesterification reactions mediated by ribonucleoprotein complexes called spliceosomes, which contain 5 RNAs known as small nuclear RNAs (snRNAs) and at least 100 polypeptides. The snRNAs mediate proper alignment of splice junctions within the spliceosome. Splicing begins with the nucleophilic attack by the 2′-OH group of an intron adenosine residue on the intron's 5′-phosphate group. The second transesterification reaction splices the 3′-OH group of the 5′ exon with the 5′ end of the 3′ exon, releasing the intron in the form of a lariat.

26. The DNA sequences of exon–intron junctions are conserved. In particular, the 5′ dinucleotide of an intron is invariably GU and the 3′ dinucleotide is invariably AG.

27. In mammals, the number of introns per gene varies from none to 234 (in the titin gene), and the total length of introns can be 4–40 times longer than that of exons. The exons in a gene can be alternatively spliced to yield either isoforms of a class of proteins (e.g., tropomyosins) or proteins whose function has changed (e.g., *Drosophila* sex determination proteins). Approximately 15% of human genetic diseases are caused by point mutations that delete splice sites or result in altered gene splicing.

28. In certain organisms, mRNA undergoes editing in which numerous U residues are inserted and removed to generate a translatable mRNA. Guide RNAs base pair with the immature mRNAs to direct these alterations. Substitutional RNA editing may result in significant protein diversity.

29. In *E. coli*, primary rRNA transcripts contain the sequences for 5S rRNA, 16S rRNA, 23S rRNA, and up to four tRNAs. These RNAs are released by the action of several specific RNases that commence processing while the primary transcript is still being synthesized.

30. Eukaryotic 45S rRNA primary transcripts contain the sequences for 5.8S rRNA, 18S rRNA, and 28S rRNA. Roughly 110 sites are methylated prior to the nucleolytic removal of spacer sequences that separate the three rRNAs. Small nucleolar RNAs (snoRNAs) guide the methyltransferases that modify the rRNAs.

31. Some eukaryotic RNA introns are self-splicing, meaning that catalysis of splicing is completely mediated by the RNA itself. Self-splicing occurs via transesterification reactions with no free energy input. Group I introns are removed by a self-splicing reaction that requires a free guanine nucleotide. Group II introns react via a lariat intermediate and require no free nucleotide; hence, this self-splicing mechanism is nearly identical to that of spliceosome-mediated pre-mRNA splicing.

32. Prokaryotic and eukaryotic transfer RNA is processed by trimming the ends of the primary transcript and modifying specific bases. Many eukaryotic pre-tRNAs also have an intron that is spliced out. In addition, tRNA nucleotidyltransferase adds two C's and an A to the 3′ end of the immature tRNA (prokaryotic pre-tRNA already contains this trinucleotide).

### *Beyond the Equations: Rates of Transcription*

Back in Chapter 11, we saw how rapid biochemical reactions can be, and indeed the complex reactions of transcription are also rapid, despite the mechanistic complexity and substrate selectivity that must be achieved. Measured eukaryotic transcription rates vary somewhat based on the context in which the measurements are made, but values around 20 nucleotides per second are a good starting point for appreciating the long time scales for the transcription of eukaryotic genes.

When the sequence of the human genome was first published (*Nature* 409, 896) some numbers for the sizes of exons and introns, and numbers of each were tabulated with the following averages:

| Gene component | Average value |
| --- | --- |
| Intron number | 6 |
| Intron size | 3,365 base pairs |
| Internal exon  size | 145 base pairs |
| 5′ UTR-containing exon | 770 base pairs |
| 3′ UTR-containing exon | 300 base pairs |

*Question*: Calculate the so-called average gene size based on the information provided in the table.

*Answer*:  An average gene based on these numbers would have 7 exons of which there are 5 internal exons along with the exons containing the 5′ and 3′ untranslated regions. Adding the numbers together, we have

5′ UTR-exon size + 3′ UTR-exon size + (5 × internal exon size) + (6 × intron size) = 770 base pairs + 300 base pairs + 725 base pairs + 20,190 base pairs = 21,985 base pairs. So at 20 nucleotides per second, we can see that transcription of an averaged-sized gene would take

21,985 nucleotides / 20 nucleotides/sec = 1099 secs = 18.32 minutes.  Since splicing and other processing reactions often occur as transcription ensues, this represents the approximate time to synthesize a mRNA for an average sized gene. Now consider the largest gene identified in the human genome.

**The *Dystrophin*** gene, with $2.3 \times 10^6$ base pairs, yields a primary, unspliced transcript that is 1,770,000 nucleotides long.

How long does it take to produce the primary transcript? How long does it take to translate this protein?

Doing the same simple calculation, we have

1,770,000 nts / (20 nucleotides/sec) = 88,500 secs = 1,475 minutes = 24.58 hours!

As you can see, size matters when it comes to macromolecular synthesis. We will play this forward in the next chapter and look at the translation of this gene and another muscle gene that is the largest polypeptide produced by vertebrate cells.

## Questions

*Prokaryotic RNA Transcription*

1.  The prokaryotic RNA polymerase holoenzyme contains the core enzyme and a σ factor. What is the function of the core enzyme? What are its limitations in terms of enzymatic activity? What is the role of the σ factor?

2.  In a binding assay, a bacterial RNA polymerase binds its 70-base promoter 100 times faster when the promoter is part of an 8 kb plasmid than when the promoter is part of a 200-bp linear DNA. What might account for this difference in binding behavior? Assume that the polymerase is incubated with an equal molar concentration of each type of DNA.

3.  What evidence suggests that newly transcribed RNA is not wound around its template DNA?

4.  You are interested in discerning the role of σ factors in prokaryotic transcription, using a system of purified core polymerase, a DNA template, a σ factor, and labeled nucleotides ($^{32}$pppN). The incorporation of $^{32}$P into RNA at 2 different core enzyme concentrations ($10^{-10}$ M and $10^{-11}$ M) and constant concentration of σ factor ($10^{-12}$ M) is shown below.
    (a)  What does the graph measure?
    (b)  What can you conclude about the behavior of σ factor?
    (c)  What can you conclude about the behavior of the core enzyme?

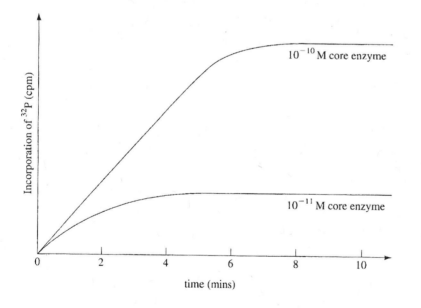

5.  What evidence indicates that a GC-rich region of DNA followed by an AT-rich region is sufficient but not necessary to promote RNA chain termination?

*Transcription in Eukaryotes*

6.  Distinguish between enhancers and promoters.

7.  You have discovered a novel eukaryotic RNA polymerase. To study this enzyme, you have developed an *in vitro* assay with an inhibitor of RNA chain elongation, 3′-deoxy-5′[α-$^{32}$P] CTP. You are surprised by the fact that no $^{32}$P is found in the transcribed RNA! What does this result tell you about the mechanism of the polymerase?

*Posttranscriptional Processing*

8.  Indicate whether the posttranscriptional modifications listed below occur in prokaryotes or in eukaryotes.

    (a)  5′ Cap
    (b)  Polyadenylation
    (c)  Methylation of nucleotide residues
    (d)  Endonucleolytic cleavage
    (e)  Splicing

9.  Which of the following is contained in the rRNA cistron of *E. coli*?
    (a) 16S rRNA
    (b) 23S rRNA
    (c) a tRNA gene
    (d) 5S RNA

10. What RNA-processing enzyme complexes contain RNA? In which does RNA participate in catalysis?

11. What kinds of posttranscriptional processing occur in eukaryotic tRNA precursors?

12. Group I introns incubated with increasing concentrations of poly(C) show Michaelis–Menten kinetics for the cleavage of poly(C); that is, in the presence of increasing amounts of poly(C), the rate increases to a maximum. Is this proof that these introns are catalysts? Explain.

13. The mRNA capping reaction involves cleavage of a methylated guanosine triphosphate to release its pyrophosphate. You wish to measure RNA chain initiation rates in isolated nuclei. Which of the isotopes shown below would you use? Explain. What assumptions are necessary to use this experimental strategy?

    $^{32}$pppA          p$^{32}$ppA          pp$^{32}$pA

14. What is the role of SXL and U2AF in generating alternatively spliced *tra* mRNA transcripts in the *Drosophila* sex-determination pathway?

15. Does GTP serve as a substrate, catalyst, cofactor, or allosteric effector in the self-splicing reaction of *Tetrahymena* pre-rRNA? Explain.

16. Describe the primer you would use to initiate reverse transcriptase–catalyzed synthesis of DNA that is complementary to a eukaryotic mRNA molecule.

# 27
# Protein Synthesis

This chapter explores the mechanisms by which messenger RNA directs the synthesis of a polypeptide. The discussion begins with an examination of the genetic code, including the arguments and experiments that led to the conclusion that an mRNA nucleotide sequence is decoded as a nonoverlapping series of triplets. Hence, in this coding scheme, there are 61 triplet combinations, or codons, that encode the 20 amino acids and three Stop codons. Because most amino acids are specified by more than one codon, the genetic code is said to be degenerate but nonrandom. Whereas some amino acids are encoded by as many as six codons, others are encoded by only one; this phenomenon may reflect the way the genetic code evolved (Box 27-1). The discussion then turns to the molecules that "translate" the genetic code: transfer RNAs. The structures of the tRNAs are explored, from their modified bases to their unusual secondary and tertiary structures. Each tRNA can covalently attach an amino acid at its 3′ end to form an aminoacyl–tRNA. Two families of aminoacyl–tRNA synthetases (aaRS), which show surprisingly little sequence similarity, mediate this aminoacylation. Structural information on the mechanisms by which aaRSs recognize their cognate tRNAs is also presented. The chapter then discusses the "wobble hypothesis," which accounts for the degeneracy of the genetic code at the level of base-pairing interactions between the codons of mRNA and the anticodons of tRNAs. Some organisms have a "21st amino acid," selenocysteine, which is incorporated into certain proteins using a unique tRNA (Box 27-2). The chapter continues with a description of ribosome structure and the mechanisms of polypeptide chain initiation, elongation, and termination. Included in this section are discussions of the nature of the peptidyltransferase activity, the role of GTP hydrolysis in protein synthesis and translational accuracy, and the mechanisms by which antibiotics interfere with translation (Box 27-3). The chapter concludes with a discussion of posttranslational processing, including proteins involved in protein folding, covalent modification, and secretion.

## *Essential Concepts*

1. Translation shares important features with DNA transcription and replication:
   (a) Translation occurs in a large multiprotein machine, which includes accessory factors that regulate the initiation, elongation, and termination of polypeptide synthesis.
   (b) Even though it is amino acids that are polymerized, translation depends on Watson–Crick base pairing between mRNA and tRNA so that amino acids are joined as genetically specified.
   (c) Accuracy is essential, so the translational machinery has proofreading capabilities.
   (d) The process is endergonic and requires the cleavage of "high-energy" phosphoanhydride bonds.

*The Genetic Code*

2. The DNA sequence encoding a protein is made up of a triplet of nucleotides called a codon. A triplet code allows more than one codon to specify an amino acid; hence, the code is said to be degenerate. For example, leucine is specified by six codons, serine by four, and glutamate by two. Codons that specify the same amino acid are termed synonyms.

3. Transfer RNA deciphers the codon. Each tRNA contains a trinucleotide sequence, called the anticodon, which is complementary to an mRNA codon specifying the amino acid attached to the tRNA.

4. The code is read sequentially and contains no internal punctuation. The sequence AUG (read 5′ to 3′ along the mRNA), or occasionally GUG, at the 5′ end of the mRNA marks the first codon to be translated. The codons UAA, UAG, and UGA are called Stop codons because they signal the termination of translation.

5. The arrangement of codons is nonrandom. Most synonyms share the first two nucleotides. Interestingly, codons with different nucleotides in the first position tend to specify chemically similar amino acids. Codons with second position pyrimidines encode mostly hydrophobic amino acids, and those with second position purines encode mostly polar amino acids. This suggests a nonrandom evolution of the genetic code that minimizes the deleterious effects of point mutations (see Box 27-1).

6. The "standard" code is not universal; exceptions have been observed in mitochondria and a few protozoans.

*Transfer RNA and Its Aminoacylation*

7. Almost all known tRNAs can be arranged in a so-called cloverleaf secondary structure, which has the following key features:
   (a) A 5′-terminal phosphate group.
   (b) Three keyhole-shaped stem-and-loop structures called the D arm, the anticodon arm, and the T or TψC arm. The D arm is named for the modified base dihydrouridine (D) that frequently occurs in the loop. The anticodon arm contains the anticodon sequence in a loop. The TψC arm is named for the sequence thymidine–pseudouridine (ψ)–cytidine that the loop usually contains.
   (c) A 3′ CCA sequence containing a free 3′ hydroxyl group. This trinucleotide is either genetically specified (in prokaryotes) or enzymatically appended to the immature tRNA (in eukaryotes).
   (d) Between the anticodon arm and the TψC arm is a variable arm whose sequence and length (3–21 nucleotides) varies among tRNAs as well as among species.
   (e) Transfer RNAs have 15 invariant positions (always the same base) and 8 semivariant positions (only a purine or only a pyrimidine) that occur mostly in the loops.

8. Up to 25% of the bases of tRNA are modified posttranscriptionally. The biological significance of many of these modifications is still mysterious, since they are not required for the structural integrity of tRNA or its proper binding to the ribosome.

9. Transfer RNA has an L-shaped tertiary structure in which structural stability is maintained by extensive base stacking interactions and base pairing within and between the helical stems.

10. Attachment of an amino acid to a tRNA is carried out in a two-reaction process by an aminoacyl–tRNA synthetase (aaRS). The first reaction involves "activation" of the amino acid by its reaction with ATP to form an aminoacyl–adenylate with the release of $PP_i$. The mixed anhydride then reacts with tRNA to form the "charged" aminoacyl–tRNA with the release of AMP. The hydrolysis of pyrophosphate ($PP_i$) ensures that the overall reaction is irreversible.

$$
\begin{aligned}
&1. \qquad \text{Amino acid} + \text{ATP} \rightleftharpoons \text{aminoacyl–AMP} + PP_i \\
&2. \qquad \text{Aminoacyl–AMP} + \text{tRNA} \rightleftharpoons \text{aminoacyl–tRNA} + \text{AMP} \\
&\overline{\qquad \text{Amino acid} + \text{tRNA} + \text{ATP} \rightleftharpoons \text{aminoacyl–tRNA} + \text{AMP} + PP_i}
\end{aligned}
$$

11. There are two structurally unrelated classes of aminoacyl–tRNA synthetases (Class I and Class II). These enzymes differ in
    (a) The mechanism by which they recognize their tRNA substrates.
    (b) The initial site of aminoacylation on the tRNA.
    (c) Amino acid specificity.

12. Each set of isoaccepting tRNAs is recognized by a single aminoacyl–tRNA synthetase (aaRS). Isoaccepting tRNAs have different anticodons but are conjugated to the same amino acid. aaRSs appear to recognize the unique structural features of the cognate tRNAs, not specific base sequences. Some aaRSs are capable of proofreading by a double-sieve mechanism, which requires a second active site for hydrolysis.

13. During protein synthesis, the proper tRNA is selected only through codon–anticodon interactions; the aminoacyl group is not involved. Many tRNAs bind to two or three of the codons that specify their cognate amino acids. This variation is due to non-Watson–Crick base pairing between the third codon position and the first anticodon position. For example, U in the first position of the anticodon can base pair with A or G at the third codon position. Some tRNAs have inosine at the first anticodon position, which can base pair with U, C, or A. The degeneracy in binding at this position is called wobble pairing.

*Ribosomes*

14. Ribosomes carry out the following functions:
    (a) They bind mRNA so that its codons are matched, with high accuracy, to tRNA anticodons.
    (b) Ribosomes contain three specific binding sites: the A site for the incoming aminoacyl–tRNA; the P site for the tRNA to which the growing polypeptide chain is attached; and the E site for the outgoing, uncharged tRNA.

(c) They mediate interactions of nonribosomal proteins that are required for polypeptide chain initiation, elongation, and termination. Many of these factors use the free energy of GTP hydrolysis to carry out their functions.

(d) Ribosomes catalyze peptide bond formation between the incoming amino acid and the growing polypeptide chain.

(e) They are responsible for the translocation of the mRNA so that codons are read sequentially.

15. Prokaryotic ribosomes are enormous ribonucleoprotein machines containing dozens of proteins and several RNA molecules. The *E. coli* ribosome has a sedimentation coefficient (which indicates its rate of sedimentation in an ultracentrifuge) of 70S and is composed of two subunits: The 30S subunit contains the 16S rRNA, while the 50S subunit contains the 23S and 5S rRNAs. The 30S subunit accommodates a single strand of mRNA (without any secondary structure); the 50S subunit harbors the growing polypeptide chain.

16. Prokaryotic ribosome structure is complex and has been pieced together through a combination of X-ray diffraction and cryoelectron microscopy. From these studies some generalizations have emerged.

(a) Both 16S and 23S rRNAs are assemblies of helices connected by loops.

(b) The four domains of 16S rRNA form distinct parts of the 30S subunit.

(c) The six domains of 23S rRNA are intertwined throughout the 50S subunit.

(d) The distribution of ribosomal proteins is not uniform. The facing surfaces of each ribosomal subunit are largely devoid of proteins; whereas the other surfaces of each subunit contains the vast majority of the ribosomal proteins.

(e) Most ribosomal proteins consist of a globular domain that is located on the surface of the ribosome and an unstructured tail that penetrates the rRNA and interacts with it via salt bridges.

17. Eukaryotic ribosomes resemble prokaryotic ribosomes in shape and function but are larger and more complex. The 80S ribosome is composed of a 40S small subunit (with 18S rRNA) and a 60S large subunit (with 5S, 5.8S, and 28S rRNAs). The 18S rRNA is homologous to the prokaryotic 16S rRNA, while the 5.8S and the 28S rRNAs are homologous to the prokaryotic 23S rRNA.

*Translation*

18. Polypeptide synthesis has the following key features:

(a) Ribosomes read mRNA in the 5′ to 3′ direction.

(b) Polypeptide synthesis proceeds from the N-terminus to the C-terminus.

(c) Chain elongation occurs by linking the growing polypeptide chain to the incoming charged tRNA.

(d) Several ribosomes can bind to a single mRNA to form a complex called a polysome or polyribosome.

19. Polypeptide chain initiation begins with Met–tRNA in eukaryotes and a modified Met–tRNA in prokaryotes, *N*-formylmethionyl–tRNA (fMet–tRNA). In prokaryotes, the AUG start codon is distinguished from other AUG codons by base pairing between the 16S

rRNA and a 5′ mRNA sequence called the Shine–Dalgarno sequence. Eukaryotes do not contain an analogous recognition sequence; initiation occurs at the first AUG of a eukaryotic mRNA.

20. Initiation of translation in *E. coli* occurs in three stages:
    (a) IF-3 binds to the 70S ribosome and promotes its dissociation into the 30S and 50S subunits.
    (b) mRNA, IF-1, and IF-2 in complex with GTP and fMet–tRNA bind to the 30S subunit.
    (c) IF-1 and IF-3 are released, allowing the 50S subunit to bind to the 30S subunit. This stimulates hydrolysis of the GTP bound to IF-2, resulting in a conformational change in the 30S subunit and the release of IF-2, which can participate in further initiation reactions.

21. Translational initiation in eukaryotes is more complex than that in prokaryotes. The three stages described for prokaryotic translational initiation also occur in eukaryotes, but up to 11 initiation factors—instead of 3—are involved. The initiation factor eIF4F binds to the 7-methylguanosine ($m^7G$) cap found on all eukaryotic mRNAs. This complex recruits several additional initiation factors, which then bind to the 43S preinitiation complex, forming the 48S initiation complex, which then scans the mRNA until it encounters the first AUG codon.

22. Elongation of a nascent polypeptide chain in prokaryotes proceeds as a three-reaction cycle (decoding, transpeptidation, and translocation) until a termination codon is encountered.

23. An incoming aminoacyl–tRNA, in a complex with EF-Tu·GTP, binds to the A site. This interaction is accompanied by the hydrolysis of GTP, which releases $P_i$ and EF-Tu·GDP from the ribosome. EF-Ts replaces the GDP bound to EF-Tu, which is then replaced by GTP, thereby regenerating the EF-Tu·GTP complex. The aminoacyl–tRNA·EF-Tu·GTP undergoes a large conformational change (a 91° domain rotation!) when GTP is hydrolyzed, which moves the aminoacyl–tRNA into a position to allow the transpeptidation reaction to occur.

24. The ribosome senses proper codon–anticodon pairing through interactions with the 16S rRNA. If there is a mispairing, the aminoacyl–tRNA·EF-Tu·GTP complex dissociates from the ribosome before GTP is hydrolyzed. In addition, a proofreading step occurs after GTP hydrolysis. GTP hydrolysis is therefore the thermodynamic price of ribosomal translation accuracy.

25. The 23S rRNA of the large ribosomal subunit acts as a ribozyme to carry out the transpeptidation reaction. A peptide bond is formed through the nucleophilic displacement of the P-site tRNA by the amino group of the aminoacyl–tRNA in the A site. Hence, the nascent polypeptide is lengthened at its C-terminus by one amino acid and transferred to the A site, a process called transpeptidation. The 23S rRNA is the catalyst in peptide bond formation, although it does not participate as an acid–base catalyst. The main catalytic mechanism appears to involve positioning and orienting the substrates for reaction.

26. Translocation of the new peptidyl–tRNA occurs in several discrete steps. After transpeptidation, the acceptor end of the new peptidyl–tRNA moves into the P site, while the acceptor end of the deacylated tRNA moves into the E site. EF-G·GTP binding to the ribosome and subsequent GTP hydrolysis allows for the complete displacement of both tRNA moieties to the E and P sites, resulting in a completely vacant A site. Hence, the alternating activities of EF-Tu and EF-G keep the ribosome cycling in a unidirectional manner. In fact, EF-G structurally mimics the EF-Tu–tRNA complex.

27. Elongation in eukaryotes is similar to that in prokaryotes, except that eEF1A and eEF1B replace EF-Tu and EF-Ts, respectively, and eEF2 replaces EF-G.

28. In *E. coli*, two release factors, RF-1 and RF-2, recognize Stop codons. Binding of either RF-1 or RF-2 to the ribosome stimulates the transfer of the peptidyl group to water, thereby terminating polypeptide chain elongation. Next, RF-3 binds and, following hydrolysis of its GTP, stimulates the release of RF-1 and RF-2. Ribosome recycling factor (RRF) and EF-G·GTP then bind. GTP hydrolysis by EF-G causes RRF to move into the P site such that the remaining tRNAs are released from the P and E sites. Finally, RRF and EF-G·GDP dissociate. Eukaryotic translational termination is similarly mediated by eRF, which carries out all the activities of RF-1, RF-2, and RF-3.

29. A mutation that converts an aminoacyl-coding ("sense") codon to a Stop codon is called a nonsense mutation and results in the premature termination of translation. Such mutations can be rescued by a nonsense suppressor tRNA, which contains an anticodon to a Stop codon but carries the amino acid of its wild-type progenitor.

30. Many commonly used antibiotics interfere with the initiation or elongation steps in prokaryotic translation. Puromycin mimics tyrosyl–tRNA and aborts transpeptidation. Streptomycin inhibits translational initiation. Chloramphenicol inhibits the peptidyltransferase reaction. Tetracycline inhibits aminoacyl–tRNA binding to the ribosome. The toxin ricin inactivates the eukaryotic ribosome by excising a base from 28S rRNA.

*Posttranslational Processing*

31. Ribosomal proteins may recruit chaperone proteins to assist with protein folding. Trigger factor binds to the L23 ribosomal protein and recognizes short hydrophobic segments emerging from the ribosome, prior to the arrival of other chaperone proteins such as DnaK, DnaJ, and the GroEL/ES chaperonins.

32. Posttranslational modifications of newly synthesized proteins are common, especially in eukaryotes. In some cases, limited proteolysis may activate a protein, which is called a proprotein. A preproprotein also contains a signal peptide that is removed in the endoplasmic reticulum, prior to excretion of the protein. Common modifications of proteins include phosphorylation (at Ser, Thr, Tyr, and His residues), glycosylation, hydroxylation of Pro and Lys residues in collagen, and the attachment of ubiquitin or small ubiquitin-related modifier (SUMO) to Lys residues.

33. The signal recognition particle (SRP) is a ribonucleoprotein containing a 300-residue RNA (in eukaryotes) or a 114-nt RNA (in prokaryotes). When an N-terminal hydrophobic region emerges from the ribosome, the SRP binds to the nascent polypeptide and facilitates the binding of the associated ribosome to the SRP receptor complex in the endoplasmic reticulum (in eukaryotes) or plasma membrane (in prokaryotes). This results in the translocation of the newly synthesized polypeptide across the membrane.

34. Proper folding of glycoproteins in the endoplasmic reticulum (ER) relies on the activity of two homologous chaperones, membrane-bound calnexin and soluble calreticulin.

## Questions

### The Genetic Code

1. Why is the genetic code degenerate?

2. How many amino acid copolymers can be specified by polynucleotides consisting of two alternating nucleotides? What are the sequences of these polypeptides?

3. A cell extract from a new species of *Amoeba* is used to translate a purified mouse hemoglobin mRNA, but only a short peptide is produced. What is the simplest explanation for this result?

4. Predict the effect on protein structure and function of an A·T to G·C transition in the first codon position for lysine.

5. Identify the polypeptide encoded by the DNA sequence below, in which the lower strand serves as the template for mRNA synthesis.

        5′  GGACCTATGATCACCTGCTCCCCGAGTGCTGTTTAGGTGGG  3′
        3′  CCTGGATACTAGTGGACGAGGGGCTCACGACAAATCCACCC  5′

6. Describe how nucleotide substitutions at different positions in codons affect the characteristics of the encoded amino acids.

### Transfer RNA and Its Aminoacylation

7. What two reactions are carried out by aminoacyl–tRNA synthetases?

8. Which of the following tRNA structural features are always involved in interactions with an aaRS?
    (a) Acceptor stem      (b) Anticodon loop         (c) TψC arm

9. A new species of yeast has just one tRNA$^{Ala}$, with the anticodon IGC. What does this suggest about codon usage in this organism?

10. Which of the following amino acids are likely to be linked to only one species of tRNA? What are their anticodons?
    (a) Phe          (b) Leu          (c) His

*Translation*

11. Reticulocyte (immature red blood cell) lysates can be made devoid of mRNA by ribonuclease treatment followed by inactivation and removal of the ribonuclease. These lysates are now capable of synthesizing protein when mRNA, GTP, and aminoacyl–tRNAs are added. How could you use this system to demonstrate that a protein is synthesized from the N-terminus to the C-terminus?

12. Match each of the functions on the left with the appropriate prokaryotic protein on the right.

    _____ Binds fMet–tRNA and GTP.                                  A. RF-1
    _____ Binds aminoacyl–tRNA and GTP.                             B. RF-2
    _____ Recognizes Stop codons.                                  C. RF-3
    _____ Binds GTP and promotes RF-1 and RF-2                     D. IF-1
          release from the ribosome
    _____ Hydrolyzes GTP to GDP.                                    E. IF-2
    _____ Promotes the transfer of peptidyl–tRNA from              F. IF-3
          the A site to the P site.
    _____ Inhibits the interaction of the 30S and 50S subunits.    G. EF-Ts
    _____ Facilitates mRNA binding to the 30S ribosome.            H. EF-Tu
    _____ Displaces GDP from EF-Tu.                                 I. EF-G

13. How could you show that the inhibition of protein synthesis in a bacterial cell extract by puromycin is not irreversible?

14. You have isolated a temperature-sensitive mutant *E. coli*. At 34°C, the growth rate is nearly normal but is reduced 10-fold at 37°C. You eventually discover that EF-Tu·GDP binding to the ribosome is two-fold higher in the mutant at 37°C. What aspect of translation is affected by the mutation?

15. Which amino acid substitutions might occur in *E. coli* opal suppressors that arise by single-base mutations?

16. The error rates of translation in some rapidly growing *E. coli* strains is as high as 2 percent. What is the percentage of correctly translated protein in such bacteria (assume the protein has 100 residues)? How can these bacteria do so well?

17. Distinguish between eukaryotic and prokaryotic polypeptide initiation, elongation, and termination.

*Posttranslational Processing*

18.   What is the adaptive significance of posttranslational cleavage of proteins such as collagen and various hormones?

19.   Proteins destined for membranes or secretion have a stretch of N-terminal hydrophobic amino acids called a signal sequence. What function does this sequence serve?

# Regulation of Gene Expression

This chapter introduces you to one of the main efforts of modern biochemistry, understanding the regulation of gene expression, which defines the phenotypic properties of cells. In eukaryotes, the regulation of gene expression is more complex than in prokaryotes, largely due to the size of the eukaryotic genome, the numerous levels of DNA packaging, the complexity of the cells, and the numerous cell types in eukaryotic organisms. This chapter begins by examining genome organization in prokaryotes and eukaryotes, focusing on gene number, gene clusters, and nontranscribed DNA. The human, yeast, and *E. coli* genomes are used to illustrate the kinds of information provided by the emerging field of genomics. Included in this discussion is an overview of the organization of DNA in eukaryotes with regard to repetitive and unique DNA sequences.

The regulation of prokaryotic gene expression is discussed next, with attention focused on the *lac* operon, *trp* operon, and riboswitches, each illustrating key features common to the regulation of all prokaryotic genes. The *lac* repressor is an example of a negative regulator regulated by an allosteric effector; the *trp* operon provides an example of attenuation. Riboswitches are interesting cases of posttranscriptional regulation via the binding of allosteric effectors to mRNA.

The chapter then turns to eukaryotic gene regulation, which can occur at many points between the exposure of promoter sites within chromatin and the translation of the mRNA. Various regulatory mechanisms involve chromatin remodeling; histone modification; DNA methylation; the binding of transcriptional activators and repressors to DNA sequences upstream from the promoter; activation of signal transduction pathways; selective degradation of RNA; phosphorylation of translation initiation factors; and DNA rearrangements (as exemplified by immunoglobulin gene diversity). Progression of cells through the cell cycle is regulated by the repression and activation of key transcriptional activators and repressors. Two key regulators of cell cycle progression are p53 and pRb, mutant forms of which cause human cancers. The chapter closes with an overview of gene expression in the development of the fruit fly, *Drosophila melanogaster*. Here, a set of genes that regulate the expression of other genes orchestrates embryogenesis. *Hox* genes encode a remarkably conserved set of transcription factors that participate in embryonic patterning in *Drosophila* and a wide variety of organisms.

## *Essential Concepts*

1. Gene expression is neither random nor fully preprogrammed. Information in an organism's genome is used in an orderly fashion during development but also responds to changes in the organism's internal or external environment.

*Genome Organization*

2. The new discipline of genomics, the study of organisms' genomes, endeavors to elucidate the functions of all genes. The vast amount of data obtained from the numerous genome projects to date is stored in computerized databases, most of which are publicly available via the Internet. Genomics also explores evolutionary relationships between genes, organelles, and organisms.

3. There is surprisingly modest correlation between the amount of DNA in an organism's genome (its C value) and its apparent morphological or metabolic complexity. This is known as the C-value paradox. However, there is a stronger correlation between gene number and morphological complexity (as measured by the number of cell types in the organism; see Figure 28-2).

4. Only ~1.4% of the 3.2 billion base pairs of the human genome encodes proteins; this proportion is typical of most vertebrate genomes sequenced to date. So far, computer analysis indicates that there are around 23,000 protein-encoding genes in mammalian genomes, of which almost 42% are of unknown function and are referred to as orphan genes. About three-quarters of all known human genes have counterparts in other species. About one-quarter are found only in other vertebrates, while another quarter are common to all prokaryotes and eukaryotes.

5. Approximately 60% of the human genome is transcribed into RNA:
   (a) Over 4000 genes encoding rRNA, tRNA, snRNA and other small RNAs; and
   (b) At least another 10,000 loci resulting in the transcription of noncoding RNAs (ncRNAs), most of which have no known function. Many of these RNAs are involved in RNA interference, a form of posttranscriptional gene silencing.

6. A gene can be identified using two broad strategies:
   (a) It can identified as an open reading frame (ORF), a sequence that is not interrupted with a Stop codon; however, this method is limited due to the short size of exons and must rely on the availability of cDNA sequences from various tissues or organisms (expressed sequence tags, ESTs).
   (b) A gene can also be identified by the presence of CpG islands that are clustered upstream of many genes.

7. In simple eukaryotes, such as yeast, it is currently possible to mutate every gene to determine its function; this information can then be applied to more complex eukaryotic organisms. In addition, human genome sequence information will make it easier to identify genes associated with disease. To this end, a catalog of single nucleotide polymorphisms (SNPs) is being compiled. A SNP occurs on average about every 1250 bp in humans, and nearly 2 million have been described so far.

8. An organism's genes are not distributed randomly throughout its genome; indeed, some genes are organized in both prokaryotic and eukaryotic genomes. Prokaryotic operons include genes related to a specific metabolic function. In eukaryotes, genes for histones and globins occur in clusters. Ribosomal RNA genes are also clustered in both prokaryotes and eukaryotes.

9. A significant portion of prokaryotic and eukaryotic genomes consists of DNA that is never transcribed. These unexpressed sequences accumulate mutations at a great rate since they experience little selective pressure; hence, they can be used to trace evolutionary relationships. In prokaryotes, these nontranscribed sequences are predominantly promoter and operator sequences adjacent to genes but also include insertion sequences and remnants of integrated viruses. In eukaryotes, nontranscribed sequences include promoters and other regulatory regions; however, the bulk of these sequences are repetitive DNA of unknown function. Some short (~10 bp) sequences are present in millions of copies, often tandemly repeated thousands of times. In humans, these short tandem repeats comprise about 3% of the genome. Eukaryotes also contain moderately repetitive sequences ($< 10^6$ copies per haploid genome) occurring in 100 to several thousand bp segments, most of which are retrotransposons. About 42% of the human genome consists of three types of retrotransposons: (a) long interspersed elements (LINEs), which are 6–8 kb long; (b) short interspersed elements (SINEs), which are 100-400 bp long; and (c) long terminal repeats (LTRs).

10. Several neurological diseases are linked to the expansion of trinucleotide repeats in certain genes. These trinucleotide repeats exhibit genetic instability: above a threshold of about 35–50 copies, they expand with successive generations (see Box 28-1). The longer the repeat, the more severe the disease and the earlier its age of onset (genetic anticipation).

## Regulation of Prokaryotic Gene Expression

11. Prokaryotic gene expression is mainly controlled at the level of transcription. Translational control is probably not necessary since mRNAs have half-lives of only a few minutes.

12. Transcription of the *lac* operon is under negative control. The *lac* repressor prevents transcription initiation by binding to the operator. A metabolite of lactose, 1,6-allolactose, acts as an inducer of transcription by binding to the *lac* repressor, which causes a conformational change in the protein that dramatically reduces its affinity for the operator.

13. The *lac* repressor is a homotetramer with four functional components that interact with two DNA segments simultaneously.
   (a) The N-terminal region contains a helix–turn–helix (HTH) motif common to several other prokaryotic DNA-binding proteins. The N-terminal fragment binds to specific operator DNA sequences;
   (b) A α-helical linker region, which acts as a hinge and also binds DNA;
   (c) A two-domain core region, which binds to inducers and allosteric effectors, such as IPTG; and
   (d) A C-terminal α helix, which is required for homotetramer formation.

14. The presence of glucose prevents the expression of genes involved in the catabolism of other sugars, a phenomenon known as catabolite repression. In the presence of cAMP (which signals the absence of glucose), catabolite gene activator protein (CAP) is a positive regulator of transcription of the *lac* operon and certain other operons. CAP–cAMP binds to the promoter region of the *lac* operon via its two HTH domains and stimulates transcription.

15. Attenuation of transcription occurs in the *E. coli trp* operon, which is also controlled by the *trp* repressor (the repressor binds to the operator when tryptophan, which is a corepressor, is also present). The 5′ end of the *trp* operon mRNA contains a leader sequence that encodes a short polypeptide containing two consecutive Trp residues. When Trp is limiting, a ribosome translating the leader peptide stalls at the Trp codons. Meanwhile, transcription of the leader sequence continues and the mRNA beyond the ribosome forms a hairpin secondary structure (the antiterminator) that allows transcription to proceed into the operon's five genes. When Trp (and therefore Trp–tRNA$^{Trp}$) is present, the ribosome proceeds rapidly through the leader sequence, which allows the formation of an alternative hairpin structure in the mRNA that terminates transcription.

16. Portions of prokaryotic mRNA act as riboswitches, such that their binding small metabolites regulates their transcription termination, translation initiation, or self-cleavage by a feedback mechanism. While this phenomenon has been most frequently described in prokaryotes, it has also been observed in fungi and plants.

*Regulation of Eukaryotic Gene Expression*

17. The regulation of gene expression in eukaryotes can occur at several levels, as outlined below. Keep in mind that no regulatory mechanism operates exclusively as an on–off switch and that several mechanisms may govern expression of a particular gene.
    (a) **Gene Availability.** Genes may be lost through transposition, rearranged through recombination, condensed into transcriptionally silent heterochromatin, or chemically modified in different tissues or at different stages of development.
    (b) **Transcriptional Control.** Regulation of the initiation, elongation, termination, or posttranscriptional processing phases of mRNA synthesis directly affects mRNA concentrations. Additional mechanisms for stabilizing and degrading mRNA influence the cellular half-life of the mRNA.
    (c) **Translational Control.** Regulation of the initiation, elongation, or termination phases of mRNA translation influences the concentrations of the corresponding polypeptides. Regulatory mechanisms that stabilize or degrade polypeptides also affect the rates of protein turnover.
    (d) **Protein Activation.** Posttranslational processing reactions, such as phosphorylation, acetylation, glycosylation, proteolysis, or assembly into multimers, may govern protein activity.
    (e) **Allosteric Effects.** Allosteric effectors that alter enzymatic properties or the assembly of multiprotein structures play a role in regulating gene expression.

18. Most of the DNA in a eukaryotic organism is packaged in a highly condensed form of chromatin called heterochromatin. Transcribed DNA is packaged more loosely as euchromatin and is more accessible to the transcription machinery.

19. Random X chromosome inactivation in female mammals results in equal amounts of X chromosome-encoded gene products in both males and females and is referred to as dosage compensation. Two mechanisms that appear to operate are (a) the binding of *Xist* mRNA to the chromosome from which it was transcribed, thereby inactivating that chromosome; and (b) methylation of CpG islands within many promoters on the inactive X chromosome.

20. Nearly all eukaryotic DNA is packaged in nucleosomes. For transcription to proceed, nucleosomes must be disassembled or remodeled to some extent so that regulatory proteins and RNA polymerase can bind to DNA. This remodeling is an ATP-driven process and is carried out by chromatin-remodeling complexes that create a dynamic state in which DNA maintains its overall packing but is exposed in various areas to allow interactions with the transcriptional machinery and regulatory transcription factors. The best characterized among these complexes are the yeast SWI/SWF and RSC complexes, which have homologs in higher eukaryotes. Other so-called architectural proteins that help regulate gene expression are the high mobility group (HMG) proteins, a group of small (<30 kD) proteins that get their name from their high electrophoretic mobilities in polyacrylamide gels.

21. Nucleosome structure may be altered by covalent modification, including:
    (a)  Acetylation/deacetylation of certain histone Lys residues
    (b)  Phosphorylation/dephosphorylation of specific Ser (and possibly Thr) residues
    (c)  Methylation of certain Lys and Arg residues

    Acetylation and phosphorylation render the side chains less positively charged, leading to weakened histone–DNA interactions, whereas methylation increases the basicity and hydrophobicity of the side chains and hence tends to stabilize nucleosomal structure. These kinds of histone modifications, which have suggested the existence of a "histone code," may act in sequence or in combination to generate specific biological outcomes or cellular phenotypes.

22. The best studied chromatin modifiers are the histone acetyltransferases (HATs), which are components of multisubunit coactivators such as TAF1 (the largest subunit of the general transcription factor TFIID). The acetate donor for the HAT reaction is acetyl-CoA. Specific patterns of histone acetylation may recruit HAT-associated transcriptional coactivators, which contain ~110-residue modules known as bromodomains that bind to specific acetylated Lys residues of histones.

23. The methylation of histones and DNA leads to the inhibition of transcription, or gene silencing. *S*-adenosylmethinine is the methyl donor for these reactions.
    (a)  Histone methyltransferases (HMTs) add methyl groups that are not readily removed and are most commonly seen in heterochromatin. Heterochromatin proteins (e.g., HP1) bind to methylated H3 Lys 9 via a protein module referred to as a chromodomain.

(b) DNA methyltransferases (DNA MTases) add a methyl group to cytosine residues to form 5-methylcytosine ($m^5C$) residues. DNA methylation appears to be self-perpetuating across cell divisions and may play a key role in genomic imprinting, which results in differential expression of maternal and paternal genes early in mammalian embryogenesis.

24. Proteins that regulate transcription are known as activators and repressors and bind to DNA targets referred to as enhancers and silencers, respectively. Since many enhancers are located upstream from the transcription start site, these regulatory proteins are also called upstream transcription factors. Many of these proteins are subject to regulation via signal transduction pathways. Certain assemblies of transcription factors that also include architectural proteins are called enhanceosomes.
    (a) Many upstream transcription factors act cooperatively to stimulate transcription in a nonstoichiometric manner such that a limited number of transcription factors can support a larger array of transcription patterns.
    (b) Mediator complexes serve as adaptors that bridge DNA-bound transcription factors with RNA polymerase II to induce or inhibit the formation of a stable preinitiation complex.
    (c) DNA sequences known as insulators limit the effects of enhancers as they define the boundaries of functional transcription units. Insulators also appear to retard the spreading of heterochromatin.

25. Several signaling systems regulate the activity of transcription factors.
    (a) Signaling pathways utilizing G protein–coupled receptors and receptor tyrosine kinases often lead to the phosphorylation of transcription factors so as to activate or inactivate them.
    (b) The sterol regulatory element binding protein (SREBP) is an inducible transcription factor that is regulated allosterically by cholesterol (see Section 20-7B).
    (c) The JAK-STAT pathway is stimulated by a family of protein growth factors called cytokines, which regulate the differentiation and proliferation of a number of cell types. Cytokine receptor dimerization, triggered by cytokine binding, allows the associated Janus tyrosine kinases (JAKs) to phosphorylate each other as well as signal transducers and activators of transcription (STAT proteins). The phosphorylated STATs dimerize and move to the nucleus, where they interact with specific DNA sequences to stimulate or repress transcription.
    (d) Members of the nuclear receptor family recognize a variety of steroid hormones and other substances and, upon ligand binding, usually dimerize and interact with specific DNA sequences called hormone response elements. The nuclear receptors consist of a ligand-binding domain, a DNA-binding domain, and a transcription activation domain.

26. Alternative mRNA splicing, which generates tissue-specific isoforms of a gene product, is a posttranscriptional regulatory mechanism discussed in detail in Section 26-3A.

27. A variety of factors can modulate the half-life of an mRNA:
    (a) Progressive deadenylation of the 3′ end by deadenylases
    (b) Removal of the 5′ cap by decapping enzyme
    (c) Protein binding to AU-rich sequences in the 3′ untranslated region of mRNA
    (d) Formation of secondary structures, usually at the 5′ or 3′ end
    (e) Presence of certain RNA-binding proteins
    (f) Presence of premature Stop codons (see Box 28-3)

28. Noncoding RNA plays an important role in the inhibition of gene expression (gene silencing). Naturally occurring RNAs of this kind are referred to as short interfering RNAs (siRNAs) or micro RNAs (miRNAs). Several hundred miRNAs have been identified in mammals, although their targets are still largely unknown.

29. Gene silencing by short noncoding RNA is referred to as RNA interference (RNAi), which was first characterized by experiments in which added RNA interfered with gene expression. Inhibition is due to base pairing between the siRNA and a complementary mRNA. The mechanism of RNAi involves (a) an ATP-dependent protein (Dicer) that cleaves dsRNA precursors into 21- to 23-nt siRNA fragments; (b) the transfer of siRNA to an RNA-induced silencing complex (RISC); and (c) the cleavage by RISC of mRNA targets that form duplexes with the siRNA. RNAi has been extremely useful in "knocking out" gene expression in numerous model systems.

30. Translational control occurs in embryonic cells, which contain stockpiled maternal mRNA. Globin synthesis is also controlled at the level of translation, via heme-regulated inhibitor (HRI), so that the protein is synthesized only when heme is also available.

31. Somatic recombination and hypermutation serve as a unique mechanism to generate antibody diversity during the development of lymphocytes. Both immunoglobulin heavy and light chains contain four segments encoded by an exon (a leader segment, $L$, a variable region segment, $V$, a joining segment, $J$, and a constant region segment, $C$). The heavy chain also contains a diversity segment, $D$, which is also encoded by one exon. The various segments occur in tandem arrays in the genome in germline DNA. During their development, clones of B lymphocytes engage in somatic recombination events that eliminate all but one randomly selected exon of each segment. As activated lymphocytes undergo cell division, the $V/D/J$ exons also undergo hypermutation at a rate of $10^{-3}$ base changes per nucleotide per cell division. Alternative splicing allows B lymphocytes to produce membrane-bound and secreted isoforms of specific immunoglobulins.

## *The Cell Cycle, Cancer, and Apoptosis*

32. The eukaryotic cell cycle is divided into four phases:
    (a) M phase (for mitosis), when mitosis and cell division occur.
    (b) $G_1$ phase (for gap), a variable but long part of the cell cycle.
    (c) S phase (for synthesis), when DNA replication occurs.
    (d) $G_2$ phase, a relatively short period of preparation for mitosis.

    Some cells enter a quiescent phase ($G_0$) after M phase in which they cease dividing.

33. Progression through the cell cycle requires the expression of certain genes, including the cyclins (whose levels rise and fall in the cell cycle) and the cyclin-dependent kinases (CDKs). CDKs are positively regulated by cyclin-dependent phosphorylation and negatively regulated by CDK inhibitors (CDKIs). Activated CDKs phosphorylate several nuclear proteins, including histone H1 and proteins that disassemble the nucleus and rearrange the cytoskeleton.

34. The tumor suppressor p53 accumulates in response to DNA damage and leads to inhibition of CDKs, thereby arresting cell division to allow DNA repair or, if repair is unsuccessful, triggering cell suicide (called programmed cell death or apoptosis). p53 is a transcriptional activator of a CDKI (p21$^{Cip1}$) that inhibits both the $G_1/S$ and $G_2/M$ transitions. The accumulation of excessive levels of p53 due to irreparable DNA lesions leads to apoptosis. p53 is also a sensor that integrates information from signaling pathways such as the Ras signaling cascade, which leads to the stabilization of p53.

35. Apoptosis is a phenomenon associated with a variety of developmental events including digit formation, development of the immune system, immune responses, and organ development and maintenance. Apoptosis is mediated by a cascade of proteases (caspases) that is triggered by either the extrinsic pathway or the intrinsic pathway. The extrinsic pathway activates the caspase cascade via the activation of so-called death receptors (Fas), while the intrinsic pathway (often initiated by the loss of key cell–cell interactions or the absence of hormones) begins with the activation of cytosolic Bcl-2 proteins. Bcl-2 associates with mitochondria to release cytochrome $c$ into the cytosol, which then binds to Apaf-1 to form a scaffold for several procaspase-9 molecules. This interaction results in the autoactivation of procaspase-9, which then activates procaspase-3 to initiate programmed cell death.

36. Another key cell cycle regulator is pRb, a tumor suppressor, which binds and inhibits a transcription factor (E2F) that stimulates the transcription of numerous S phase genes. Phosphorylation of pRb by various CDKs leads to the dissociation of pRb from E2F, allowing the transcription factor to activate S phase genes.

37. Embryonic development in *Drosophila melanogaster* begins with rapid nuclear divisions in the fertilized egg followed by a cellularization process that leads to formation of a cell monolayer called the blastoderm. During this development, the embryo begins a segmentation process that ultimately specifies the differentiation of adult body structures.

38. A hierarchy of gene interactions orchestrates early patterning of the *Drosophila* embryo:
    (a) Maternal-effect genes define the embryo's anteroposterior and dorsoventral axes, even before embryogenesis begins, via the deposition of mRNA in different portions of the egg. The resulting protein gradients subsequently influence the expression of embryonic genes.
    (b) Segmentation genes, which specify the correct number and polarity of embryonic body segments, include gap genes, pair-rule genes, and segment polarity genes. The products of these genes activate and repress other genes, producing sharp stripes of proteins that define the embryonic body segments.

(c) Homeotic selector genes specify segment identity; their mutation changes one body part into another. The developmental fate of a body segment depends on its position in a gradient of homeotic gene products.

39. Many homeotic genes contain a conserved 60-residue polypeptide sequence called the homeodomain or homeobox. Homeodomains occur in many developmentally important genes, called *Hox* genes, in vertebrates and invertebrates. The homeodomain encodes an HTH DNA-binding motif, and *Hox* proteins are transcription factors that regulate the expression of other genes usually involved in defining key positions along the anteroposterior axis of the embryo.

## Questions

### Genome Organization

1. From Figure 28-2, select the two types of organisms that most dramatically illustrate the C-value paradox.

2. Based on the information in Figure 28-3, what percentage of known human genes are involved in signal transduction?

3. Can the *D. melanogaster* histone genes be considered a type of moderately repetitive DNA?

4. How might recombination during meiosis result in the expansion of trinucleotide repeats?

### Regulation of Prokaryotic Gene Expression

5. What observation indicates that transcription of the *lac* operon is never completely repressed?

6. Match the molecules or DNA sequences on the left with the terms on the right.
   _____ Attenuator          A.  Tryptophan
   _____ Inducer             B.  TPP
   _____ Allosteric inhibitor  C.  3·4 terminator hairpin
   _____ Leader              D.  IPTG
   _____ Co-repressor        E.  TrpL

7. A mutation in which element of the *lac* operon (*P, O, Z, Y, A*) and/or the *lac* repressor gene (*I*) will lead to constitutive expression of the structural genes of the operon? Explain.

8. Imagine a culture of *E. coli* grown for many generations in the presence of glucose and absence of lactose. How do these bacteria "turn on" the *lac* operon if the permease gene has been "off" for so long?

9. What mechanisms have been observed in riboswitches for the inhibition of gene expression?

*Regulation of Eukaryotic Gene Expression*

10. Match the structural elements on the left with their functions on the right.

    \_\_\_\_ Promoter                  A. Transcription factor binding site
    \_\_\_\_ Silencer                    B. Transcription terminator
    \_\_\_\_ Operator                   C. Protein coding region
    \_\_\_\_ Attenuator                D. Inducer binding site
    \_\_\_\_ Enhancer                 E. RNA polymerase binding site
    \_\_\_\_ Structural gene         F. Repressor binding site

11. Which of the following DNA structures are likely to be transcriptionally active: Barr body, euchromatin, heterochromatin, DNase I–sensitive regions, and highly methylated DNA.

12. Match the following proteins or protein domains on the left with the functions on the right.

    \_\_\_\_ Enhanceosome    A. Binds acetylated Lys residues on histones
    \_\_\_\_ Bromodomain     B. Acidic region found in upstream transcription factors
    \_\_\_\_ Mediators            C. Defines the boundaries of functional transcription units
    \_\_\_\_ Chromodomain   D. Part of chromatin proteins that bind methylated histones
    \_\_\_\_ Insulator            E. Complex of architectural proteins containing coactivators and/or corepressors
    \_\_\_\_ Activation domain   F. Adaptor proteins that bind upstream transcription factors and RNAPII

13. Given its mechanisms, explain how RNAi might act catalytically.

14. Not counting variation due to hypermutation and the addition and deletion of nucleotides at the junctions between *V*, *D*, and *J* segments, what is the total possible number of different antibodies that a human can produce? Assume that the number of possible λ chains is the same as the number of possible κ chains.

*The Cell Cycle, Cancer, and Apoptosis*

15. Which phase of the cell cycle is likely to be missing in the early embryonic cells of the frog *Xenopus laevis*, which divide about every 20–30 minutes?

16. Why is a mutated p53 gene present in so many cancers?

17. How do mutations resulting in nonfunctional ATM predispose individuals to cancer?

18.   Inactivation of the eukaryotic proto-oncogene *c-myc* (see Box 13-3 on p. 421) involves premature termination of its RNA transcripts near the promoter. What feature of this phenomenon is likely to be similar to attenuation in prokaryotes? What feature must be different?

19.   Why is it important that cells other than reticulocytes not synthesize heme-controlled repressor?

20.   Match the following genes involved in *Drosophila* development with the statements below.

   A. Homeotic genes       B. Pair-rule genes
   C. Gap genes            D. Segment polarity genes

   _____   Gradients of these gene products define the polarity of body segments.
   _____   A deletion in one of these genes converts an antenna into a leg.
   _____   The spatial expression of these genes is regulated by the distribution of a maternal mRNA.
   _____   Mutations in these genes result in deletion of portions of every second body segment.
   _____   The products of these genes contain helix–turn–helix motifs.

21.   How do Bicoid and Nanos proteins determine the differential expression of the *giant, Krüppel*, and *knirps* genes?

22.   Do homeotic genes encode morphogens? Explain.

# Answers to Questions

## Chapter 1

1.  Carbon, nitrogen, oxygen, hydrogen, phosphorus, calcium, potassium, and sulfur account for 98% of the dry weight of living cells.

2.  The primitive earth's atmosphere was a relatively reducing atmosphere lacking appreciable amounts of $O_2$. Much recent controversy revolves around just how much $NH_3$ and $CH_4$ existed in the young earth's atmosphere. Relatively low amounts would indicate that the atmosphere was both nonreducing and nonoxidizing. In either case, the absence of significant levels of $O_2$ indicates that the atmosphere was nonoxidizing compared to the present atmosphere.

3.  The Miller and Urey experiments were designed to ask whether biological molecules could be generated by mimicking the atmospheric conditions of the early earth. The early atmosphere was thought then to consist of $H_2O$, $CH_4$, $NH_3$, $SO_2$, and possibly $H_2$. Miller and Urey subjected these molecules, except for $SO_2$, to electrical discharges that were meant to simulate discharges believed to be prevalent in the earth's early atmosphere. The generation of diverse organic acids showed that the precursors for larger biological molecules could be spontaneously produced in the early atmosphere, paving the way for subsequent chemical evolution.

4.  Photosynthesis.

5.  Shown below is the condensation of galactose and glucose to form lactose. The atoms involved in the elimination of water are circled.

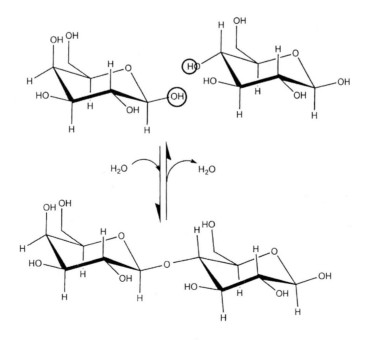

6.  Shown below are condensation reactions of two pairs of functional groups found in Table 1-2. The atoms involved in the elimination of water are circled.

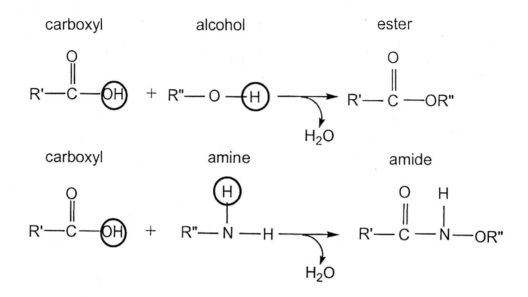

7.  See Figure 1-8.

8.  (a) Eukaryotes are defined by their elaborate internal membrane system and the enclosure of genomic DNA inside a double-membrane compartment, the nucleus, which is lacking in prokaryotes.
    (b) The cytoplasm is the contents of a prokaryotic cell and also refers to the cellular contents outside of the nucleus in eukaryotes. The cytosol is the cellular contents minus all of the membranous compartments of the cell (e.g., mitochondria, endoplasmic reticulum, Golgi apparatus, chloroplasts, and vacuoles).
    (c) The endoplasmic reticulum is the extensive network of internal membranes in eukaryotic cells (which is topologically continuous with the nuclear membrane). The cytoskeleton is the extensive network of protein filaments in the cytosol of eukaryotic cells.

9.  All living organisms can be classified into three domains: archaea (archaebacteria), bacteria (eubacteria), and eukarya (eukaryotes).

10. The most compelling evidence that these organelles were once symbiotic bacteria is the presence of distinct genetic material and protein synthesis machinery inside of these organelles. The RNA and proteins that make up the protein synthesis machinery of these organelles is much more similar to that of bacteria than that of eukaryotes (see Chapter 27).

11. Genetic variation results from mutations that persist in a population and have not been eliminated by natural selection because they did not significantly decrease fitness. Evolution of a population occurs when variations that increase an individual's chances for survival and reproduction spread throughout the population.

12. Enthalpy is defined as $H = U + PV$. Hence $\Delta H = \Delta U + \Delta(PV)$. The enthalpy of a system is equal to its energy change when $\Delta(PV) = 0$, which occurs at the constant pressure and volume conditions typical of living things.

13. If $q$ is positive, then heat has been transferred *to* the system, increasing its internal energy. If $w$ is positive, then work has been done *by* the system, decreasing its internal energy.

14. In this example, urea and the water are the system, and the vessel and beyond are the surroundings. As urea dissolves, the enthalpy increases (an endothermic process). Heat is absorbed into the interactions between the urea and water, making the solution cooler.

15. **Energy** is measured as the heat absorbed by a system minus the work done by the system on its surroundings. **Enthalpy**, or heat content, is the amount of heat generated or absorbed by a system when a process occurs at constant pressure, as in biological systems, and no work is done other than the work of expansion or contraction ($\Delta V$) of the system. **Entropy** is a measure of the heat absorbed or generated by a system at constant temperature and reflects the number of equivalent ways of arranging a system with no change in its internal energy. **Gibbs free energy** is the energy available to do work; it is a combination of enthalpy and entropy ($\Delta H - T\Delta S$) and an indicator of the spontaneity of a process at constant pressure and temperature.

16. (a) The beaker containing the solution with both salts has higher entropy. The molecules in this solution can be arranged in many more ways with respect to each other than each salt solution alone.
    (b) The set of dice with the 6's showing on the side faces has more entropy since there are many more ways to obtain this configuration.
    (c) The symmetric molecule has greater entropy, since it can polymerize by joining reactions involving either end. Note that the information content of the asymmetric molecule is higher, however. All biological polymers are formed from asymmetric subunits.

17. When enthalpy and entropy are both positive, $\Delta G$ decreases with increasing temperature, and the temperature at which the reaction occurs spontaneously must be high enough that the $T\Delta S$ term is larger than the $\Delta H$ term in the equation $\Delta G = \Delta H - T\Delta S$. For instance, dissolving crystalline urea in water is endothermic; however the process is spontaneous when it is carried out at room temperature. When enthalpy and entropy are both negative, $\Delta G$ decreases with decreasing temperature, and for the reaction to be spontaneous, the temperature must be low enough that the $T\Delta S$ term is not more negative than the $\Delta H$ term.

18. The expression for the Gibbs free energy is $\Delta G = \Delta H - T\Delta S$. In this reaction, the entropy decreases since the number of molecules decreases and the product is therefore more ordered than the reactants. Hence, in order for this reaction to be spontaneous the enthalpy term ($\Delta H$) must be more negative than the entropy term ($T\Delta S$). In many cases such as this, the enthalpy of the reaction is negative and the reaction releases heat to the surroundings.

19. This statement is incorrect. An enzyme does not alter the spontaneity of a reaction; rather, it increases the rate at which a reaction reaches equilibrium.

20. Catalysis, replication, and mutability have been argued to be the minimum criteria for life. In addition, an organism must be able to maintain a novel chemical environment that is not in equilibrium with its surroundings and to resist the environmental fluctuations that might disturb its ability to carry out the other three essential functions of living systems—this steady state condition is called homeostasis.

21. Since the synthesis of glucose requires more ATP input than the amount produced during glucose degradation into pyruvate, we know that more chemical work is required for glucose synthesis than its degradation. Hence, glucose synthesis is endergonic (non-spontaneous) and its degradation is exergonic (spontaneous).

    The figure below shows one representation of this process ($P_i = HPO_4^{2-}$)

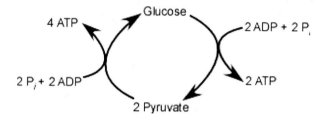

    Remember that the second law of thermodynamics tells us that every transformation of energy results in an increase in entropy of the universe, so that all the free energy released during glucose degradation to pyruvate cannot supply the total energy needed for the reverse process. We can see this in the relationship $\Delta G = \Delta H - T\Delta S$, which, when rearranged, gives $\Delta H = \Delta G + T\Delta S$. Hence, the total energy contained in the bonds of two pyruvate molecules is less than that of the bonds in glucose, so that that an additional input of energy is required. This is a biochemical example of the idea that one cannot have a perpetual motion machine.

## Chapter 2

1.

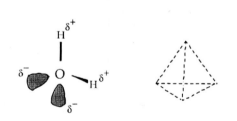

    The molecule forms a tetrahedron.

**AQ-5**

2. (a) Carbon dioxide (O=C=O), which is more polarizable than oxygen (O=O). (b) Ammonia ($NH_3$), which is more polar than nitrogen (N≡N). (c) Hydrogen sulfide (HS), which is more polar than methane ($CH_4$).

3. Water expands upon freezing. Full bottles of beer have little room for expansion and the frozen liquid expands beyond the boundaries of the glass and cap. Similarly, bottles of wine lose their corks in the freezer.

4. Hydrophilic substances are polar compounds that are readily soluble in water. Hydrophobic substances are nonpolar compounds that are insoluble in water. Amphipathic substances have both polar and nonpolar segments; they form micelles or bilayers in aqueous solution in which the polar groups face the water and the nonpolar groups exclude water and face each other (see Figure 2-11).

5. When the olive oil and vinegar (which is an aqueous solution) are mixed, the entropy of the solution decreases (the action of mixing is an input of energy that drives this process). Mechanical mixing results in an ordering of water around the thousands of oil droplets. Once the mixing has stopped, the system moves toward equilibrium, which maximizes entropy. As the oil droplets fuse, the surface-to-volume ratio of all of the oil decreases, thereby decreasing the amount of ordered water. Consequently, the entropy of the solution increases.

6. $Na^+$ and $Cl^-$ ions diffuse out of the bag down their concentration gradient across the cellulose membrane. Water moves into the bag via osmosis since its concentration inside the bag is relatively lower due to the presence of the protein that cannot traverse the membrane (and initially due to the higher [NaCl] inside the bag—which eventually equalizes with the [NaCl] outside the bag). The bag expands as water enters. This setup is called dialysis and is used to remove or add salts to solutions of macromolecules.

7. $CO_2$, $BF_3$, $Zn^{2+}$, HCOOH, $NH_4^+$, and $H_2O$ are Lewis acids; however, only HCOOH, $NH_4^+$, and $H_2O$ are Brønsted–Lowry acids. $NH_3$, $OH^-$ and $Cl^-$ are considered bases under both definitions.

8. p$K$ is the negative logarithm of the dissociation constant of a weak acid. pH is the negative logarithm of the $[H^+]$. The Henderson–Hasselbalch equation is

$$pH = pK + \log \frac{[\text{conjugate base}]}{[\text{acid}]}$$

9.  pK is the negative logarithm of the dissociation constant, $K$, so the smaller the value of pK, the larger the value of $K$. The larger the value of $K$, the greater the dissociation of the acid in water. Therefore, trichloroacetic acid is the stronger acid since it has a lower pK than acetic acid.

    For trichloroacetic acid, $pK = -\log K$

    Therefore, $K = 10^{-0.7} = 0.2$

    Similarly for acetic acid,

    $$\log K = -4.76$$
    $$K = 10^{-4.76} = 1.74 \times 10^{-5}$$

10. HCl is a strong acid and completely dissociates in water, so [HCl] = [H+] = 0.05 M. The pH is given by

    $$pH = -\log [H^+]$$
    $$= -\log 0.05$$
    $$= 1.3$$

    Since $[H^+][OH^-] = K_w$,

    $$[OH^-] = pK_w / [H^+]$$
    $$= 10^{-14} / 0.05$$
    $$= 2 \times 10^{-13} \text{ M}$$

11. The amount of NaOH needed to titrate equivalent amounts of all monovalent acids is the same for both acid solutions. In both cases, 10 mL of 1 M NaOH will bring the titration of the acid to its end point. (Note: 10 mL of 1 M NaOH provides 0.01 moles of $OH^-$ to neutralize the 0.01 moles of $H^+$.)

12. The equilibrium constant for a weak acid is

    $$K = \frac{[H^+][A^-]}{[HA]}$$

    The concentration of each of the dissociated ions is 0.05% of 0.01 M, or $0.0005 \times 0.01$ M $= 5 \times 10^{-6}$ M. Because only 0.05% of the acid is ionized, we can assume that [HA] does not change. We then calculate

    $$K = \frac{(5 \times 10^{-6})(5 \times 10^{-6})}{0.01} = 2.5 \times 10^{-9}$$

    $$pK = -\log (2.5 \times 10^{-9}) = 8.6$$

AQ-7

13. A buffer is a solution of a weak acid and its conjugate base. A buffer works by "absorbing" base or acid equivalents within about one pH unit of its p$K$. For example, when a small amount of $OH^-$ is added, it reacts with HA to form water and $A^-$, with little change in pH. Similarly, a small amount of $H^+$ reacts with $A^-$ to form HA. In biological systems, protons produced during catabolic reactions must be buffered by the cell. Phosphate and carbonate ions, as well as proteins, nucleic acids, and fatty acids, serve as buffers in cells.

14. Since p$K = 6.27$, $K = 10^{-6.27} = 5.37 \times 10^{-7}$. For the dissociation of cacodylic acid

$$\text{cacodylic acid} \leftrightarrows H^+ + \text{cacodylate}^-$$

we can approximate $[H^+] = [\text{cacodylate}^-] = x$.

We approximate [cacodylic acid] by $0.1 - x$. Therefore,

$$K = \frac{[H^+][\text{cacodylate}^-]}{[\text{cacodylic acid}]}$$

$$5.37 \times 10^{-7} = \frac{x^2}{0.1 - x}$$

$$0 = x^2 + (5.37 \times 10^{-7})x - (5.37 \times 10^{-8})$$

Solve for $x$ using the quadratic equation

$$x = \frac{-b \pm \sqrt{b^2 - 4ac}}{2a}$$

$$= \frac{-5.47 \times 10^{-7} \pm \sqrt{(5.47 \times 10^{-7})^2 + [4(5.47 \times 10^{-8})]}}{2}$$

$$x = -2.32 \times 10^{-4} \text{ or } 2.31 \times 10^{-4}$$

The result can only be $2.31 \times 10^{-4}$ since there is no such thing as a negative concentration.

Therefore, $x = [H^+] = 2.31 \times 10^{-4}$ M and

pH = 3.64

Note that for weak acids with pK > 5, one can approximate the pH by ignoring the change in acid concentration due to its dissociation. Thus, [cacodylic acid] = 0.1 M and

$$5.37 \times 10^{-7} = \frac{x^2}{0.1}$$

Therefore, $x = 2.32 \times 10^{-4}$ M

15. The added acetate anion reacts with the $H^+$ present in the solution to form undissociated acid, thereby causing a decrease in $[H^+]$ and an increase in pH.

16. The final concentrations are [lactic acid] = 0.01 M and [lactate] = 0.025 M. Use the Henderson–Hasselbalch equation to calculate the pH:

$$pH = pK + \log \frac{[\text{lactate}]}{[\text{lactic acid}]}^{[K15]}$$
$$= 3.86 + \log 2.5$$
$$= 3.86 = 0.398$$

$$= 4.26$$

17. Use the Henderson–Hasselbalch equation to calculate the concentration of acetic acid $(x)$ necessary to obtain a solution of pH 5 that contains 3 M $K^+$ (3 M KOAc). The pK of acetic acid is 4.76 (Table 2-4).

$$pH = pK + \log \frac{[\text{KOAc}]}{[\text{HOAc}]}$$
$$5 = 4.76 + \log \frac{3}{x}$$
$$^{[K17]}0.24 = \log 3 - \log x$$
$$0.24 - 0.48 = -\log x$$
$$0.24 = \log x$$
$$x = 1.73 \text{ M}$$

Therefore, for a 500 mL solution, the amount of acetic acid added should be 1.73/2 or 0.86 moles. The amount of KOAc should be 3/2 or 1.5 moles.

$$x = \frac{3}{100.24}$$
$$= 1.73 \text{ M}$$

18. At pH 7, $H_2PO_4^-$ is in equilibrium with $HPO_4^{2-}$ at nearly a 1:1 ratio since pH 7 is near $pK_2$ (6.82). See Figure 2-18 and Table 2-4.

19. Atmospheric carbon dioxide reacts with the distilled water in the beaker to form carbonic acid. The carbonic acid dissociates into protons and bicarbonate with a $pK$ of about 6. These protons make the distilled water slightly acidic.

$$CO_2 + H_2O \leftrightarrows H_2CO_3$$

$$H_2CO_3 \leftrightarrows HCO_3^- + H^+$$

20. The pH is 7. Note here that the dissociation of water ($H_2O \leftrightarrows OH^- + H^+$), produces a concentration of $10^{-7}$ M $H^+$, compared to the added $10^{-12}$ $H^+$ from HCl. Hence, HCl at this concentration contributes negligibly little $H^+$.

21. (a) The low ambient partial pressure of $CO_2$ in air coming into the lung alveoli provides a steep concentration gradient across the endothelium of the alveolar and vascular lining for the incoming blood, where the partial pressure of $CO_2$ is close enough to its solubility limit to favor diffusion of $CO_2$ into the lung air space. The loss of $CO_2$ from the left-hand side of this equilibrium will favor the reaction shifting to the left to re-establish equilibrium. Hence, the [$H^+$] will decline and the pH of the blood will rise (from 7.2 to about pH 7.4).

    (b) Respiration in the tissues produces $CO_2$, such that the elevation of $CO_2$ in the capillaries will drive this equilibrium to the right to re-establish equilibrium, thereby decreasing the pH of the blood (back to 7.2).

### Answers to Graphical Analysis Questions

1. The weak acids are listed as strongest to weakest from top to bottom down the table. The stronger the acid, the higher the value of Ka and the lower the value of pKa an acid will have.

2. $HPO_4^{2-}$ has the highest affinity for protons while Oxalic acid has the lowest affinity for protons.

3. The useful pH range of a buffer = the pKa of the buffer $\pm$ 1. Thus, the buffering capacity needed to buffer the range of pH of the typical cytoplasm is 7.1 – 7.8 $\pm$ 1, or 6.1 – 8.8. Thus, buffers made from $H_2CO_3$, PIPES, $H_2PO_4-$, MOPS, HEPES, and Tris would be appropriate, depending on which end of the pH range is needed. The best buffer would be one with a pKa between the two pH extremes of the cytoplasm, so that the whole range falls between $\pm$ 1 of the pKa of the buffer. From the above list, HEPES would be the best choice.

4.

| Acid | Ka | pKa | pH of environment | [A$^-$]/[HA] |
|---|---|---|---|---|
| Oxalic acid | 5.37 X 10$^{-2}$ | 1.27 (pk$_1$) | 3.25 | **100/1** |
| H$_3$PO$_4$ | 7.08 X 10$^{-3}$ | 2.15 (pk$_1$) | **4.15** | 100/1 |
| Formic acid | 1.78 X 10$^{-4}$ | 3.75 | 3.75 | **1/1** |
| Succinic acid | 6.17X 10$^{-5}$ | 4.21 (pk$_1$) | 1.21 | **1/1000** |
| Oxalate$^-$ | 5.37 X 10$^{-5}$ | 4.27 (pk$_2$) | **8.27** | 10,000/1 |
| Acetic acid | 1.74 X 10$^{-5}$ | 4.76 | **3.76** | 1/10 |
| Succinate$^-$ | 2.29 X 10$^{-6}$ | 5.64 (pk$_2$) | 3.65 | **1/100** |
| MES | 8.13 X 10$^{-7}$ | 6.09 | 7.1 | **10/1** |
| H$_2$CO$_3$ | 4.47 X 10$^{-7}$ | 6.35 (pk$_1$) | **4.35** | 1/100 |
| PIPES | 1.74 X 10$^{-7}$ | 6.76 | **2.76** | 1/10,000 |
| H$_2$PO$_4^-$ | 1.51 X 10$^{-7}$ | 6.82 (pk$_2$) | 10.80 | **10,000/1** |
| MOPS | 7.08 X 10$^{-8}$ | 7.15 | 1.15 | **1/1,000,000** |
| HEPES | 3.39 X 10$^{-8}$ | 7.47 | 10.47 | 1,000/1 |
| Tris | 8.32 X 10$^{-9}$ | 8.08 | 5.1 | **1/1,000** |
| NH$_4^+$ | 5.62 X 10$^{-10}$ | 9.25 | **4.25** | 1/100,000 |
| Glycine (amino group) | 1.66 X 10$^{-10}$ | 9.78 | 2.8 | **1/10,000,000** |
| HCO$_3^-$ | 4.68 X 10$^{-11}$ | 10.33 (pk$_2$) | **10.33** | 1/1 |
| Piperidine | 7.58 X 10$^{-12}$ | 11.12 | 5.1 | **1/1,000,000** |
| HPO$_4^{2-}$ | 4.17 X 10$^{-13}$ | 12.38 (pk$_3$) | **9.38** | 1/1,000 |

## Answers to Play It Forward Questions

1.  The process of dissolving sucrose in water is spontaneous (exergonic) and is therefore characterized by a negative value of ΔG. How do we know this? Because it is happening!

2.  If the glass or water with the sucrose represents a thermodynamic system, then the spoon and the person stirring it can be considered the surroundings. The stirring of the solution with the spoon increases the rate at which the sucrose crystals come into contact with water, enhancing the rate at which the sucrose dissolves in water. However, stirring is not absolutely necessary. As long as the amount of sugar added does not exceed the solubility limit of sucrose in water at room temperature, then theoretically, the sugar should eventually dissolve on its own without stirring using the input of energy provided by the ambient temperature in the room.

3.  Adding the sucrose to water that is warmer than room temperature, or alternatively heating the water-sugar solution after the sugar has been added will add energy into the system, increasing both the rate at which the sugar dissolves and the solubility limit of the sucrose.

4.    The cold glass implies that as the dissolving takes place, energy is being extracted from the surroundings into the system to drive the process. Thus, this process is endothermic and is characterized by a positive value for ΔH.

5.    In order for sucrose to dissolve in water, a number of molecular bonds must be broken and formed. For example, as the sucrose becomes solvated and water forms concentric solvation cages around the individual molecules of sugar, the interactions between water molecules that will form the hydration cages and the water of the bulk solvent must be broken. Conversely, the water molecules that will form the innermost hydration cage must interact with the sugar via hydrogen bonds. The multiple hydroxyl groups of the sucrose can serve as both a hydrogen bond donor and a hydrogen bond acceptor. The same is true of water. Multiple water molecules may be necessary to interact with each hydroxyl group on each sugar molecule in order to solvate them. In addition, the crystal lattice of the solid sucrose must be disrupted. Ultimately, energy must be utilized by the system to break existing solvent bonds, break existing solute-solute bonds, and make new bonds between the solvent and solute to allow for solvation to proceed. The enthalpy absorbed in breaking hydrogen bonds in the solvent and the crystal lattice is greater that the enthalpy released during the formation of new hydrogen bonds. Therefore the overall process is endothermic.

6.    In order for this process to be spontaneous (negative value for ΔG) but still be endothermic, it must be entropically driven. That is, the value of the TΔS term in the Gibbs Free Energy equation must be sufficiently positive so that its subtraction from the positive ΔH term results in the value of ΔG being negative:

$$\overset{-}{\Delta G} = \overset{+}{\Delta H} - \overset{+}{T\Delta S}$$

In other words, ΔS must be sufficiently positive for TΔS to overcome the unfavorable effect of a positive ΔH.

## Chapter 3

1.    (a) Pyrimidine base; (b) pyrimidine nucleoside; (c) purine nucleotide; (d) purine nucleoside; (e) pyrimidine base; (f) purine nucleotide; (g) pyrimidine nucleotide; (h) purine base.

2.    (i) Chargaff's rules, (ii) knowledge that nitrogenous bases are predominantly in the keto tautomeric form, (iii) X-ray crystallographic data indicating that DNA has a double-helical structure, (iv) X-ray crystallographic data suggesting that the planar aromatic bases form a stack parallel to the fiber axis.

3.  (a) The nucleotides of a DNA strand are linked between the 3′ and 5′ ribose carbons by a phosphate group, thereby giving the strand polarity. The two strands of DNA form a duplex in which the strands run in opposite directions, that is, in an antiparallel arrangement.

    (b) Each nitrogenous base of one strand is hydrogen bonded to a different specific nitrogenous base on the opposing strand to form a planar base pair. This qualitatively specific hydrogen bonding between base pairs is called complementary base pairing. In the Watson–Crick structure, adenine specifically pairs with thymine, and guanine pairs with cytosine.

4.  Samples (a) and (e), since by Chargaff's rules, 27.3% T implies 27.3% A, so that G + C = 45.4% and C = 22.7%. Samples (c) and (d), since 13.1% C implies 13.1% G and 36.9% each of A and T.

5.  The gene in the first strain contains the sequence CTGCAG, which is recognized and cleaved by *Pst*I but not by *Pvu*II. In the second strain, the site has mutated so that it cannot be recognized by *Pst*I. Its sequence must be CAGCTG, since it is recognized and cleaved by *Pvu*II.

6.  Shown below is one of two possible restriction maps of the recombinant plasmid (in the other possible map, the *Sal*I site would be 0.5 kb from the other *Bam*HI site). To determine the orientation of the insert, it would be necessary to perform a restriction enzyme digest with *Sal*I and *Pst*I (or *Sal*I and *Eco*RI) to reveal the location of the *Sal*I site. For the map shown below, the *Sal*I/*Pst*I digestion products would be 2.25 kb and 6.25 kb.

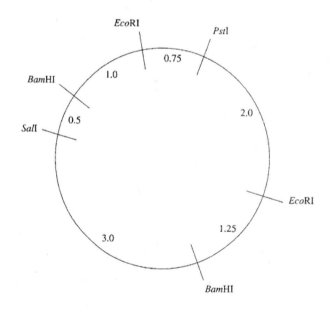

7.  At pH 7.5, purified DNA is negatively charged; hence, the DNA fragments will migrate from the cathode toward the anode.

8. (a) T7 DNA polymerase is likely to give a longer ladder, since it is less likely to terminate DNA chain elongation.

   (b) The *E. coli* polymerase I fragment is better suited for obtaining sequences close to the primer, since it terminates DNA chain elongation more often and hence produces more short fragments than does T7 DNA polymerase.

9. The bands on the gel identify the 3′ ends of DNA fragments synthesized by the DNA polymerase in the presence of dideoxynucleotides. Their sequence, read from bottom to top, is

   5′-T-C-G-A-C-T-C-G-A-A-G-T-C-A-G- 3′

   This is complementary to the DNA of interest, which has the sequence

   5′- C-T-G-A-C-T-T-C-G-A-G-T-C-G-A- 3′

10. The human genome contains $3.2 \times 10^9$ base pairs (bp), which means we need sequence information for $6.4 \times 10^9$ bases. So, with 15 systems producing 550,000 bases per day, we have $8.2 \times 10^6$ bases read every day.

    Therefore, it would take

    $$\frac{6.4 \times 10^9 \text{ bases in human genome}}{8.25 \times 10^6 \text{ bases per day}} = 775.75 \text{ days} = 2.13 \text{ years}$$

11. The smallest prokaryotic genome in Table 3-3 is *Mycoplasma genitalium* with 580,000 base pairs; the largest shown in the table is *Escherichia coli* with 4,649,000 base pairs. Dividing each of these genome lengths by 1000 base pairs or 1400 base pairs per gene, respectively, shows that one can expect about 580 genes in *M. genitalium* and 3,321 genes in *E. coli*.

12. The average chromosome length is

    $$\frac{3.2 \times 10^9 \text{ bp in human genome}}{23 \text{ chromosomes}} = 1.39 \times 10^8 \text{ bp per chromosome}$$

    The length of the DNA is

    $$(1.39 \times 10^8 \text{ bp})(34 \times 10^{-10} \text{ m/turn})/10 \text{ bp/turn} = 0..47 \text{ m} = 47 \text{ cm } !!$$

    So, try to imagine how 23 chromosomes with an average length of 47 cm of DNA are packaged into a human cell nucleus with an average diameter of only ~20 μm. Molecular biologists do not yet have a satisfying answer to this enormous packaging problem.

13. Bacteria containing the plasmids will be resistant to ampicillin, since the pUC18 vector contains the $amp^R$ gene, and so can grow in the presence of ampicillin. Bacteria lacking the vector will not grow in the presence of ampicillin. Cells containing the foreign gene at the polylinker site, which interrupts the *lacZ* gene, will not produce any β-galactosidase and so can be detected as white colonies in the presence of a β-galactosidase substrate that turns blue when hydrolyzed. Cells containing the vector without the foreign DNA will appear blue because they express functional β-galactosidase.

14. The frequency (*f*) of the gene of interest is $10/(3.2 \times 10^6) = 3.12 \times 10^{-6}$. The probability (*P*) of finding this gene is

$$P = 1 - (1 - f)^N$$

where *N* is the number of clones. Hence, $P = 1 - (1 - 3.12 \times 10^{-6})^{100,000} = 1 - 0.73 = 0.27$. This low probability inspires no confidence that a single-copy gene can be easily found. The probability may be even lower, since not all genomic DNA fragments will be incorporated into the cloning vector with equivalent efficiency. Using the relationship above, the probability increases to 0.96 when the library contains $10^6$ clones. Most investigators will shoot for a probability of 0.999 when constructing a recombinant DNA library to ensure the best chance of identifying a single-copy gene.

15. The finite amounts of nucleotides, primers, and enzyme will become rate-limiting. In the early rounds of PCR, these elements are at or near saturating levels (that is, their concentrations are much higher than that of the DNA template). Consequently, the rate of DNA production is directly proportional to the amount of template, which accumulates in an exponential fashion. But as the template DNA accumulates, its concentration approaches that of the enzyme, primers, and nucleotides. This slows the rate at which these elements come together to generate additional DNA, so the rate of DNA accumulation becomes linear.

16. Looking at the products shown in the second cycle (Figure 3-27), one sees a "hybrid-length" DNA in which one strand is of unit length and the other is much longer (and of varying length). This DNA molecule will dissociate into two strands in the third round, producing two products: a new "hybrid-length" DNA molecule and another DNA molecule in which each strand is unit length. This latter unit-length DNA duplex will accumulate exponentially in subsequent rounds. The "hybrid-length" molecule will accumulate in a linear fashion as a new pair is generated with the replication of each strand of the original strands of DNA.

17. These hormones require posttranslational processing that cannot be carried out in the bacteria.

18. What has been cloned is the cDNA corresponding to the mRNA sequences of tissue A that are absent from tissue B. The hybridization reaction eliminates the mRNA sequences shared by the tissues, including mRNAs encoding proteins required for common metabolic activities. The cloned cDNA therefore represents the genes that are active (i.e. that are transcribed into mRNA) in tissue A but not tissue B.

19. During in vitro DNA replication, one half of the population of DNA molecules will be the mutated version and another will be the original DNA sequence (the so-called wild-type sequence). Upon transformation, the replicating bacteria will form three populations (as shown in the rectangles below), where the fractions on the right show their contribution to the population. Note that only 50% of the population will carry the desired mutation.

Hence, the molecular biologist must collect many colonies and sequence them, with the anticipation that one-half will be the desired mutated form, or she must devise a vector or host that will selectively allow only the transformed bacteria with the mutated form to be easily recognized or survive. (Note that the heterozygous plasmid will express the mutant protein only if the DNA sequence is inserted in the proper orientation with respect to transcription.)

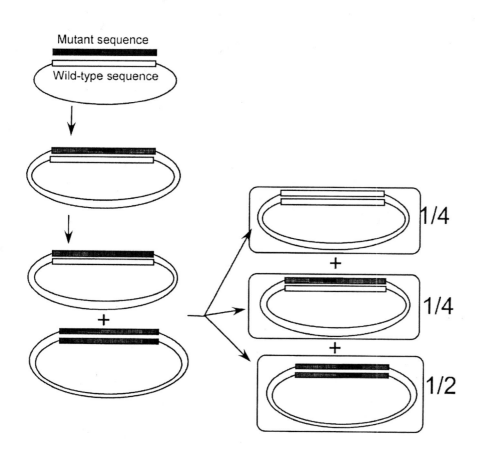

**Chapter 4**

1.

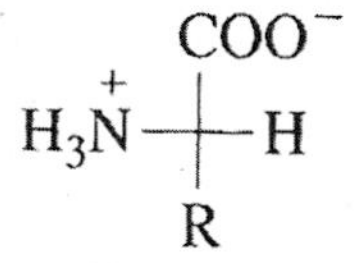

$$H_3\overset{+}{N}\underset{R}{\overset{COO^-}{\rule{1cm}{0.4pt}}}H$$

or

$$H_3\overset{+}{N}\underset{H}{\overset{R}{\rule{1cm}{0.4pt}}}CO$$

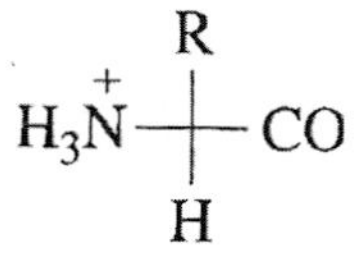

Remember, in Fischer projections, horizontal bonds extend out from the paper while vertical bonds extend behind the paper.

2. (a) Nonpolar: **A**; uncharged, polar: **C** and **E** (because **E** has a p$K$ of 6.0, it is largely uncharged at pH 7); charged, polar: **B** and **D** (at pH 7, the ε-amino group of **B** is positively charged, and the γ-carboxyl group of **D** is negatively charged.)
   (b) **B** and **D** cannot exist at any pH in aqueous solution. In **B**, the ε-amino group will become deprotonated at lower pH values than the α-amino group since its p$K$ is lower. Similarly, the γ-carboxyl group of **D** is more acidic than the α-amino group.
   (c) **A** is tryptophan; **B** is lysine; **C** is tyrosine; **D** is glutamate; and **E** is histidine.

3. Glutamine and asparagine can be converted to glutamate and aspartate, respectively.

4. Alanine gives rise to serine, and phenylalanine gives rise to tyrosine.

5. Histidine and cysteine are the most likely to be sensitive in the physiological pH range since the p$K$ values of their R groups are 6.04 and 8.37, respectively.

6.

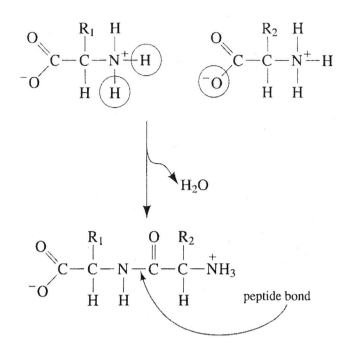

7.  Use the Henderson–Hasselbalch equation and rearrange terms to find the percentage of protonated species:

$$pH = pK + \log\frac{[A^-]}{[HA]} \quad [\text{K11}]$$

$$\frac{[A^-]}{[HA]} = 10^{(pH - pK)} \quad [\text{K12}]$$

$$= 10^{(7.2 - 6.04)}$$
$$= 10^{1.16} = 14.45$$

To obtain the percentage of the protonated species:

$$\% \, HA = \frac{[HA]}{[HA] + [A^-]} \times 100$$

$$= \frac{1}{1 + 14.45} \times 100$$

$$= 6.47\%$$

Hence, 6.47% is in the protonated form.

8.  Use the Henderson–Hasselbalch equation and rearrange terms to find the percentage of deprotonated species:

$$pH = pK + \log \frac{[A^-]}{[HA]}$$

$$\frac{[A^-]}{[HA]} = 10^{(pH - pK)}$$

$$= 10^{(7.6 - 8.37)}$$
$$= 10^{-0.77} = 0.17$$

$$\% \, A^- = \frac{[A^-]}{[HA] + [A^-]} \times 100$$

$$= \frac{0.17}{1 + 0.17} \times 100$$

$$= 14.53\%$$

Therefore, about 14.53% of the sulfhydryl groups are deprotonated.

9.  (a)  The intracellular pH is also near 7, so about half the imidazole groups will be protonated. In other words, the imidazole groups can easily abstract a proton or give one up, as is required for catalysis, and remain unchanged at the end of the reaction.
    (b)  It is more likely that the pK of these groups is near the pK of free histidine (6.04) so that they are available to form hydrogen bonds with potential substrates.

10.  (a)  Lysine; tyrosine is similar but its $pK_2$ is closer to 9.2.
     (b)  Proline; it has only two pK's, one at about 2 and the other well above 10. The only other amino acid with a $pK_2$ above 10 is cysteine, but it has three pK's.

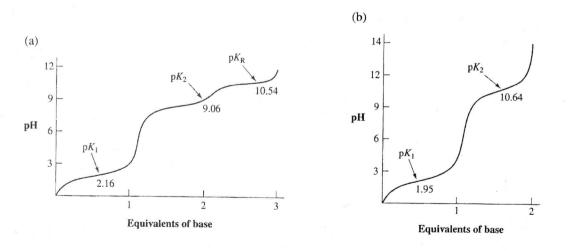

11. A, C, and D contain peptide bonds not commonly found in proteins. In A, the γ-carboxylate group of the first amino acid participates in an amide linkage (the second and third amino acids are linked by a conventional peptide bond). C contains a linkage between the carboxylate of a β-amino acid and the amino group of an α-amino acid. In D, the side chain of the first amino acid is linked to the amino group of the second. Tripeptide B can be referred to as serylvalylglycine.

12. The positively charged amino group of glycine electrostatically stabilizes the nearby carboxylate ion. Hence, the carboxylic acid group of glycine dissociates more readily to form the carboxylate ion than does the carboxylic acid group of acetic acid.

13. The p$I$ is approximately midway between the p$K$'s of the two ionizations involving the neutral species. In amino acids with ionizable side chains, the relevant p$K$'s may be p$K_1$ and p$K_R$ or p$K_R$ and p$K_2$. For aspartate, the neutral species results from ionization of the α-COOH (p$K_1$) and the β-COOH (p$K_R$):

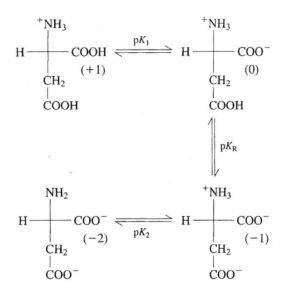

p$I$ = (p$K_1$ + p$K_R$)/2 = (1.99 + 3.90)/2 = 2.95.

For serine, which has no ionizable R group, p$I$ = (p$K_1$ + p$K_2$)/2 = (2.19 + 9.21)/2 = 5.7.

For lysine, the p$I$ lies between p$K_2$ and p$K_R$:

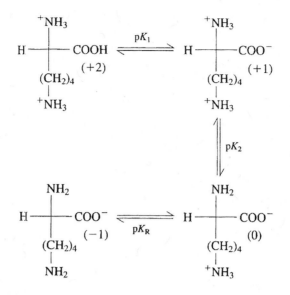

Hence, p$I$ = (p$K_2$ + p$K_R$)/2 = (9.06 + 10.54)/2 = 9.8.

14. The p$I$ of the tripeptide Asp-Lys-Ser cannot be precisely predicted and must be measured. However, one can use the p$K$ values in Table 4-1 to make a good estimate. The p$I$ of the tripeptide is 6.9, while the average of the three p$I$'s of the individual amino acids is 6.15. The graph below shows the titration of this peptide, including the predicted charge at the beginning of the titration, when all ionizable groups are protonated, and at each equivalence point. Each equivalence point is estimated as the average between adjacent p$K$'s. As you can see, when the peptide is fully protonated, it carries a net charge of +2. The complete ionization of the second carboxyl group in the side chain of the aspartyl residue occurs at pH 6.9, resulting in a net charge of 0.

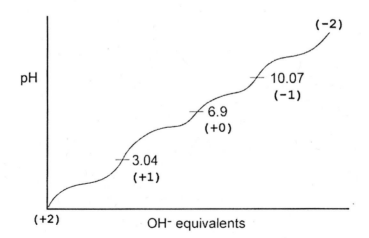

15. In free amino acids, the neighboring α-carboxyl and α-amino groups affect each other's p$K$ values. For instance, the negatively charged carboxyl group stabilizes the protonated amino group, making it a weaker acid and thus raising its p$K$ value. Similarly, the positively charged amino group stabilizes the anionic carboxyl group, making it a stronger acid and thereby lowering its p$K$ value. In proteins, the N-terminal and C-terminal residues participate in peptide bonds rather than having a free carboxyl group and a free ammonium group, respectively. Neutral peptide groups have little effect on the p$K$ values of nearby ionizable groups.

16. Yes. The formation of the peptide bond between each amino acid in the intact protein eliminates two ionizable groups. However, at physiological pH (pH 5–9), the α-carboxyl and the α-amino groups of amino acids are poor buffers, because their respective p$K$ values are far from the physiological pH range. However, at physiological pH, a protein containing a number of histidines and cysteines, for instance, may have a buffering capacity similar to that of an equimolar solution of its constituent amino acids; the side groups of these amino acids are free to ionize regardless of whether they are free or in a protein. Other ionizable R groups have p$K$'s too far from physiological pH to be effective buffers.

17. The α-COOH of GABA does not come under the stabilizing influence of its amino group as it does in glutamate. Hence, its p$K$ can be expected to be greater than that of the α-COOH of glutamate. Similarly, the amino group of GABA does not experience the stabilizing effect of the ionized form of its α-COOH and so has a lower p$K$ than the cognate amino group in glutamate.

18. Lysine solid is of necessity neutral and therefore has the structure

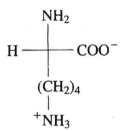

Neutral species are not as soluble as charged species. Adding base causes lysine's ε-NH$_3^+$ group to be converted to the neutral ε-NH$_2$, giving the molecule a net negative charge and higher solubility.

19. (a)  α-Amino-β-hydroxybutyric acid (later named threonine)
    (b)  The chirality of the β-carbon.

20.

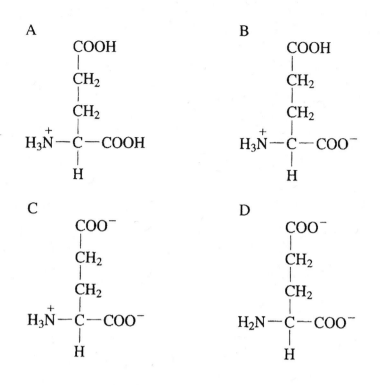

21. Glycine; its $C_\alpha$ has two identical substituents, H.

22. In a Fischer projection, the vertical bonds by convention point into the paper while the horizontal bonds point out from the paper. When the carbon groups are aligned vertically as drawn below, the amino group is on the left in an L-amino acid; it is on the right in a D-amino acid. To identify $R$ or $S$ configurations, swing the α-carbon hydrogen behind the α carbon. Then assign a priority number based on the atomic mass of the atom or group bound to the α carbon. The configuration is $R$ or $S$ according to whether the groups with decreasing priority have a clockwise or counterclockwise orientation, respectively. If two substituent atoms are the same (C, for example), the other atoms attached to them are used to assign priority. Note that for cysteine, sulfur has a higher atomic mass than oxygen, which reverses the priority of the groups. (See figure on next page.)

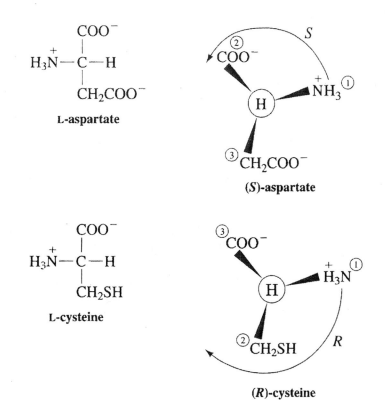

L-aspartate

(S)-aspartate

L-cysteine

(R)-cysteine

23. The amino acids with two or more prochiral centers are Arg, Gln, Glu, Leu, Lys, Met, and Pro.

24. *N*-formylmethionine and *N,N,N*-trimethylalanine cannot occur in the interior of a polypeptide, because they lack free amino groups to form peptide bonds.

25. (a)      GABA is derived from glutamate via decarboxylation of the α-carboxyl group.
    (b)      Histamine is derived from histidine also by decarboxylation of the α-carboxyl group.
    (c)      Thyroxine is derived from tyrosine by addition of a phenyl group and iodination.
    (d) Dopamine is derived from tyrosine via hydroxylation of the phenyl group and decarboxylation of the α-carboxyl group.

26. Six (see Box 4-3).

27. Below is one possible Venn diagram showing the chemical and structural relationships between the R groups of the 20 genetically encoded amino acids in Table 4-1. Note that glycine is shown here as an outlier to all of the groups.

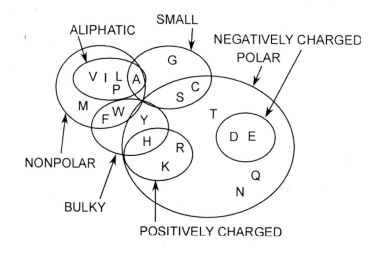

## Chapter 5

1.
- (a) Proteins at low concentrations can be adsorbed by glass or some plastic surfaces, thus significantly decreasing their recovery.
- (b) Sudsing of protein solutions indicates increased contact with the air–water interface, which can denature the proteins.
- (c) Proteins can be eaten or degraded by contaminating bacteria and fungi.
- (d) Protease inhibitors are essential for limiting protein degradation by proteases released from lysosomes during cell lysis.
- (e) Many proteins are only marginally stable and slowly denature over time at temperatures above 25°C.
- (f) Protein structure as well as enzymatic activity are strictly dependent on pH.

2. (a) If the protein is an enzyme, incubate it with a substrate that generates an easily detected (e.g., colored) product or a product that can be converted to a detectable product by a second enzyme.
   (b) If an antibody to the protein is available, the protein can be detected using an immunoassay such as an RIA or ELISA.

3. Immunoaffinity chromatography, immunoassay, and Western blotting could work. One protocol would be to coat a surface with antibodies to protein X, then incubate it with a cell lysate to allow any protein X present to bind to the immobilized antibodies. Wash away unbound proteins, then incubate the surface-bound antibody–protein X complex with the anti-phosphotyrosine antibody to which a readily assayed enzyme has been attached. Wash away unbound antibody and assay the enzyme. The appearance of the enzyme's reaction product indicates the presence of phosphorylated protein X. Unphosphorylated protein X will not bind the second antibody, and lysates that contain no protein X will not cause the second antibody to bind. See Figure 5-3.

4. Following salting out, the precipitated protein can be redissolved in a very small volume, thus concentrating it. A highly concentrated protein solution may be necessary for a subsequent gel filtration step or other analytical procedure. Concentrated protein solutions are also easier to handle, and they minimize protein loss by adsorption to surfaces.

5. Adjust the pH of the solution so that it approaches the p$I$ of one of the proteins. As that protein's net charge approaches 0, it will become less soluble and may precipitate, whereupon it can be eliminated by centrifugation or filtration. Alternatively, add a water-soluble organic solvent to selectively precipitate one of the proteins.

6. In pepsin, aspartate and glutamate must predominate because they have acidic R groups. Lysine, arginine, and tyrosine, all with basic R groups, must predominate in lysozyme.

7.
   (a) At pH 7.0, the side chains of Asp and Glu are ionized. Although both peptides contain ionized N- and C-terminal groups, $(Glu-Asp)_{10}$ contains more onized groups than $Gly_{20}$ and is therefore the more soluble peptide.
   (b) Although Cys would be ionized at higher pH, neither peptide has ionized R groups at pH 5.0. However, $(Cys-Ser-Ala)_5$ has more polar groups and is therefore more soluble than $(Pro-Ile-Leu)_5$.
   (c) $Leu_3$ is less hydrophobic since it has two charged groups (its N- and C-termini) per three residues, compared to two charged groups per 10 residues in $Leu_{10}$. The shorter peptide is more soluble since it requires less disruption of water structure to dissolve.

8. Bovine histone is highly basic (p$I$ = 10.8) and therefore positively charged at pH 7.0. It would therefore bind to the anionic matrix of a cation exchanger (e.g., carboxymethyl cellulose).

9. Above pH 9, the amino groups of the DEAE matrix lose their protons and are therefore unable to bind anionic substances.

10. Some pre-purification before immunoaffinity chromatography may be necessary because the cell lysate may contain other proteins that compete with the desired protein for binding to the immobilized antibody. Although such proteins would bind with lower affinity to the antibody, their greater numbers would diminish the amount of desired protein that binds specifically to the antibody.

11. Two methods for determining molecular mass are gel filtration chromatography and SDS-PAGE. SDS-PAGE is more accurate. Gel filtration is sensitive to the native shape of a protein, whereas SDS-PAGE is performed on an SDS-denatured polypeptide so that its folded shape is not an issue. Note that SDS-PAGE reveals the molecular masses of individual polypeptide chaina; gel filtration is appropriate for determining the mass of a protein consisting of multiple subunits.

12.
  (a) The step that provides the greatest purification is the step that gives the largest fold purification from the previous step, or the greatest increase in specific activity from the previous step. In this example, the step is substrate affinity chromatography, with a 10-fold purification from the previous step.

  (b) The enzyme may be pure, or there may yet be other proteins present. SDS-PAGE provides the most powerful technique for ascertaining whether the final protein preparation contains a single polypeptide. However, if SDS-PAGE reveals equivalent amounts of more than one polypeptide, it is possible that the native enzyme is a multimer of nonidentical subunits.

13.
  __C__ An alkylating agent that reacts with Cys residues.
  __D__ A reducing agent that carries out reductive cleavage of disulfide bonds.
  __A, B__ A reagent used to identify N-terminal residues.
  __E__ A reagent that cleaves polypeptides into smaller fragments.
  __B__ A reagent used in sequencing polypeptides from the N-terminus.

14. If the insulin is reduced and alkylated to eliminate disulfide bonding between Cys residues, then endopeptidase V8, which cleaves after Glu residues, will generate six peptides. If the disulfide bonds between the two chains of insulin are intact, the peptidase will generate four fragments.

| | Total Protein (mg) | Activity (nkat) | Specific Activity (nkat/mg) | Fold Purification |
|---|---|---|---|---|
| Crude extract | 50,000 | 10,000,000 | 200 | 1 |
| Ammonium sulfate precipitation | 5,000 | 3,750,000 | 750 | 3.75 |
| DEAE-cellulose, stepwise KCl gradient | 500 | 500,000 | 1,000 | 5 |
| DEAE-cellulose, KCL linear gradient | 250 | 500,000 | 2,000 | 10 |
| Gel filtration | 25 | 250,000 | 10,000 | 50 |
| Substrate affinity chromatography | 1 | 100,000 | 100,000 | 500 |

15. Since the average mass of an amino acid residue is ~110 D, a peptide with a mass of 4500 D contains ~44 residues. Thus, both peptides are small enough to be directly sequenced. The two peptides are linked by a disulfide bond, so determining the protein's complete sequence requires the following steps: (a) reduce the disulfide bond; (b) alkylate the free thiol groups; (c) separate the peptides; and (d) sequence each peptide. If either peptide contains more than one Cys residue, additional fragmentation and sequencing steps would be necessary to identify the position of the disulfide bond in the intact protein.

16. Yes, there is enough information. The overlap of the relevant fragments is

MLYCRGM
    GMNIK
      NIKGLM
        GLMR
          RFM
           FMK

Thus, the peptide sequence is **MLYCRGMNIKGLMRFMK**. To confirm the deduced sequence, perform one round of Edman degradation to verify that the N-terminus is methionine.

17. Since there are no cysteines, there is only one N-terminus in this polypeptide. Therefore, treatment (b) reveals that His or Lys must be at the N-terminus (the ε-amino group of Lys can also react with dansyl chloride). Treatment (c) indicates that there is a Lys at the C-terminus, and since trypsin treatment did not cleave the polypeptide, the **YSK** peptide must represent the C-terminal peptide. Since peptide A contains **F** and must end in **F** due to the specificity of chymotrypsin, **T** must precede **F**. We can now deduce the following:

HSE
  EGT...

    FTS
      TSD
        DYS
          YSK

From the masses of the two peptides produced by chymotrypsin treatment, we know that peptide A is about 6 amino acids long (~110 D per residue) and peptide B is 4 amino acids long. Hence, the sequence of peptide A must be **HSEGTF** and that for peptide B **TSDY**. Since **YSK** is the C-terminal peptide, we can deduce that the sequence of the polypeptide is **HSEGTFTSDYSK**.

18. The table below shows the percent difference in mass when examining different pairs of peaks. See Sample Calculation 5-1 for the determination of $M$.

| $m/z$ pair | $M$ | % Difference |
|---|---|---|
| 1884.7 / 1696.3 | 16,950.3 D | 0.004 |
| 1542.3 / 1414.0 | 16,975.0 D | 0.14 |
| 893.3 / 848.7 | 16,959.7 D | 0.051 |

19. The rate of protein evolution depends on the ability of a protein to accept mutations without significantly altering its function (neutral drift), which varies among proteins. Hence, proteins with slow rates of evolution cannot accept many amino acid substitutions without loss of function, thereby eliminating mutant forms by natural selection.

20. From inspection of the last row of Table 5-6, it is seen that residue 56 has three different amino acids in the table, 61 has four, 74 has one, 85 has two, and 89 has nine. Hence their order, from least to most conserved is: 89, 61, 56, 85, 74.

21. The fibrinopeptides are evolving too rapidly to show the more ancient relationships among mammals, since over periods greater than 100 million years, many residues will have mutated two or more times. Histone H4 has evolved too slowly to reveal variations in mammalian species (they all have nearly identical histone H4's). However, cytochrome $c$ and hemoglobin evolve at rates that would be useful for assessing phylogenetic relationships among mammalian species, with hemoglobin, the faster-evolving protein of the two, providing a mores sensitive measure.

22. In the comparison between these two proteins, the follow amino acid pairs are listed as positives:

| | |
|---|---|
| H:Y | – Histidine and tyrosine |
| S:N | – Serine and glutamine |
| L:V | – Leucine and valine |
| D:N | – Glutamate and glutamine |
| I:V | – Isoleucine and valine |
| L:I | – Leucine and isoleucine |
| S:A | – Serine and alanine |
| M:L | – Methionine and leucine |

The following pairs are considered positives since they all have nonpolar R groups: L:V, I:V, L:I, and M:L. Similarity based on polarity can be seen in the S:N and D:N pairs being categorized as positives in BLAST. Similarity in size or bulkiness accounts for the H:Y and S:A pairings being scored as positives, with H and Y both being bulky, and S and A having small R groups.

23. Shown below is an alignment file that can be obtained at the Clustal W site:

```
CAP_Ecoli                      -MVLGKPQTDPTLEWFLSHCH----------------------IHKY  24
CAP_Kpneumoniae                -MVLGKPQTDPTLEWFLSHCH----------------------IHKY  24
CAP_Ypestis                    -MVLGKPQTDPTLEWFLSHCH----------------------IHKY  24
Crp_Shewanella                 MTIEQNKNRRPAASGCAIHCHDCSMGTLCIPFTLNANELDQLDDIIERKK  50
fnr_Aeromonas_hydrophila       MIPEKKPGRRIQSGGCAIHCQDCSISQLCIPFTLNDNELDQLDSIIERKK  50
                                 :          **:        *:                  :*

CAP_Ecoli                      PSK--STLIHQGEKAETLYYIVKGSVAVLIKDEEGKEMILSYLNQGDFIG  72
CAP_Kpneumoniae                PSK--STLIHQGEKAETLYYIVKGSVAVLIKDEEGKEMILSYLNQGDFIG  72
CAP_Ypestis                    PSK--STLIHQGEKAETLYYIVKGSVAVLIKDEEGKEMILSYLNQGDFIG  72
Crp_Shewanella                 PIQKGEQIFKSGDPLKSLFAIRSGTVKSYTITEQGDEQITGFHLAGDVIG 100
fnr_Aeromonas_hydrophila       PIQKGEELFKAGDELKSLYAIRSGTIKSYTITEQGDEQITAFHLAGDLVG 100
                               * :   . ::: *:  ::*:  * .*::     *:*.*  * .:    **.:*

CAP_Ecoli                      ELGLFEEGQERSAWVRAKTACEVAEISYKKFRQLIQVNPDILMRLSAQMA 122
CAP_Kpneumoniae                ELGLFEEGQERSAWVRAKTACEVAEISYKKFRQLIQVNPDILMRLSSQMA 122
CAP_Ypestis                    ELGLFEEGQERSAWVRAKTACEVAEISYKKFRQLIQVNPDILMRLSSQMA 122
Crp_Shewanella                 FDGIHAQSHQ--SFAQALETSMVCEIPFNILDELSGSMPKLRQQIMRLMS 148
fnr_Aeromonas_hydrophila       FDAIHKQAHP--SFAQALETAMVCEIPFDVLDDLSGKMPKLRQQIMRLMS 148
                               .:. :.:   ::.:*  :. *.**.:. : :*    *.: ::    *:

CAP_Ecoli                      RRLQVTSEKVGNLAFLDVTGRIAQTLLNLAKQP-DAMTHPDGMQIKITRQ 171
CAP_Kpneumoniae                RRLQVTSEKVGNLAFLDVTGRIAQTLLNLAKQP-DAMTHPDGMQIKITRQ 171
CAP_Ypestis                    NRLQITSEKVGNLAFLDVTGRIAQTLLNLAKQP-DAMTHPDGMQIKITRQ 171
Crp_Shewanella                 NEIMSDQEMILLLSKKNAEERLAAFISNLANRFGNRGFSPKEFRLTMTRG 198
fnr_Aeromonas_hydrophila       NEIMGDQEMILLLSKKNAEERLAAFLHNLSVRFSERGFSAKEFRLSMTRG 198
                               ..:   .* : *: :. *:* : **: :  :   .. :::.:**

CAP_Ecoli                      EIGQIVGCSRETVGRILKMLEDQNLISAHGKTIVVYGTR---------- 210
CAP_Kpneumoniae                EIGQIVGCSRETVGRILKMLEDQNLISAHGKTIVVYGTR---------- 210
CAP_Ypestis                    EIGQIVGCSRETVGRILKMLEDQNLISAHGKTIVVYGTR---------- 210
Crp_Shewanella                 DIGNYLGLTVETISRLLGRFQKSGLIEVKGKYITILDHHELNLLAGNARI 248
fnr_Aeromonas_hydrophila       DIGNYLGLTVETISRLLGRFQKSGLISVKGKYITVLDHVALGVMAGATRP 248
                               :**: :* : **:.*:*  ::...**..:** *.:  .

CAP_Ecoli                      ---
CAP_Kpneumoniae                ---
CAP_Ypestis                    ---
Crp_Shewanella                 AR- 250
fnr_Aeromonas_hydrophila       PCG 251
```

(a) The N-terminal and C-terminal amino acids show the most divergence across all 5 species; however, note that even in these areas, some species show some similarities between each other. This may be a reflection of the broad range of genes, ranging from sugar catabolism to nitrogen metabolism, which this family of proteins may control.

(b) Indeed, most of this protein seems to be conserved.

(c) Two small domains are identified, the CAP_ED domain and the HTH_CRP domain. The CAP_ED domain is a ligand-binding domain that serves as the sensor for the DNA-binding activity for this protein. Notice that current information suggests a broad range of ligands (cAMP, CoA, iron–sulfur clusters), depending on the polypeptide. The HTH_CRP domain is the DNA-binding domain. The colors indicate degrees of probability that the matches shown are not due to chance. However, since the lengths of these amino acids sequences are similar and not very long, the e-values can be deceptive. Note that in the CAP_ED domain, 55 of 90 amino acids are either identical or similar in chemistry (61.1%), while in the HTH_CRP domain, 50 of 68 amino acids are identical or similar (73.5%), which makes a little more biochemical sense given the putative biological roles of each domain.

*Chapter 6*

1. The C—N bond has a partial double-bond character due to resonance interactions. This limits rotation around the C—N bond, constraining the carbonyl and amide groups to lie in the same plane.

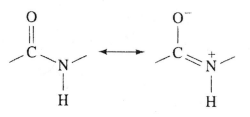

2.

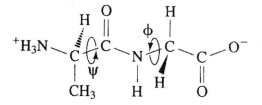

Note that both $\phi$ and $\psi$ increase clockwise when viewed from $C_\alpha$.

3. In the structure shown below, the thicker lines represent the peptide bonds. The polypeptide backbone contains three single bonds capable of free rotation between each peptide bond (versus two in a standard polypeptide). This greater flexibility allows for more possible conformations of the peptide chain, so that the polymer is more likely to assume a random, fluctuating conformation rather than a regular secondary structure.

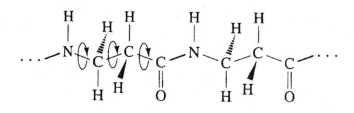

4. When $\phi = 180°$ and $\psi = 0°$, the van der Waals spheres of the amide hydrogens overlap.

5. Rotation around $\phi$ will cause the van der Waals sphere of the carbonyl oxygen of residue $n$ to collide with the van der Waals sphere of the $C_\beta$ of residue $n + 1$. This can be clearly seen in Kinemage 3-1. Turn on AAs 1–3, PH1, planes, O1, and CB. Then try rotating around $\phi$ to observe collision of van der Waals spheres.

6. Shown below is the dipole of a peptide unit. In an α helix, all the hydrogen bonds point in the same direction, so the dipoles of each peptide unit sum up to an overall dipole moment for the entire helix. Since the carbonyl groups all point from the N-terminus toward the C-terminus of the helix, the N-terminus of the helix has a partial positive charge and the C-terminus has a partial negative charge.

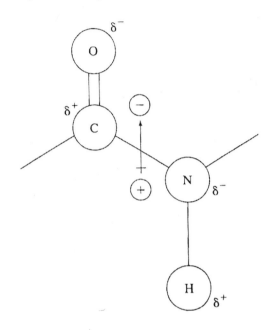

7. Antiparallel β sheets are more stable because the hydrogen bonds connecting adjacent polypeptide strands are not distorted, as they are in parallel β sheets (see Figure 6-9).

8. The first and fourth residues of α keratin's seven-residue repeat (residues *a* and *d* of the repeating sequence *a-b-c-d-e-f-g*) are primarily nonpolar residues that form a hydrophobic strip along one side of the α helix (which has 3.6 ≈ 7/2 residues per turn). When the hydrophobic strips of two keratin helices interact, the helices incline slightly, causing them to coil around each other (see Figure 6-15).

9. The steaming process stretches α keratin into a pleated β sheet conformation. When subsequently heated, the keratin in the wool is converted back to an α-helical conformation, which is more compact.

10. The repeating sequence is Gly-X-Y, where X is often Pro and Y is often Hyp. The glycine residues occur in the interior of the triple helix, where there is no room to accommodate any side chain larger than a hydrogen atom. The highly constrained Pro and Hyp side chains confer rigidity on the helical structure.

11. Scurvy is caused by a deficiency of ascorbic acid (vitamin C), which is necessary for the activity of prolyl hydroxylase.

12. Hydroxyproline is not one of the 20 standard amino acids used in the synthesis of proteins from mRNA. Specific proline residues in collagen are oxidized by prolyl hydroxylase to form 4-hydroxyproline only after peptide synthesis. Hence, only the administration of $^{14}$C-labeled proline can yield radioactively labeled collagen fibers.

13. Glu, Asp, and Pro are the least likely to occur in a β sheet as judged by their low propensity for appearing in the β sheets of known proteins (see Table 6-1 on p. 141.)

14. Several pairs of amino acids give indistinguishable electron density maps: Asp and Asn, Glu and Gln, Thr and Val. Other amino acids that may have similar shapes are Ser and Cys, although the S atom of Cys has a much greater electron density than the O atom of Ser.

15. At 6 Å resolution, larger elements of protein structure can be identified, most notably helices, which have a diameter of several Å and are visible as rodlike shapes.

16. The advantages of NMR include (a) structural information from proteins that fail to crystallize and (2) information about protein folding and dynamics since protein movements can be traced over relatively long time scales. The primary disadvantage of NMR is that the protein must be no larger than ~40 D.

17. When interpreting an electron density map, knowledge of the primary structure allows the identification of specific amino acids along the polypeptide chain. This would otherwise be quite difficult if not impossible for R groups with similar shapes (see Question 14.)

18. Charged or polar amino acids are commonly on the outside of the molecule, where they are exposed to the aqueous solvent, whereas uncharged or nonpolar amino acids are often buried the interior of the protein, out of contact with the aqueous solvent, due to hydrophobic effects.

19. As protein size increases, the surface-to-volume ratio decreases. Recall that the surface of a sphere increases as the square of the radius ($r^2$), while the volume of a sphere increases as the cube of the radius ($r^3$). Hence, the ratio of hydrophilic to hydrophobic residues decreases as the molecular mass increases since the interior (volume) of the protein increases more rapidly than its surface area.

20. (a)  On the MMDB Summary Page, one can obtain
    (i) The currently accepted name for the protein;
    (ii) Taxonomic information about the organism from which the protein was obtained;
    (iii) An overview of structural and sequence information; and
    (iv) Links to other data files, particularly the PDB page, which contains a wealth of information about the biochemistry of the protein, and the VAST page, which allows one to examine proteins with similar structural domains.

(b) From the Sequence Details page, one can see that β sheets dominate domain 1 and alpha helices dominate domain 2. Note, however, that the catalytic domain includes both of these structural domains, although domain 2 is where the regulatory function, the binding of $Ca^{2+}$-calmodulin, occurs (see the Structural Genomics Consortium at http://www.sgc.utoronto.ca/SGC-WebPages/StructureDescription/2V7O.php).

(c) While domain 2 has VAST "hits" mostly to protein kinases, domain 1 has some similarity to different proteins, including a secretion chaperone, interferon regulatory factor (part of a transcription regulatory complex), and an archeabacterial endonuclease.

21. A tetrameric protein can have $C_4$ or $D_2$ symmetry. A pentameric protein can have only $C_5$ symmetry.

22. (a) The protein is probably a trimer of 40 kD subunits. The gel shows cross-linked dimers and trimers in addition to free monomers.

(b) The apparent masses of the faint bands are multiples of 120 kD and therefore probably represent protein trimers that have been chemically cross-linked to other trimers.

(c) The protein is a trimer of two 20 kD subunits and one 50 kD subunit. Some of the 90-kD oligomers have been cross-linked to form larger structures of 180 kD (a dimer), 270 kD (a trimer), and 540 kD (a hexamer).

23. Recall that $\Delta G° = \Delta H° - T\Delta S°$. At 4°C, $T\Delta S°$ is negative and its absolute value is greater than that of $\Delta H$, so $\Delta G°$ is positive, indicating that the equilibrium favors monomer formation (see Table 1-3, p. 15). At 45°C, $T\Delta S°$ is positive and greater than $\Delta H$, so $\Delta G°$ is negative, indicating that the equilibrium now favors the formation of dimers. The cytoskeletal elements called microtubules behave in such a manner.

24. The favorable free energy of formation of an ion pair is nearly equivalent to the loss of solvation free energy that occurs when the charged groups form the ion pair.

25. Heat, pH changes, detergents, and chaotropic agents.

26. The probability for each monomer is $1/5 \times 1/3 \times 1/1 = 1/15 = 0.067$ or 6.7%. The probability that a native dimer can form is $0.067 \times 0.067 = 0.0044$ or 0.44%.

27. Thermostable proteins are very similar in amino acid composition to mesophilic proteins. One distinguishing feature in some is that many such proteins have networks of salt bridges on their surfaces.

28. $t = 10^n/10^{13}$, where $t$ is the time in seconds, and $n$ is the number of amino acid residues. Rearranging terms,

$$n = \log 10^{13} + \log t$$

Here, $t = (60 \text{ s/min})(60 \text{ min/h})(24 \text{ h/day}) = 86{,}400 \text{ s}$, so that

$$n = 13 + \log 86{,}400$$
$$= 13 + 4.9$$
$$= 17.9$$

Therefore, the peptide could have no more than 18 residues.

29. For a 200-residue protein, $t = 10^n/10^{13} = 10^{200}/10^{13} = 10^{187}$ s or about $3 \times 10^{179}$ years. If the eight $\alpha$ helices and six $\beta$ sheets act as nucleating elements for the formation of tertiary structure, $n = 200/(8+6) \approx 14$ so that $t = 10^{14}/10^{13} = 10$ s.

30. A 70 kD polypeptide would be composed of roughly 625 amino acids (the 70,000 molecular mass of the polypeptide divided by 112, the molecular mass of an amino acid), which means that there are 624 peptide bonds. Therefore, 2496 ATP equivalents (4 ATP equivalents per peptide bond × 624 peptide bonds) are required to synthesize this polypeptide. Folding of this average-sized polypeptide in the GroEL/ES complex would require 168 ATP molecules (7 ATP per cycle × 24 cycles for an average-sized polypeptide). Hence, properly folding the polypeptide costs about 6.7% of the ATP energy needed to synthesize it. Note that this calculation ignores all the energy required to activate the gene and transcribe it as well as the considerable energy required to synthesize amino acids (Section 21-5).

31. Under physiological conditions, most polypeptides display marginal stability, averaging ~0.4 kJ/mol amino acids (see page 156). Furthermore, some native proteins in *E. coli* appear to repeatedly bind to GroEl/ES complexes long after they have been initially folded into their native structure, which suggests that native proteins frequently denature under physiological conditions (see page 169).

## Chapter 7

1.  
| | |
|---|---|
| E | A component of cytochromes |
| B, C, D | Binds $O_2$ |
| A | Contains iron in the Fe(III) state |
| B | Found in muscle only |
| D | Forms filaments in the deoxy state |

2. The globin prevents the oxidation of its bound heme to the Fe(III) state, and in the case of hemoglobin, permits cooperative $O_2$ binding, which is responsible for the efficient transport of $O_2$ from the lungs to the tissues.

3. High metabolic activity generates $CO_2$, which reacts with water to form $H_2CO_3$, which in turn ionizes to yield $HCO_3^- + H^+$. The protons preferentially bind to T-state hemoglobin (the Bohr effect) and thereby cause $O_2$ to be released.

4. A lower $p_{50}$, or higher affinity of hemoglobin for $O_2$, would have almost no effect on $O_2$ uptake in the lungs, since hemoglobin is nearly saturated at arterial $pO_2$. However, the lower $p_{50}$ would result in less $O_2$ released at the peripheral tissues, where the $pO_2$ is ~30 torr. The $p_{50}$ of normal hemoglobin (26 torr) is an evolutionary compromise that allows hemoglobin to become saturated with $O_2$ in the lungs but to be deoxygenated in the oxygen-poor peripheral tissues.

5. The $O_2$ that is released in the capillaries diffuses to the mitochondria, where it is reduced to water. However, the solubility of $O_2$ in aqueous solution is too low to support its required rate of diffusion in rapidly respiring muscle. The presence of myoglobin, in effect, increases the solubility of $O_2$ in muscle tissue. Thus, the $O_2$ released by hemoglobin is passed from myoglobin molecule to myoglobin molecule in a kind of molecular bucket brigade until it is taken up by the mitochondria. In this way, myoglobin increases the rate that $O_2$ can diffuse through the tissues.

6. $CO_2$ reacts preferentially with amino groups in deoxyhemoglobin to form carbamates. This helps transport $CO_2$ and facilitates the release of $O_2$ by stabilizing the deoxy state.

7. None; $CO_2$, $H^+$, and BPG all bind preferentially to deoxyhemoglobin to reduce the affinity of hemoglobin for $O_2$.

8. CO binds to hemoglobin with so much higher affinity than $O_2$ that CO cannot be displaced by high concentrations of oxygen. For practical purposes, CO binding to hemoglobin is therefore irreversible (at least in the short term). Therefore, a fresh blood transfusion is required to counteract the effects of this poison.

9.

| | |
|---|---|
| A | Forms a coordination bond with Fe(II) |
| A, C | Involved in the binding of heme |
| E | Partially occludes the $O_2$-binding site |
| G | Conformations are nearly superimposable |
| B | Forms a hydrogen bond with $O_2$ |
| I | Is associated with a rotational shift of the hemoglobin $\alpha_1\beta_1$ dimer with respect to the $\alpha_2\beta_2$ dimer. |
| D | Disrupts the N- and C-terminal salt bridges in hemoglobin |
| J | Stabilizes the T state |
| F, I | Involved in the interactions of hemoglobin subunits |

10. The decrease in p$K$ promotes deprotonation of the imidazole group at pH 7.4. The release of $H^+$ contributes to the Bohr effect.

11. BPG binds preferentially to the T state and stabilizes it through the formation of salt bridges between BPG and the β subunits. This makes the shift to the R state less energetically favorable, thereby decreasing hemoglobin's ability to bind $O_2$.

12. The symmetry model cannot account for negative cooperativity, in which the binding of a ligand to one subunit decreases the binding affinity of the other subunits.

13. (a) Only two conformational states are possible, either R or T, since all subunits change conformation simultaneously.
    (b) Four conformations are possible, corresponding to a trimer in which 0, 1, 2, or 3 subunits have bound ligand.

14. (a) Hb Kansas. Individuals with this variant might be relatively symptom-free in the Andes, since this mutation, which destabilizes oxyhemoglobin, would allow more oxygen to be given up to tissues.
    (b) All the variants listed, except for Hb Kansas and Hb Yakima, are likely to be unstable, since all will affect key electrostatic interactions within the globin polypeptides or the quaternary structure of the variant hemoglobin.
    (c) Hb Yakima. The mutation in this variant destabilizes the T state, which leads to a decrease in oxygen dissociation in the peripheral tissues. The body thereby compensates by increasing the amount of hemoglobin.
    (d) All the variants listed may show changes in the Hill constant, since cooperativity will be affected by all the changes described in Table 7-1.

15. The low $pO_2$ of the capillaries, which leads to an increase in the concentration of deoxyhemoglobin, promotes polymerization of hemoglobin S.

16. Fetal hemoglobin (which contains γ globin chains rather than β chains) dilutes the hemoglobin S in erythrocytes, so that the deoxyhemoglobin S in the venous blood is less likely to achieve the critical concentration for fiber formation.

17. Myosin moves along the actin filament via repeated cycles of conformational changes in the head region of the myosin molecule. This cycle of conformational changes is unidirectional because it is coupled to the irreversible hydrolysis of ATP. When the myosin head binds ATP, it releases the actin filament to which it is bound. ATP hydrolysis follows, resulting in a change in the conformation of the myosin head to the "cocked" or high-energy state. The myosin head then binds weakly to the actin filament at a new position. The strength of this binding interaction increases upon the release of $P_i$. The myosin head then undergoes a second major conformational change, producing the power stroke that causes the translocation of the actin filament relative to the myosin filament. Upon completion of the power stroke, the myosin head releases ADP but remains bound to the actin filament. It can then bind another ATP molecule to begin the cycle anew. The thick filament's multiple myosin heads, each undergoing multiple ATP-driven reaction cycles, cause the thick filament to "walk" along the thin filament. As the thick and thin filaments slide past each other, the sarcomere decreases in length and the muscle thereby contracts.

18. The thin filaments of striated muscle contain actin in complex with tropomyosin and troponin. At resting $Ca^{2+}$ concentrations, the muscle is relaxed because tropomyosin blocks the myosin binding sites on the actin filament. When the intracellular $Ca^{2+}$ concentration increases in response to a nerve impulse, troponin C (a subunit of troponin) binds $Ca^{2+}$, causing a conformational change in troponin. This results in the movement of tropomyosin, which uncovers the myosin binding sites, and thus permits the myosin heads to interact with the thin filament.

19. (a) During phase A, most of the actin exists in the form of small oligomers. Only a few oligomers nucleate microfilament growth. During phase B, rapid growth occurs at the ends of growing filaments. During phase C, the rate of filament growth is equal to the rate of dissociation of ADP–G-actin from the filaments, so there is no net increase in F-actin.
    (b) If the ATP concentration becomes limiting as it is hydrolyzed in actin subunits, the entire population of actin subunits will eventually contain bound ADP. ADP–G-actin can still form filaments, but since ADP–F-actin is more likely to disassemble than ATP–F-actin, the F-actin concentration will decrease until a new equilibrium is established, as shown below.

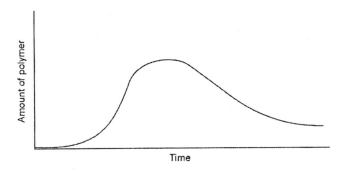

20. A secondary immune response is greater than the first (see Figure 7-37), so more antibody can be recovered.

21. (a) IgM        (b) IgE        (c) IgA        (d) IgA, IgM        (e) IgG

22. (a) Antibody-producing lymphocytes have a limited proliferative capacity and are therefore unsuitable for growing in large numbers in culture to produce large amounts of antibody. Myeloma cells, like all cancer cells, have an unlimited proliferative capacity and impart their immortal phenotype to the hybrid cells.
    (b) All the lymphocytes harvested from the immunized animal can potentially fuse with the myeloma cells. The resulting hybridomas will therefore secrete a variety of different antibodies, only a small fraction of which are specific for protein X.

(c) Unfused lymphocytes have no metabolic deficiency but grow for only a limited time, and then die out. In the selective medium, unfused myeloma cells cannot synthesize purines, which are necessary for DNA replication and cell division. Only the fused cells can both synthesize purines and proliferate without limit.

23. (a) Affinity chromatography using immobilized protein X would be the best procedure.

(b) It is likely that protein X contains several antigenic features that are recognized by B cells and elicit antibody production. The anti-protein X antibodies isolated from the rabbit are therefore a mixture of different IgG molecules that recognize different antigenic features on protein X. Even antibodies that recognize the same portion of the antigen may have different amino acid sequences and different antigen-binding affinities. In contrast, each mouse monoclonal antibody preparation is homogeneous. Because monoclonal antibodies are all the products of identical B cells, they all have the same unique sequence and antigen-binding specificity.

(c) One of the mouse monoclonal antibodies would be most suitable, since it would yield identical Fab fragments that would be likely to form regular crystals. In contrast, Fab fragments isolated from a heterogeneous population of rabbit antibodies could not form such crystals.

24. The loops can accommodate a wide variety of amino acid sequences because they are on the surface of the protein. Such sequence variation in the β sheet core of the protein would likely disrupt its structure.

## Chapter 8

1. (a) aldose, pentose, reducing sugar
   (b) aldose, hexose, reducing sugar
   (c) ketose, hexose, reducing sugar
   (d) aldose, hexose, uronic acid, reducing sugar
   (e) hexose, alditol
   (f) aldose, hexose, deoxy sugar, reducing sugar

2.

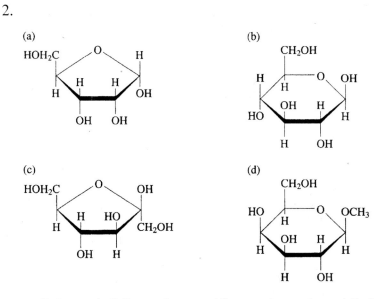

(a)

(b)

(c)

(d)

Only methyl-β-D-galactose (d) contains a glycosidic bond.

3. The three other forms are the open chain form of glucose, α-D-glucofuranose and β-D-glucofuranose.

4. The α anomer of glucose is less soluble than the β anomer and therefore comes out of solution more readily. As it crystallizes, the β-glucose remaining in solution interconverts with α-glucose thereby maintaining the 36% α–63% β equilibrium ratio. Thus, the α anomer is continually generated and deposited in the crystal.

5.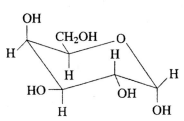

The most stable conformation has its bulky $CH_2OH$ group and two OH groups in equatorial positions, which minimizes their steric interference.

6.

   D   Differs in configuration at the anomeric carbon
   A   Polyhydroxy alcohol
   E   Product of condensation of anomeric carbon with an alcohol
   B   Differs in configuration at one carbon atom.
   C   Polymer of monosaccharides

7. (a) Contains fructose, contains an α anomeric bond.
   (b) Is a reducing sugar.
   (c) Is a reducing sugar.
   (d) Contains an α anomeric bond, is a reducing sugar.
   (e) Contains fructose, contains an α anomeric bond.

8.

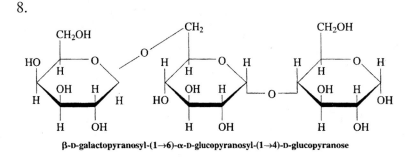

β-D-galactopyranosyl-(1→6)-α-D-glucopyranosyl-(1→4)-D-glucopyranose

**β-D-galactopyranosyl-(1→6)-α-D-glucopyranosyl-(1→4)-D-glucopyranose**

9. The β(1→4) glycosidic bonds that join glucose groups in cellulose give rise to extended polymers that line up in parallel and are stabilized by extensive intrachain and interchain hydrogen bonds. This gives cellulose fibers unusual strength and renders them water insoluble. In contrast, the α(1→4) glycosidic bonds linking glucose residues in starch give rise to a totally different structure, namely a linear chain that assumes a relatively open helical conformation.

10. Three advantages are (1) the reduction in osmotic pressure in a cell when many residues are combined into a single polymeric molecule: (2) the rapid mobilization of glucose when it is removed from the ends of many branches simultaneously; and (3) the resulting globular structure enables the glycogen molecule to pack snugly within storage granules.

11. Lectins are proteins that bind specific sugars with high affinity. Thus, a lectin attached to an immobile support can be used to selectively adsorb polysaccharides or glycoproteins via affinity chromatography. The adsorbed molecules can then be eluted by applying a solution containing an excess of the free sugar to which the lectin preferentially binds.

12. (a) The core oligosaccharide is (mannose)$_3$(GlcNAc)$_2$.
    (b) Typical terminal sugars are galactose, *N*-acetylneuraminic acid (sialic acid), and fucose.

13. Swainsonine causes an accumulation of high-mannose *N*-linked oligosaccharides by interfering with processing of the core oligosaccharide.

14. In the form of tPA with three *N*-linked oligosaccharide chains, the presence of the additional bulky chain impedes access of plasmin to the site of proteolysis, thereby protecting tPA from rapid cleavage.

15. Each disaccharide unit bears a negative charge (from the ionized COOH group of *N*-acetylneuraminic acid). Because of the strong repulsion between charged groups, the protein assumes a rigid elongated shape, thereby accounting for the high viscosity of the solution. Removal of the *N*-acetylneuraminic acid residues by sialidase abolishes the large net negative charge and permits the protein to assume a more compact shape. As a result, the viscosity of the solution decreases.

16. Type AB individuals have both type A and type B carbohydrate structures on their cell surfaces, so they do not recognize either type A or type B blood cells as foreign. Consequently, they can receive either type A or type B blood. Type A individuals synthesize antibodies to type B carbohydrate antigens, so transfused blood cells bearing the type B carbohydrate antigen will agglutinate in the blood vessels. Similarly, type B individuals synthesize antibodies to the type A carbohydrate antigen and therefore cannot receive type A blood.

## Chapter 9

1.

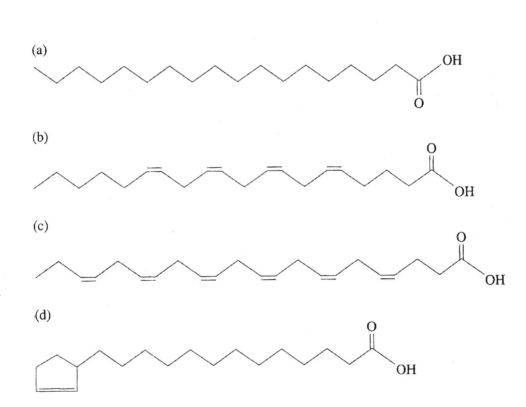

2.  (a)  The melting point increases with increasing hydrocarbon chain length. Fully saturated hydrocarbon chains have the least amount of steric hindrance and thus can pack together closely.

    (b)  The melting point decreases with an increasing number of double bonds.

    (c)  Fatty acid residues that have a cis double bond exhibit a 30° bend in their hydrocarbon chain that interferes with close packing. In contrast, trans double bonds distort the acyl chain conformation much less, allowing closer packing of molecules. Hence a fatty acid with trans double bonds has a higher melting point than the corresponding cis fatty acid of the same chain length and with the same number of double bonds.

3.  For example, 1-palmitoyl-2-palmitoleoyl-3-oleoyl-glycerol

$$
\begin{array}{ccc}
CH_2 & CH & CH_2 \\
| & | & | \\
O & O & O \\
| & | & | \\
C{=}O & C{=}O & C{=}O \\
| & | & | \\
CH_2 & CH_2 & CH_2 \\
| & | & | \\
CH_2 & CH_2 & CH_2 \\
| & | & | \\
CH_2 & CH_2 & CH_2 \\
| & | & | \\
CH_2 & CH_2 & CH_2 \\
| & | & | \\
CH_2 & CH_2 & CH_2 \\
| & | & | \\
CH_2 & CH_2 & CH_2 \\
| & | & | \\
CH_2 & CH_2 & CH_2 \\
| & \| & \| \\
CH_2 & CH & CH \\
| & \| & \| \\
CH_2 & CH & CH \\
| & | & | \\
CH_2 & CH_2 & CH_2 \\
| & | & | \\
CH_2 & CH_2 & CH_2 \\
| & | & | \\
CH_2 & CH_2 & CH_2 \\
| & | & | \\
CH_2 & CH_2 & CH_2 \\
| & | & | \\
CH_2 & CH_2 & CH_2 \\
| & | & | \\
CH_3 & CH_3 & CH_2 \\
& & | \\
& & CH_2 \\
& & | \\
& & CH_3 \\
\end{array}
$$

4.  Triacylglycerols serve as metabolic energy sources and thermal insulators.

5.  Glycerophospholipids are amphiphilic because they are composed of a hydrophobic diacylglycerol "tail" attached to a hydrophilic phosphoryl derivative "head."

6.

(a)

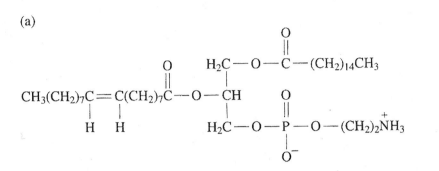

(b)

7.  A plasmalogen differs from a diacylglycerophospholipid in that a hydrocarbon chain is attached to the C1 position of glycerol via an α,β-unsaturated ether bond in the cis configuration instead of an ester bond.

8.  Sphingomyelin and gangliosides both contain a sphingosine moiety to which is attached a fatty acyl group in an amide linkage. They differ in that sphingomyelin has a phosphocholine (or less commonly a phosphoethanolamine) head group, whereas the ganglioside head group is an oligosaccharide that includes one or more sialic acid residues.

9.  This compound is a sterol that closely resembles testosterone and estradiol (shown in Figure 9-11). It is most likely to exert its effects by influencing physiological processes that depend on the sex hormones. In fact, it is dehydroepiandrosterone (DHEA), a metabolic precursor of androgens and estrogens.

10. Although eicosanoids, like hormones, are synthesized by a variety of cells and exert their effects on other cells, they are considered local mediators rather than hormones because they act near their sites of synthesis rather than being carried throughout the body via the bloodstream, and because they decompose quickly.

11. (b)

12. Vitamin A, or retinol, is oxidized to retinal, which plays an essential role as a photoreceptor. A further oxidation product, retinoic acid, functions much like a hormone. Vitamin E ($\alpha$-tocopherol) is an important physiological antioxidant. Vitamin D exists in two inactive forms, $D_2$ and $D_3$. Both forms are activated by enzymatic hydroxylation. The active vitamin aids the deposition of $Ca^{2+}$ in bones and teeth. Vitamin K is required for the carboxylation of several proteins, an essential modification to maintain their function in blood clotting.

13. The difference arises from geometrical considerations. Single-tailed amphiphiles, such as fatty acid anions, form micelles because of their tapered shape. Their hydrated head groups are wider than their tails, enabling them to pack efficiently into a spheroidal micelle. The cylindrical shape of two-tailed amphiphiles, such as glycerophospholipids, prevents their packing into a spheroidal micelle. Instead they pack together to form disklike micelles, which are really extended bilayers.

14. In order for a polar solute to enter a lipid bilayer, its interactions with surrounding water molecules must first be disrupted (i.e. it must lose its hydration shell), and new interactions with the hydrophobic bilayer constituents would have to form. This would require an increase in free energy for the system; that is, the process is unfavorable.

15. When a pure phospholipid bilayer in an orderly gel-like state is warmed, it is converted (at its transition temperature) to a liquid crystal form that is more fluid. The presence of cholesterol in a phospholipid bilayer both decreases its fluidity and broadens the transition temperature range. These effects occur because the sterically rigid cholesterol molecules insert between the fatty acyl chains of the phospholipids and thus restrict their motion.

16. Integral proteins are strongly associated with membranes through hydrophobic effects and can typically be extracted only by agents such as detergents, organic solvents, or chaotropic agents, which disrupt membrane structure. Some integral proteins bind lipids so tightly that they are dissociated from them only under denaturing conditions. Integral proteins are amphiphilic molecules that are oriented in the membrane such that the regions buried in the interior of the membrane have a surface composed mainly of hydrophobic residues, whereas the portions exposed to the aqueous environment have mostly polar residues. These polar regions may occur on one side of the membrane or the other, or may asymmetrically extend from both sides (transmembrane proteins). Integral proteins can diffuse laterally in the plane of the membrane, but their flip-flop rate is nil.

17. Lipids linked covalently to proteins often anchor the proteins in the membrane.

18. The three main types of lipid moieties that can be covalently attached to proteins are isoprenoid groups, fatty acyl groups, and glycosylphosphatidylinositol (GPI) groups.

19.  The principal lipids in lipid rafts are cholesterol and sphingolipids, including sphingomyelin and glycosphingolipids. A phospholipid enriched in docosahexaenoic acid, a polyunsaturated fatty acid with six double bonds, would not be present in lipid rafts, but in highly disordered membrane lipid microdomains. These domains have in fact been called "ultimate non-raft domains".

20.  The cysteine residue in the C-terminal sequence is the site where $p21^{c\text{-}ras}$ is prenylated. Since this event permits the protein to insert into membranes, it appears that $p21^{c\text{-}ras}$ must be membrane-bound to promote tumor formation.

21.  Fusion of the two membranes could not occur at low temperature because the fluidity of the membranes was too low to allow the free diffusion of lipids and proteins (presumably, at the lower temperature they became gel-like solids.) At the higher temperature, the membranes were more fluid, making the lateral movement of lipids and proteins possible.

22.  In normal erythrocytes, the cross-linking of spectrin and other membrane proteins creates a submembrane network of proteins, somewhat akin to a geodesic dome, that gives the cells a biconcave disklike shape. A cell that contains too little spectrin to assemble a complete membrane skeleton is not constrained in shape and becomes a sphere, the simplest possible shape for a membrane enclosing the cell contents.

23.  Growing bacteria were exposed briefly to radioactive phosphate so that newly synthesized phospholipids (mostly phosphatidylethanolamine in *E. coli*) would become labeled. Trinitrobenzenesulfonic acid (TNBS), a membrane-impermeable reagent that reacts with phosphatidylethanolamine (PE), was then added to the cells. At various times, samples of cells were tested for the presence of $^{32}$P-labeled PE that also contained the TNB group. The appearance of doubly labeled PE after three minutes indicated that PE initially synthesized on the cytoplasmic side of the membrane (where it was labeled with $^{32}$P) traversed the membrane and reacted with TNBS in the extracellular medium.

24.  Biological membranes differ from artificial lipid membranes in having protein constituents (e.g., flipases and phospholipid translocases) that greatly accelerate the rate at which phospholipids move from one side of the bilayer to the other.

25.

| | |
|---|---|
| G | free ribosome |
| D | signal peptide |
| H | signal recognition particle (SRP) |
| B | core glycosylation |
| A | membrane anchor |
| I | translocon |
| C | signal peptidase |
| E | membrane-bound ribosomes |
| F | SRP receptor |

26. (a) The membrane protein was synthesized on ribosomes bound to the endoplasmic reticulum and passed partially through the membrane into the ER lumen. Treatment with Triton X-100 disrupted the membrane structure, which prevented the nascent protein from being glycosylated by the glycosyltransferases that are normally present in the ER lumen. Thus, posttranslational glycosylation requires an intact ER membrane.

    (b) Because glycosylation is initiated as the protein is being synthesized, the protein will almost certainly have been glycosylated by the ER enzymes before the membrane was disrupted by the detergent.

27. Disruption of posttranslational protein modification reactions would likely also disrupt protein-targeting mechanisms, many of which depend on modifications, such as glycosylation, that occur in the Golgi apparatus.

28. Most likely, these proteins bring the two membranes sufficiently close to overcome the repulsion energy of the charged phospholipids of each membrane.

## Chapter 10

1. The free energy change is not the same. For glucose $Z_A$ is zero, since glucose has no charge. Hence, the free energy change for glucose movement relative to $Na^+$ movement is reduced by $Z_A \mathcal{F}$

2.
$$
\begin{aligned}
\Delta \bar{G} &= RT \ln ([Ca^{2+}]_{in}/[Ca^{2+}]_{out}) + Z\mathcal{F}\Delta\Psi \\
&= RT \ln (0.001/1) + (2)\mathcal{F}\Delta\Psi \\
&= (8.3145 \text{ J·K}^{-1}\text{·mol}^{-1})(298 \text{ K})(-6.91) + (2)(96,485 \text{ J·V}^{-1}\text{·mol}^{-1})(-0.1\text{V}) \\
&= -17,121 \text{ J·mol}^{-1} - 19297 \text{ J·mol}^{-1} \\
&= -36,418 \text{ J·mol}^{-1} = -36.4 \text{ kJ·mol}^{-1}
\end{aligned}
$$

3.    **1, 2, 3.** The charged molecule is least able to penetrate the hydrophobic core of the bilayer. Although compounds 2 and 3 bear the same polar groups, the larger hydrocarbon chain of compound 3 lets it diffuse more easily through the bilayer.

4. Compound B enters the cell by mediated transport since its rate of entry falls off as [B] increases. This is consistent with a carrier molecule that becomes saturated with B. Compound A enters by a nonmediated process, since its flux is directly related to its concentration at all values of [A].

5. C. A carrier ionophore must diffuse through the membrane. The gel-like state of the lipids at low temperatures slows ionophore movement. The steep portion of the curve corresponds to the transition of the membrane to a more fluid state that allows free ionophore diffusion.

6. The hydration shells around the $K^+$ and $Na^+$ ions must be stripped in order for the ions to enter the channel. In $K^+$ channels, the selectivity filter can accommodate dehydrated $K^+$ ions but is not flexible enough to coordinate dehydrated $Na^+$ ions, which are smaller. Consequently, the $Na^+$ ions remain hydrated and do not enter the $K^+$ channel.

7. Aquaporin is lacking. The presence of aquaporin would allow water to enter the eggs, which would cause the eggs to swell and burst.

8. Probably not, since this transporter allows equilibration of glucose concentration across the membrane in either direction.

9. The cell interior becomes more negative since 3 Na$^+$ exit the cell for every 2 K$^+$ that enter.

10. (a) uniport
    (b) uniport
    (c) antiport
    (d) symport
    (e) symport
    (f) antiport

    The active transport systems are c, d, e, and f.

11.

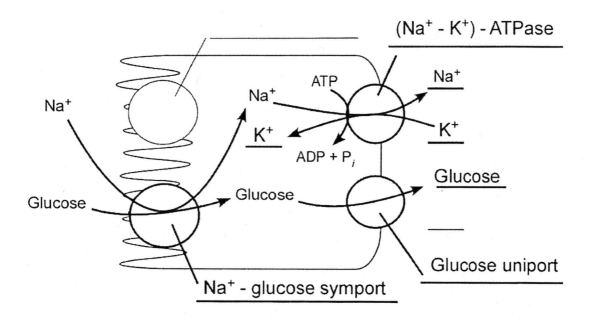

*Chapter 11*

1. The enzyme commission (EC) number is unique to each enzyme. An enzyme is assigned an EC number according to the type of reaction it catalyzes.

2. The protein (or RNA) enzyme is a chiral molecule whose binding clefts and catalytic residues are arranged in a specific three-dimensional asymmetric array. Hence, only substrates with the appropriate stereochemistry can bind to the enzyme, and the enzyme transforms the substrate to product according to the spatial arrangement of interacting functional groups.

3. An apoenzyme is the protein portion of an enzyme that has lost its cofactor (a metal ion or coenzyme). A holoenzyme is an active enzyme containing both the protein and the cofactor.

4. Many coenzymes are synthesized from precursors that are vitamins (substances, that an animal cannot synthesize and must obtain from its diet). However, not all coenzymes have vitamin precursors and not all vitamins are precursors of coenzymes.

5. Protein-modifying reagents can provide clues to the identities of catalytic residues. If chemical modification of a residue does not result in loss of activity, that residue can be ruled out as essential for catalysis. Loss of enzymatic activity on modification of a residue may indicate that the modified residue plays an essential role. However, chemical modification of residues at sites other than the active site may interfere with catalysis nonspecifically by disrupting protein structure.

6. The rate-determining step is the slowest of the steps in the reaction mechanism, the step with the greatest free energy of activation.

7.
   (a) Yes. $\Delta G$ is the difference in free energy between reactants and products, whereas $\Delta G^{\ddagger}$ is the difference in free energy between the reactants and the transition state.
   (b) No. By definition, a catalyst decreases $\Delta G^{\ddagger}$ of a reaction.
   (c) No. The free energy of the intermediate may be greater than that of the reactant. The reaction will proceed as long as the $\Delta G$ of the overall reaction A→P is negative.
   (d) No. The rate-determining step is the one whose $\Delta G^{\ddagger}$ is greatest. This does not always correspond to the step with the highest free energy, since $\Delta G^{\ddagger}$ depends on the difference in free energies between a reactant or an intermediate and the following transition state, not just on the free energy of this transition state.

8. $\Delta G^{\ddagger}$ is largely independent of temperature. An increase in temperature increases the rate of a reaction by increasing the number of reacting molecules that reach the transition state.

9.

    (a) The rate enhancement is

$$e^{\Delta\Delta G^{\ddagger}/RT}$$

$$= e^{(13\ kJ\cdot mol^{-1})/(8.3145\ J\cdot K^{-1}\cdot mol^{-1})(298\ K)}$$

$$= e^{5.25}$$

$$= 190$$

    The enzyme-catalyzed reaction therefore proceeds 190 times faster than the uncatalyzed reaction.

    (b) When the rate enhancement is $10^5$,

$$100{,}000 = e^{\Delta\Delta G^{\ddagger}/RT}$$
$$\Delta\Delta G^{\ddagger} = RT\ \ln 100{,}000$$
$$= (8.3145\ J\cdot K^{-1}\cdot mol^{-1})(298K)(11.5)$$
$$= 28.5\ kJ\cdot mol^{-1}$$

10. Aspartate, cysteine, glutamate, histidine, lysine, and tyrosine are likely to participate in general acid–base catalysis. Glycine does not have an ionizable side chain.

11. As the pH approaches the $pK_R$, the His become progressively deprotonated and less effective as an acid catalyst. When pH $= pK_R$, 50% of the His residues are deprotonated and the enzyme is half-active. When the pH significantly exceeds $pK_R$, the His is almost completely deprotonated and the enzyme is inactive.

12. (a) His 12 has a $pK$ of 5.4, since it is active when unprotonated. It would lose activity when the pH decreases.
      His 119 has a $pK$ of 6.4 since it is active when protonated. It would lose activity when the pH increases.

(b)

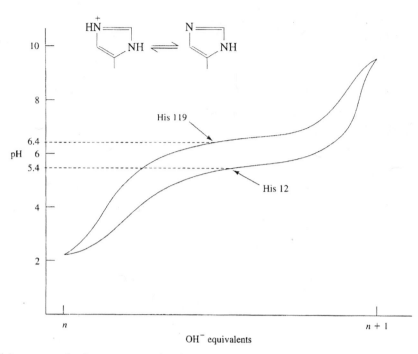

OH⁻ equivalents

13. In general base catalysis, a proton is abstracted from the substrate, whereas in nucleophilic catalysis, a covalent bond forms.

14. Amino acid side chains that have highly delocalized electrons or high polarizabilities are good covalent catalysts. The hydroxyl group of serine, the carboxyl group of aspartate, and the thiol group of cysteine are highly polarizable groups, and the electrons are highly delocalized in the imidazole group of histidine.

15.
    (a) nucleophile
    (b) electrophile
    (c) electrophile
    (d) nucleophile
    (e) nucleophile

16. (a)

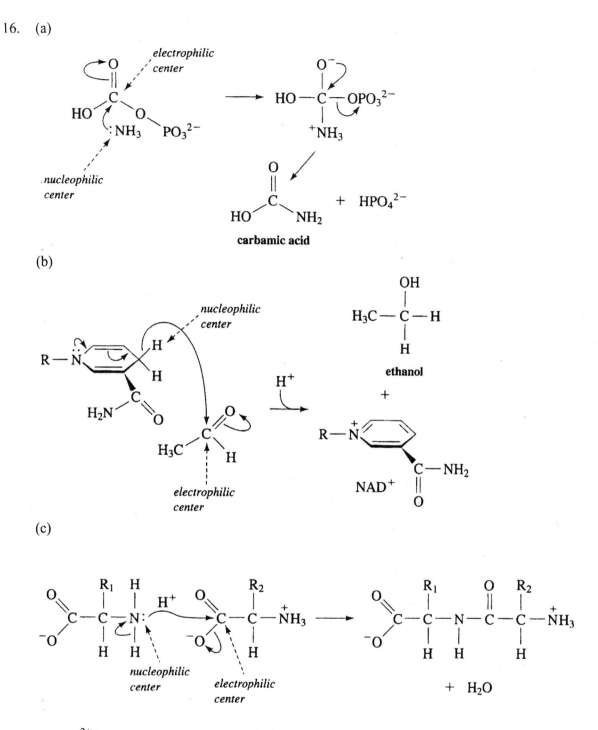

(b)

(c)

17. The $Zn^{2+}$ in carbonic anhydrase polarizes a water molecule and thereby causes it to ionize. The resulting $Zn^{2+}$—$OH^-$ is the nucleophile that attacks carbon dioxide, converting it to $HCO_3^-$. The metal ion stabilizes the negative charge of the $OH^-$, which would otherwise not form at neutral pH.

18. Glu 35 gives up its proton to O1 of the D ring of the substrate (see Figure 11-21), which leads to the creation of the oxonium ion transition state. The ionized hydroxyl group of Asp 52 attacks the electron-poor C1 of the oxonium ion to form a covalent intermediate. Water regenerates the original form of the enzyme by its nucleophilic attack on the Asp 52–substrate covalent intermediate, releasing the substrate.

19. The lactyl side chain of NAM residues sterically prevents NAM binding to subsites C and E. Hence only NAG residues bind to these subsites.

20. $NAG_6$ would be better because the C and E residues of $NAM_6$ would not bind to their subsites.

21. The backbone NH group of Val 109 forms a hydrogen bond with O6 of the D-site residue, helping stabilize its half-chair conformation. *N*-acetylxylosamine lacks O6 and therefore cannot form this hydrogen bond.

22.

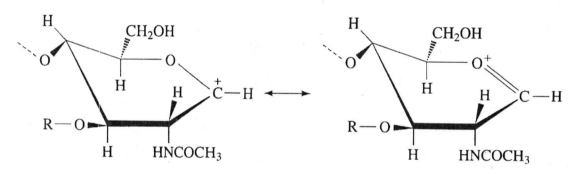

23.
   (a) Asp 102 polarizes the imidazole ring of His 57.
   (b) His 57 extracts a proton from the nearby hydroxyl group of Ser 195.
   (c) Ser 195 provides the ionized hydroxyl group that will act as a nucleophile to break the nearby amide bond of the substrate.
   (d) Gly 193 stabilizes the oxyanion tetrahedral intermediate of the reaction.
   (e) Asp 102, His 57, and Ser 195 form the "catalytic triad."

24. Ser 195 was identified through the use of diisopropylphosphofluoridate (DIPF), which reacts with the Ser —OH group at the active site of chymotrypsin and irreversibly inactivates the enzyme. His 57 was identified through affinity labeling using the substrate analog tosyl-L-phenylalanine chloromethylketone. DIPF also reacts with the active-site Ser in other serine proteases and therefore would label this residue. However, these other enzymes, not all of which share chymotrypsin's specificity for Phe-containing substrates, would require chloromethylketone analogs that incorporated residues corresponding to their substrate specificities, in order to react with their active-site His residues.

25. Site-directed mutations often change (or fail to change) enzymes in unexpected ways. For example, site-directed mutation of trypsin's Asp 189 to Ser did not change trypsin's specificity to that of chymotrypsin. Several other changes involving amino acids found on the surface loops surrounding the binding pocket were required for trypsin to emulate chymotrypsin.

26. Chymotrypsin acts as an esterase, attacking the carbonyl C of the substrate's ester bond to form a tetrahedral intermediate. This is followed by decomposition to an acyl–enzyme intermediate with release of the first product, *p*-nitrophenolate.

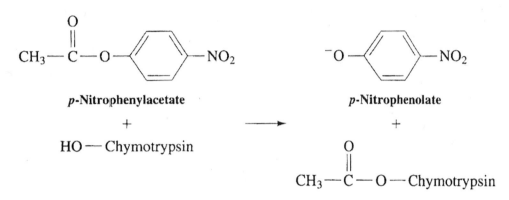

Release of the second product, acetate, via hydrolysis of the acyl–enzyme intermediate, is much slower. Consequently, every active site attacks the substrate and releases the first product in a stoichiometric fashion, accounting for the initial burst of product formation. Because the second phase of the reaction is slow, the enzyme is only slowly regenerated and made available to catalyze additional rounds of the esterolytic reaction.

27. Stabilization of the transition state via the oxyanion hole is responsible for the largest portion of chymotrypsin's rate enhancement.

28. Bovine pancreatic trypsin inhibitor binds trypsin so tightly that the complex is too rigid to allow formation of the tetrahedral intermediate, to allow release of the first product, or to allow water to enter the active site of trypsin.

29. Trypsinogen's catalytic activity is too low to activate other trypsinogen molecules at a biologically significant rate. Enteropeptidase cleaves trypsinogen, thereby generating a small amount of trypsin that then commences autocatalytic activation. The role of enteropeptidase is to initiate this process in a controlled manner.


## *Chapter 12*

1.
   (a) $v = -d[A]/dt = k[A]$   This is a first-order reaction.
   (b) $v = -d[A]/dt = -d[B]/dt = k[A][B]$   This is a second-order reaction (A and B must collide to form product).

(c) $v = -d[A]/dt = k[A]^2$  This is also a second-order reaction (A must collide with another molecule of A for the reaction to proceed).

2.  The progress of a reaction can be followed by measuring the rates of the appearance of the product(s) or the disappearance of the reactant(s). In practice, any physical property, such as light absorbance, pH, or an NMR signal, can be followed, provided that it changes in proportion to the concentration(s) of the reactant(s) or product(s).

3.

(a)  For a first-order reaction, $t_{1/2} = 0.693/k$ (Equation 12-9). Therefore, $k = 0.693/20$ min $= 0.035$ min$^{-1}$.

(b)  Since $\ln([A]/[A]_o) = -kt$ (from Equation 12-6), $t = \ln([A]/[A]_o)/-k$. When 20% of A has been converted to product, $[A]/[A]_o = 0.8$ and $t = (\ln 0.8)/(-0.035$ min$^{-1}) = (-0.22)/(-0.035$ min$^{-1}) = 6.4$ min.

(c)  When 80% of A has been converted to product, $[A]/[A]_o = 0.2$ and $t = (\ln 0.2)/(-0.035$ min$^{-1}) = (-1.61)/(-0.035$ min$^{-1}) = 46$ min.

(d)  When $t = 15$ min, $\ln([A]/[A]_o) = (-0.035$ min$^{-1})(15$ min$) = -0.525$. Since $e^{-0.525} = 0.59$, 59% of A remains at 15 minutes.

(e)  For the decomposition of $^{32}$P, $k = (0.693/14$ days$)$ $(1$ day$/1440$ min$) = 3.4 \times 10^{-5}$ min$^{-1}$. This is ~1000 times slower than the reaction described above.

4.

(a)  $K^{\ddagger} = [X^{\ddagger}]/[A]$.

(b)  $\Delta G^{\ddagger} = -RT \ln K^{\ddagger}$.

(c)  As described in Box 12-3, $d[P]/dt = k[A] = k'[X^{\ddagger}]$ where $k'$ is the rate constant for the decomposition of $X^{\ddagger}$ to from products. $k'$ can be expressed in terms of the Boltzman constant ($k_B$) and Planck's constant ($h$): $k' = k_B T/h$. Thus, since $[X^{\ddagger}] = K^{\ddagger}[A]$

$$e^{-\Delta G^{\ddagger}/RT},$$

$$\frac{d[P]}{dt} = \frac{k_B T}{h} e^{-\Delta G^{\ddagger}/RT} [A]$$

and

$$k = \left(\frac{k_B T}{h}\right) e^{-\Delta G^{\ddagger}/RT}$$

5.

(a)  The enzyme–substrate (ES) complex is the species formed by the interaction between an enzyme and its substrate.

(b)  The rate equation for the net formation of ES is the rate of formation of ES minus the rate of ES degradation: $d[ES]/dt = k_1[E][S] - k_{-1}[ES] - k_2[ES]$.

(c)  The rate of product formation, $d[P]/dt = v = k_2[ES]$.

(d) Increasing substrate concentration when all the enzyme is in the ES complex does not increase the rate of the reaction since then $v = V_{max} = k_2[E]_T$ where $[E]_T$ is the total enzyme concentration; that is, the enzyme is working at its maximal rate; it can work no faster.

6.

(a) The steady state assumption assumes that during the course of an enzyme-catalyzed reaction, the concentration of the ES complex does not change.

(b) $K_M$ is the Michaelis constant: $K_M = (k_{-1} + k_2)/k_1$. $K_M$ is the substrate concentration at which the reaction velocity is half-maximal.

(c) The $k_{cat}$ is the maximum velocity divided by the total enzyme concentration: $k_{cat} = V_{max}/[E]_T$. It is the number of reaction processes (turnovers) that each active site catalyzes per unit time. For the simple kinetic scheme used to derive the Michaelis–Menten equation, $k_{cat} = k_2$.

(d) The turnover number is the same as $k_{cat}$.

(e) Catalytic efficiency, calculated as $k_{cat}/K_M$, is the apparent second-order rate constant for the reaction of E + S and indicates how often the enzyme catalyzes a reaction upon encountering its substrate.

(f) If an enzyme catalyzes a reaction every time it collides with its substrate, it has reached catalytic perfection and the rate is controlled by how often the molecules collide, that is, by their rate of diffusion. At this point, the rate is said to have reached its diffusion-controlled limit.

7.

(a) To calculate $v_0$ for Reaction 1, for example,
$v_0 = (26 \text{ nmol}/5 \text{ min})/(1.0 \text{ mL}) \times (10^3 \text{ mL}/1 \text{ L}) \times (0.001 \text{ μmol/nmol}) = 5.2 \text{ μM·min}^{-1}$

| Reaction | $v_0$ (μM·min$^{-1}$) |
|---|---|
| 1 | 5.2 |
| 2 | 5.8 |
| 3 | 7.8 |
| 4 | 8.6 |
| 5 | 11.2 |
| 6 | 12.4 |
| 7 | 14.2 |

(b) First, calculate [S] for each reaction. In Reaction 1, for example, [A] = (0.008 mL)(5 mM)/(1 mL) × (1000 μM/1 mM) = 40 μM. Next, convert the data to values of 1/[S] and 1/$v$.

| Reaction | [S] (μM) | 1/[S] (μM$^{-1}$) | $v$ (μM·min$^{-1}$) | 1/$v$ (min·μM$^{-1}$) |
|---|---|---|---|---|
| 1 | 40 | 0.025 | 5.2 | 0.192 |
| 2 | 50 | 0.02 | 5.8 | 0.172 |
| 3 | 75 | 0.0133 | 7.8 | 0.128 |
| 4 | 100 | 0.010 | 8.6 | 0.0116 |
| 5 | 200 | 0.005 | 11.2 | 0.089 |
| 6 | 300 | 0.0033 | 12.4 | 0.081 |
| 7 | 500 | 0.002 | 14.2 | 0.070 |

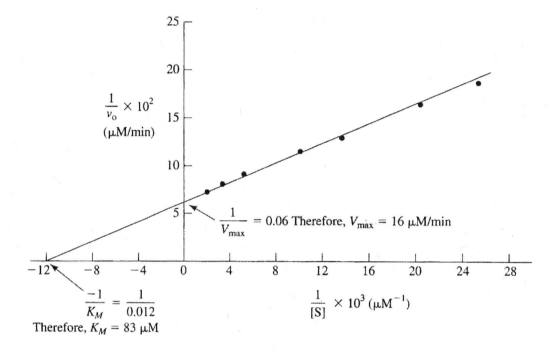

(c) First calculate $[E]_T$ = (0.002 mL)(10 μM Aase)/(1 mL) = 0.02 μM.
Using the value of $V_{max}$ determined above and Equation 12-27, $k_{cat}$ = $V_{max}$/$[E]_T$ = (16 μM·min$^{-1}$)/(0.02 μM) = 800 min$^{-1}$.

8. Kinetics can support but cannot prove a particular reaction mechanism. A single kinetic model may explain several mechanisms, so additional experiment must be performed to establish a particular mechanism. However, kinetic data can rule out mechanisms that are inconsistent with the observed behavior.

9. In a Sequential reaction, both substrates must bind before any product is released. Hence it is possible for the two substrates to bind in an Ordered or Random fashion. In a Ping Pong reaction, a group is transferred from the first substrate to the enzyme to form the first product; then a second substrate binds and is converted to a second product. Consequently, the second product cannot bind first or yield the second product until the first substrate has bound and been converted to the first product.

10.

(a) We can approximate $\Delta G^{\ddagger}$ by assuming we have conditions where $k_2 = k_{cat}$ and $k_{cat} = k$ in the Arrhenius equation, $k = Ae^{-\Delta G^{\ddagger}/RT}$. We can substitute and convert this equation to a logarithmic form in the form $y = mx + b$.

$$k_{cat} = Ae^{-\Delta G^{\ddagger}/RT}$$

$$\ln k_{cat} = \ln A - \Delta G^{\ddagger}/RT$$

$$\ln k_{cat} = (-\Delta G^{\ddagger}/R)(1/T) + \ln A$$

Hence, now we can plot data in graphical form as shown below.

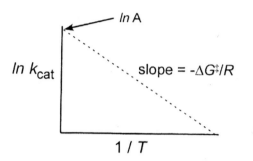

(b) Since our enzyme is in highly purified form, we can calculate our total enzyme concentration in an experiment as a molar quantity. $k_{cat}$ can be calculated after we experimentally obtain $V_{max}$. We then do kinetic experiments at several temperatures at which the enzyme is stable so that we can determine $k_{cat}$ at each temperature. These data are then plotted as shown above so that the slope provides $\Delta G^{\ddagger}$.

(c) A variety of assumptions need to be made:
  (i) Small errors in $[E]_T$ can lead to substantive differences in the $\Delta G^{\ddagger}$;
  (ii) We assume that $[E]_T = [X^{\ddagger}]$; which may only be an approximation for enzymes with low values for $k_{cat}$ and one transition-state complex (see Box 12-2.)
  (iii) The enzyme's structure, and hence its catalytic activity, is not significantly altered at the temperatures tested.

11. The three general mechanisms of inhibition are competitive, uncompetitive, and mixed inhibition. In competitive inhibition, the inhibitor competes with the substrate for binding to the active site, increasing the apparent $K_M$. The inhibitor binds only to the free enzyme and not to the ES complex. In uncompetitive inhibition, the inhibitor binds only to the ES complex to inhibit the enzyme. Although the inhibitor does not directly affect substrate binding, it decreases the apparent $K_M$. In mixed inhibition, the inhibitor can bind to both the free enzyme and the ES complex. Binding of a mixed inhibitor may alter the binding of the substrate and therefore may alter the apparent $K_M$. If the inhibitor does not alter the apparent $K_M$, inhibition is said to be noncompetitive.

12. Aside from extremely low or high pH's, which can denature an enzyme, changes in pH may affect the protonation/deprotonation of residues involved in substrate binding and/or catalysis. Constructing a Lineweaver–Burk plot at several different pH values should yield a set of lines that indicate whether $K_M$ (substrate binding) and/or $V_{max}$ (catalytic activity) is affected (just as Lineweaver–Burk plots can reveal the different types of enzyme inhibition).

13.

(a) The process obeys Michaelis–Menten (saturation) kinetics:

$$\text{Glucose}_{out} + \text{T}_{out} \ \rightarrow \ \text{glucose}_{out} \cdot \text{T}_{out} \ \rightarrow \ \text{glucose}_{in} \cdot \text{T}_{in} \ \rightarrow \ \text{glucose}_{in} + \text{T}_{in}$$

(b) Yes. All three inhibitory modes are possible. An inhibitor could compete with glucose for binding to the transport protein (competitive inhibition); it could bind to the transporter–glucose complex and interfere with the conformational change that exposes glucose to the other side of the membrane (uncompetitive inhibition); or it could interfere with the transporter's function with glucose bound or not bound (mixed inhibition).

14. These substitutions would weaken substrate binding to lysozyme by removing hydrogen bonds that hold the substrate in place. The $K_M$ would increase, but $V_{max}$ would remain nearly the same, since the catalytic residues Asp 52 and Glu 35 are not changed.

15. The catalytic trimer does not display cooperativity in the absence of the regulatory subunits. Hence the rate profile is hyperbolic, much like the binding of $O_2$ to myoglobin. Intact ATCase exhibits a sigmoidal curve characteristic of cooperative substrate binding, which is possible when the regulatory subunits mediate conformational changes between catalytic subunits.

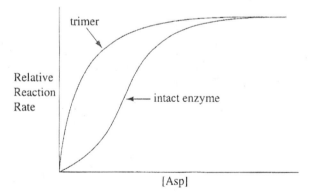

16. Carbamoyl phosphate, aspartate, and ATP have higher affinity for and stabilize the R (more active) state, whereas CTP has higher affinity for and stabilizes the T (less active) state.

17. The intrinsic biochemical parameters that are likely to vary among different allelic isozymes of cytochrome P450 include:
    (a) The range of specificities for different substrates;
    (b) The $K_M$ values for different substrates;
    (c) The $V_{max}$ values for different substrates;
    (d) The concentration of the isozyme.

    A key biological parameter that may vary among allelic isozymes includes the steady-state concentration of the isozyme, which is a function of the rate of synthesis of the protein (as determined by the steady-state concentration of its mRNA), and its intracellular stability (as determined by its amino acid sequence, especially at the N-terminus).

18. A lead compound for a drug that interacts with a target protein can come from several sources:
    (a) Natural products, most often from plants, including potential candidates from so-called folk remedies;
    (b) Rational or structure-based design using the three-dimensional structure of the protein to explore chemical structures that may fit in an active or allosteric site;
    (c) Combinatorial libraries of peptides and other compounds (see Fig. 12-18, for example), which can generate a variety of chemical structures that may bind effectively to the target protein. This has been recently facilitated by automated high-throughput screening technologies; and
    (d) A combination of combinatorial libraries, high-throughput screening, and structure-based design.

    The most time-consuming (and expensive) step is the identification of adverse side reactions during clinical trials.

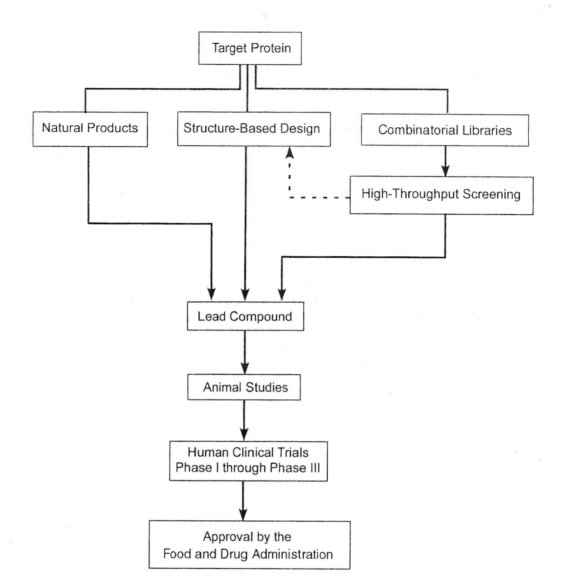

## Chapter 13

1.  Both receptor systems lead to the mobilization of energy reserves, shunting them to key tissues. The physiological responses include relaxation of smooth muscle for rapid blood delivery as well as stimulation of skeletal muscle contraction to increase mobility, most commonly for the flight or fight response.

2.  Steroid–receptor complexes act as transcriptional regulators, while soluble hormones act more indirectly by activating intermediary proteins that trigger cascades of enzymatic activity affecting multiple pathways.

3.  The SH2 and SH3 domains function in the assembly of multiprotein signaling complexes that linkphosphorylated receptor tyrosine kinases to the monomeric G protein Ras. For example, the SH2 domain in the Grb2/Sem-5 protein binds the phosphorylated cytoplasmic

domain of the receptor tyrosine kinase, while its SH3 domains recognize and bind proline-rich sequences in Sos protein. This complex induces Ras to release GDP and bind GTP in order to become active (see Figure 13-7).

4. Ligand binding to a receptor tyrosine kinase leads to the activation of Ras via the formation of a complex of SH2- and SH3-containing proteins. Activated Ras triggers a kinase cascade consisting of Raf, MEK, and MAPK, which are activated in series. Activated MAPK migrates to the nucleus, where it phosphorylates a variety of proteins, including transcription factors. The activated transcription factors stimulate gene expression.

5. Scaffold proteins serve as intracellular organizing sites that frequently tether or spatially restrict the movement of component proteins in a signaling pathway. In this way, they help to confer specificity on a particular pathway and limit or prevent unwanted crosstalk between pathways. The best understood example of a pathway regulated in this manner is the MAP kinase cascade.

6. G protein–coupled receptors are transmembrane proteins that contain seven membrane-spanning segments. Ligand binding leads to interactions between the receptor and an associated G protein, which is anchored to the membrane by its lipid groups. Activation of the G protein leads to production of second messengers such as cAMP (if the G protein stimulates adenylate cyclase) and $IP_3$ and $Ca^{2+}$ (if the G protein stimulates phospholipase C).

   Receptor tyrosine kinases are usually monomeric proteins with a single membrane-spanning segment. Ligand binding induces dimerization of the receptor, which leads to autophosphorylation at one or more tyrosine residues in its cytoplasmic domain. These phosphorylated residues provide binding sites for target proteins with SH2 domains, which are generally activated upon binding to the receptor tyrosine kinase.

7. When the G protein binds to a hormone–receptor complex, the $G_\alpha$ subunit exchanges GDP for GTP. Binding of GTP induces a conformational change in $G_\alpha$ that causes it to dissociate from $G_{\beta\gamma}$. GTP binding also increases the affinity of $G_\alpha$ for adenylate cyclase, which is thereby activated.

8. $AlF_3$ has an atomic structure that resembles a phosphate group. Thus when it is present in X-ray crystallographic studies of inactive G proteins (such as Ras-GDP or a $G_\alpha$-GDP), it will tend to substitute for the $\gamma$ phosphate group of GTP. In fact the inclusion of $AlF_3$ yields G protein preparations that are considered to mimic the transition state leading to GTP hydrolysis. Consequently, information can be gained concerning the conformational changes that accompany conversion of a G protein between its inactive and active states.

9.  Among possible mechanism by which a β-adrenergic signaling pathway can be "turned off" include:
    (a) Internalization of the receptor by endocytosis
    (b) Dissociation of the ligand from the receptor
    (c) Increased cAMP phosphodiesterase activity
    (d) Increased rate of GTP hydrolysis or inhibition of GTP exchange for GDP
    (e) Increased protein phosphatase activity to counteract the actions of PKA
    (f) Proteolytic degradation of key enzymes of the signaling pathway

    Note that (a) and (b) "turn off" the pathway because the activated heterotrimeric G protein is short-lived. In general, however, regulatory mechanisms result in the cell becoming refractory to stimulatory or inhibitory effects on a signaling pathway.

10. Both hormones probably act via the adenylate cyclase signaling pathway: XGF activates a stimulatory G protein to release a $G_{s\alpha}$ subunit, while YGF activates an inhibitory G protein that that releases a $G_{i\alpha}$ subunit. These activated G proteins both affect the activity of adenylate cyclase to modulate the amount of cAMP that is synthesized and available to stimulate PKA activity.

11. All of these drugs antagonize adenosine receptors that act through inhibitory G proteins. This creates an imbalance in the regulation of adenylate cyclase activity that is modulated in drug-responsive cells by signaling systems generating stimulatory and inhibitory G proteins.

12. Activation of phospholipase C leads to the hydrolysis of phosphatidylinositol-4,5-bisphosphate ($PIP_2$) to inositol-1,4,5-trisphosphate ($IP_3$) and 1,2-diacyglycerol (DAG). Both of these products are second messengers. $IP_3$ moves through the cytosol and binds to the $IP_3$ receptor, a $Ca^{2+}$ channel on the endoplasmic reticulum. Binding of $IP_3$ to its receptor triggers the release of $Ca^{2+}$ from the ER, increasing the intracellular $[Ca^{2+}]$. $Ca^{2+}$ can stimulate many cellular activities either by interacting directly with target proteins or through the formation of the $Ca^{2+}$-calmodulin complex, which regulates many other target proteins. The DAG is retained within the membrane, where it activates protein kinase C. This enzyme then phosphorylates and activates additional target proteins. Members of the phospholipase C family can be activated in two ways. Some forms are activated by interacting with a $G_\alpha$–GTP complex derived from a G protein following ligand binding to a G protein–coupled receptor. Other isoforms are activated by binding, via their SH2 domains, to a receptor tyrosine kinase that has undergone ligand binding, dimerization, and autophosphorylation.

13. When it binds insulin, the insulin receptor undergoes a conformational change that activates its tyrosine kinase domains, resulting in autophosphorylation. The RTK can then phosphorylate the target proteins IRS-1 and IRS-2. These proteins can interact with SH2-containing proteins, which in turn may activate G proteins, act as kinases, or act as phosphatases. Phosphorylation of IRS proteins may also lead to activation of protein kinase C. The insulin receptor is thereby linked to kinase cascades, which is typical of RTKs, and to elements of the phosphoinositide pathway.

14. All other things being equal, Li$^+$ is likely to counteract the action of phorbol-13-acetate, a DAG mimic that activates protein kinase C, as the pool of available IP$_3$ is hydrolyzed to IP$_2$.

15. Both IP$_3$ and cAMP synthesis are stimulated by stimulatory G proteins and both second messengers are inactivated by hydrolysis of a phosphoester bond.

## Chapter 14

1. The fundamental processes by which living organisms acquire and use free energy depend on oxidation–reduction reactions. For example, the oxidation of a metabolic fuel generates a reduced cofactor, such as NADH, whose subsequent reoxidation generates ATP. Rapid (i.e., enzyme-catalyzed) oxidation–reduction reactions therefore would be consistent with (but would not prove) the presence of life on Mars.

2.

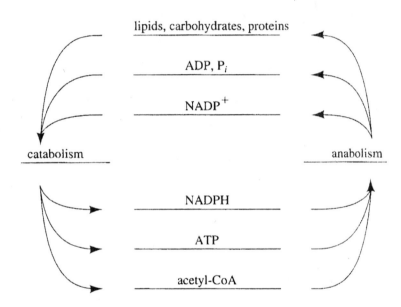

3. PFK operates far from equilibrium, so it is not sensitive to changes in substrate concentrations. It most likely catalyzes a rate-determining step of a metabolic pathway. (In fact, it is part of the glycolytic pathway, and its activity is under allosteric control.)

4. $\Delta G$ would become more negative because, without the shielding effect of the divalent cations, the phosphate groups in ATP would experience more repulsion.

5. Processes that maintain the cellular concentration of ATP are oxidative phosphorylation (the synthesis of ATP from ADP + P$_i$ as driven by the free energy of dissipation of a transmembrane proton concentration gradient); substrate-level phosphorylation (direct transfer of a phosphoryl group from a "high-energy" compound such as phosphoenolpyruvate to ADP); and reactions catalyzed by kinases such as nucleoside diphosphate kinase and adenylate kinase.

6. These "high-energy" compounds have large negative values for $\Delta G$ of hydrolysis. Therefore, their breakdown is thermodynamically spontaneous (exergonic). However, the kinetics of their breakdown depends on the availability and activity of enzymes to catalyze such reactions. In the absence of the appropriate enzymatic activity, the compounds are quite kinetically stable.

7. The reaction catalyzed by creatine kinase

$$\text{ATP + creatine} \rightleftharpoons \text{phosphocreatine + ADP}$$

is at equilibrium and is freely reversible in the cell. Hence, the concentration of phosphocreatine is directly sensitive to changes in [ATP]. When ATP is plentiful, phosphocreatine is formed by the forward reaction. When [ATP] drops, the reaction proceeds in reverse, so that phosphocreatine can transfer its phosphoryl group to ADP to produce more ATP. The extent of this reaction depends on the decrease in [ATP].

8. (a) At equilibrium, $\Delta G^{\circ\prime} = -RT \ln K_{eq}$,

Hence, 
$$K_{eq} = \frac{[G6P]}{[\text{Glucose}][P_i]} = e^{-\Delta G^{\circ\prime}/RT}$$
$$= e^{-(14,000 \text{ J·mol}-1)/(8.3145 \text{ J·K}-1\cdot\text{mol}-1)(298K)}$$
$$= e^{-5.65}$$
$$= 0.0035$$

(b) The phosphorylation of glucose coupled to ATP hydrolysis is the sum of the following reactions:

| | | |
|---|---|---|
| $\text{ATP + H}_2\text{O} \rightleftharpoons \text{ADP + P}_i$ | $\Delta G^{\circ\prime} = -30.5 \text{ kJ·mol}^{-1}$ (Table 13-2) | |
| $\text{glucose + P}_i \rightleftharpoons \text{glucose-6-phosphate + H}_2\text{O}$ | $\Delta G^{\circ\prime} = 14 \text{ kJ·mol}^-$ | |

$$\text{ATP + glucose} \rightleftharpoons \text{glucose-6-phosphate + ADP} \qquad \Delta G^{\circ\prime} = -16.5 \text{ kJ·mol}^{-1}$$

At equilibrium, $\Delta G = 0$ and $\Delta G^{\circ\prime} = -RT \ln K_{eq}$. Thus,

$$K_{eq} = e^{-\Delta G^{\circ\prime}/RT} = \frac{[\text{ADP}][G6P]}{[\text{ATP}][\text{Glucose}]}$$

Since [ATP]/[ADP] = 12,

$$\frac{[G6P]}{[\text{Glucose}]} = e^{-\Delta G^{\circ\prime}/RT}$$

$$= 12e^{-(-16,500 \text{J·mol}^{-1})/(8.3145 \text{K·K}^{-1}\cdot\text{mol}^{-1})(310K)}$$
$$= 12e^{6.40} = 72.33$$

9.    Use Equation 14-1:

$$\Delta G = \Delta G^{\circ\prime} + RT \ln\left(\frac{[GAP][DHAP]}{[FBP]}\right)$$

$$\frac{\Delta G - \Delta G^{\circ\prime}}{RT} = \ln\left(\frac{[GAP][DHAP]}{[FBP]}\right)$$

$$\frac{(-5900 \text{kJ} \cdot \text{mol}^{-1})(22{,}800 \text{kJ} \cdot \text{mol}^{-1})}{(8.3145 \text{J} \cdot \text{K}^{-1} \cdot \text{mol}^{-1})(310 \text{K})} = \ln\left(\frac{[GAP][DHAP]}{[FBP]}\right)$$

$$-11.13 = \ln\left(\frac{[GAP][DHAP]}{[FBP]}\right)$$

$$e^{-11.13} = \frac{[GAP][DHAP]}{[FBP]}$$

$$1.46\times10^{-5} = \frac{[GAP][DHAP]}{[FBP]}$$

10.

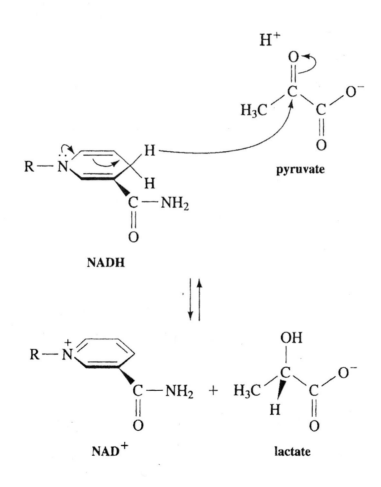

11. Electrons flow from substances of lower reduction potential to substances of higher reduction potential. $\Delta\mathscr{E}° = -0.250 \text{ V} - (-0.763 \text{ V}) = 0.513 \text{ V}$.

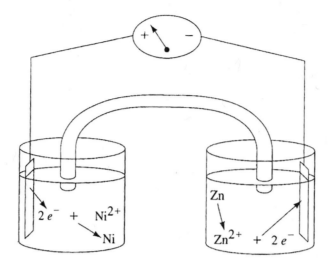

12. At pH 7.0, $\mathscr{E}°' = 0.815 \text{ V}$ (Table 14-4). Therefore, reduction is more favorable at pH 0 (where $\mathscr{E}° = 1.23 \text{ V}$) since the more positive the reduction potential, the more negative the $\Delta G$ (Equation 14-7). The law of mass action dictates that an increase in the concentration of one of the reactants ($H^+$) will shift the equilibrium toward product. Thus, decreasing the pH, which increases $[H^+]$, will favor the reduction of oxygen.

13. The reduction of acetoacetate by NADH is a coupled redox reaction, where the half-reactions (shown in Table 14-4) are

$$\text{Acetoacetate} + 2\,H^+ + 2\,e^- \rightleftharpoons \beta\text{-hydroxybutyrate} \qquad \mathscr{E}°' = -0.346 \text{ V}$$
$$NAD^+ + H^+ + 2\,e^- \rightleftharpoons NADH \qquad \mathscr{E}°' = -0.315 \text{ V}$$

For the overall reaction direction specified, the $NAD^+$/NADH half-reaction is the electron donor, and the acetoacetate/$\beta$-hydroxybutyrate half-reaction is the electron acceptor. Use the Nernst equation to calculate $\Delta E$:

$$\Delta\mathcal{E} = \Delta\mathcal{E}^{\circ\prime} - \frac{RT}{n\mathcal{F}} \ln\left(\frac{[\beta\text{-Hydroxybutyrate}]\left[NAD^+\right]}{[\text{Acetoacetate}]\left[NADH\right]}\right)$$

$$= \mathcal{E}^{\circ\prime}_{e^-\,acceptor} - \mathcal{E}^{\circ\prime}_{e^-\,donor} - \frac{\left(8.3145\;J\cdot K^{-1}\cdot mol^{-1}\right)(298K)}{(2)\left(96,485\;J\cdot V^{-1}\cdot mol^{-1}\right)}\ln\left[\frac{(0.001)(0.001)}{(0.01)(0.01)}\right]$$

$$= (-0.346\;V\;+\;0.315\;V)\;-\;(0.01284\;V)\ln(0.01)$$
$$= -0.031\;V\;-\;(0.01284\;V)(-4.605)$$
$$= 0.028\;V$$

Next, use Equation 14-7 to calculate $\Delta G$:

$$\Delta G\;=\;n\mathcal{F}\Delta\mathcal{E}$$
$$=\;-(2)(96,485\;J\cdot V^{-1}\cdot mol^{-1})(0.028\;V)$$
$$=\;-5403\;J\cdot mol^{-1}$$
$$=\;-5.4\;kJ\cdot mol^{-1}$$

14. Radioactive tracers can be used to determine the order of metabolic transformations in a pathway. An inhibitor that blocks a step of the pathway causes earlier intermediates to accumulate but does not necessarily reveal their order, and cannot reveal information about the order of metabolites following the blocked step.

15. To study the metabolic steps between compounds A and Z, first generate mutants by irradiating the cells or treating them with a chemical mutagen. Next, screen cells for their inability to grow in the absence of compound Z, the end product of the pathway. This would yield cells with defects in the A→Z pathway. To identify mutations in enzymes that catalyze individual steps of the pathway, screen the mutant cells for their inability to grow in the absence of each of the compounds suspected to be intermediates in the transformation of A to Z.

16.
(a) Since Mutant 1 requires two compounds for growth, the pathway for their synthesis is probably branched, such that both are derived from a common precursor lacking in Mutant 1. Mutant 2 is blocked in the pathway leading to X, while Mutants 3 and 4 are blocked in the pathway leading to Z. The step blocked in Mutant 4 precedes that blocked in Mutant 3, since Mutant 3 accumulates compound W, which supports the growth of Mutant 4. Since compound Y, which accumulates in Mutant 4, supports the growth of Mutant 1, it must be the common precursor of X and Z. Therefore, the most likely pathway is

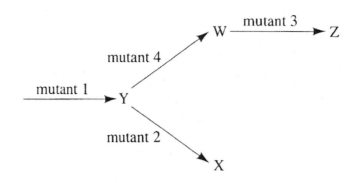

(b) The first committed step in the synthesis of Z is the step that converts Y to W, the step missing in Mutant 4.

17.

(a) DNA chip data only allow investigators to assess the *relative* concentrations of specific mRNAs from cells under two different physiological states, since mRNA populations from each source are quantitatively compared to each other.

(b) DNA chip data only show gene expression at the point of mRNA synthesis. The relative amounts of proteins made from these mRNAs cannot be accurately assessed, since the rates of translation of different mRNAs may differ under different physiological states. The protein diversity and content (i.e. the proteome) at a systems level requires the tools of proteomics, such as the ICAT method for quantitative proteome analysis.

## Chapter 15

1. In the energy investment phase, glucose is phosphorylated and cleaved to yield two molecules of glyceraldehyde-3-phosphate. Two ATPs are consumed in this stage so that the second stage can produce "high-energy" compounds whose breakdown drives ATP synthesis. The initial hexose is split, in the first stage, to two trioses. These undergo the reactions of the second stage, thereby generating four ATPs for a net "profit" of two ATPs per glucose. The equation for Stage I is

$$\text{Glucose} + 2\,\text{ATP} \rightarrow 2\,\text{GAP} + 2\,\text{ADP} + 2\,\text{H}^+$$

The equation for Stage II is

$$2\,\text{GAP} + 2\,\text{NAD}^+ + 4\,\text{ADP} + 2\,\text{HPO}_4^{2-} \rightarrow$$
$$2\,\text{pyruvate} + 2\,\text{NADH} + 4\,\text{ATP} + 2\,\text{H}_2\text{O} + 2\,\text{H}^+$$

2. ATP and NADH

3.
   (a)  **A** Fructose-1,6-bisphosphate
       **B** Glucose-6-phosphate
       **C** Phosphoenolpyruvate
       **D** 3-Phosphoglycerate
       **E** Glyceraldehyde-3-phosphate
   (b)  **B, A, E, D, C**
   (c)  **C**
   (d)  **A** Aldolase
       **B** Phosphoglucose isomerase
       **C** Pyruvate kinase
       **D** Phosphoglycerate mutase
       **E** Glyceraldehyde-3-phosphate dehydrogenase
   (e)  Fructose-6-phosphate, dihydroxyacetone phosphate, 1,3-bisphosphoglycerate, and 2-phosphoglycerate.

4.  Hexokinase, phosphofructokinase, phosphoglycerate kinase, and pyruvate kinase.

5.  There are <u>three</u> isomerization reactions in glycolysis. The enzymes that catalyze them are <u>phosphoglucose isomerase, triose phosphate isomerase, and phosphoglycerate mutase</u>.

6.  Triose phosphate isomerase catalyzes the interconversion of GAP and DHAP. Its $\Delta G^{\circ\prime} = 7.9$ kJ·mol$^{-1}$ and $\Delta G = 4.4$ kJ·mol$^{-1}$ (Table 15-1). Since $\Delta G = -RT \ln K$, $\ln K = -\Delta G/RT$ and $K = e^{-\Delta G/RT}$. Under standard conditions,

$$K = \frac{[\text{GAP}]}{[\text{DHAP}]} = e^{-\left(7900 \text{ J·mol}^{-1}\right)/\left(8.3145 \text{ J·K}^{-1}\text{·mol}^{-1}\right)(310\,\text{K})}$$

$$= 0.047$$

In the cell, where the reaction is not at equilibrium,

$$\Delta G = \Delta G^{\circ\prime} + RT \ln \frac{[\text{GAP}]}{[\text{DHAP}]}$$

$$\frac{[\text{GAP}]}{[\text{DHAP}]} = e^{(\Delta G - \Delta G^{\circ\prime}/RT)}$$

$$\frac{[\text{GAP}]}{[\text{DHAP}]} = e^{(4400 - 7900 \text{ J·mol}^{-1})/(8.3145 \text{ J·K}^{-1}\text{·mol}^{-1})(310 \text{ K})}$$

$$= 0.26$$

Therefore, the cellular ratio [GAP]/[DHAP] is nearly 6 times larger than under standard conditions, so that in the cell, formation of GAP is favored.

7. This isotope exchange reaction is consistent with the existence of an acyl–enzyme intermediate through the following series of events: For exchange to occur, 1,3-bisphosphoglycerate (1,3-BPG) must react with GAPDH to form an acyl–enzyme intermediate and eliminate $P_i$. (Figure 15-9, Reaction 5 in reverse). In this reaction, the $^{32}$P-labeled $P_i$ reacts with the acyl-enzyme intermediate to yield $^{32}$P-labeled 1,3-BPG.

8. 2,3-BPG is occasionally released by bisphosphoglycerate mutase. The resulting enzyme is inactive because it requires a phospho-His residue to phosphorylate 2PG. The presence of trace amounts of 2,3-BPG permits this substance to rebind to the enzyme and thereby activate it by forming the phospho-His residue.

9. The combined glyceraldehyde-3-phosphate dehydrogenase/phosphoglycerate kinase reaction is

$$GAP + NAD^+ + ADP + P_i \rightarrow 3PG + NADH + ATP + H^+$$

$\Delta G^{\circ\prime} = -16.7\ kJ \cdot mol^{-1}$ and $\Delta G = -1.1\ kJ \cdot mol^{-1}$ (Table 15-1).

Since $\Delta G = \Delta G^{\circ\prime} + RT \ln Q$, where

$$Q = \frac{[3PG][NADH][ATP][H^+]}{[GAP][NAD^+][ADP][P_i]}$$

$$Q = e^{(\Delta G - \Delta G^{\circ\prime}/RT)}$$
$$= e^{(-1100 + 16,700\ J \cdot mol^{-1})/(8.3145\ J \cdot K^{-1} \cdot mol^{-1})(310\ K)}$$
$$= 425$$

Since $\quad Q = \dfrac{[3PG][NADH][ATP][H^+]}{[GAP][NAD^+][ADP][P_i]} = \dfrac{[3PG]}{[GAP]} \times \dfrac{1}{100} \times 10 \times 1 = 425$

$$\frac{[GAP]}{[3PG]} = \frac{1}{4250} = 2.35 \times 10^{-4}$$

10.

   (a) Inhibition of enolase causes its substrate, 2PG, to accumulate. Because the preceding reaction (catalyzed by phosphoglycerate mutase) is near equilibrium, 2PG equilibrates with 3PG so that [3PG] increases as [2PG] increases.
   (b) The formation of 3PG from 1,3-BPG is endergonic, so this reaction (catalyzed by phosphoglycerate kinase) does not proceed in reverse. Therefore, an increase in [3PG] does not cause [1,3-BPG] to increase.

11.
  (a)  A
  (b)  C
  (c)  A
  (d)  C

12.  During the catabolic breakdown of glucose, $NAD^+$ is reduced to NADH, thereby lowering the $[NAD^+]/[NADH]$ ratio. Under anaerobic conditions, pyruvate can be converted to lactate with the concomitant oxidation of NADH to $NAD^+$. Under aerobic conditions, oxidative phosphorylation regenerates the $NAD^+$. Both of these processes will increase the $[NAD^+]/[NADH]$ ratio.

13.  C3 or C4 of glucose must be labeled with $^{14}C$.

14.  The rate of ATP production during anaerobic fermentation can be as much as 100 times greater than during oxidative phosphorylation. Therefore, anaerobic fermentation provides the bulk of ATP during rapid bursts of muscle activity, even though the yield of ATP per glucose is much lower than in oxidative phosphorylation.

15.
  (a)  The phosphofructokinase reaction is the rate-determining step.
  (b)  The first reaction of glycolysis, catalyzed by hexokinase, is not a suitable control point for the overall pathway because significant amounts of sugar enter glycolysis as glucose-6-phosphate, the product of this reaction. For example, galactose is converted to glucose-6-phosphate to enter glycolysis. Not all this glucose-6-phosphate proceeds through glycolysis: Some is shunted to the pentose phosphate pathway. Furthermore, mannose and, in muscle, fructose enter glycolysis as fructose-6-phosphate, which freely interconverts with glucose-6-phosphate. Therefore, the production of fructose-1,6-bisphosphate by PFK is the first committed step of glycolysis and the most effective point for regulation.

16.  Substrate cycles were originally referred to as futile cycles because of their apparent futile waste of free energy.

17.  Mannose.

18.  No additional NADH is produced. The $NAD^+$ is probably reduced and then reoxidized in the sequential oxidation and reduction of C4 during epimerization of the hexose.

19.  The primary sugar in human milk is lactose, a disaccharide of glucose and galactose.

20.

(a) Glycerol can be converted to pyruvate through reactions catalyzed by glycerol kinase, glycerol phosphate dehydrogenase, triose phosphate isomerase, and the enzymes of Stage II of glycolysis (see Figure 15-27):

$$\text{Glycerol} + \text{ATP} \rightarrow \text{glycerol-3-phosphate} + \text{ADP}$$
$$\text{Glycerol-3-phosphate} + \text{NAD}^+ \rightarrow \text{DHAP} + \text{NADH} + \text{H}^+$$
$$\text{DHAP} \rightarrow \text{GAP}$$
$$\text{GAP} + \text{NAD}^+ + 2\,\text{ADP} + \text{P}_i \rightarrow \text{pyruvate} + \text{NADH} + 2\,\text{ATP} + \text{H}_2\text{O} + 2\,\text{H}^+$$

$$\text{Glycerol} + 2\,\text{NAD}^+ + \text{ADP} + \text{P}_i \rightarrow \text{pyruvate} + 2\,\text{NADH} + \text{ATP} + \text{H}_2\text{O} + 3\,\text{H}^+$$

(b) The net yield of ATP is the same for the oxidation of 1 glucose or 2 glycerol to pyruvate.

(c) For every pyruvate formed from glycerol, 2 $\text{NAD}^+$ are reduced to NADH. Alcoholic fermentation regenerates only 1 $\text{NAD}^+$ per pyruvate, so this pathway does not maintain the redox balance of the cell.

21.

(a) $5C_a$ is ribose-5-phosphate, and $5C_b$ is xylulose-5-phosphate.
(b) 6C is fructose-6-phosphate.
(c) 3C is glyceraldehyde-3-phosphate.
(d) A and C are transketolase reactions, and B is a transaldolase reaction.

22.

(a) Two
(b) One directly, although four more can be produced via glycolysis from the two F6P molecules also produced.

23. The conversion of Ru5P to Xu5P is a racemization that inverts the configuration at the C3 chiral center. The isomerization of Ru5P to R5P converts a ketose to an aldose by shifting the position of a double bond.

24.

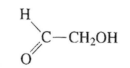

**2-hydroxy-ethanol (hydroxy-acetaldehyde)**

As in the pyruvate decarboxylase reaction, an aldehyde might be expected to be released from the TPP cofactor of transketolase in the absence of the second substrate.

25.  (3); (ketose); (aldose); (ketose and an aldose)

26.  (2); (ketose); (aldose); (ketose and an aldose)

27.  The committed step is the exergonic reaction mediated by G6PD, which is regulated by the availability of $NADP^+$.

28.

(1)  The 5 G6P are converted to 5 F6P by phosphoglucose isomerase.

(2)  One F6P is converted to FBP, which requires ATP for the reaction catalyzed by phosphofructokinase.

(3)  The FBP is converted to 2 GAP by aldolase and triose phosphate isomerase.

(4)  The transaldolase and transketolase reactions of the pentose phosphate pathway operate in reverse to convert 2 F6P + 1 GAP to 2 Xu5P and 1 R5P. Since the starting materials are 4 F6P and 2 GAP, the result is 4 Xu5P and 2 R5P.

(5)  The 4 Xu5P can be isomerized to 4 R5P, for a total yield of 6 R5P from the original 5 G6P.

## Chapter 16

1.  Glycogen phosphorylase, glycogen debranching enzyme, and phosphoglucomutase.

2.  A narrow crevice on the surface of glycogen phosphorylase, which connects the glycogen storage site and the active site, can accommodate up to five residues of an unbranched glycogen chain but cannot accommodate a branched chain.

3.  The phosphoryl group of PLP participates in general acid–base catalysis by donating a proton to the anionic $P_i$ that reacts with glycogen to release G1P.

4.  The reaction product would be a mixture of $\alpha$ and $\beta$ anomers of G1P because the reaction intermediate is an oxonium ion whose C1 can react with phosphate approaching either face of the sugar residue.

5.

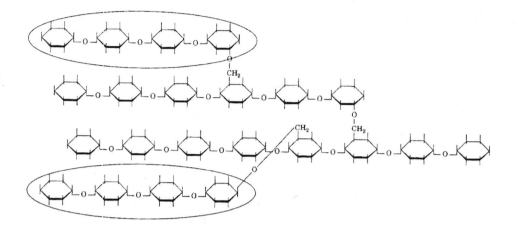

Glycogen debranching enzyme has two catalytic functions: (1) It transfers three or four $\alpha(1\rightarrow4)$-linked glucose residues from a "limit branch" of glycogen to the nonreducing end of another branch, and (2) it hydrolyzes the $\alpha(1\rightarrow6)$ bond of the remaining residue to form free glucose. In the diagram above, the top and bottom branches serve as substrates in which the trisaccharide is transferred and the remaining glucose residue is released. Once these branches have been eliminated, continued phosphorolysis would produce an additional "limit branch" substrate for debranching enzyme.

6. Highly branched glycogen molecules have more nonreducing ends available for phosphorylase to produce G1P, so the rate of G1P release remains high until a limit branch is encountered.

7.

$$K_{eq} = \frac{[glycogen_{n-1}][G1P]}{[glycogen_n][P_i]} \approx \frac{[G1P]_{eq}}{[P_i]_{eq}}$$

$$\Delta G = \Delta G^{o\prime} + RT \ln \frac{[G1P]}{[P_i]}$$

In order for $\Delta G$ to be less than zero, $-RT \ln ([G1P]/[P_i])$ must be at least as great as $\Delta G^{o\prime}$.

$-RT \ln ([G1P]/[P_i]) \geq \Delta G^{o\prime}$

$-\ln ([G1P]/[P_i]) \geq \Delta G^{o\prime}/RT$

$\ln ([G1P]/[P_i]) \leq -\Delta G^{o\prime}/RT$

$[G1P]/[P_i] \leq e^{-\Delta G^{o\prime}/RT}$

$[G1P]/[P_i] \leq e^{-(3100 \text{ J·mol}^{-1})/(8.3145 \text{ J·K}^{-1}\text{·mol}^{-1})(310 \text{ K})}$

$[G1P]/[P_i] \leq 0.30$

Therefore, $[G1P]/[P_i]$ can be no greater than 0.30, so $[P_i]/[G1P]$ must be at least $1/0.30 = 3.33$.

8.  A "high-energy" compound is required to bypass the exergonic glycogen phosphorylase reaction. The formation of UDP–glucose from G1P and UTP yields PP$_i$, whose hydrolysis provides the thermodynamic "pull" to add another glucose residue to glycogen.

9.

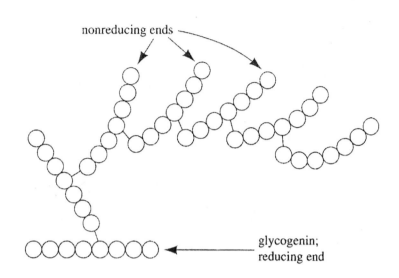

Branching enzyme has three constraints: (1) Branches must be separated by four glucosyl residues; (2) seven residues are transferred at a time to make a branch; and (3) the donating chain must be at least 11 residues long. The first branch is made by moving a seven-residue segment from the reducing end. The most highly branched structure results when branches are added to branches for a final product with six $\alpha(1\rightarrow6)$ branch points, as shown above. There are seven nonreducing ends (three are labeled).

10.

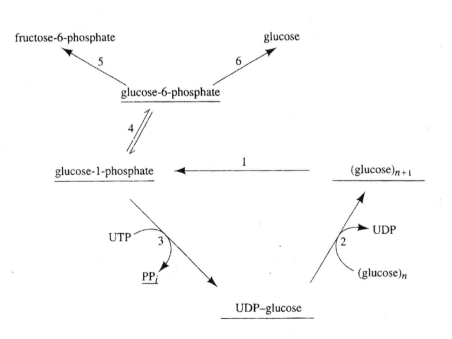

    1   Glycogen phosphorylase
    2   Glycogen synthase
    3   UDP–glucose pyrophosphorylase
    4   Phosphoglucomutase
    5   Phosphoglucose isomerase
    6   Glucose-6-phosphatase

11.  In the T form, the active site is less accessible to its substrates compared to the R form due to the presence of a loop that covers the T state active site so as to prevent access of substrate to it. In the R state, the tower helices have tilted and pulled apart relative to their positions in the T state, thereby inducing a ~10° counter-rotation of the two subunits. This also displaces and disorders the loop covering the active site, thus making the active site accessible to substrate. In addition, the side chain of Arg 569, which is located in the active site, rotates in such a way that it increases the R-state enzyme's affinity for its $P_i$ substrate.

12.  $Ca^{2+}$ released during muscle contraction activates phosphorylase kinase via binding of $Ca^{2+}$ to calmodulin (the δ subunit of phosphorylase kinase). Phosphorylase kinase is therefore active (although not maximally so) even without PKA-catalyzed phosphorylation of its α and β subunits.

13.  Epinephrine-dependent stimulation of PFK-2 leads to an increase in [F2,6P], which activates PFK-1. The result is increased flux through glycolysis.

14.  The full activity of phosphorylase kinase requires both the presence of $Ca^{2+}$ (whose concentration increases in response to hormone binding to α-adrenoreceptors) and phosphorylation by PKA (which is activated by the binding of the cAMP second messenger whose synthesis is stimulated by hormone binding to β-adrenoreceptors).

15.  Only Type 0 (glycogen synthase deficiency) causes a decrease in stored glycogen.

16.  The $H^{14}CO_3^-$ is added to pyruvate to yield oxaloacetate via the pyruvate carboxylase reaction. The $^{14}C$ is then released as $CO_2$ by the PEP carboxykinase reaction, which converts oxaloacetate to PEP + $CO_2$.

17.

*gluconeogenesis:*

$$\begin{array}{l} \text{pyruvate} + HCO_3^- + ATP \rightleftharpoons \text{oxaloacetate} + ADP + P_i \\ \underline{\text{oxaloacetate} + GTP \rightleftharpoons PEP + GDP + CO_2} \\ \text{pyruvate} + HCO_3^- + ATP + GTP \rightleftharpoons PEP + ADP + GDP + P_i + CO_2 \end{array}$$

*glycolysis:*

$$H^+ + PEP + ADP \rightleftharpoons ATP + \text{pyruvate}$$

The formation of PEP from pyruvate requires the investment of two ATP equivalents ("high-energy" phosphoanhydride bonds), whereas PEP's reaction to form pyruvate yields only one ATP.

18. Glycerol enters the gluconeogenic pathway as dihydroxyacetone phosphate (DHAP; see Figure 15-27 on p. 503.) Glycerol is a substrate for glycerol kinase, which catalyzes the reaction

Glycerol + ATP $\rightleftharpoons$ glycerol-3-phosphate + ADP

Glycerol phosphate dehydrogenase then catalyzes the reaction

Glycerol-3-phosphate + NAD$^+$ $\rightleftharpoons$ DHAP + NADH + H$^+$

19.

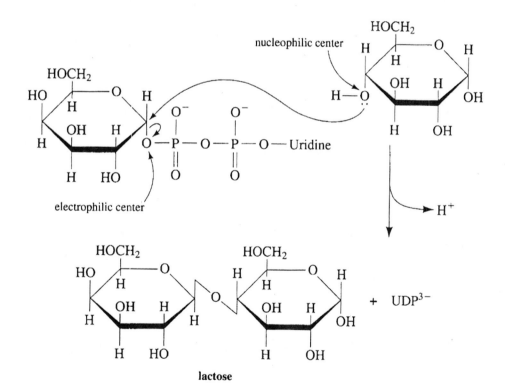

lactose

## Chapter 17

1. The first phase involves the net oxidation of acetyl-CoA to two molecules of $CO_2$ and the regeneration of CoA (Reactions 1–5 of the cycle).

Acetyl-CoA + oxaloacetate + GDP + 2 NAD$^+$ + H$_2$O + P$_i$ →
succinate + GTP + CoASH + 2 NADH + 2 CO$_2$ + 2 H$^+$

The second phase involves the regeneration of oxaloacetate from succinate (Reactions 6–8).

$$\text{Succinate} + \text{FAD} + \text{NAD}^+ + \text{H}_2\text{O} \rightarrow$$
$$\text{oxaloacetate} + \text{NADH} + \text{FADH}_2 + \text{H}^+$$

2.  To demonstrate that succinate is an intermediate in glucose oxidation, show that, in the presence of malonate, cell extracts containing newly added $^{14}$C-labeled acetyl-CoA (labeled at either of the acetyl C atoms) and any of the other citric acid cycle intermediates yield $^{14}$C-labeled succinate.

3.  Pyruvate dehydrogenase ($E_1$), dihydrolipoyl transacetylase ($E_2$), and dihydrolipoyl dehydrogenase ($E_3$).

4.

| Reaction | Enzyme | Cofactor |
| --- | --- | --- |
| (a) Oxidative formation of an enzymatic disulfide bond | $E_3$ | FAD, $\text{NAD}^+$ |
| (b) Transfer of hydroxyethyl group bound to TPP acid | $E_2$ | lipoic |
| (c) Liberation of $CO_2$ | $E_1$ | TPP |
| (d) Oxidation of dihydrolipoamide | $E_2$, $E_3$ | |
| (e) Formation of acetyl-CoA | $E_2$ | CoA, lipoic acid |

5.  FAD functions as an electron conduit to $\text{NAD}^+$ as the enzyme reoxidizes the $E_3$ sulfhydryl groups that were reduced during the regeneration of lipoamide on $E_2$. The FAD and $\text{NAD}^+$ therefore participate in regenerating the functional groups of $E_3$.

6.  The data suggest that in preparation B, $E_1$ (pyruvate dehydrogenase) is somehow unable to transfer its bound hydroxyethyl carbanion to the lipoamide cofactor of $E_2$ (dihydrolipoyl transacetylase). Instead, free acetaldehyde is released from the TPP group of $E_1$ (as occurs in the pyruvate decarboxylase reaction; Section 15-3B).

7.  (b) and (c) would yield $^{14}CO_2$ after the pyruvate dehydrogenase reaction since the C3 and C4 atoms of glucose both become C1 of glyceraldehyde-3-phosphate and then pyruvate. Pyruvate dehydrogenase liberates C1 of pyruvate as $CO_2$.

8.  The large, negative free energy of hydrolysis of the thioester bond provides the thermodynamic force for the otherwise endergonic condensation of the acetyl group and oxaloacetate under conditions of low oxaloacetate concentration. The oxaloacetate concentration is low because of the endergonic nature of the preceding malate dehydrogenase reaction.

9.

(a) For the reaction succinate + FAD → fumarate + FADH$_2$, succinate is the electron donor and FAD is the electron acceptor. $\mathscr{E}^{\circ\prime}$ for the succinate → fumarate half-reaction is 0.031 V (Table 14-5). Therefore,

$$\Delta\mathscr{E}^{\circ\prime} = \mathscr{E}^{\circ\prime}_{FAD} - \mathscr{E}^{\circ\prime}_{succinate} = 0.00\ V - 0.031\ V = -0.031\ V$$

$$\begin{aligned}
\Delta G^{\circ\prime} &= -n\mathscr{F}\Delta\mathscr{E}^{\circ\prime} \\
&= -(2)(96{,}485\ J \cdot V^{-1} \cdot mol^{-1})(-0.031\ V) \\
&= 5982\ J \cdot mol^{-1} = 6.0\ kJ \cdot mol^{-1}
\end{aligned}$$

(b) When NAD$^+$ is the electron acceptor ($\mathscr{E}^{\circ\prime} = -0.315\ V$),

$$\mathscr{E}^{\circ\prime} = -0.315\ V - 0.031\ V = -0.346\ V$$
$$\begin{aligned}
\Delta G^{\circ\prime} &= -(2)(96{,}485\ J \cdot V^{-1} \cdot mol^{-1})(-0.346\ V) \\
&= 66768\ J \cdot mol^{-1} = 67\ kJ \cdot mol^{-1}
\end{aligned}$$

Therefore, under standard biochemical conditions, the oxidation of succinate by FAD is slightly disfavored ($\Delta G^{\circ\prime} > 0$). However, the oxidation of succinate by NAD$^+$ is strongly disfavored ($\Delta G^{\circ\prime} \gg 0$).

(c) Succinate lacks the reducing power (has a reduction potential that is too high) to reduce NAD$^+$ to NADH but has sufficient reducing power to reduce FAD to FADH$_2$.

10. It is lost as CO$_2$ in the isocitrate dehydrogenase reaction in the first turn of the cycle (see Figure 17-2).

11. The rate of glucose oxidation would decrease because succinyl-CoA inhibits its own synthesis by the α-ketoglutarate dehydrogenase reaction, and it inhibits the citrate synthase reaction via feedback inhibition. The rapid accumulation of succinyl-CoA would also deplete the mitochondrial pool of CoA, thereby slowing the production of acetyl-CoA from glucose-derived pyruvate.

12. Glucose synthesis (gluconeogenesis) utilizes oxaloacetate; lipid biosynthesis utilizes acetyl-CoA derived from citrate; amino acid biosynthesis utilizes α-ketoglutarate and oxaloacetate, and porphyrin biosynthesis utilizes succinyl-CoA (Figure 17-17).

13. Acetyl-CoA activates pyruvate carboxylase. Increases in [acetyl-CoA] are indicative of an inability of the citric acid cycle to oxidize acetyl-CoA as fast as it is being produced (from glycolysis and fatty acid oxidation). Hence, activating pyruvate carboxylase, which adds more oxaloacetate to the pool of citric acid cycle intermediates will catalytically accelerate acetyl-CoA oxidation.

14. Two rounds of the glyoxylate cycle are necessary to yield two oxaloacetate, which give rise to one glucose by gluconeogenesis.

*glyoxylate cycle:*
$$4\,\text{Acetyl-CoA} + 4\,NAD^+ + 2\,FAD + 6\,H_2O \rightarrow$$
$$2\,\text{oxaloacetate} + 4\,CoA + 4\,NADH + 2\,FADH_2 + 8\,H^+$$

*gluconeogenesis:*
$$2\,\text{Oxaloacetate} + 2\,GTP + 2\,ATP + 2\,NADH + 4\,H_2O + 6\,H^+ \rightarrow$$
$$\text{glucose} + 2\,GDP + 2\,ADP + 2\,NAD^+ + 4\,P_i + 2\,CO_2$$

*net:*
$$4\,\text{Acetyl-CoA} + 2\,GTP + 2\,ATP + 2\,NAD^+ + 2\,FAD + 10\,H_2O \rightarrow$$
$$\text{glucose} + 4\,CoA + 2\,CO_2 + 2\,GDP + 2\,ADP + 2\,NADH + 2\,FADH_2 + 2\,H^+ + 4\,P_i$$

15. *Isocitrate lyase*: isocitrate $\rightarrow$ succinate + glyoxylate
    *Malate synthase*: acetyl-CoA + glyoxylate $\rightarrow$ malate + CoA

16. Aspartate and α-ketoglutarate move from the mitochondrion to the glyoxysome, and succinate and glutamate move from the glyoxysome to the mitochondrion.

## Chapter 18

1.

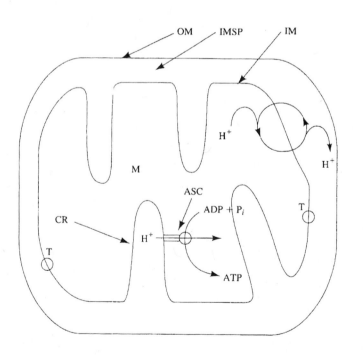

2.

    __E__ Pyruvate dehydrogenase
    __A__ 3-Phosphoglycerate dehydrogenase
    __C__ Flavoprotein dehydrogenase
    __E, A__ Malate dehydrogenase
    __D__ Cytochrome $c$
    __C__ Cytochrome $c_1$
    __E__ Fatty acid oxidation enzymes
    __E__ Mitochondrial DNA
    __C__ ADP–ATP translocator
    __B__ Mitochondrial porin

3. The protein's molar concentration would be

$$\frac{1}{40\,000\ \text{g·mol}^{-1}} \times \frac{1.37\ \text{g}}{\text{mL}} \times \frac{1000\ \text{mL}}{\text{L}} \times 50\%$$
$$= 0.0171\ \text{mol·L}^{-1}$$
$$= 17.1\ \text{mM}$$

4.

(a) The negatively charged phosphate groups of NADH (Figure 11-4) prevent its diffusion across the inner mitochondrial membrane, and there are no NADH transport proteins to facilitate its transport.

(b) The malate–aspartate shuttle allows the indirect import of NADH reducing equivalents (Figure 16-20). The reduction of oxaloacetate to malate by NADH followed by the facilitated transport of malate across the inner mitochondrial membrane yields NADH in the mitochondrial matrix when the malate is reoxidized to oxaloacetate. Oxaloacetate then returns to the cytosol by being converted to aspartate, for which there is a transporter. In insects, the glycerophosphate shuttle utilizes NADH to convert dihydroxyacetone phosphate to 3-phosphoglycerate by a flavoprotein dehydrogenase that donates electrons to the electron-transport chain in a manner similar to succinate dehydrogenase (Fig. 18-5).

(c) Cytosolic $NAD^+$ is required for the glyceraldehyde-3-phosphate dehydrogenase reaction of glycolysis. Limited $[NAD^+]$ would shut down glycolysis.

5. Isocitrate dehydrogenase, α-ketoglutarate dehydrogenase, and malate dehydrogenase produce NADH, which transfers its electrons to Complex I. Succinate dehydrogenase, whose FAD group is reduced in the oxidation of succinate to fumarate, is a component of Complex II.

6.

  (a)  The percent reduction is

$$\frac{[\text{Reduced}]}{[\text{Reduced}] + [\text{Oxidized}]} \times 100 \qquad \text{or}$$

$$\frac{[\text{FADH}_2]}{[\text{FADH}_2] + [\text{FAD}]} \times 100$$

For the $\text{FADH}_2/\text{FAD}$ half-cell, the Nernst equation is

$$\mathscr{E} = (-0.219\text{V}) - \left( \frac{(8.3145\text{J} \cdot \text{K}^{-1} \cdot \text{mol}^{-1})(298\text{K})}{(2)(96,486\text{J} \cdot \text{V}^{-1} \cdot \text{mol}^{-1})} \right) \ln \left( \frac{[\text{FADH}_2]}{[\text{FAD}]} \right)$$

$$\mathscr{E} = (-0.219\text{V}) - (0.0218\text{V}) \ln \left( \frac{[\text{FADH}_2]}{[\text{FAD}]} \right)$$

| [FADH₂]/[FAD] | % Reduction | E (V) |
|---|---|---|
| 100 | 99.0 | − 0.278 |
| 10 | 90.9 | − 0.248 |
| 5 | 83.3 | − 0.240 |
| 2 | 66.7 | − 0.228 |
| 1 | 50.0 | − 0.219 |
| 0.5 | 33.3 | − 0.210 |
| 0.2 | 16.7 | − 0.198 |
| 0.1 | 9.1 | − 0.190 |
| 0.01 | 0.99 | − 0.160 |

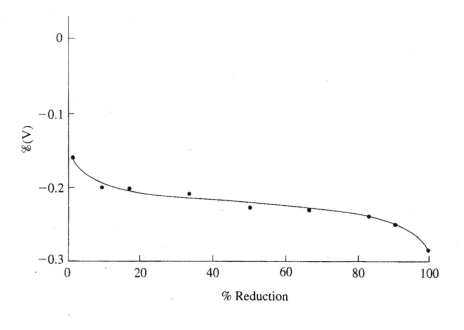

(b)    For cytochrome $c$,

$$\mathscr{E} = (0.235 \text{ V}) - \left( \frac{(8.3145 \text{J} \cdot \text{K}^{-1} \cdot \text{mol}^{-1})(298 \text{K})}{(1)(96,485 \text{ J} \cdot \text{V}^{-1} \cdot \text{mol}^{-1})} \right) \ln \left( \frac{\text{cyto } c \,[\text{Fe}^{2+}]}{\text{cyto } c \,[\text{Fe}^{3+}]} \right)$$

$$\mathscr{E} = (0.235 \text{ V}) - (0.0275 \text{ V}) \ln \left( \frac{\text{cyto } c \,[\text{Fe}^{2+}]}{\text{cyto } c \,[\text{Fe}^{3+}]} \right)$$

| [cyto $c$ (Fe²⁺)]/[cyto $c$ (Fe³⁺)] | % Reduction | E (V) |
|---|---|---|
| 100 | 99.0 | 0.117 |
| 10 | 90.9 | 0.176 |
| 5 | 83.3 | 0.194 |
| 2 | 66.7 | 0.217 |
| 1 | 50.0 | 0.235 |
| 0.5 | 33.3 | 0.253 |
| 0.2 | 16.7 | 0.276 |
| 0.1 | 9.1 | 0.294 |
| 0.01 | 0.99 | 0.353 |

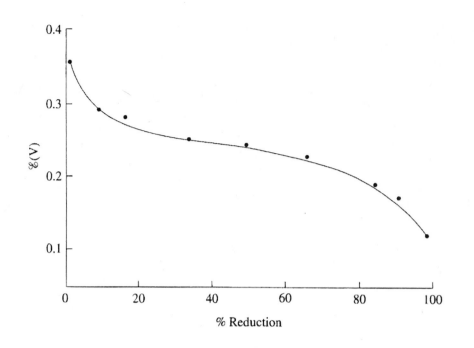

$\mathscr{E}$ (V) vs % Reduction

(c)  When [FADH$_2$]/[FAD] = 10, $\mathscr{E}_{FAD}$ = –0.248 V
When [cyto $c$ (Fe$^{2+}$)]/[cyto $c$ (Fe$^{3+}$)] = 0.1, $\mathscr{E}_{cyto\ c}$ = 0.294 V
According to Equation 13-10,

$$\Delta\mathscr{E} = \mathscr{E}_{(e-\text{acceptor})} - \mathscr{E}_{(e-\text{donor})}$$
$$= \mathscr{E}_{cyto\ c} - \mathscr{E}_{FAD}$$
$$= 0.294\ \text{V} - (-0.248\ \text{V})$$
$$= 0.542\ \text{V}$$

7.

(a)  Stigmatellin inhibits the electron flow from CoQH$_2$ to Complex III. Hence cytochrome $a$ (a component of Complex IV) cannot obtain electrons and would be largely in its oxidized state. Similarly, cytochrome $b_L$ (part of Complex II) and cytochrome $c$ (which links Complexes III and IV) would be largely in their oxidized states.

(b)  Antimycin A blocks electron transport in Complex III from heme $b_H$ to CoQ or CoQ·$^-$. Thus, cytochrome $b_L$ would be reduced since it could not pass its electrons on to cytochrome $b_H$. However, cytochromes $a$ and $c$ would be oxidized since they are both downstream of the block.

(c)  Rotenone blocks electron transport in Complex I. However, since electrons are being introduced into the electron-transport chain at Complex II, electron transport can proceed all the way to O$_2$. Therefore, all three cytochromes would be predominantly in their reduced forms.

8.  The most abundant type of electron carrier in Complex I is iron–sulfur clusters.

9. Only two electron carriers in the mitochondrial electron-transport chain can also carry protons, FMN and ubiquinone (CoQ). The paucity of proton carriers compared to electron carriers suggests that transmembrane proton transport involves protein-mediated proton pumping (a proton wire) rather than only direct transport by the proton carriers.

10. Since cytochrome $c$ contains positively charged Lys residues around its heme crevice, its redox partners must have negatively charged residues such as Glu and Asp at their cytochrome $c$–binding sites.

11. The $Cu_A$ center accepts the first electron from cytochrome $c$. Oxygen binds to the partially reduced Fe(II)–Cu(I) form of the cytochrome $a_3$–$Cu_B$ binuclear complex.

12. Key observations that support the chemiosmotic hypothesis are: (a) Oxidative phosphorylation requires an intact inner membrane; (b) the inner mitochondrial membrane is impermeable to ions such as $H^+$, $OH^-$, $K^+$, and $Cl^-$; (c) electron transport results in the transport of $H^+$ out of the mitochondrial matrix; and (d) compounds that increase proton permeability across the membrane uncouple phosphorylation from electron transport and inhibit ATP synthesis.

13. A pH gradient can exist without $\Delta\Psi$. If a counterion such as $Cl^-$ moved in the same direction as the $H^+$, or if $K^+$ moved in the opposite direction, a $[H^+]$ gradient could form without altering the net charge on either side of the membrane. Similarly, the transmembrane movement of ions other than $H^+$ could generate $\Delta\Psi$ without generating a proton concentration gradient.

14.
   (a) The P/O ratio is a measure of how many ATPs are synthesized for every O atom reduced.
   (b) Oxygen consumption stops when the production of ATP from ADP cannot be carried out. This is because electron transport and proton translocation cannot take place independently. Hence, when the proton gradient has built up to the point that electron transport has insufficient free energy to translocate additional protons across the inner mitochondrial membrane, electron transport must stop.
   (c) When ADP is added, ATP can be synthesized. $O_2$ consumption then resumes because the dissipation of the proton concentration gradient by ATP synthase reduces the free energy of proton translocation across the inner mitochondrial membrane to the point that electron transport can continue.

15.
   (a) The pH gradient would remain unchanged. However, the $\Delta\Psi$ would collapse because $K^+$ would equilibrate across the membrane in response to the $\Delta\Psi$.
   (b) Nigericin would dissipate the proton gradient by exchanging protons for $K^+$ and hence arrest ATP synthesis. The rate of electron transport would increase because there would be no buildup of a proton gradient to hold electron transport in check. However, $\Delta\Psi$ would remain intact because there would be no net change in charge across the membrane.

(c) Gramicidin would cause the collapse of both the $\Delta\Psi$ and the pH gradient, so ATP synthesis would stop but electron transport would accelerate.

16. In solution, the ATP-forming reaction is $ADP + P_i \rightarrow ATP + H_2O$. Recall that an enzymatic reaction may progress through several steps, but it cannot alter the $\Delta G$ for a reaction. For the ATP synthase–catalyzed reaction, the binding of substrates and release of products are just two of the steps in the overall process. Letting E represent the enzyme, the reaction can be written as

$$ADP + P_i + E \rightarrow ADP \cdot P_i \cdot E \rightarrow ATP \cdot H_2O \cdot E \rightarrow ATP + H_2O + E$$

where each step has a different $\Delta G$ value. However, the overall reaction remains

$$ADP + P_i + E \rightarrow ATP + H_2O + E$$

and hence the overall free energy change is still 30.5 kJ·mol$^{-1}$.

17. The reaction of DCCD with ATP synthase inhibits ATP formation by reacting with Asp 61, preventing its protonation and deprotonation, actions required for inducing the rotation of the $c$ subunit. The proton gradient cannot be dissipated and therefore builds up to the point that it arrests further electron transport.

18. The $H^+/P$ ratio is a measure of the number of protons transported across the inner membrane for each ATP molecule synthesized. This could be difficult to determine since the measurement of pH would not take into account $\Delta\Psi$, which also contributes to the protonmotive force. In addition, the measurement would be highly sensitive to cytosolic pH changes that were not involved in mitochondrial electron transport.

19. $\Delta G^{\circ\prime} = +30.5$ kJ·mol$^{-1}$ for the synthesis of ATP from $ADP + P_i$. The conversion of glucose to lactate by glycolysis is accompanied by the synthesis of 2 ATP. Thus, the efficiency of ATP production under standard conditions is $[(2 \times 30.5 \text{ kJ·mol}^{-1})/(196 \text{ kJ·mol}^{-1})] \times 100 = 31\%$. When glucose is completely oxidized, the yield is 32 ATP. The efficiency of this process is $[(32 \times 30.5 \text{ kJ·mol}^{-1})/(2823 \text{ kJ·mol}^{-1})] \times 100 = 35\%$. Therefore, not only does oxidative phosphorylation yield more ATP than glycolysis, but at least under standard conditions, it does so with greater efficiency.

20.
(a) The energy is supplied by 7000 kJ/30.5 kJ·mol$^{-1}$ = 230 moles of ATP.
(b) 230 moles × 507 g·mol$^{-1}$ = 11700 g = 117 kg
(c) Since 230 moles are needed each day, 0.1 mol of ADP must recycled 230/0.1 = 2300 times.

21. The irreversible step in electron transport is the formation of water from oxygen. The rate of cytochrome $c$ oxidase is controlled by the ratio of reduced to oxidized cytochrome $c$, which is in turn controlled by the [NADH]/[NAD$^+$] and [ATP]/[ADP][P$_i$] ratios.

22. The change in redox potential of the Fe atom promotes its ability to accept the electrons from NADPH ($\mathscr{E}^{\circ\prime} = -0.320$). Recall that electrons flow spontaneously from a substance with a more negative redox potential to a substance with a more positive redox potential.

23. Rats that consume large quantities of metabolic fuels have a higher rate of oxidative metabolism and hence generate more oxygen radicals than rats that consume less food and have lower rates of oxidative metabolism. The cumulative oxidative damage would be greater in the cafeteria-fed rats, which therefore die sooner.

## Chapter 19

1. The light reactions depend on light for activity and include the reactions of the photosystems and the electron transport chain. The dark reactions are those of the Calvin cycle, which convert $CO_2$ to GAP.

    The dark reactions do not require light for their mechanisms. However, the light reactions regulate the dark reactions to ensure that the cell maintains adequate levels of ATP and NADPH. Thus the dark reactions do not actually occur in the dark.

2.

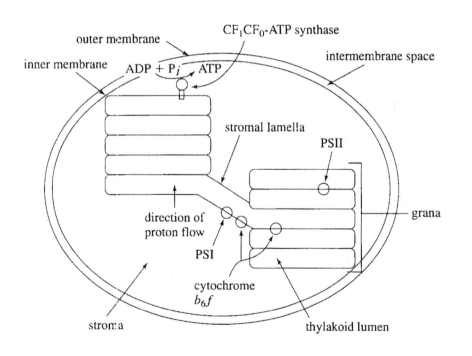

3.  Use Planck's law multiplied by Avogadro's number:

$$E = \frac{hc}{\lambda} N$$

$= (6.626 \times 10^{-34} \text{ J·s}) (2.998 \times 10^{8} \text{ m·s}^{-1}) (6.0221 \times 10^{23} \text{ mol}^{-1}) / \lambda$
$= (0.1196 \text{ J·m·mol}^{-1}) / \lambda$

(a) $E = (0.1196 \text{ J·m·mol}^{-1}) / (4 \times 10^{-7} \text{ m}) = 3.0 \times 10^{5} \text{ J·mol}^{-1} = 300 \text{ kJ·mol}^{-1}$
(b) $E = (0.1196 \text{ J·m·mol}^{-1}) / (5 \times 10^{-7} \text{ m}) = 2.4 \times 10^{5} \text{ J·mol}^{-1} = 240 \text{ kJ·mol}^{-1}$
(c) $E = (0.1196 \text{ J·m·mol}^{-1}) / (6 \times 10^{-7} \text{ m}) = 2.0 \times 10^{5} \text{ J·mol}^{-1} = 200 \text{ kJ·mol}^{-1}$
(d) $E = (0.1196 \text{ J·m·mol}^{-1}) / (7 \times 10^{-7} \text{ m}) = 1.7 \times 10^{5} \text{ J·mol}^{-1} = 170 \text{ kJ·mol}^{-1}$

4.  The pigments in purple photosynthetic bacteria absorb radiation with longer wavelengths than visible light. This is the most intense radiation in the environments that they inhabit, the murky bottoms of stagnant ponds.

5.  The chlorophyll molecules in a reaction center have a slightly lower-energy excited state than that of the antenna chlorophyll molecules. This allows excitation energy to be transferred from the antennae molecules to the reaction center, where photooxidation occurs.

6.  (a) The electron ejected from the excited special pair must be transferred away so that it cannot return immediately to the special pair, which would allow the excitation energy to be released as heat (or possibly in a way that would damage the reaction center).
    (b) After its reduction by an electron, the quinone (now a semiquinone) must await a second excitation event and electron transfer to become fully reduced to a quinol so that it can transfer both electrons to the membrane-bound quinol pool.

7.  Electrons from the bacterial reaction center flow to ubiquinone and then to cytochrome $bc_1$ and cytochrome $c_2$ before returning to the special pair. This cyclic electron flow does not yield reduced $NADP^+$, but it translocates four protons per electron pair to the periplasmic space.
    In the chloroplast, electrons flow in a linear fashion (the Z-scheme) from water to $NADP^+$, so that 2 NADPH are produced for every 2 $H_2O$ oxidized to $O_2$. The 4 electrons pass from PSII to cytochrome $b_6f$ and then to PSI before they reach ferredoxin–$NADP^+$ reductase. This results in the transmembrane movement of 12 protons. In cyclic electron flow, electrons cycle from PSI back to cytochrome $b_6f$ and hence no NADPH is produced. Instead, this increases the number of protons translocated without affecting the stoichiometry of the $H_2O \rightarrow O_2$ reaction.

8.  The standard reduction potential for the reaction $O_2 + 4\,e^- + 4\,H^+ \rightarrow 2\,H_2O$ is 0.815 V. The reduction potential of the P870 bacterial reaction center is ~0.500 V (Figure 19-10), which is not sufficient to oxidize water (electrons spontaneously flow to centers with more positive reduction potentials). The two-reaction center Z-scheme (Figure 19-12) spans a redox range that allows both the oxidation of water and the reduction of $NADP^+$.

9.  The longer wavelength activates only PSI (which contains P700), whereas the shorter wavelength activates both PSI and PSII (which contains P680). Since PSII feeds electrons to PSI, both photosystems must operate together for the redox reactions to proceed most efficiently. When only 700 nm light is available, PSII is unable to extract the electrons from $H_2O$ necessary to form $O_2$. However, in the presence of 700 nm and 500 nm light, PSII can supply electrons to PSI, which can energize them with 700 nm light, thereby driving $O_2$ production.

10. DCMU blocks electron flow between PSII and PSI. This would cause the excitation energy of PSII to be dissipated by a mechanism other than photooxidation (since the electrons have nowhere to go). Some of the absorbed energy is released as fluorescence. Plants in which electron flow was blocked by DCMU can be detected by their fluorescence.

11. The $S$ states in the OEC are chemical intermediates in the five-step reaction in which $H_2O$ is oxidized to $O_2$. Dark-adapted chloroplasts are in the $S_1$ state, since $O_2$ is generated on the third flash of light (Figure 19-16).

12. The cytochrome $b_6f$ complex, via the Q cycle, pumps the majority of the protons that make up the proton gradient required for ATP synthesis. Eight protons are translocated for every 2 $H_2O$ oxidized; more during cyclic electron flow. The membrane-bound cytochrome $b_6f$ complex resembles mitochondrial Complex III (cytochrome $bc_1$).

13. The Cu(II)/Cu(I) half-reaction normally has a standard redox potential of 0.158 V, but plastocyanin's standard redox potential is 0.370 V. In the protein, the Cu(II) atom is strained toward the tetrahedral coordination geometry of Cu(I), which promotes its reduction.

14. Both photosystems are membrane-bound protein complexes, contain special pairs of chlorophyll where photooxidation occurs, and have near-symmetrical arrangements of pigment molecules at the reaction center.
    The photosystems differ in overall structure, their redox groups, and the pathway of electron transfer.

15. The reduction potential for P680 should be greater than that of the $O_2/H_2O$ half-reaction (0.815 V; Table 14-4) since electrons flow from $H_2O$ to P680. The standard reduction potential of ferredoxin should be less than that for the $NADP^+/NADPH$ half-reaction (−0.320 V) since electrons flow from ferredoxin to $NADP^+$.

16. Under bright sun (high proportion of short-wavelength light), PSII is more active than PSI. As a result, reduced plastoquinone accumulates. This activates a protein kinase to phosphorylate the LHCs, which then move to the stromal lamellae, where they associate with and funnel more light energy to PSI.
    Under shady light (high proportion of long-wavelength light), PSI is more active than PSII. Oxidized plastoquinone therefore accumulates. This leads to dephosphorylation of the LHCs, which move to the grana to associate with and funnel light energy to PSII.

17. 3-Phosphoglycerate is the first stable sugar that incorporates the $^{14}CO_2$. This suggests that $^{14}CO_2$ is added to a 2-carbon compound. Ribulose-5-phosphate levels increase after the removal of $^{14}CO_2$, which instead suggests that ribulose-5-phosphate is the $^{14}CO_2$ acceptor.

18. Shown below are the 13 reactions of the Calvin cycle:

    (1)  $3 \text{ Ru5P} + 3 \text{ ATP} \rightarrow 3 \text{ RuBP} + 3 \text{ ADP}$
    (2)  $3 \text{ RuBP} + 3 \text{ CO}_2 \rightarrow 6 \text{ 3PG}$
    (3)  $6 \text{ 3PG} + 6 \text{ ATP} \rightarrow 6 \text{ BPG} + 6 \text{ ADP}$
    (4)  $6 \text{ BPG} + 6 \text{ NADPH} \rightarrow 6 \text{ GAP} + 6 \text{ P}_i + 6 \text{ NADP}^+$
    (5)  $2 \text{ GAP} \rightleftharpoons 2 \text{ DHAP}$
    (6)  $\text{GAP} + \text{DHAP} \rightleftharpoons \text{FBP}$
    (7)  $\text{FBP} \rightarrow \text{F6P} + \text{P}_i$
    (8)  $\text{F6P} + \text{GAP} \rightleftharpoons \text{Xu5P} + \text{E4P}$
    (9)  $\text{E4P} + \text{DHAP} \rightleftharpoons \text{SBP}$
    (10) $\text{SBP} \rightarrow \text{S7P} + \text{P}_i$
    (11) $\text{S7P} + \text{GAP} \rightleftharpoons \text{Xu5P} + \text{R5P}$
    (12) $\text{R5P} \rightleftharpoons \text{Ru5P}$
    (13) $2 \text{ Xu5P} \rightleftharpoons 2 \text{ Ru5P}$

    Stage I (Reactions 1–4) is the production phase or energy-requiring phase and consists of carboxylation, phosphorylation, and reduction steps. Stage II (Reactions 5–13) is the rearrangement or recovery phase, which regenerates ribulose-5-phosphate.

    Stage I:
    $$3 \text{ Ru5P} + 3 \text{ CO}_2 + 9 \text{ ATP} + 6 \text{ NADPH} \rightarrow 6 \text{ GAP} + 9 \text{ ADP} + 6 \text{ P}_i + 6 \text{ NADP}^+$$
    Stage II:
    $$5 \text{ GAP} \rightarrow 3 \text{ Ru5P} + 2 \text{ P}_i$$

19. The interconversion of fructose-1,6-bisphosphate (FBP) and fructose-6-phosphate (F6P) is a potential futile cycle. The relevant enzymes are phosphofructokinase (PFK, for glycolysis) and fructose bisphosphatase (FBPase, for the Calvin cycle). To control these enzymes, a redox-sensing protein, thioredoxin, activates FBPase and deactivates PFK in the light, and activates PFK and deactivates FBPase in the dark.

20. Photorespiration is a side reaction of ribulose bisphosphate carboxylase in which $O_2$ (which competes with $CO_2$ for binding to the active site) reacts with RuBP to form 3PG and 2-phosphoglycolate. The 2-phosphoglycolate is eventually converted back to 3PG by a multistep pathway that consumes NADH and ATP and yields $CO_2$. Photorespiration therefore wastes some of the free energy captured in the photosynthetic light reactions as well as "unfixing" some of the $CO_2$ fixed by photosynthesis.

21. $C_4$ plants have an advantage when $CO_2$ is relatively scarce, since they concentrate $CO_2$ in mesophyll cells. However, this process consumes 2 ATP equivalents. The more energy-efficient $C_3$ plants therefore have an advantage when $CO_2$ is not limiting. Thus, a higher atmospheric concentration of $CO_2$ would favor $C_3$ plants over $C_4$ plants.

## Chapter 20

1. The energy content of fats is greater because the carbon atoms of fatty acids are more reduced and therefore release more free energy upon oxidation than the carbons of carbohydrates or proteins. In addition, fats can be stored in large amounts in anhydrous form. In contrast, proteins cannot be stored in large amounts, and carbohydrates require a large volume for hydration.

2. The gall bladder stores the bile acids secreted by the liver and releases them on demand. Bile acids are essential for emulsifying triacylglycerols and thereby promoting their hydrolysis. In addition, bile acids and released fatty acids are components of mixed micelles that are absorbed across the intestinal wall. The liver secretes bile acids more or less continuously so that if large amount of fats are rapidly ingested, insufficient bile acids will be available to emulsify them.

3. Both pancreatic lipase and phospholipase $A_2$ hydrolyze their substrates at the lipid–water interface, where the enzymes undergo interfacial activation. Pancreatic lipase binds to the interface only when complexed with pancreatic colipase. In the presence of a lipid micelle, the lipase undergoes a structural change that exposes the active site, forms an oxyanion hole, and, in conjunction with colipase, creates a large hydrophobic surface near the active site that helps to bind the lipase–colipase complex to the lipid.

   Phospholipase $A_2$ does not alter its conformation but instead has a hydrophobic channel that enables a micellar phospholipid to gain access to the enzyme active site without having to pass through the aqueous phase.

4.
   | | |
   |---|---|
   | C | Bile acid |
   | F | Intestinal fatty acid–binding protein |
   | E | Albumin |
   | B | Phospholipase $A_2$ |
   | A | Colipase |
   | D | Chylomicrons |

5.

_5_ Triacylglycerols are removed from circulating VLDL by lipoprotein lipase.
_2_ Chylomicrons are transported through the lymphatic system and enter the bloodstream.
_1_ Chylomicrons are formed in the intestinal mucosa.
_6_ Cells take up cholesterol via receptor-mediated LDL endocytosis.
_3_ Chylomicrons are degraded by lipoprotein lipase to chylomicron remnants.
_7_ LDL components are rapidly degraded by lysosomal enzymes.
_4_ VLDL are synthesized in liver.

6. Cholesteryl esters are highly hydrophobic and therefore occupy the nonaqueous interior of the lipoprotein. Cholesterol, with an OH group, is weakly polar and therefore can interact with water molecules at the surface of the lipoprotein.

7. ApoB-100 is the LDL protein component that is recognized by the LDL receptor. Increased binding of LDL by its receptor would increase the cellular uptake of cholesterol-rich LDL. Consequently, the level of cholesterol in the serum would be lower than normal.

8. (a) Hippuric acid (a benzoic acid derivative). (b) Phenylaceturic acid (a phenylacetic acid derivative). This experiment suggested that fatty acids are degraded by oxidation through removal of successive two-carbon fragments.

9. The patient probably has a carnitine palmitoyl transferase I deficiency. As a result, acyl-CoA transport across the mitochondrial membrane is inadequate. Fatty acids released from adipose tissue stores would accumulate in the liver. Glucose levels would fall because it would be the primary metabolic fuel in the absence of oxidizable fatty acids. Treatment with high doses of carnitine would elevate tissue carnitine levels and hence enable some acyl-carnitine to form and enter the mitochondrion so that the acyl group could be oxidized by β oxidation.

10. (a) and (b) are correct. (c) is incorrect because the product is 3-L-hydroxyacyl-CoA.

11.  120 ATP. Stearic acid (an 18:0 fatty acid) is first activated by conversion to stearoyl-CoA, with the consumption of 2 equivalents of ATP. Stearoyl-CoA is then degraded by eight rounds of β oxidation to form 9 acetyl-CoA, 8 $FADH_2$, and 8 NADH. Oxidation of each acetyl-CoA by the citric acid cycle yields GTP, NADH, and $FADH_2$. Reoxidation of the $FADH_2$ and NADH, with the transfer of electrons to $O_2$ to form $H_2O$, yields ATP by oxidative phosphorylation. The stoichiometry of ATP production can be summarized as follows:

| Process | Reduced coenzymes formed | ATPs formed |
|---|---|---|
| Fatty acid activation | | −2 |
| 8 rounds of β oxidation | 8 $FADH_2$ | 12 |
| | 8 NADH | 20 |
| 9 rounds of citric acid cycle | | 9 (GTP) |
| | 27 NADH | 67.5 |
| | 9 $FADH_2$ | 13.5 |
| Total | | 120 |

12.  The missing terms are: cis $\Delta^9$; third; cis $\Delta^3$; enoyl-CoA hydratase; enoyl-CoA isomerase; cis $\Delta^3$; trans $\Delta^2$.

13.  The missing terms are: cis $\Delta^{12}$; 2,4-dienoyl-; enoyl-CoA hydratase; 2,4-dienoyl-CoA; *trans*-$\Delta^3$-enoyl-CoA; 3,2-enoyl-CoA; *trans*-$\Delta^2$-enoyl.

14.  Cobalamin is essential for converting (R)-methylmalonyl-CoA to succinyl-CoA. The cobalt alternates between the Co(III) and Co(II) states and therefore functions as a free radical generator. The weak carbon—cobalt(III) bond of the coenzyme undergoes homolytic cleavage such that the carbon and cobalt each acquire one electron. This yields a deoxyadenosyl radical that can abstract a hydrogen atom from methylmalonyl-CoA, which then rearranges to form succinyl-CoA.

15. Since citric acid cycle intermediates function as catalysts and are continuously regenerated, the net oxidation of succinyl-CoA can occur only if it is converted to another compound that is oxidized by the operation of the cycle, such as pyruvate or acetyl-CoA. This is accomplished by transforming succinyl-CoA to malate and then transporting the malate to the cytosol, where it is oxidatively decarboxylated to pyruvate and $CO_2$ in a reaction catalyzed by malic enzyme. Pyruvate can then be transported into the mitochondrion and oxidized via pyruvate dehydrogenase and the citric acid cycle.

16. The first step in β oxidation in peroxisomes is catalyzed by acyl-CoA oxidase rather than acyl-CoA dehydrogenase as occurs in mitochondria. In the acyl-CoA oxidase reaction, electrons from acyl-CoA are transferred directly to $O_2$ to form $H_2O_2$ and therefore do not pass through the electron-transport chain with concomitant formation of ATP. For this reason, peroxisomal β oxidation yields less ATP than mitochondrial β oxidation.

17. The large intake of milk means that fat supplies a major portion of the calories consumed by infants prior to weaning. The oxidation of fatty acids produces abundant acetyl-CoA which is partly converted to ketone bodies in the liver. The ketone bodies, like glucose, are water-soluble and easily transported. They are therefore readily available as metabolic fuels to support the rapid growth and development of the infant.

18.
(a) The net production of ATP from the complete oxidation of linoleic acid can be calculated as in Problem 11. The total yield is 116.5 ATP. This is 3.5 less than for the oxidation of stearate because the first double bond of linoleate does not need an FAD to reduce it ($FADH_2$ is equivalent to 1.5 ATP) and the reduction of its second double bond consumes an NADPH (which is equivalent to ~2.5 ATP).

(b) If the 9 acetyl-CoA generated by the β oxidation of linoleic acid were instead converted to ketone bodies, 4.5 molecules of acetoacetate would be formed. The transformation of acetoacetate back into acetyl-CoA does not directly consume ATP. However, it involves the consumption of 4.5 succinyl-CoA, which would otherwise provide the energy for a substrate level phosphorylation. Therefore, the net yield of ATP for this metabolic route can be considered to be $116.4 - 4.5 = 112$ ATP.

19. Ketogenesis:
     2 Acetyl-CoA + $H_2O$ → acetoacetate + 2 CoASH

    Ketone body breakdown:
     Acetoacetate + succinyl-CoA + CoASH → 2 acetyl-CoA + succinate

    Combined synthesis and breakdown:
     Succinyl-CoA + $H_2O$ → succinate + CoASH

20.
(a) $H^{14}CO_3^-$ is incorporated into malonyl-CoA by the action of acetyl-CoA carboxylase, but the labeled carboxyl group is subsequently eliminated as $^{14}CO_2$ during condensation of malonyl-ACP with the growing fatty acyl chain. Therefore, the biosynthesized palmitate will be unlabeled.
(b) The radioactive methyl carbon in acetate will label each even-numbered carbon atom in the completed fatty acid.

21. The proximity of multiple enzyme activities involved in fatty acid synthesis on a single polypeptide chain may enhance the efficiency of the process. Moreover, groups that are anchored to the ACP on one subunit are mainly processed by the enzymatic activities on the opposite subunit.

22. (a) and (c) are correct. (b) is incorrect because the tricarboxylate transport system is irreversible, that is, unidirectional. This is because both the ATP-citrate lyase reaction (in the cytosol) and the pyruvate carboxylase reaction (in the matrix) involve ATP hydrolysis. Thus, the net result of one turn of the cycle is the hydrolysis of 2 ATP to 2 ADP + 2 $P_i$ and the transport of acetyl-CoA from the matrix to the cytosol (assuming that the NADH and NADPH in the malate dehydrogenase and malic enzyme reactions are equivalent).

23.
| A | Citrate |
| C | cAMP-dependent phosphorylation |
| D | Palmitate |
| B | Malonyl-CoA |

24. AMP-dependent protein kinase (AMP-PK) is activated by AMP and inhibited by ATP. It is thought to act as a sensor of available energy supply in the cell. In fatty acid metabolism, AMP-PK acts to phosphorylate acetyl-CoA carboxylase, the rate-limiting enzyme in fatty acid biosynthesis, which is inactive when phosphorylated. In conditions of energy scarcity, AMP will tend to accumulate as ATP is consumed. The resulting activation of AMP-PK will therefore inhibit fatty acid formation and at the same time promote fatty acid oxidation by diminshing the level of malonyl-CoA, the product of the acetyl-CoA carboxylase reaction. In the absence of malonyl-CoA, which blocks carnitine palmitoyltransferase I, fatty acyl-CoA will more readily enter the mitochondrion and be oxidized, providing ATP.In contrast, when ATP levels are high and AMP-PK is inhibited, acetyl-CoA carboxylase undergoes dephosphorylation and becomes active, increasing the level of malonyl-CoA and enhancing fatty acid synthesis, while causing the blockade of fatty acid oxidation.

25.
| E | CDP–diacylglycerol |
| D | Ceramide |
| A | Phosphatidylglycerol |
| C | Plasmalogen |
| B | CDP–Choline |

26. Propranolol primarily inhibits phosphatidic acid phosphatase, which converts phosphatidic acid to 1,2-diacylglycerol. Since diacylglycerol is an intermediate in the synthesis of phosphatidylcholine and phosphatidylethanolamine, propranolol decreases the production of these lipids. Inhibition of phosphatidic acid phosphatase also increases the concentration of phosphatidic acid, which is then converted in greater amounts to phosphatidylinositol and phosphatidylglycerol.

27. Fumonisin B will block the conversion of sphinganine to dihydroceramide. Since dihydroceramide is converted to ceramide and then to cerebroside by the addition of a hexose moiety, cerebroside synthesis will be depressed. It is also possible that dihydrosphinganine, the immediate precursor of dihydroceramide, will accumulate.

28. Tay–Sachs disease results from a defect in the enzyme hexosaminidase A, which breaks down ganglioside $G_{M2}$ to ganglioside $G_{M3}$.

29. In lipid metabolism, the production of prostaglandins is blocked by aspirin, which inhibits the reaction catalyzed by prostaglandin $H_2$ synthase (COX).

30. Cholesterol can be obtained either by synthesis *de novo* from acetyl-CoA or from circulating lipoproteins, primarily LDL, that enter cells by receptor-mediated endocytosis.

31. The order is: HMG-CoA, mevalonate, geranyl pyrophosphate, farnesyl pyrophosphate, squalene, 2,3-oxidosqualene, lanosterol.

32. HMG-CoA reductase catalyzes the first unique step in the synthesis of cholesterol, the conversion of HMG-CoA to mevalonate. This reaction is followed by three reactions that consume ATP. By regulating the activity of HMG-CoA reductase so that it operates only when cholesterol is needed, the cell avoids wasting metabolic energy on the production of cholesterol precursors.

33. (a) and (c) are correct. (b) is incorrect because inhibition of cholesterol biosynthesis tends to increase the synthesis of LDL receptors.

34.
| C | Cholesterol |
|---|---|
| A | Site-1 protease |
| D | Site-2 protease |
| E | WD repeat |
| B | bHLH |

## Chapter 21

1. (a) False, (b) true, (c) false, (d) false, (e) true.

2.  (a)

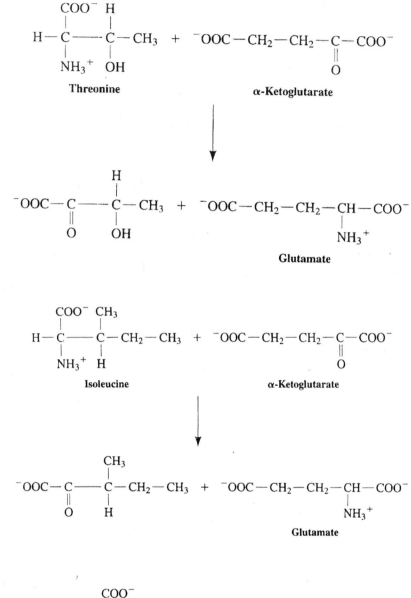

(b)

(c)

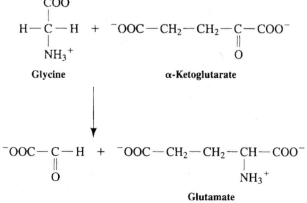

3. Prior to reacting with an amino acid, PLP is covalently bound to an enzyme Lys side chain via an imine linkage (Schiff base). With the release of the α-keto acid, PLP is converted to PMP, which is no longer covalently attached to the enzyme.

4.

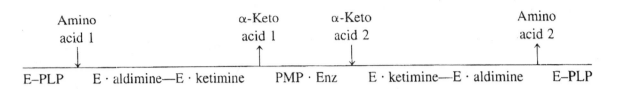

5. Flux in this reaction would favor the production of α-ketoglutarate. Starved individuals break down protein (mostly muscle protein) and oxidize the resultant amino acids for energy. The α-ketoglutarate would be needed as an amino-group acceptor in the transamination of amino acids to be used as fuel.

6. $NH_3 + HCO_3^- + \text{aspartate} + 3\,ATP + H_2O \rightarrow$
$$\text{Urea} + \text{fumarate} + 2\,ADP + AMP + 4\,P_i$$

The total free energy cost is four ATP equivalents, since the $PP_i$ produced in Step 3 is hydrolyzed by inorganic pyrophosphatase.

7. Carbamoyl phosphate synthetase and ornithine transcarbamoylase are mitochondrial enzymes. The reaction catalyzed by carbamoyl phosphate synthetase is the first committed step.

8.

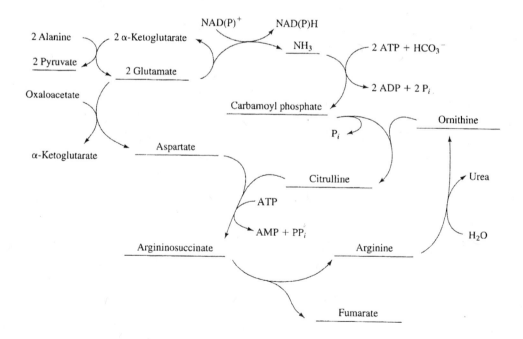

9.  Aspartate and glutamate yield oxaloacetate and α-ketoglutarate, respectively.

10. Bond *a*: Transamination reactions such as occurs in the transfer of glutamic acid's amino group to oxaloacetate to yield α-ketoglutarate and aspartate.
    Bond *b*: The decarboxylation of α-amino acids such as occurs in the synthesis of physiologically active amines.
    Bond *c*: A C—C bond cleavage such as occurs in the serine hydroxymethyltransferase reaction.

11. All citric acid cycle intermediates can form oxaloacetate, which can then be decarboxylated and phosphorylated by PEP carboxykinase. The resulting PEP can then be converted to glucose by gluconeogenesis.

12.
    (a) Reaction 2
    (b) Reaction 3

13. In such cases, the defect is poor binding of thiamine pyrophosphate (TPP) to the E1 subunit of branched-chain α-keto acid dehydrogenase. The administration of large doses of the vitamin precursor of TPP improves the catalytic efficiency of the defective enzyme, which breaks down isoleucine, valine, and leucine.

14. Biotin, *S*-adenosylmethionine, and tetrahydrofolate.

15. Sulfonamides are structural analogs of the *p*-aminobenzoate group of folate and thereby interfere with folate synthesis. Unlike bacteria and plants, most animals cannot synthesize folate and hence are unaffected by these drugs.

16. After SAM donates its methyl group, the resulting *S*-adenosylhomocysteine is hydrolyzed to adenosine and homocysteine. Homocysteine is then methylated by $N^5$-methyltetrahydrofolate to form methionine, which is then adenylated by ATP to regenerate SAM (see Figure 21-18).

17. (a) 12.5 ATP. Aspartate is converted to oxaloacetate by transamination. Oxaloacetate is decarboxylated by PEP carboxykinase and the resulting PEP is converted to pyruvate and then to acetyl-CoA for entry into the citric acid cycle.

| Reaction | Reduced coenzymes produced | ATP equivalents |
|---|---|---|
| PEP carboxykinase | | −1 (GTP) |
| Pyruvate kinase | | 1 |
| Pyruvate dehydrogenase | 1 NADH | 2.5 |
| Citric acid cycle | | 1 (GTP) |
| | 3 NADH | 7.5 |
| | 1 FADH$_2$ | 1.5 |
| Total | | 12.5 |

(b) 22.5 ATP. Glutamine is converted to glutamate and then oxidized to α-ketoglutarate (Figure 21-17), which is then converted to oxaloacetate. Oxidation of oxaloacetate proceeds as outlined in part *a*. (Alternatively, α-ketoglutarate can be converted to malate by the citric acid cycle. Malic enzyme converts malate to pyruvate in a reaction that produces NADPH. The net ATP yield is the same.)

| Reaction | Reduced coenzymes produced | ATP equivalents |
|---|---|---|
| Glutamate dehydrogenase | 1 NADPH | 2.5 |
| α-Ketoglutarate dehydrogenase | 1 NADH | 2.5 |
| Succinyl-CoA synthetase | | 1 (GTP) |
| Succinate dehydrogenase | 1 FADH$_2$ | 1.5 |
| Malate dehydrogenase | 1 NADH | 2.5 |
| PEP carboxykinase | | −1 (GTP) |
| Pyruvate kinase | | 1 |
| Pyruvate dehydrogenase | 1 NADH | 2.5 |
| Citric acid cycle | | 1 (GTP) |
| | 3 NADH | 7.5 |
| | 1 FADH$_2$ | 1.5 |
| Total | | 22.5 |

(c) 28 ATP. Lysine is converted to acetoacetate in 11 reactions (Figure 21-22), several of which yield reduced coenzymes. The acetoacetate is converted to 2 acetyl-CoA at the expense of 1 GTP (since a CoA group is donated by succinyl-CoA).

| Reaction | Reduced coenzymes produced | ATP equivalents |
|---|---|---|
| Lysine degradation | | |
| Reaction 1 | $-1$ NADPH | $-2.5$ |
| Reaction 2 | 1 NADH | 2.5 |
| Reaction 3 | 1 NAD(P)H | 2.5 |
| Reaction 5 | 1 NADH | 2.5 |
| Reaction 6 | 1 FADH$_2$ | 1.5 |
| Reaction 9 | 1 NADH | 2.5 |
| 3-Ketoacyl-CoA transferase | | $-1$ (succinyl-CoA) |
| 2 Citric acid cycle | | 2 (GTP) |
| | 6 NADH | 15 |
| | 2 FADH$_2$ | 3 |
| Total | | 28 |

18. The oxidative deamination of glutamate by glutamate dehydrogenase yields ammonia and α-ketoglutarate. Hence rising levels of α-ketoglutarate signal rising levels of free ammonia. The activation of glutamine synthetase by α-ketoglutarate increases the production of glutamine to help prevent the accumulation of ammonia, which is toxic in high concentrations.

19.

(a) Arginine

(b) Glucose is degraded to 2 pyruvate by glycolysis. One pyruvate is converted to oxaloacetate by the pyruvate carboxylase reaction. The other pyruvate, converted to acetyl-CoA by pyruvate dehydrogenase, combines with the oxaloacetate to form citrate, then isocitrate and then α-ketoglutarate (the first three reactions of the citric acid cycle). Glutamate dehydrogenase, which operates close to equilibrium, converts α-ketoglutarate and NH$_3$ to glutamate. Three additional reactions convert glutamate to ornithine (Figure 21-30). Since ornithine is a urea cycle intermediate, the reactions of the urea cycle convert it to the other cycle intermediates, including arginine.

20. Chorismate is also the precursor of phenylalanine and tyrosine, whose biosynthesis branches from the tryptophan pathway at chorismate.

21. Channeling, the transfer of an intermediate directly from one active site to another, increases the efficiency of the overall reaction by preventing the loss of the intermediate, which could then diffuse away or be degraded. Channeling is important for indole, a nonpolar molecule that might otherwise be lost from the cell.

22. Eight succinyl-CoA units are required to synthesize the porphyrin ring:

   8 Succinyl-CoA + 8 glycine $\rightarrow$ 8 ALA $\rightarrow$ 4 BPG $\rightarrow$ uroporphyrinogen III

23.
   (a) None. The carboxyl group of glycine is lost as carbon dioxide during condensation of glycine with succinyl-CoA.
   (b) Two. Uroporphyrinogen III contains all eight succinyl-derived carboxyl groups. However, subsequent steps remove six of these (Figure 21-37).

24. Heme oxygenase converts heme to the linear biliverdin and releases $Fe^{3+}$ and CO.

25. Oxidized bilirubin is biliverdin, which is reduced by NADPH to regenerate bilirubin.

26. The reduction potentials of NADH (−0.315 V) and $FADH_2$ (−0.219 V) are not sufficiently negative for these coenzymes to reduce $N_2$ ($\mathscr{E}°' = -0.340$ V).

## Chapter 22

1. Note that animals cannot undertake the net conversion of acetyl units to glucose.

| Fate | Pathway | Key Enzyme |
| --- | --- | --- |
| 1. $CO_2$ and $H_2O$ | Citric acid cycle | Citrate synthase |
| 2. Acetoacetate and/or β-hydroxybutyrate | Ketogenesis | Mitochondrial HMG-CoA synthase |
| 3. Fatty acids | Fatty acid synthesis | Acetyl-CoA carboxylase |
| 4. Cholesterol | Cholesterol synthesis | Cytosolic HMG-CoA synthase |
| 5. Amino acids | Pathways utilizing citric acid cycle intermediates | Varies; often catalyzes an exergonic reaction coupled to ATP hydrolysis, e.g., glutamine synthetase |

AQ-103

2. Note that the conversion of lactate to pyruvate is reversible.

| Fate | Pathway | Key Enzyme |
|---|---|---|
| 1. $CO_2$ and $H_2O$ | Pyruvate dehydrogenation/ citric acid cycle | Pyruvate dehydrogenase |
| 2. Oxaloacetate for citric acid cycle | Anaplerotic reactions | Pyruvate carboxylase |
| 3. Glucose | Gluconeogenesis | Phosphofructokinase-2, which alters the levels of F2,6P, a key allosteric regulator of PFK-1 and FBPase-1 |
| 4. Amino acids | Transamination and pathways utilizing citric acid cycle intermediates | Varies; the transamination of pyruvate to alanine is a reversible transamination |

3.

| Fate | Pathway | Key Enzyme |
|---|---|---|
| 1. Glucose | Glucose transport (liver) | Glucose-6-phosphatase |
| 2. Glycogen | Glycogen synthesis | Glycogen synthase |
| 3. Acetyl-CoA | Glycolysis/pyruvate dehydrogenation | PFK-1 and pyruvate dehydrogenase |
| 4. Ribose-5-phosphate and NADPH | Pentose phosphate pathway | G6P dehydrogenase |

4.

A  Glycolysis
A  Lactic fermentation
C  Pyruvate dehydrogenation
C  Citric acid cycle
B  Oxidative phosphorylation of ADP
A  Pentose phosphate pathway
F  Fatty acid elongation and desaturation

A  Palmitoyl-CoA oxidation
A  Palmitate synthesis
E  Lignoceric (24:0) acid synthesis
A,C  Amino acid degradation
A,C  Urea cycle
A,C  Gluconeogenesis
A  Cholesterol synthesis

5.    (a) B; (b) C, F; (c) D; (d) E; (e) A; (f) B; (g) D; (h) A, F

6.    The brain (and the rest of the central nervous system) has a high energy demand, and under normal conditions, its sole source of metabolic fuel is blood-delivered glucose. Hexokinase has a high affinity for glucose, so that glucose that enters a brain cell via facilitated diffusion can be rapidly converted to glucose-6-phosphate, thereby allowing additional glucose to enter the cell.

7.    Both cycles operate under conditions of metabolic stress. In both cycles, metabolites arriving from skeletal muscle are converted by the liver to glucose, which is returned to the bloodstream. The Cori cycle involves the synthesis of glucose from lactate produced by lactic fermentation in muscle during sustained vigorous exercise. The glucose–alanine cycle involves the synthesis of glucose from alanine produced in the muscle. It is thereby a mechanism of transporting nitrogen from muscle to liver. However, under conditions of starvation, the cycle is broken and the muscle-derived glucose is supplied to other tissues.

8.    The degradation of proteins yields amino acids whose further breakdown, in many cases [e.g., glycine, cysteine, serine, and threonine (Figure 21-14); and glutamine (Figure 21-17), yields ammonia. The ammonia is then transferred to α-ketoglutarate via the glutamate dehydrogenase reaction to yield glutamate, and the resulting α-amino group is transferred to pyruvate by a transaminase to yield alanine. The alanine is then transported to the liver where, via transaminases, glutamate dehydrogenase, and urea cycle, its α-amino group is transferred to urea for excretion by the kidneys.

    The elevation of the $NH_4^+$ level in muscle is indicative of increased protein breakdown (e.g., due to starvation). The activation by $NH_4^+$ of muscle phosphofructokinase (PFK) stimulates glycolysis (PFK is the major flux-determining enzyme) and thereby increases the rate of pyruvate production to accommodate the increased need to synthesize alanine.

9.    Increasing levels of blood glucose favor triacylglycerol synthesis as the rate of glucose uptake by adipocytes increases. The three-carbon backbone of triacylglycerols is glycerol-3-phosphate, which is obtained by the reduction of dihydroxyacetone phosphate. A high level of glycolytic activity in adipocytes results in a sufficiently large pool of glycerol-3-phosphate to favor fatty acid esterification.

10.   See Table 22-1 on p. 777.

11.   The pancreas is the sensor organ for blood glucose concentration, while the liver serves as the primary effector organ that regulates blood glucose concentration. For the pancreas, it makes more biological sense to have a transporter that is constitutively expressed, such as the GLUT2 transporter, which is in fact the case.

12.   Recall from Section 15-4A that ATP, ADP, and AMP concentrations are rapidly equilibrated by adenylate kinase (2 ADP ⇆ ATP + AMP). Small (<10%) decreases in [ATP] consequently result in rapid, large increases in [ADP] and at least 4-fold increases in [AMP].

13. The activity of AMPK promotes metabolic pathways that rapidly produce ATP while inhibiting energy-storing biosynthetic pathways.

| Tissue | AMPK activity |
|---|---|
| Cardiac Muscle | Activation of glycolysis |
| Skeletal Muscle | Activation of fatty acid oxidation and glucose uptake |
| Liver | Inhibition of lipogenesis and gluconeogenesis |
| Adipose Tissue | Inhibition of lipolysis |

14. Decreases in leptin act as a long-term regulator of neuropeptide Y production, while ghrelin and $PYY_{3-26}$ appear to act as short-term regulators of neuropeptide Y production.

15. Phosphocreatine, glycogen, fatty acids, protein.

16. The order of usage of these metabolic fuels is glycogen, muscle protein, fatty acids, and nonmuscle protein. Glucose is the primary metabolic fuel of muscle and the central nervous system. After the first day of a fast, the glycogen stores of the liver (and muscle) are exhausted. In order to maintain circulating glucose levels, muscle protein is broken down into amino acids, many of which are glucogenic. Extended fasting alters the metabolic machinery of the liver to favor fatty acid degradation into ketone bodies, thereby conserving critical muscle mass. Severe, extended starvation eventually leads to loss of nonmuscle protein, which can severely compromise organ function and can be life-threatening.

17. During a fast, glucagon stimulates glycogen degradation via PKA, which phosphorylates and activates glycogen phosphorylase (glycogen phosphorylase *a*). The addition of a large amount of glucose shifts the T $\rightleftharpoons$ R equilibrium toward the T state. Stabilization of the T state by glucose binding promotes dephosphorylation of Ser 14, which converts the enzyme to the less active glycogen phosphorylase *b*.

18. A certain level of activity facilitates the transition from metabolism of glucose to metabolism of fatty acids. In a sedentary fasting individual, inactive muscle would provide more metabolic fuel via protein degradation than active muscle; hence, a completely sedentary individual might experience significant loss of muscle mass without a major loss of adipose tissue mass, the primary target for weight loss. Therefore, it is important to be active rather than sedentary while dieting.

## *Chapter 23*

1.    (b) and (c) are correct

2.    Glutamine, $N^{10}$-formyltetrahydrofolate, aspartate, $CO_2$, and glycine.

3.    Aspartate; fumarate; $NAD^+$; glutamine

4.    (c) and (d) are correct

5.    The synthesis of the purine ring is energetically expensive (7–9 ATP equivalents). In contrast, the salvage pathways in which a pre-existing purine base reacts with PRPP to reform a nucleotide requires only the energy of the activated PRPP.

6.    This molecule is cytosine arabinoside, an analog of cytidine in which arabinose replaces ribose.

7.    $HCO_3^-$ and glutamine; aspartic acid; dihydroorotate; orotic acid; PRPP; uridylic acid (UMP), glutamine (second use).

8.    In both pathways in mammals, multiple enzyme activities are located on a single polypeptide chain. This increases the rate of the multistep process and allows intermediates to be channeled from one active site to the next without being lost to the surrounding medium.

9.    Orotic aciduria is caused by a deficiency of the bifunctional enzyme that converts orotic acid to UMP. Uridine can be phosphorylated to UMP, thereby bypassing the metabolic block and acting as precursor for both uridine and cytidine nucleotides. Uridine is preferable to UMP because, being an uncharged molecule, it can readily cross cell membranes.

10.   The pentose phosphate pathway also produces NADPH, which is the ultimate electron donor for the reduction of NDPs to dNDPs via thioredoxin reductase, thioredoxin, and ribonucleotide reductase.

11.   
    D    dATP
    E    1 subunit
    A    2 subunit
    B    dTTP
    F    ATP
    C    thioredoxin

12. 5-Fluorouracil is readily taken up by cancer cells and converted to 5-fluorodeoxyuridylate (FdUMP), which is an inactivator of thymidylate synthase. The rapid proliferation of tumor cells requires a constant supply of dTTP for DNA synthesis. Most slow-growing normal cells are relatively insensitive to 5-fluorouracil treatment. FdUMP inactivates thymidylate synthase only after going partially through the catalytic cycle. Since the enzyme actually generates the species that inactivates it, FdUMP is termed a mechanism-based inhibitor or a suicide substrate.

13. Methotrexate is a dihydrofolate (DHF) analog that binds avidly to dihydrofolate reductase and consequently prevents the reduction of dihydrofolate to tetrahydrofolate (THF). Rapidly growing cancer cells, which require THF for the synthesis of dTTP, are particularly sensitive to methotrexate. A high dose of methotrexate is intended to rapidly kill cancer cells. Subsequent administration of a tetrahydrofolate derivative and thymidine supplies sufficient amounts of folate and a dTTP precursor to enable the more slowly growing normal cells to survive.

14. In the absence of ADA activity, deoxyadenosine accumulates and is phosphorylated. The elevated concentration of dATP inhibits ribonucleotide reductase. Inadequate production of deoxyribonucleotides inhibits DNA synthesis and hence cell proliferation. Lymphocytes exhibit especially active phosphorylation of deoxyadenosine and are therefore particularly vulnerable to ADA deficiency.

15. In the catabolism of pyrimidines, uracil is degraded in several steps to malonyl-CoA (Figure 23-24), which is a substrate for fatty acid synthesis (Section 20-4C). C5 of the pyrimidine ring becomes the methylene carbon of malonyl-CoA and can therefore be incorporated into fatty acyl chains.

16. Uric acid excretion requires little water since it is eliminated as insoluble crystals. Urea is water-soluble and requires large volumes of water to be excreted. Thus uric acid excretion is advantageous over urea excretion in environments where water is scarce.

## Chapter 24

1. The A form of DNA, favored under dehydrating conditions, is the most likely form to occur in desiccated cysts or spores or in such things as brine shrimp eggs, which survive for years in a desiccated form.

2. Hydrogen bonding between base pairs and the limited rotation of the bases with respect to the sugar residues impose limitations on the rotation of the bonds in the ribose–phosphate backbone of double-stranded DNA. However, in single-stranded DNA, the ribose–phosphate bonds have greater freedom to rotate, allowing the polymer to take up a greater range of conformations.

3. The ribose pucker governs the relative orientations of the phosphate groups to each sugar residue. Residues in B-DNA have the C2′-*endo* conformation; in A-DNA, they have the C3′-*endo* conformation; and in Z-DNA, purine nucleotides are C3′-*endo* and pyrimidine nucleotides are C2′-*endo*.

4. The fifteen bands represent the native, negatively supercoiled plasmid (migrating the fastest because it is the most compact) and progressively less supercoiled DNA molecules (with the most relaxed DNA migrating the slowest because it is the least compact). Because Type IA topoisomerase relaxes DNA by making single-strand cuts, the linking numbers of the 15 DNA bands change by increments of one from bottom to top.

5. DNA does not contain a 2′-OH group that can be deprotonated and then serve as a nucleophile to attack the phosphate group at the 3′ position.

6. Chaotropic agents such as urea and formamide tend to disrupt the structure of water. Hence, in their presence, water's ability to solvate DNA's anionic phosphate groups is reduced. Consequently, the phosphate groups repel one another more strongly than they do in the absence of the chaotropic agents, which induces the DNA to take up an extended rodlike conformation.

7.
(a) Monovalent salts increase the $T_m$ because they attenuate the repulsions between the negatively charged phosphate groups.
(b) A large decrease in pH disrupts hydrogen bonding between base pairs by protonating some of the bases and therefore decreases Tm.
(c) A large increase in pH disrupts hydrogen bonding between base pairs by deprotonating some of the base pairs and therefore decreases Tm.
(d) Formamide disrupts water structure and thereby promotes denaturation, leading to a decrease in $T_m$.

8. An increase in ionic strength reduces the repulsions between phosphate groups and hence promotes base stacking. UV absorbance, which increases when the bases melt apart, would therefore decrease.

9. The tertiary structure of tRNA is stabilized by non-Watson–Crick hydrogen bonding and by stacking interactions.

10. Ethanol decreases the $T_m$ of the DNA as well as decreasing its solubility, so it is important to keep the solution cold to keep the duplex intact. Shorter DNA molecules are more vulnerable to melting and hence strand separation, since $T_m$ depends in part on the length of the DNA.

11. Gel A best represents the expected banding pattern. With increasing ethidium bromide, the plasmid unwinds, becoming progressively less supercoiled until it is a relaxed open circle. The relaxed open circle is less compact than the native plasmid; therefore, it migrates more slowly during electrophoresis. Further increases in ethidium bromide induce positive supercoils, so the plasmid again becomes more compact and migrates faster.

12.
    (a) The bases are more exposed in the major groove of B-DNA, which is wider than the minor groove, and are therefore more accessible to binding proteins. In addition, more base-specific hydrogen bonding donors and acceptors are exposed in the major groove.
    (b) In the major groove, the groups available for hydrogen bonding are N7 and N6 of adenine, N7 and O6 of guanine, N4 of cytosine, and O4 of thymine. In the minor groove, the groups available for hydrogen bonding are N3 of adenine, N3 and N2 of guanine, O2 of cytosine, and O2 of thymine.

13. Both kinds of protein exhibit a twofold symmetry that is reflected in the palindromic DNA sequence at their binding sites. The α helices of HTH proteins contact the major groove of DNA directly or indirectly via H$_2$O bridges. In *met* repressor-like proteins, β strands contact the major groove of DNA.

14. Studies using micrococcal nuclease digestion indicated that the nucleosome contains ~200 bp of DNA. Further digestion trims this DNA to ~146 bp, leaving the nucleosome core particle.

15. Histone H1 is probably absent from the preparation shown in Figure 24-41; compare with Figure 24-45*b*. H1 appears to compact the DNA by binding to the ends of the DNA entering and leaving the nucleosome core.

16. The uncondensed DNA would have a length of $(6 \times 10^9 \text{ bp})(0.34 \text{ nm/bp}) = 2.04$ m. The packing ratio is therefore 2.04 m/200 μm = ~10,000.

17. The bacterial nuclease does not cleave the DNA of the nucleosome; however, DNase I appears to be able to cleave nucleosome-bound DNA, cutting it once per helical turn.

## Chapter 25

1. The largely symmetric labeling of each replication fork suggests that both strands of DNA are synthesized simultaneously. However, this implies that one strand is synthesized from 5' to 3' and the other strand is synthesized from 3' to 5'. A polymerase that acts in the 3'→5' direction has never been found. Okazaki demonstrated that one of the new strands is synthesized continuously in the 5'→3' direction and one strand is synthesized discontinuously in the 5'→ 3' direction, after which the DNA fragments are ligated together.

2.  The polymerase reaction is irreversible *in vivo* because the pyrophosphate released on incorporation of a nucleotide into a DNA strand is quickly cleaved to 2 $P_i$ by inorganic pyrophosphatase.

3.  The $3'\rightarrow5'$ exonuclease activity, located near the polymerase active site, recognizes and removes mismatched bases. The $5'\rightarrow 3'$ exonuclease site of Pol I binds to DNA at single-strand breaks and initiates nick translation to replace RNA primers with DNA.

4.

| | |
|---|---|
| G | Required for the initiation of DNA synthesis |
| D | Essential for the condensation of Okazaki fragments |
| E | A helicase required for unwinding the DNA duplex |
| H | Involved in the termination of DNA strand replication |
| C | The enzyme that is strictly involved in DNA replication |
| I | The enzyme critical for relieving buildup of positive supercoils |
| A,B | The enzyme that participates in DNA repair |
| F | Prevents the reannealing of DNA at the replication fork |
| A | Excision and replacement of RNA primers |
| B, C | Incapable of nick translation |
| D | Requires $NAD^+$ in *E. coli* |
| A | The most abundant DNA polymerase |

5.  A trace amount of DNase is necessary to produce single-strand nicks that are the starting point for the $5'\rightarrow3'$ exonuclease and polymerase activities of Pol I.

6.

(a) The single-stranded nucleic acid is likely to be composed entirely of RNA, suggesting that replication of the phage genome begins with the synthesis of a "genomic" RNA.

(b) The appearance of only two bands of phage DNA suggests that the "genomic" RNA serves as a template to generate new double-stranded DNA. In this scenario, $^{15}N$ nucleosides are incorporated into a DNA strand complementary to the RNA and then into a second DNA strand complementary to the first. Hence, both strands of newly synthesized DNA are labeled with $^{15}N$. During infection, when progeny phage are produced, there is much more of the heavy new phage DNA than there is of the original light $^{14}N$-containing DNA.

7.

(a) DNA polymerase δ
(b) DNA polymerase δ
(c) DNA polymerase γ
(c) DNA polymerase α

8.  The host tRNA serves as an RNA primer for the synthesis of the DNA. The 3' end is the most likely section of the molecule to serve as the primer, since it provides the 3'-OH group to which deoxynucleotides can be attached.

9.  The 3′ end of the newly synthesized cDNA strand folds back to prime the synthesis of a complementary DNA strand, presumably on initiation of RNase H activity, which digests away the original mRNA template. This folded DNA then serves as a template for the synthesis of a complementary strand by DNA polymerase.

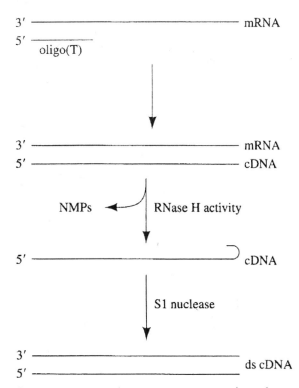

10. The cells that give rise to eggs and sperm must retain telomerase activity in order to produce immortal gametes.

11.

    (a) The mutation is an insertion of an A·T base pair in the third position.
    (b) The mutation is a transversion at the fifth position (T·A to G·C).

12. 5-Bromouracil (5BU) is a base analog that can substitute for thymine and base pair with adenine. However, its enol tautomer, which forms more readily than that of T, can also base pair with guanine, thereby generating a T·A → C·G transition. When 5BU substitutes for cytosine, it generates a C·G → T·A transition.

13.

    | | |
    |---|---|
    | C | Acridine orange |
    | A | Nitrous acid |
    | B | Ethylnitrosourea |
    | B | Dimethyl sulfate |
    | C | Ethidium bromide |

14.

    _B_   Removes 7-methyladenine residues from damaged DNA

    _A_   Serves as a sink for methyl residues abstracted from $O^6$-methylguanine residues

    _C_   Removes deaminated cytosine residues

    _D_   Removes damaged DNA via recombination repair

    _B_   Removes thymine dimers

    _D_   Initiates the SOS response

15. There is no transfer of a DNA segment from a donor to a recipient; instead, the transposon is replicated during insertion so that both the donor and recipient ultimately contain transposon sequences.

16.

(a) The transposon may contain a defective resolvase so that it can integrate into the host DNA but cannot complete site-specific recombination that lets the original plasmid be recovered from the host DNA.

(b) RecA (provided by the host cell) can resolve the cointegrate at a low level, so that the cointegrate is resolved in at least 1 in 100 cases.

## Chapter 26

1. The core enzyme carries out RNA synthesis but it cannot initiate transcription. The σ factor binds specifically to a promoter; this allows the DNA to melt apart so that RNA polymerase can begin transcribing.

2. The polymerase binds nonspecifically to the plasmid and finds its promoter in a one-dimensional search, which proceeds faster than when the enzyme must locate its promoter through a three-dimensional search.

3. *E. coli* topoisomerase mutants show a buildup of positive or negative supercoils during transcription, which would not occur if RNA polymerase simply followed the helical path of the DNA template. In fact, electron micrographs such as Figure 26-9 indicate that the RNA extends away from the DNA as transcription proceeds.

4.

(a) Since radioactivity from $^{32}$pppN is incorporated only into the 5′ nucleotide, the graph measures RNA chain initiations.

(b) Since more initiations occur when more core enzyme is present despite constant σ factor concentration, the σ factor can dissociate from the core enzyme after initiation and bind to a new core enzyme to initiate a new transcript.

(c) Since the curves flatten at both core enzyme concentrations, the core enzyme does not dissociate from the DNA template (another factor must facilitate dissociation of the core enzyme from the template). Consequently, once all the core enzyme binds to DNA and initiates transcription, no new RNA chain initiations are observed.

5.  Termination resulting from the presence of GC- and AT-rich regions does not always occur. In some cases, particularly where these sequences are missing, Rho factor–dependent termination occurs.

6.  Enhancers are eukaryotic transcriptional control elements where specific proteins bind and thereby promote RNA polymerase binding at the promoter. Enhancers are located at variable distances from the gene and in different orientations. Promoters, in contrast, are located near the transcription start site. A promoter is sufficient for transcription initiation and can operate with or without an enhancer. An enhancer cannot operate without a promoter.

7.  Normal RNA chain elongation requires a 3′-OH group. Because the inhibitor lacks this functional group, it should normally be incorporated into the RNA but prevent further polymerization. The absence of $^{32}$P in this experiment suggests that RNA chain elongation may proceed in the 3′→5′ direction. In this case, the RNA chain would grow through nucleophilic attack by the 3′-OH group of the incoming nucleotide, a reaction in which the 3′-deoxy compound cannot participate.

8.
   (a) 5′ Cap: eukaryotes
   (b) Polyadenylation: eukaryotes
   (c) Methylation of nucleotide residues: eukaryotes and prokaryotes
   (d) Endonucleolytic cleavage: eukaryotes and prokaryotes
   (e) Splicing: eukaryotes

9.  All of these elements are contained in the rRNA primary transcript.

10. RNA-processing enzyme complexes that contain RNA include RNase P, snRNPs, the mRNA-editing machinery, and the rRNA-methylating machinery. Only the RNase P RNA participates in catalysis.

11. Nucleolytic removal of nucleotides, splicing of two exons, addition of CCA residues at the 3′ end, and chemical modification of certain nucleotides.

12. No. It must be demonstrated that the intron is regenerated in the course of the reaction. It is possible that the intron is irreversibly cleaved in the course of poly(C) cleavage.

13. The β-labeled compound p$^{32}$ppA should be used because the $^{32}$P will appear only in the first incorporated nucleotide (in subsequent residues, the β and γ phosphates are eliminated). If the α-labeled compound were used, all A residues in the mRNA would be labeled. During cap addition, two phosphates of the methylated GTP and one terminal mRNA phosphate are removed, so only a β label would remain. The assumptions in this experiment are that the initial residue is A rather than G and that the capping reaction is not rate limiting, so that the rate of appearance of β-labeled nucleotide indicates the rate of RNA chain initiation.

$$p^{32}ppApNpNpNpNpN-$$

$$\Big| + p\text{-}p\text{-}p\text{-}G$$

$$Gp^{32}ppApNpNpNpNpN-$$
$$+ PP_i + P_i$$

14. The role of U2AF is to bind to the splice junction of exon 2 in the *tra* pre-mRNA to orchestrate splicing of exon 1 with exon 2. In the absence of SXL (in males), U2AF binds to the proximal 3′ splice site to yield an mRNA that contains an exon with a premature Stop codon, resulting in a truncated, nonfunctional TRA protein. In the presence of SXL (in females), this splice site is blocked by SXL; U2AF binds to a distal 3′ splice site to join the 3′ portion of exon 2, which no longer has the premature Stop codon and thereby produces a functional TRA protein.

15. GTP serves as a substrate since it becomes part of the excised intron and is not regenerated.

16. The primer would be a poly(T) sequence, because this sequence can pair with the poly(A) sequence at the 3′ end of the mRNA. The reverse transcriptase can then extend the poly(T) sequence to generate a cDNA.

## Chapter 27

1. The genetic code, which consists of 64 codons, is degenerate because the codons are used in a redundant manner. Only 21 codons would be strictly required to code for 20 amino acids and a Stop codon.

2. The six possible polynucleotides are
   AUAUAUAUAU⋯
   ACACACACAC⋯
   AGAGAGAGAG⋯
   CGCGCGCGCG⋯
   GUGUGUGUGU⋯
   CUCUCUCUCU⋯
   Each encodes a polypeptide that repeats two amino acids: Poly(AU) yields poly(Ile–Tyr) since it alternates AUA codons and UAU codons; poly(AC) yields poly(Thr–His) since it alternates ACA and CAC codons; poly(AG) yields poly(Arg–Glu) since it alternates AGA codons and GAG codons; poly(CG) yields poly(Arg–Ala) since it alternates CGC codons and GCG codons; poly(GU) yields poly(Val–Cys) since it alternates GUG codons and UGU codons; and poly(CU) yields poly(Leu–Ser) since it alternates CUC codons and UCU codons. The first amino acid in the polymer can be either of the two since the translation reading frame can begin with either codon, so technically, 12 copolymers are possible.

3.   An *Amoeba* tRNA interprets a mouse sense codon as a Stop codon, thereby halting translation of the mouse RNA.

4.   An A to G transition would change the lysine AAA and AAG codons to GAA and GAG, both of which encode glutamate. This would result in an acidic residue replacing a basic residue in the protein. If the change were on the protein surface, it might have no effect on protein structure or function, but if the change involved a buried ion pair, a catalytic residue, or some other essential residue, its effect might be significant.

5.   The mRNA includes an AUG start codon and a UAG stop codon and has the sequence

   5′  GGACCU**AUG**AUCACCUGCUCCCCGAGUGCUGUU**UAG**GUGGG  3′

   The encoded polypeptide has the sequence

   $H_3N^+$–Met–Ile–Thr–Cys–Ser–Pro–Ser–Ala–Val–$COO^-$

6.   See Table 27-1:
   (a)   Changes in the third position of a codon infrequently change the identity of the amino acid. In half of these exceptions, the amino acids are of very similar chemistry (e.g., Asp → Glu when GA(U/C) → GA(A/G)); others result in Stop codons or amino acids of very different chemical character (e.g., Ser → Arg when AG(U/C) → AG(A/G)).
   (b)   Changes in the first codon position usually change the amino acid to another of similar chemistry.
   (c)   Changing the second position of a codon to a pyrimidine generally specifies a hydrophobic amino acid, while changing it to a purine generally specifies a polar amino acid.

7.   The first reaction "activates" an amino acid via condensation with ATP to form an aminoacyl–AMP + $PP_i$. The second reaction condenses this mixed anhydride with its cognate tRNA to form an aminoacyl–tRNA + AMP.

8.   Only the acceptor stem is involved in all tRNA–aaRS interactions.

9.   This yeast uses Ala codons GCA, GCC, and GCU but not GCG, since I can pair with A, C, or U but not G.

10.   Only Phe and His are likely to be linked to one tRNA species. The anticodon of tRNA$^{Phe}$ is GAA, so it can bind both UUU and UUC by wobble pairing (see Table 27-3). Similarly, the anticodon of tRNA$^{His}$ is GUG, so it can bind codons CAC and CAU. tRNA$^{Leu}$ must have at least three anticodons (UAA, IAG, and CAG or UAG) to accommodate the six Leu codons (UUA, UUG, CUU, CUC, CUA, and CUG).

11. Add the mRNA for a known protein to the lysate. After a minute or two, add puromycin to terminate translation prematurely. Analyze the sequences of the resulting peptides to determine whether they correspond to the C- or the N-terminus of the protein. Alternatively, add to the lysate an isotopically labeled amino acid (as an aminoacyl–tRNA) that occurs only near one terminus of the protein and determine whether the label becomes incorporated into the peptides.

12.

| | |
|---|---|
| E | Binds fMet–tRNA and GTP. |
| H | Binds aminoacyl–tRNA and GTP. |
| A, B | Recognizes Stop codons. |
| C | Binds GTP and promotes RF-1 and RF-2 release from the ribosome. |
| C, E, H, I | Hydrolyzes GTP to GDP. |
| I | Promotes the transfer of peptidyl–tRNA from the A site to the P site. |
| F | Inhibits the interaction of the 30S and 50S subunits. |
| E | Facilitates mRNA binding to the 30S ribosome. |
| G | Displaces GDP from EF-Tu. |

13. Puromycin, which resembles the 3′ end of Tyr–tRNA, competes with aminoacyl–tRNAs for binding to ribosomes. Increasing the concentrations of aminoacyl–tRNAs in the extract should at least partially overcome the effects of puromycin and restore protein synthesis.

14. EF-Tu· GDP dissociation is essential for polypeptide elongation. Hence, in the mutant at 37°C, slower dissociation of the EF-Tu·GDP complex slows translation. The dramatic decrease in growth rate may be a function of kinetic proofreading in which an increase in the time of EF-Tu·GDP binding to the ribosome increases the chance of misincorporating an aminoacyl–tRNA.

15. The anticodon of an opal suppressor tRNA must be UCA. The anticodons that can mutate to UCA via a single-base change are UCC, UCG, UCU, UAA, UGA, UUA, ACA, CCA, and GCA, which correspond to amino acids Arg, Cys, Gly, Leu, Ser, and Trp (UUA corresponds to the Stop codon UAA).

16. The probability that any one amino acid is incorporated correctly is 0.98. Hence, the probability that a 100-residue protein is synthesized correctly is $0.98^{100}$, or 13.26%! The bacteria survive because many translation errors do not alter protein function. In addition, protein turnover is sufficiently rapid to overcome the effects of abnormal proteins. In fact, many nonfunctional proteins have shorter-than-normal half-lives.

17. **Initiation:** Eukaryotes lack Shine–Dalgarno binding interactions, and they have many more initiation factors, including an mRNA cap–binding protein.

   **Elongation:** Eukaryotic eEF1 carries out the functions of prokaryotic EF-Tu and EF-Ts, while eEF2 functions similarly to EF-G.

   **Termination:** Eukaryotic polypeptide termination requires only one factor, eRF, which carries out all the functions of the three prokaryotic RFs.

18. Mature collagen self-assembles into large polymeric structures. Such assembly inside the cell would likely destroy the cell. Hence, the immature procollagen, which is incapable of self-assembly, exits the cell before being proteolytically converted to its mature form. In the case of peptide hormones, a mature hormone present in the cell might find its receptor and activate a signal transduction cascade that is inappropriate for that cell. This is prevented by extracellular activation, via posttranslational proteolysis, of the prohormone.

19. Signal sequences recruit signal recognition particles (SRPs), in which the resulting complex binds to an SRP receptor in the ER membrane. Binding of the signal peptide to an SRP halts translation until the SRP–ribosome–nascent polypeptide complex binds to the SRP receptor.

## Chapter 28

1. Some algae (with less than five cell types) have more than 1000 times more DNA than some arthropods (with dozens of different cell types).

2. At least 12.2% of the human genome appears to encode genes involved in signal transduction (1.2% are signaling molecules, 5% receptors, 2.8% kinases, 3.2% regulatory molecules). However, other genes important to signal transduction are not apparent in this pie chart, notably phosphatases.

3. The *Drosophila* genome contains ~100 copies of the histone genes, which can therefore be classified as moderately repetitive sequences.

4. A recombination event that was not in register would result in two daughter cells with differing numbers of trinucleotide repeats, as diagrammed below for a CAG repeat (only one DNA strand is shown). The longer the repeat, the more out-of-register the recombination is likely to be.

```
· · · · · CAGCAGCAGCAGCAGCAGCAGCAG · · · · ·
              | | |
· · · · · CAGCAGCAGCAGCAGCAGCAGCAG · · · · ·
```

meiotic recombination and
cell division

**daughter 1**
· · · · · CAGCAGCAGCAGCAGCAGCAGCAGCAGCAGCAG · · · · ·

**daughter 2**
· · · · · CAGCAGCAGCAG · · · · ·

5. The natural inducer of the *lac* operon is 1,6-allolactose, which is produced from lactose by the action of β-galactosidase. The *lac* operon must be transcribed at a low level in order to generate a small amount of β-galactosidase as well as galactoside permease, which allows lactose to enter the cell.

6.
C Attenuator
D Inducer
B Allosteric inhibitor
E Leader
A Co-repressor

7. Mutations in *I* or *O* can lead to constitutive expression of the *Z*, *Y*, and *A* genes of the *lac* operon. A mutation in *I* (encoding the *lac* repressor) that prevents binding to the operator will allow expression of the *Z*, *Y*, and *A* genes in the absence of inducer. Similarly, mutations in the operator that prevent binding of the *lac* repressor will also lead to constitutive expression of the *Z*, *Y*, and *A* genes.

8. Rarely is the activation or repression of transcription an all-or-none affair. In order for the bacteria to be responsive to lactose, the bacteria must occasionally transcribe the *lac* operon and produce a few molecules of the permease to allow a relatively rapid response to the presence of lactose. The presence of lactose allows for a rapid and high level production of the proteins encoded in the *lac* operon.

9. Riboswitches with bound metabolites have been found to inhibit gene expression via (a) the masking of the Shine–Dalgarno sequence, (b) formation of a premature transcriptional termination structure, and (c) activation of a self-cleaving reaction.

10.
E Promoter
A Silencer
F Operator
B Attenuator
A Enhancer
C Structural gene

11. The transcriptionally active structure is euchromatin. DNase I–sensitive regions only mark potentially active loci. The Barr body, heterochromatin, and highly methylated DNA are transcriptionally inactive.

12.
E Enhanceosome
A Bromodomain
F Mediators
D Chromodomain
C Insulator
B Activation domain

13. After the RISC–siRNA complex has bound an mRNA and degraded it, the complex can interact with another mRNA, and so on, so that a single RISC–siRNA can degrade multiple mRNA molecules.

14. Antibodies combine a heavy chain and a light chain. Since there are ~2000 different κ and λ light chains, the number of different combinations is

    $(2 \times 10^3$ possible κ light chains$)(8.4 \times 10^6$ possible heavy chains$) + (2 \times 10^3$ possible λ light chains$)(8.4 \times 10^6$ possible heavy chains$) = 3.36 \times 10^{10}$ different antibody molecules

    (When one factors in hypermutation and the addition and deletion of nucleotides at the junctions between $V$, $D$, and $J$ segments, this number increases to an estimated $10^{18}$ possible antibodies, although this is far greater than the number of antibodies that an individual produces in a lifetime.)

15. These cells progress directly from M phase to S phase, essentially skipping $G_1$ phase.

16. p53 normally prevents cell division when DNA is damaged. Loss of this function leads to mutations that contribute to uncontrolled cell proliferation.

17. Mutations resulting in nonfunctional ATM do not allow cells to stabilize p53 via phosphorylation initiated by DNA damage, which normally stimulates ATM-mediated phosphorylation of p53. Unphosphorylated p53 binds to Mdm2 and so becomes targeted for proteolytic degradation by proteosomes; hence, the steady-state level of p53 cannot rise sufficiently to stimulate DNA repair.

18. The eukaryotic transcript likely forms a transcription termination structure similar to that in prokaryotic attenuation. The obvious difference is that in eukaryotes, ribosomal binding and translation cannot be involved in transcription termination.

19. Heme-controlled repressor phosphorylates eIF2 so that eIF2·GTP cannot be regenerated. This prevents translation initiation and would inhibit protein synthesis in all cells. The inhibition of protein synthesis does not harm reticulocytes, which synthesize little protein other than hemoglobin.

20.
      D   Gradients of these gene products define the polarity of body segments.
      A   A deletion in one of these genes converts an antenna into a leg.
      C   The spatial expression of these genes is regulated by the distribution of a maternal mRNA.
      B   Mutations in these genes result in deletion of portions of every second body segment.
      A   The products of these genes contain helix–turn–helix motifs.

21. Bicoid and Nanos proteins are translated from mRNAs that are maternally deposited in the anterior and posterior poles, respectively, of the unfertilized egg. They are therefore present at different concentrations along the length of the early embryo (see Fig. 28-49). The gradients of these proteins establish a gradient of expression of Hunchback protein that decreases nonlinearly from anterior to posterior. Specific concentrations of Hunchback protein regulate the level of transcription of the *giant, Krüppel,* and *knirps* genes so that these proteins form bands across the embryonic body.

22. Homeotic genes encode transcription factors, which cannot be morphogens because they do not diffuse between cells or between nuclei in a syncytium.

# SOLUTIONS TO PROBLEMS

## Chapter 1

1. **A** Thiol (sulfhydryl) group

   **B** Carbonyl group

   **C** Amide linkage

   **D** Phosphoanhydride (pyrophosphoryl) linkage

   **E** Phosphoryl group ($P_i$)

   **F** Hydroxyl group

2. The cell membrane must be semipermeable so that the cell can retain essential compounds while allowing nutrients to enter and wastes to exit.

3. Concentration = (number of moles)/(volume)

   Volume = $(4/3)\pi r^3 = (4/3)\pi(5 \times 10^{-7}\text{ m})^3$

   $\qquad = 5.24 \times 10^{-19}\text{ m}^3 = 5.24 \times 10^{-16}\text{ L}$

   Moles of protein = (2 molecules)/($6.022 \times 10^{23}$ molecules $\cdot$ mol$^{-1}$)

   $\qquad = 3.32 \times 10^{-24}$ mol

   Concentration = ($3.32 \times 10^{-24}$ mol)/($5.24 \times 10^{-16}$ L)

   $\qquad = 6.3 \times 10^{-9}$ M = 6.3 nM

4. Number of molecules = (molar conc.)(volume)

   $\qquad\qquad (6.022 \times 10^{23}$ molecules $\cdot$ mol$^{-1}$)

   $= (1.0 \times 10^{-3}$ mol $\cdot$ L$^{-1}$)($5.24 \times 10^{-16}$ L)

   $\qquad\qquad (6.022 \times 10^{23}$ molecules $\cdot$ mol$^{-1}$)

   $= 3.2 \times 10^5$ molecules

5. (a) Liquid water; (b) ice has less entropy at the lower temperature.

6. (a) Decreases; (b) increases; (c) increases; (d) no change.

7. (a) $T = 273 + 10 = 283$ K

   $\Delta G = \Delta H - T\Delta S$

   $\Delta G = 15$ kJ $- (283$ K$)(0.05$ kJ $\cdot$ K$^{-1}$)

   $\quad = 15 - 14.15$ kJ $= 0.85$ kJ

   $\Delta G$ is greater than zero, so the reaction is not spontaneous.

   (b) $T = 273 + 80 = 353$ K

   $\Delta G = \Delta H - T\Delta S$

   $\Delta G = 15$ kJ $- (353$ K$)(0.05$ kJ $\cdot$ K$^{-1}$)

   $\quad = 15 - 17.65$ kJ $= -2.65$ kJ

   $\Delta G$ is less than zero, so the reaction is spontaneous.

8. $\Delta G = \Delta H - T\Delta S$

   $\Delta G = -7000$ J $\cdot$ mol$^{-1}$ $- (298$ K$)(-25$ J $\cdot$ K$^{-1}$ $\cdot$ mol$^{-1}$)

   $\Delta G = -7000 + 7450$ J $\cdot$ mol$^{-1}$ $= 450$ J $\cdot$ mol$^{-1}$

   The reaction is not spontaneous because $\Delta G > 0$. The temperature must be decreased in order to decrease the value of the $T\Delta S$ term.

9. In order for $\Delta G$ to have a negative value (a spontaneous reaction), $T\Delta S$ must be greater than $\Delta H$.

   $T\Delta S > \Delta H$

   $T > \Delta H/\Delta S$

   $T > 7000$ J $\cdot$ mol$^{-1}$/20 J $\cdot$ K$^{-1}$ $\cdot$ mol$^{-1}$

   $T > 350$ K or 77°C

10. (a) False. A spontaneous reaction only occurs in one direction.
    (b) False. Thermodynamics does not specify the rate of a reaction.
    (c) True. (d) True. A reaction is spontaneous so long as $\Delta S > \Delta H/T$.

11. $\Delta G^{\circ\prime} = -RT\ln\dfrac{[C]}{[A][B]}$

    $= -(8.314$ J $\cdot$ K$^{-1}$ $\cdot$ mol$^{-1}$)$(298$ K$)\ln\dfrac{(9)}{(2)(3)}$

    $= -1000$ J $\cdot$ mol$^{-1}$ $= -1$ kJ $\cdot$ mol$^{-1}$

12. $\Delta G^{\circ\prime} = -RT\ln K_{eq} = -RT\ln([C][D]/[A][B])$

    $= -(8.314$ J $\cdot$ K$^{-1}$ $\cdot$ mol$^{-1}$)$(298$ K$)\ln[(3)(5)/(10)(15)]$

    $= 5700$ J $\cdot$ mol$^{-1}$ $= 5.7$ kJ $\cdot$ mol$^{-1}$

    Since $\Delta G^{\circ\prime}$ is positive, the reaction is endergonic under standard conditions.

13. $K_{eq} = e^{-\Delta G^{\circ\prime}/RT} = e^{-(-20,900\text{ J} \cdot \text{mol}^{-1})/(8.314\text{ J} \cdot \text{K}^{-1} \cdot \text{mol}^{-1})(298\text{ K})}$

    $= 4.6 \times 10^3$

14. From Eq. 1-17, $K_{eq} = [G6P]/[G1P] = e^{-\Delta G^{\circ\prime}/RT}$

    $[G6P]/[G1P] = e^{-(-7100\text{ J} \cdot \text{mol}^{-1})/(8.314\text{ J} \cdot \text{K}^{-1} \cdot \text{mol}^{-1})(298\text{ K})}$

    $[G6P]/[G1P] = 17.6$

    $[G1P]/[G6P] = 0.057$

15. (a) Because $K_{eq} = \dfrac{[R]}{[Q]} = 25$, at equilibrium, the concentration of R is 25 times greater than the concentration of Q. When equal concentrations of Q and R are mixed, molecules of Q will be converted to molecules of R.

    (b) Let $x$ = amount of Q converted to R, so that [R] will be 50 $\mu$M $+ x$ and [Q] will be 50 $\mu$M $- x$. Since $K_{eq} = \dfrac{[R]}{[Q]} = 25$,

    $50 + x = 25(50 - x)$

    $50 + x = 1250 - 25x$

    $26x = 1200$

    $x = 46.15$

    $[R] = 50\ \mu$M $+ 46.15\ \mu$M $= 96.15\ \mu$M

    $[Q] = 50\ \mu$M $- 46.15\ \mu$M $= 3.85\ \mu$M

16. At 10°C = 283 K ($1/T = 0.00353$), $K_{eq} = 100$ and ln $K = 4.61$. At 30°C = 303 K ($1/T = 0.00330$), $K_{eq} = 10$ and ln $K = 2.30$.

    These two points generate a line on a van't Hoff plot (ln $K_{eq}$ versus $1/T$) with a positive slope that is equal to $(-\Delta H^\circ/R)$. $\Delta H$ must therefore be negative, indicating that enthalpy decreases during the reaction (heat is given off).

17. This strategy will NOT work because Reaction 1 has a negative enthalpy change, releasing heat, and will therefore become more favorable with decreasing temperature, whereas Reaction 2, which has a positive enthalpy change, will become less favorable. Thus decreasing the temperature will favor Reaction 1, not Reaction 2. In order to make Reaction 2 more favorable, the temperature must be raised.

    To calculate the amount that the temperature must be raised, Equation 1-18 may be used as follows:

    $$\ln K_{eq} = \frac{-\Delta H^\circ}{R}\left(\frac{1}{T}\right) + \frac{\Delta S^\circ}{R}$$

    $$\ln\frac{K_1^{T_1}}{K_1^{T_2}} = \frac{-\Delta H_1^\circ}{R}\left(\frac{1}{T_1} - \frac{1}{T_2}\right)$$

    $$\ln\frac{K_2^{T_1}}{K_2^{T_2}} = \frac{-\Delta H_2^\circ}{R}\left(\frac{1}{T_1} - \frac{1}{T_2}\right)$$

On subtraction of the previous two equations, and taking into account that $\dfrac{K_2^{T_1}}{K_1^{T_1}} = 1$, we get

$$\ln\left[\frac{K_1^{T_1} K_2^{T_2}}{K_1^{T_2} K_2^{T_1}}\right] = \ln\frac{K_2^{T_2}}{K_1^{T_2}} = \frac{\Delta H_2^\circ - \Delta H_1^\circ}{R}\left(\frac{1}{T_1} - \frac{1}{T_2}\right)$$

We would like $\dfrac{K_2^{T_2}}{K_1^{T_2}} = 10$, Substituting in all values and solving for $T_2$ we get

$$\ln\frac{K_2^{T_2}}{K_1^{T_2}} = \ln 10 = 2.3 = \frac{28{,}000 + 28{,}000}{8.31}\left(\frac{1}{298} - \frac{1}{T_2}\right)$$

Solving for $T_2$ we get

$$T_2 = \frac{1}{\dfrac{1}{298} - \dfrac{2.3 \times 8.31}{56{,}000}} = 332 \text{ K}$$

Hence to increase $K_2/K_1$ from 1 to 10, the temperature must be raised from 298 K to 332 K.

# Chapter 2

1. (a) Donors: NH1, NH2 at C2, NH9; acceptors: N3, O at C6, N7. (b) Donors: NH$^+$, NH2 at C4; acceptors: O at C2, N3. (c) Donors: NH$_3^+$ group, OH group; acceptors: COO$^-$ group, OH group.

2. A protonated (and therefore positively charged) nitrogen would promote the separation of charge in the adjacent C—H bond so that the C would have a partial negative charge and the H would have a partial positive charge. This would make the H more likely to be donated to a hydrogen bond acceptor group.

3. (a) Water; (b) water.

4. (a) Micelle, with the polar carboxylate group on the surface; (b) in the interior of the micelle.

5. From most soluble (most polar) to least soluble (least polar): c, b, e, a, d.

6. The waxed car is a hydrophobic surface. To minimize its interaction with the hydrophobic molecules (wax), each water drop minimizes its surface area by becoming a sphere (the geometrical shape with the lowest possible ratio of surface to volume). Water does not bead on glass, because the glass presents a hydrophilic surface with which the water molecules interact. This allows the water to spread out.

7. Water molecules move from inside the dialysis bag to the surrounding seawater by osmosis. Ions from the seawater diffuse into the dialysis bag. At equilibrium, the compositions of the solutions inside and outside the dialysis bag are identical. If the membrane were solute-impermeable, essentially all the water would leave the dialysis bag.

8. (a) Water will move out of the cell by osmosis, from an area of high concentration (low solute concentration) to an area of low concentration (high solute concentration). (b) Salt ions would undergo a net movement by diffusion from the surrounding solution (high salt concentration) into the cell (low salt concentration).

9. The high solute concentration of honey tends to draw water out of microorganisms by osmosis, thereby preventing their growth.

10. Option 2 would reduce the NaCl concentration more effectively. For option 1, the final NaCl concentration in the sample would be

$$\frac{\text{initial amount of NaCl}}{\text{total volume}} = \frac{(0.005 \text{ L})(0.5 \text{ M})}{4.005 \text{ L}} = 0.000624 \text{ M} = 0.624 \text{ mM}$$

For option 2, the NaCl concentration after the first step would be

$$\frac{(0.005 \text{ L})(0.5 \text{ M})}{1.005 \text{ L}} = 0.0025 \text{ M}$$

After the second step, the concentration would be

$$\frac{(0.005 \text{ L})(0.0025 \text{ M})}{1.005 \text{ L}} = 0.0000124 \text{ M} = 0.012 \text{ mM}$$

11. The high concentration of bicarbonate in the dialysate means that some bicarbonate will diffuse from the dialysate across the dialysis membrane into the patient's blood, where it will combine with and neutralize excess protons.

12. With a p$K$ value of 9.25, ammonia exists in the blood (pH 7.4) as NH$_4^+$. The ammonium ion is charged, so it will not easily diffuse across a hydrophobic membrane.

13. (a)
```
      COO⁻
       |
       CH
       ||
       HC
       |
      COO⁻
```
(b)
```
      COO⁻
       |
    H—C—H
       |
      NH₃⁺
```

14. (a)
```
         COO⁻
          |
      H—C —H
          |
         NH₃⁺
```
(b)
```
         COO⁻
          |
      H—C—CH₂—COO⁻
          |
         NH₃⁺
```

15. pH 4, NH$_4^+$; pH 8, NH$_4^+$; pH 11, NH$_3$.

16. pH 4, H$_2$PO$_4^-$; pH 8, HPO$_4^{2-}$; pH 11, HPO$_4^{2-}$.

17. The increase in [H$^+$] due to the addition of HCl is (50 mL) (1 mM)/(250 mL) = 0.2 mM = $2 \times 10^{-4}$ M. Because the [H$^+$] of pure water, $10^{-7}$ M, is relatively insignificant, the pH of the solution is equal to $-\log(2 \times 10^{-4})$ or 3.7.

18. (a) (0.010 L)(5 mol · L$^{-1}$ NaOH)/(1 L) =
$$0.05 \text{ M NaOH} \equiv 0.05 \text{ M OH}^-$$
[H$^+$] = $K_w$/[OH$^-$] = $(10^{-14})/(0.05) = 2 \times 10^{-13}$ M
pH = $-\log$[H$^+$] = $-\log(2 \times 10^{-13})$ = 12.7
(b) (0.020 L)(5 mol · L$^{-1}$ HCl)/(1 L) =
$$0.1 \text{ M HCl} \equiv 0.1 \text{ M H}^+$$
Since the contribution of 0.010 L × 100 mM/(1 L) = 1 mM glycine is insignificant in the presence of 0.1 M HCl,
$$\text{pH} = -\log[\text{H}^+] = -\log(0.1) = 1.0$$
(c) pH = p$K$ + log([acetate]/[acetic acid])
[acetate] = (5 g)(1 mol/82 g)/(1 L) = 0.061 M
[acetic acid] = (0.010 L)(2 mol · L$^{-1}$)/(1 L) = 0.02 M
pH = 4.76 + log(0.061/0.02) = 4.76 + 0.48 = 5.24

19. The p$K$ corresponding to the equilibrium between H$_2$PO$_4^-$ (HA) and HPO$_4^{2-}$ (A$^-$) is 6.82 (Table 2-4). The concentration of A$^-$ is (50 mL)(2.0 M)/(200 mL) = 0.5 M, and the concentration of HA is (25 mL)(2.0 M)/(200 mL) = 0.25 M. Substitute these values into the Henderson–Hasselbalch equation (Eq. 2-9):

$$\text{pH} = \text{p}K + \log\frac{[\text{A}^-]}{[\text{HA}]}$$

$$\text{pH} = 6.82 + \log\frac{0.5}{0.25}$$

$$\text{pH} = 6.82 + \log 2$$

$$\text{pH} = 6.82 + 0.30 = 7.12$$

20. Use the Henderson–Hasselbalch equation (Eq. 2-9) and solve for $pK$:

$$pH = pK + \log\frac{[A^-]}{[HA]}$$

$$pK = pH - \log\frac{[A^-]}{[HA]}$$

$$pK = 6.5 - \log\frac{0.2}{0.1}$$

$$pK = 6.5 - 0.3 = 6.2$$

21. Let HA = sodium succinate and $A^-$ = disodium succinate.

$[A^-] + [HA] = 0.05$ M, so $[A^-] = 0.05$ M $- [HA]$.

From Eq. 2-9 and Table 2-4,

$\log([A^-]/[HA]) = pH - pK = 6.0 - 5.64 = 0.36$

$[A^-]/[HA] = $ antilog $0.36 = 2.29$

$(0.05$ M $- [HA])/[HA] = 2.29$

$[HA] = 0.015$ M

$[A^-] = 0.05$ M $- 0.015$ M $= 0.035$ M

grams of sodium succinate $= (0.015$ mol $\cdot$ L$^{-1})(140$ g $\cdot$ mol$^{-1}) \times$ $(1$ L$) = 2.1$ g

grams of disodium succinate $= (0.035$ mol $\cdot$ L$^{-1})(162$ g $\cdot$ mol$^{-1})$ $\times (1$ L$) = 5.7$ g

22. At pH 4, essentially all the phosphoric acid is in the $H_2PO_4^-$ form, and at pH 9, essentially all is in the $HPO_4^{2-}$; form (Fig. 2-18). Therefore, the concentration of $OH^-$ required is equivalent to the concentration of the acid: $(0.100$ mol $\cdot$ L$^{-1}$ phosphoric acid)$(0.1$ L$) = 0.01$ mol NaOH required $= (0.01$ mol$)(1$ L/5 mol NaOH$) = 0.002$ L $= 2$ mL

23. The standard free energy change can be calculated using Eq. 1-16 and the value of $K$ from Table 2-4.

$\Delta G^{\circ\prime} = -RT\ln K$

$= -(8.314$ J $\cdot$ K$^{-1} \cdot$ mol$^{-1})(298$ K$)$ $\ln(3.39 \times 10^{-8})$

$= 42,600$ J $\cdot$ mol$^{-1} = 42.6$ kJ $\cdot$ mol$^{-1}$

24. (a) Succinic acid; (b) ammonia; (c) HEPES.

25. The dissociation of TrisH$^+$ to its basic form and H$^+$ is associated with a large, positive enthalpy change. Consequently, heat is taken up by the reactant on dissociation. When the temperature is lowered, there is less heat available for this process, shifting the equilibrium constant toward the associated form (the effect of temperature on the equilibrium constant of a reaction is given by Eq. 1-18). To avoid this problem, the buffer should be prepared at the same temperature as its planned use.

26. (a) Carboxylic acid groups are stronger acids than ammonium groups and therefore lose their protons at lower pH values. This can be seen in Fig. 2-17, where the carboxylic acid group of $CH_3COOH$ is 50% dissociated to $CH_3COO^- + H^+$ at pH 4.7 while it is not until pH 9.25 that the ammonium ion is 50% dissociated to $NH_3$.

(b) $H_3N^+CH_2COOH \rightleftharpoons H_3N^+CH_2COO^- + H^+ \rightleftharpoons$ $H_2NCH_2COO^- + H^+$

(c) The $pK$ values of glycine's two ionizable groups are sufficiently different so that the Henderson–Hasselbalch equation (Section 2-2B) adequately describes the behavior of the solution of the diacid and the monodissociated species.

$$pH = pK + \log\frac{[A^-]}{[HA]}$$

$$2.65 = pK + \log\frac{0.02}{0.01}$$

$$pK = 2.65 - 0.3$$

$$pK = 2.35$$

(d)

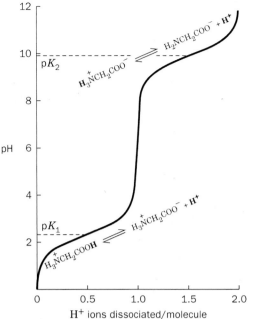

## Chapter 3

1. Guanosine 5'-diphosphate

2. Adenosine-3',5'-cyclic monophosphate (cyclic AMP)

3.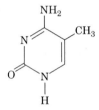

**5-Methylcytosine**

4. The resulting base is uracil.

5. (a) Yes; (b) no.

6. (a) No; (b) yes.

7. Since the haploid genome contains 21% G, it must contain 21% C (because G = C) and 58% A + T (or 29% A and 29% T, because A = T). Each cell is diploid, containing 90,000 kb or $9 \times 10^7$ bases. Therefore,

$A = T = (0.29)(9 \times 10^7) = 2.61 \times 10^7$ bases

$C = G = (0.21)(9 \times 10^7) = 1.89 \times 10^7$ bases

8. The DNA contains 40 bases in all. Since G = C, there are 7 cytosine residues. The remainder $(40 - 14 = 26)$ must be adenine and thymine. Since A = T, there are 13 adenine residues. There are no uracil residues (U is a component of RNA but not DNA).

9.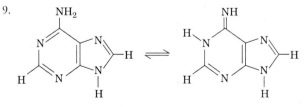

10.

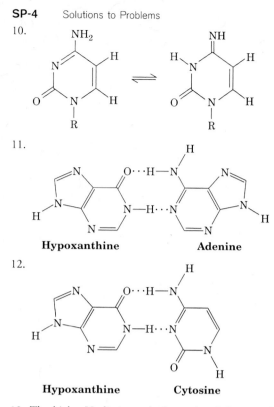

11.

**Hypoxanthine**          **Adenine**

12.

**Hypoxanthine**          **Cytosine**

13. The high pH eliminates hydrogen bonds between bases, making it easier to separate the strands of DNA.

14. At higher NaCl concentrations, there are more $Na^+$ ions to shield the negatively charged DNA backbones, thus reducing electrostatic repulsions and requiring more energy to separate the strands.

15. According to the central dogma, DNA serves as a template for RNA synthesis. The HIV enzyme, called reverse transcriptase, works in reverse by synthesizing DNA from an RNA template.

16. The number of possible sequences of four different nucleotides is $4^n$ where $n$ is the number of nucleotides in the sequence. Therefore, (a) $4^1 = 4$, (b) $4^2 = 16$, (c) $4^3 = 64$, and (d) $4^4 = 256$.

17. 5′–ACGT–3′       +       5′–CGAATC–3′
   3′–TGCAGC–5′              3′–TTAG–5′

18. (a) AluI, EcoRV, HaeIII, PvuII; (b) HpaII and MspI; (c) BamHI and BglII; HpaII and TaqI; SalI and XhoI.

19. (a) Newly synthesized chains would be terminated less frequently, so the bands representing truncated fragments on the sequencing gel would appear faint.

   (b) Chain termination would occur more frequently, so longer fragments would be less abundant.

20. (a) The amount of DNA synthesis would decrease and the resulting gel bands would appear faint.

   (b) No effect.

21. In *O. tauri*, the gene density is 8000 genes/13,000 kb = 0.62, a value that is somewhat lower than that of the prokaryote *E. coli* (4300 genes/4639 kb = 0.93) but less than that of the plant *A. thaliana* (25,500 genes/119,200 kb = 0.21).

22. The desired clones are colorless when grown in the presence of ampicillin and X-gal. Nontransformed bacteria cannot grow in the presence of ampicillin, because they lack the $amp^R$ gene carried by the plasmid. Clones transformed with the plasmid only are blue, since they have an intact *lacZ* gene and produce β-galactosidase, which cleaves the chromogenic substrate X-gal. Clones that contain the plasmid with the foreign DNA insert are colorless because the insert interrupts the *lacZ* gene.

23. The *C. elegans* genome contains 97,000 kb, so
$$f = 5/97,000 = 5.2 \times 10^{-5}$$
Using Eq. 3-2,
$$N = \log(1 - P)/\log(1 - f)$$
$$N = \log(1 - 0.99)/\log(1 - 5.2 \times 10^{-5})$$
$$N = -2/(-2.24 \times 10^{-5}) = 8.91 \times 10^{4}$$

24. Use Equation 3-1 to calculate $P$, given $N = 5000$ and $f = 250$ kb/2,500,000 kb = $10^{-4}$.
$$P = 1 - (1 - f)^N$$
$$= 1 - (1 - 10^{-4})^{5000}$$
$$= 0.39$$
The probability that you have cloned the desired DNA segment is less than 40%.

25. The genomic library contains DNA sequences corresponding to all the organism's DNA, which includes genes and nontranscribed sequences. A cDNA library represents only the DNA sequences that are transcribed into mRNA.

26. Different cell types express different sets of genes. Therefore, the populations of mRNA molecules used to construct the cDNA libraries also differ.

27. (a) Only single DNA strands of variable length extending from the remaining primer would be obtained. The number of these strands would increase linearly with the number of cycles rather than geometrically.

   (b) PCR would yield a mixture of DNA segments whose lengths correspond to the distance between the position of the primer with a single binding site and the various sites where the multispecific primer binds.

28. (a) The first cycle of PCR would yield only the new strand that is complementary to the intact DNA strand, since DNA synthesis cannot proceed when the template is broken. However, since the new strand has the same sequence as the broken strand, PCR can proceed normally from the second cycle on.

   (b) DNA synthesis would terminate at the breaks in the first cycle of PCR.

29. ATAGGCATAGGC and CTGACCAGCGCC.

30. (a) If an individual is homozygous (has two copies of the same allele) at a locus, then only one peak will appear in the electrophoretogram (for example, the D3S1358 locus from Suspect 2).

   (b) Suspect 3, whose alleles exactly match those from the blood stain, is the most likely source of the blood.

   (c) Analysis of each of the three STR loci in this example shows a match between the sample and Suspect 3, and no matches with the other suspects. In practice, however, multiple loci are analyzed in order to minimize the probability of obtaining a match by chance.

   (d) The peak heights are lower for Suspect 1 compared to Suspect 4, suggesting that less DNA was available for PCR amplification from Suspect 1.

## Chapter 4

1. Gly and Ala, Ser and Thr; Val, Leu and Ile; Asn and Gln; Asp and Glu.

2. Arginine, glutamine, and proline.

3. $^+H_3N—CH_2—CH_2—COO^-$

4. (a) Taurine lacks a carboxylate group at the carbon atom where the amino group is attached. (b) Taurine is derived from cysteine. The

sulfhydryl group of the Cys side chain has been oxidized to a sulfonic acid group, and the α-carboxylate group has been removed.

5.
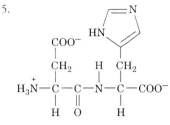

6. The first residue can be one of five residues, the second one of the remaining four, etc.

$$N = 5 \times 4 \times 3 \times 2 \times 1 = 120$$

7. Hydrogen bond donors: α-amino group, amide nitrogen. Hydrogen bond acceptors: α-carboxylate group, amide carbonyl.

8. The negatively charged aspartate —COO⁻ group helps "pull" a hydrogen ion from the hydroxyl group of the serine side chain.

9. The polypeptide would be even less soluble than free Tyr, because most of the amino and carboxylate groups that interact with water and make Tyr at least slightly soluble are lost in forming the peptide bonds in poly(Tyr).

10. (a) The net charge is zero (the N-terminus is positively charged and the C-terminus is negatively charged). (b) The tripeptide has one positive and one negative charge. Hydrolysis increases the total number of charges to 6: one positively charged ammonium group and one negatively charged carboxylate group for each of the three alanines.

11. (a) +1; (b) 0; (c) −1; (d) −2.

12. (a) +2; (b) +1; (c) 0; (d) −1.

13. (a) $pI = (2.35 + 9.87)/2 = 6.11$
    (b) $pI = (6.04 + 9.33)/2 = 7.68$
    (c) $pI = (2.10 + 4.07)/2 = 3.08$

14. The relevant ionizable groups for the neutral species are the His side chain ($pK_R = 6.04$) and the Ser amino group ($pK_2 = 9.21$).

$$pI = (6.04 + 9.21)/2 = 7.62$$

This value is only an estimate because the $pK$ values of ionizable groups in free amino acids are not the same as the $pK$ values of the groups in amino acid residues.

15.

H₃N⁺—CH—C(=O)—NH—CH—C(=O)—NH—CH—C(=O)—NH—CH—C(=O)—NH—CH—COO⁻

(a) The $pK$'s of the ionizable side chains (Table 4-1) are 3.90 (Asp) and 10.54 (Lys); assume that the terminal Lys carboxyl group has a $pK$ of 3.5 and the terminal Ala amino group has a $pK$ of 8.0 (Section 4-1D). The $pI$ is approximately midway between the $pK$'s of the two ionizations involving the neutral species (the $pK$ of Asp and the N-terminal $pK$):

$$pI \approx \tfrac{1}{2}(3.90 + 8.0) \approx 5.95$$

(b) The net charge at pH 7.0 is 0 (as drawn above).

16. At position A8, duck insulin has a Glu residue, whereas human insulin has a Thr residue. Since Glu is negatively charged at physiological pH and Thr is neutral, human insulin has a higher $pI$ than duck insulin. (The other amino acids that differ between the proteins do not affect the $pI$ because they are uncharged.)

17. (a) chiral; (b) nonchiral; (c) chiral; (d) chiral; (e) nonchiral; (f) chiral; (g) chiral.

18.
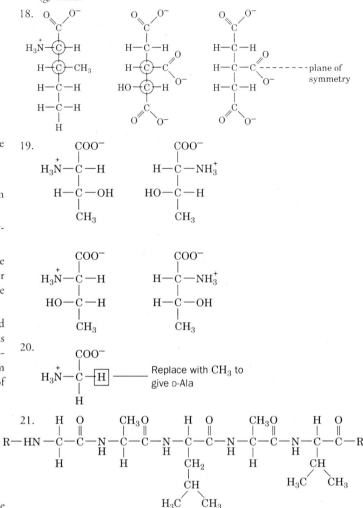

19.

20. Replace with CH₃ to give D-Ala

21.

22. (2S,3S)-Isoleucine

23. (a) Glutamate; (b) aspartate

24. Decarboxylation of L-DOPA yields dopamine (see Fig. 4-15).

25. (a) Serine (N-acetylserine); (b) lysine (5-hydroxylysine); (c) methionine (N-formylmethionine)

26. Phosphorylation of Ser, carboxylation of Glu, and acetylation of Lys would lower the $pI$. Hydroxylation of Pro and methylation of His would not greatly affect the $pI$.

27. Isopeptide bonds can form from the side chain amino group of Lys and from the side chain carboxylate groups of Asp and Glu.

28.

⁺H₃N—CH(COO⁻)—CH₂—CH₂—CH₂—CH₂—NH—C(=O)—CH(NH₃⁺)—CH₃

# Chapter 5

1. There are $8^{10}$ or about 1 billion possibilities.

2. Since there are four Cys residues on the A chain and two on the B chain, there are $4 \times 2 = 8$ possible ways to form a single Cys—Cys linkage between the two chains. For two disulfide bonds, there are four ways of forming the first bond and then three ways of forming the second bond so there are $4 \times 3 = 12$ possible ways to form two disulfide bonds. Thus, there are a total

of 8 + 12 = 20 different ways of linking the A and B chains via disulfide bonds.

3. Peptide B, because it contains more Trp and other aromatic residues.

4. Since $A = \varepsilon cl$, $A = (0.4 \text{ mL} \cdot \text{mg}^{-1} \cdot \text{cm}^{-1})(2.0 \text{ mg/mL})(1 \text{ cm})$ = 0.8

5. Lowering the pH from 7.0 to 5.0 would promote the precipitation of protein Q because the protein will be least soluble when its net charge is zero (when pH = p$I$).

6. (a) The supernatant contains 2 M ammonium sulfate, a salt concentration that is too high for the cells to survive. (b) Placing the solution inside dialysis tubing in a large volume of a low-salt solution would allow the ammonium sulfate to diffuse out (see Fig. 2-14), leaving the desired protein in a low-salt solution that would be compatible with the cells used in the assay.

7. (a) Leu, His, Arg. (b) Lys, Val, Glu.

8. At neutral pH, serum albumin (p$I$ = 4.9) is negatively charged, and ribonuclease A (p$I$ = 9.4) is positively charged. Serum albumin will therefore flow through a cation exchange column and can be recovered, while ribonuclease A will interact with the carboxymethyl groups and can be recovered later by increasing the salt concentration or increasing the pH.

9. The protein behaves like a larger protein during gel filtration, suggesting that it has an elongated shape. The mass determined by SDS-PAGE is more accurate since the mobility of a denatured SDS-coated protein depends only on its size.

10. The protein contains two 60-kD polypeptides and two 40-kD polypeptides. Each 40-kD chain is disulfide bonded to a 60-kD chain. The 100-kD units associate noncovalently to form a protein with a molecular mass of 200 kD.

11. Because protein 1 has a greater proportion of hydrophobic residues (Ala, Ile, Pro, Val) than do proteins 2 and 3, hydrophobic interaction chromatography could be used to isolate it.

12. (a)

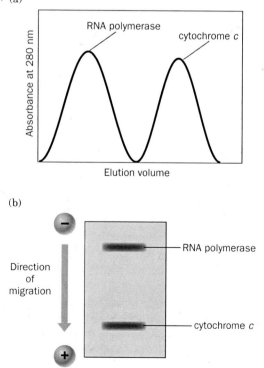

(b)

13. The protein aggregates at the higher salt concentration.

14. The protein appears to be a tetramer of 200-kD subunits. The subunits separate under the denaturing conditions of SDS-PAGE (giving an apparent mass of 200 kD). Both gel filtration and ultracentrifugation are performed under nondenaturing conditions, so the protein behaves as an 800-kD particle.

15. (a)

| Purification step | mg total protein | μmol Mb | Specific activity (μmol Mb/mg total protein) | % yield | Fold purification |
|---|---|---|---|---|---|
| 1. Crude extract | 1550 | 0.75 | $4.8 \times 10^{-4}$ | 100 | 1 |
| 2. DEAE-cellulose chromatography | 550 | 0.35 | $6.4 \times 10^{-4}$ | 47 | 1.3 |
| 3. Affinity chromatography | 5.0 | 0.28 | $5.6 \times 10^{-2}$ | 80 from affinity chromatography (37 overall) | 117-fold overall, (87-fold from affinity chromatography) |

(b) The DEAE chromatography step results in only a 47% yield, while the affinity chromatography step results in an 80% yield from the step before it. The DEAE chromatography step therefore results in the greatest loss of Mb.

(c) The DEAE chromatography step results in a 1.3-fold purification, while the affinity chromatography step results in a 87-fold purification from the previous step. The affinity chromatography step therefore results in the greatest purification of Mb.

(d) The affinity chromatography step is the best choice for a one-step purification of Mb in this example.

16. (a) The extract contains 32 mg $\cdot$ mL$^{-1}$ $\times$ 50 mL = 1600 mg protein and has a specific activity of (0.14 μmol $\cdot$ min$^{-1}$ $\cdot$ 10 μL$^{-1}$) / 1600 mg = $8.75 \times 10^{-6}$ μmol $\cdot$ min$^{-1}$ $\cdot$ μL$^{-1}$ $\cdot$ mg$^{-1}$. The purified fraction contains 50 mg $\cdot$ mL$^{-1}$ $\times$ 10 mL = 500 mg protein and has a specific activity of (0.65 μmol $\cdot$ min$^{-1}$ $\cdot$ 10 μL$^{-1}$) / 500 mg = $1.3 \times 10^{-4}$ μmol $\cdot$ min$^{-1}$ $\cdot$ μL$^{-1}$ $\cdot$ mg$^{-1}$. The fold purification is $(1.3 \times 10^{-4}$ μmol $\cdot$ min$^{-1}$ $\cdot$ μL$^{-1}$ $\cdot$ mg$^{-1})$ / $(8.75 \times 10^{-6}$ μmol $\cdot$ min$^{-1}$ $\cdot$ μL$^{-1}$ $\cdot$ mg$^{-1})$ = 14.9.

(b) The total enzyme activity in the crude extract is (0.14 μmol $\cdot$ min$^{-1}$ $\cdot$ 0.010 mL$^{-1}$) $\times$ 50 mL = 700 μmol $\cdot$ min$^{-1}$, and the total enzyme activity in the purified fraction is (0.65 μmol $\cdot$ min$^{-1}$ $\cdot$ 0.010 mL$^{-1}$) $\times$ 10 mL = 650 μmol $\cdot$ min$^{-1}$, so the % yield is (650 μmol $\cdot$ min$^{-1}$) / (700 μmol $\cdot$ min$^{-1}$) $\times$ 100 = 93%.

17. Dansyl chloride reacts with primary amino groups, including the ε-amino group of Lys residues.

18. (a) Gly; (b) Thr; (c) none (the N-terminal amino group is acetylated and hence unreactive with Edman's reagent).

19. Thermolysin would yield the most fragments (9) and endopeptidase V8 would yield the fewest (2).

20. (a) There is one Met, so CNBr would produce two peptides.

(b) There are four possible sites for chymotrypsin to hydrolyze the peptide: following Phe, Tyr (twice), and Trp. This would yield five peptides unless one of the residues was at the C-terminus, in which case it would yield four peptides.

(c) Four Cys residues form two disulfide bonds.

(d) Arbitrarily choosing one Cys residue, there are three ways it can make a disulfide bond with the remaining three Cys residues. After choosing one of them, there is only one way that the remaining two Cys residues can form a disulfide bond. Thus there are $3 \times 1 = 3$ possible arrangements of the disulfide bonds.

21. (a) The positive charges are caused by the protonation of basic side chains (H, K, and R) and the N-terminal amino groups of the protein.

    (b) There are 1 H, 6 K, 11 R, and 1 $NH_2$ at the N-terminus. Therefore, the maximum number of positive charges that can be obtained is 19.

22. (a) From Sample Calculation 5-1 we see that
$$M = (p_2 - 1)(p_1 - 1)/(p_2 - p_1)$$
Therefore
$$M = (1789.2 - 1)(1590.6 - 1)/(1789.2 - 1590.6)$$
$$= (1788.2)(1589.6)/198.6$$
$$= 14{,}312.8$$

    (b) Peak 5 is $p_1$ in our calculation. From Sample Calculation 5-1,
$$p_1 = (M + z)/z$$
$$1590.6 = (14312.8 + z)/z$$
$$1590.6z - z = 14312.8$$
$$z(1590.6 - 1) = 14312.8$$
$$z = 14312.8/1589.6 = 9.00$$
The charge on the fifth peak in the mass spectrum is 9.00.

23. Gln–Ala–Phe–Val–Lys–Gly–Tyr–Asn–Arg–Leu–Glu

24. Asp–Met–Leu–Phe–Met–Arg–Ala–Tyr–Gly–Asn

25.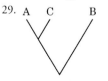

26. Arg–Ile–Pro–Lys–Cys–Arg–Lys–Phe–Gln–Gln–Ala–Gln–His–Leu–Arg–Ala–Cys–Gln–Gln–Trp–Leu–His–Lys–Gln–Ala–Asn–Gln–Ser–Gly–Gly–Gly–Pro–Ser

27. Because the side chain of Gly is only an H atom, it often occurs in a protein at a position where no other residue can fit. Consequently, Gly can take the place of a larger residue more easily than a larger residue, such as Val, can take the place of Gly.

28. (a) Position 6 (Gly) and Position 9 (Val) appear to be invariant.

    (b) Conservative substitutions occur at Position 1 (Asp and Lys, both charged), Position 10 (Ile and Leu, similar in structure and hydrophobicity), and Position 2 (all uncharged bulky side chains). Positions 5 and 8 appear to tolerate some substitution.

    (c) The most variable positions are 3, 4, and 7, where a variety of residues appear.

29.

30. There are several possibilities, since the lengths of the branches matter but not their positions. For example,

## Chapter 6

1.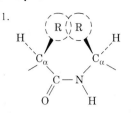

2. There are 10 peptide bonds.

3. (a) $3.6_{13}$; (b) steeper.

4. The large side chains of the amino acids all project from the side of the helix but would still sterically interfere with each other.

5. (100 residues)(1 α-helical turn/3.6 residues)(5.1 Å/keratin turn) = 142 Å

6. (a) The first and fourth side chains of the two helices of a coiled coil form buried hydrophobic interacting surfaces, but the remaining side chains are exposed to the solvent and therefore tend to be polar or charged.

    (b) Although the residues at positions 1 and 4 in both sequences are hydrophobic, Trp and Tyr are much larger than Ile and Val and would therefore not fit as well in the area of contact between the two polypeptides in a coiled coil.

7. The reducing conditions promote cleavage of the disulfide bonds that cross-link α keratin molecules. This helps the larvae digest the wool clothing that they eat.

8. The residues are 5-hydroxylysine (*left*) and methionine (*right*).

9. Collagen's primary structure is its amino acid sequence, which is a repeating triplet of mostly Gly–Pro–Hyp. Its secondary structure is the left-handed helical conformation characteristic of its repeating sequence. Its tertiary structure is essentially the same as its secondary structure, since most of the protein consists of one type of secondary structure. Collagen's quaternary structure is the arrangement of its three chains in a right-handed triple helix.

10. Because collagen has such an unusual amino acid composition (almost two-thirds consists of Gly and Pro or Pro derivatives), it contains relatively fewer of the other amino acids and is therefore not as good a source of amino acids as proteins containing a greater variety of amino acids.

11. A fibrous protein such as α keratin does not have a discrete globular core. Most of the residues in its coiled coil structure are exposed to the solvent. The exception is the strip of nonpolar side chains at the interface of the two coils.

12. Yes, although such irregularity should not be construed as random.

13. Peptide c is most likely to form an α helix with its three charged residues (Lys, Glu, and Arg) aligned on one face of the helix. Peptide a has adjacent basic residues (Arg and Lys), which would destabilize a helix. Peptide b contains Gly and Pro, both of which are helix-breaking (Table 6-1).

14. The presence of Gly and Pro in peptide b would inhibit the formation of β strands, so peptide b is least likely to form a β strand.

15. In a protein crystal, the residues at the end of a polypeptide chain may experience fewer intramolecular contacts and therefore tend to be less ordered (more mobile in the crystal). If their disorder prevents them from generating a coherent diffraction pattern, it may be impossible to map their electron density.

16. (a) Gln; (b) Ser; (c) Ile; (d) Cys. See Table 6-1.

17. (a) $C_4$ and $D_2$; (b) $C_6$ and $D_3$.

18. $D_6$

19. (a) Phe. Ala and Phe are both hydrophobic, but Phe is much larger and might not fit as well in Val's place.

    (b) Asp. Replacing a positively charged Lys residue with an oppositely charged Asp residue is likely to be more disruptive.

    (c) Glu. The amide-containing Asn would be a better substitute for Gln than the acidic Glu.

    (d) His. Pro's constrained geometry is best approximated by Gly, which lacks a side chain, rather than a residue with a bulkier side chain such as His.

20. A polypeptide synthesized in a living cell has a sequence that has been optimized by natural selection so that it folds properly (with hydrophobic residues on the inside and polar residues on the outside). The random sequence of the synthetic peptide cannot direct a coherent folding process, so hydrophobic side chains on different molecules aggregate, causing the polypeptide to precipitate from solution.

21. No.

22. Hydrophobic effects, van der Waals interactions, and hydrogen bonds are destroyed during denaturation. Covalent cross-links are retained.

23. At physiological pH, the positively charged Lys side chains repel each other. Increasing the pH above their $pK$ ($>10.5$) would neutralize the side chains and allow an $\alpha$ helix to form.

24. Intrinsically disordered polypeptide segments would contain relatively more hydrophilic residues because such structures would be extended and exposed to aqueous solution. Hydrophobic groups would tend to aggregate with each other or with other hydrophobic substances in the cell.

25. The molecular mass of $O_2$ is 32 D. Hence the ratio of the masses of hemoglobin and 4 $O_2$, which is equal to the ratio of their volumes, is $65,000/(4 \times 32) = 508$. The 70-kg office worker has a volume of 70 kg $\times$ 1 $cm^3/g$ $\times$ (1000 g/kg) $\times$ (1 m/100 cm)$^3$ = 0.070 $m^3$. Hence the ratio of the volumes of the office and the office worker is $(4 \times 4 \times 3)/0.070 = 686$. These ratios are similar in magnitude, which you may not have expected.

26. (a) Each codon corresponds to an amino acid. The probability that the first residue is correct is $1 - 5 \times 10^{-4} = 0.9995$. Hence the probability that the entire 500-residue polypeptide is correctly synthesized is $(0.9995)^{500} = 0.78$, so that the fraction of polypeptides containing at least one incorrect amino acid is $1 - 0.78 = 0.22$.

    (b) For a 2000-residue polypeptide the fraction that is correctly synthesized is $(0.9995)^{2000} = 0.37$, so that the fraction of polypeptides containing at least one incorrect residue is $1 - 0.37 = 0.63$.

27. At elevated temperatures, more protein denaturation is likely, so it is advantageous for a cell to increase the rate at which it can eliminate these nonfunctional proteins.

28. Oxidation can cause a protein to unfold, so cells increase the production of heat shock proteins to help the damaged proteins refold. Under the same conditions, the level of reduced glutathione (GSH) decreases and the level of oxidized glutathione (GSSG) increases as oxidized proteins are restored to a reduced state.

29. The brains of the Alzheimer's-prone mice were already burdened with accumulated A$\beta$, so the PrP$^{Sc}$ aggregation augmented the brain damage, leading to earlier onset of symptoms than in mice whose brains were not already damaged by A$\beta$ accumulation. (There is also some evidence that A$\beta$ oligomers may interact directly with PrP.)

30. The formation of hydrogen-bonded $\beta$ strands in and between the proteins makes it difficult to solubilize individual protein molecules. Furthermore, this stable but nonnative structure may not revert easily to a native or functional structure.

## Chapter 7

1.

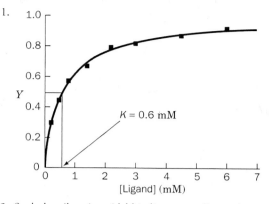

2. Set b describes sigmoidal binding to an oligomeric protein and hence represents cooperative binding.

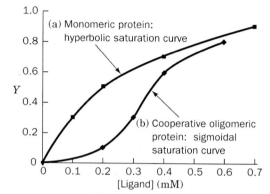

3. According to Eq. 7-6,

$$Y_{O_2} = \frac{pO_2}{K + pO_2}$$

When $pO_2 = 10$ torr,

$$Y_{O_2} = \frac{10}{2.8 + 10} = 0.78$$

When $pO_2 = 1$ torr,

$$Y_{O_2} = \frac{1}{2.8 + 1} = 0.26$$

The difference in $Y_{O_2}$ values is $0.78 - 0.26 = 0.52$. Therefore, in active muscle cells, myoglobin can transport a significant amount of $O_2$ by diffusion from the cell surface to the mitochondria.

4. Using Equation 7-8 and a $p_{50}$ value of 26 torr to calculate a fractional saturation shows that when $pO_2 = 10$ torr,

$$Y_{O_2} = \frac{10}{26 + 10} = 0.28$$

When $pO_2 = 1$ torr,

$$Y_{O_2} = \frac{1}{26 + 1} = 0.04$$

The difference in these quantities, $0.27 - 0.03 = 0.24$, is relatively small [about half of the 0.52 difference exhibited by normal myoglobin ($p_{50} = 2.8$ torr) under the same conditions; see Problem 3] and hence this hypothetical myoglobin would be relatively ineffective in facilitating the diffusion of $O_2$.

5. (a) For hemoglobin, $p_{50} = 26$ torr. Let the Hill coefficient, $n$, equal 3.

$$Y_{O_2} = \frac{(pO_2)^n}{(p_{50})^n + (pO_2)^n}$$

$$= \frac{(20)^3}{(26)^3 + (20)^3} = \frac{8000}{17,576 + 8000} = 0.31$$

(b) $Y_{O_2} = \dfrac{(40)^3}{(26)^3 + (40)^3} = \dfrac{64{,}000}{17{,}576 + 64{,}000} = 0.78$

(c) $Y_{O_2} = \dfrac{(60)^3}{(26)^3 + (60)^3} = \dfrac{216{,}000}{17{,}576 + 216{,}000} = 0.92$

6. Let the Hill coefficient, $n$, equal 3 and use Equation 7-8 to solve for $p_{50}$.

$$Y_{O_2} = \frac{(pO_2)^n}{(p_{50})^n + (pO_2)^n}$$

$$(p_{50})^n + (pO_2)^n = \frac{(pO_2)^n}{Y_{O_2}}$$

$$(p_{50})^n = \frac{(pO_2)^n}{Y_{O_2}} - (pO_2)^n$$

$$(p_{50})^3 = \frac{(25)^3}{0.82} - (25)^3 = \frac{15{,}625}{0.82} - 15{,}625$$

$$(p_{50})^3 = 3429.9$$

$$p_{50} = 15 \text{ torr}$$

7. (a) Vitamin O is useless because the body's capacity to absorb oxygen is not limited by the amount of oxygen available but by the ability of hemoglobin to bind and transport $O_2$. Furthermore, oxygen is normally introduced into the body via the lungs, so it is unlikely that the gastrointestinal tract would have an efficient mechanism for extracting oxygen.

(b) The fact that oxygen delivery in vertebrates requires a dedicated $O_2$-binding protein (hemoglobin) indicates that dissolved oxygen by itself cannot attain the high concentrations required. Moreover, a few drops of vitamin O would make an insignificant contribution to the amount of oxygen already present in a much larger volume of blood.

8. High-altitude adaptation includes the production of additional red blood cells, which would increase the oxygen-delivery capacity of the body at both high and low altitude. The production of red blood cells requires several weeks, so a day or two at high elevation would not provide much advantage for the runner.

9.

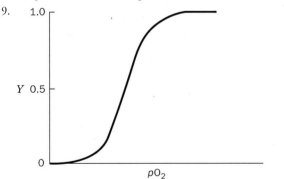

10. The Hill coefficient most likely has a value between 1 (no cooperativity) and 2 (perfect cooperativity between the two subunits).

11. (a) Hyperventilation eliminates $CO_2$, but it does not significantly affect the $O_2$ concentration, since the hemoglobin in arterial blood is already essentially saturated with oxygen.

(b) The removal of $CO_2$ also removes protons, according to the reaction

$$H^+ + HCO_3^- \rightleftharpoons H_2O + CO_2$$

The resulting increase in blood pH would increase the $O_2$ affinity of hemoglobin through the Bohr effect. The net result would be that less oxygen could be delivered to the tissues until the $CO_2$ balance was restored. Thus, hyperventilation has the opposite of the intended effect (note that since hyperventilation suppresses the urge to breathe,

doing so may cause the diver to lose consciousness due to lack of $O_2$ and hence drown).

12. As the crocodile remains under water without breathing, its metabolism generates $CO_2$ and hence the $HCO_3^-$ content of its blood increases. The $HCO_3^-$ preferentially binds to the crocodile's deoxyhemoglobin, which allosterically prompts the hemoglobin to assume the deoxy conformation and thus release its $O_2$. This helps the crocodile stay under water long enough to drown its prey.

13. (a) Lower; (b) higher. The Asp $99\beta \rightarrow$ His mutation of hemoglobin Yakima disrupts a hydrogen bond at the $\alpha_1$–$\beta_2$ interface of the T state (Fig. 7-9a), causing the $T \rightleftharpoons R$ equilibrium to shift toward R state (lower $p_{50}$). The Asn $102\beta \rightarrow$ Thr of hemoglobin Kansas causes the opposite shift in the $T \rightleftharpoons R$ equilibrium by abolishing an R-state hydrogen bond (Fig. 7-9b).

14. (a) Because the mutation destabilizes the T conformation of hemoglobin Rainier, the R (oxy) conformation is more stable. Therefore, the oxygen affinity of hemoglobin Rainier is greater than normal.

(b) The ion pairs that normally form in deoxyhemoglobin absorb protons. The absence of the ion pairs in hemoglobin Rainier decreases the Bohr effect (in fact, the Bohr effect in hemoglobin Rainier is about half that of normal hemoglobin).

(c) Because the R conformation of hemoglobin Rainier is more stable than the T conformation, even when the molecule is not oxygenated, $O_2$-binding cooperativity is reduced. The Hill coefficient of hemoglobin Rainier is therefore less than that for normal hemoglobin.

15. The increased BPG helps the remaining erythrocytes deliver $O_2$ to tissues. However, BPG stabilizes the T conformation of hemoglobin, so it promotes sickling and therefore aggravates the disease.

16. (a) A Lys residue is positively charged and therefore would not bind in the hydrophobic pocket to which the side chain of Val 6 in hemoglobin S binds (as is also the case for the negatively charged Glu 6 in normal hemoglobin). Hence, hemoglobin C does not polymerize as does hemoglobin S. (b) The shorter lifetime of red blood cells containing Hb C would reduce the time that *Plasmodium* can spend inside the cell, thereby helping to limit infection by the parasite.

17. Myosin is both fibrous and globular. Its two heads are globular, with several layers of secondary structure. Its tail, however, consists of a lengthy, fibrous coiled coil.

18. Each cell of striated muscle is a muscle fiber with numerous sarcomeres positioned end-to-end. If cytokinesis occurred more frequently, individual muscle cells would be much shorter. Sarcomeres located in separate small cells would likely not align optimally, and the overall shortening of the muscle would be less.

19. Because many myosin heads bind along a thin filament where it overlaps a thick filament, and because the myosin molecules do not execute their power strokes simultaneously, the thick and thin filaments can move past each other by more than 100 Å in the interval between power strokes of an individual myosin molecule.

20. In the absence of ATP, each myosin head adopts a conformation that does not allow it to release its bound actin molecule. Consequently, thick and thin filaments form a rigid cross-linked array.

21. Microfilaments consist entirely of actin subunits that are assembled in a head-to-tail fashion so that the polarity of the subunits is preserved in the fully assembled fiber. In keratin filaments, however, successive heterodimers align in an antiparallel fashion, so that in a fully assembled intermediate filament, half the molecules are oriented in one direction and half are oriented in the opposite direction (Fig. 6-16).

22. The newly synthesized microfilaments would have a higher proportion of actin subunits that have not yet hydrolyzed their bound ATP.

Older microfilaments would contain relatively more ADP in the nucleotide-binding sites of the actin subunits.

23. (a) 150–200 kD; (b) 150–200 kD; (c) ~23 kD and 53–75 kD

24. The fish IgM molecule has the structure (H₂L₂)₄J.

Total mass = 4[(2 × 70 kD) + (2 × 25 kD)] + 15 kD = 775 kD

25. The loops are on the surface of the domain, so they can tolerate more amino acid substitutions. Amino acid changes in the β sheets would be more likely to destabilize the domain.

26. (a) 12; (b) 60

27. The antigenic site in the native protein usually consists of several peptide segments that are no longer contiguous when the tertiary structure of the protein is disrupted.

28. (a) Fab fragments are monovalent and therefore cannot cross-link antigens to produce a precipitate. (b) A small antigen has only one antigenic site and therefore cannot bind more than one antibody to produce a precipitate. (c) When antibody is in great excess, most antibodies that are bound to antigen bind only one per immunoglobulin molecule. When antigen is in excess, most immunoglobulins bind to two independent antigens.

29. Cleaving the IgA molecules into separate Fab fragments destroys their ability to cross-link antigens. Although the Fab fragments can still bind to the bacteria, the bacteria will not become trapped in a cross-linked network and can still initiate an infection.

30. (a) Normally DNA is intracellular and not accessible to B cells, so no anti-DNA antibodies are produced. For substances such as phospholipids, which are present on cell surfaces, the body's self-tolerance mechanisms prevent antibody production. (b) In an autoimmune disease such as SLE, there is an essentially unlimited supply of antigens, so the resulting antigen–antibody complexes overwhelm the mechanisms that normally clear the complexes from the body.

# Chapter 8

1. (a) 4; (b) 8; (c) 16

2. (a) Yes; (b) no (its symmetric halves are superimposable); (c) no.

3. (a) and (c)

4. b

5.

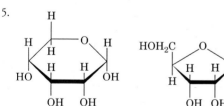

6. Tagatose is derived from galactose.

7.

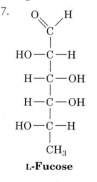

L-**Fucose**

L-Fucose is the 6-deoxy form of L-galactose.

8. Rhamnose is a deoxy sugar (6-deoxy-L-mannose).

9. 19

10. There are 39 in addition to lactose.

11. (a) α-D-glucose-(1 → 1)-α-D-glucose or α-D-glucose-(1 → 1)-β-D-glucose. (b) The numerous hydrogen-bonding —OH groups of the disaccharide act as substitutes for water molecules. Because trehalose is a nonreducing sugar, it is unlikely to participate in oxidation–reduction reactions with other biomolecules when it is present at high concentrations.

12.

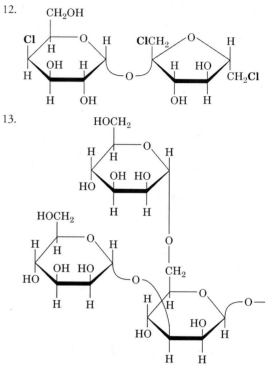

13.

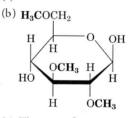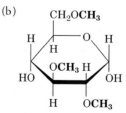

14. Galactose is linked to fructose by a β(1→6) bond.

15. (a) One

(b)

16. (a) There are four types of methylated glucose molecules, corresponding to (1) the residue at the reducing end of the glycogen molecule, (2) residues at the nonreducing ends, (3) residues at the α(1→6) branch points, and (4) residues from the linear α(1→4)-linked segments of glycogen. Type 4 is the most abundant type of residue.

(b)

17. One

18. Amylose (it has only one nonreducing end from which glucose can be mobilized).

19. Glucosamine is a building block of certain glycosaminoglycan components of proteoglycans (Fig. 8-12). Boosting the body's supply of glucosamine might slow the progression of the disease osteoarthritis, which is characterized by the degradation of proteoglycan-rich articular (relating to a joint) cartilage.

20.
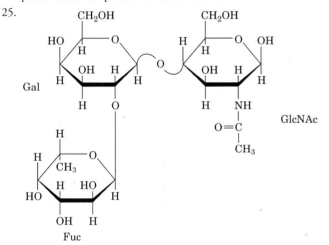

21. $-200$

22. The growth factor is rich in Lys and Arg. These positively charged groups can interact with the negatively charged groups of glycosaminoglycans.

23. The cationic $Ca^{2+}$ ions shield the negative charges of the glycosaminoglycans in the proteoglycans, so that they can be stored in a relatively small volume inside the cell. When the $Ca^{2+}$ ions are pumped out, the repulsion between the glycosaminoglycan chains causes them to expand in the extracellular space.

24. The pores in the peptidoglycan structure allow nutrients to diffuse through the cell wall to the cell surface, where they can be transported across the plasma membrane into the cell.

25.

26. Each population of glycoprotein molecules is a mixture of glycoforms that differ slightly in size and possibly charge due to the heterogeneity in the number and structure of their oligosaccharide groups. In contrast, the nonglycosylated protein molecules are all identical and therefore exhibit no variation in their electrophoretic mobility.

## Chapter 9

1. *trans*-Oleic acid has a higher melting point because, in the solid state, its hydrocarbon chains pack together more tightly than those of *cis*-oleic acid.

2. The triacylglycerol containing the stearic acid residues yields more energy since it is fully reduced.

3.

4. 1-myristoyl-2-palmitoleoyl-3-phosphatidylserine

5. Of the $4 \times 4 = 16$ pairs of fatty acid residues at C1 and C3, only 10 are unique because a molecule with different substituents at C1 and C3 is identical to the molecule with the reverse substitution order. However, C2 may have any of the four substituents for a total of $4 \times 10 = 40$ different triacylglycerols.

6. (a) Palmitic acid and 2-oleoyl-3-phosphatidylserine;

   (b) oleic acid and 1-palmitoyl-3-phosphatidylserine;

   (c) phosphoserine and 1-palmitoyl-2-oleoyl-glycerol;

   (d) serine and 1-palmitoyl-2-oleoyl-phosphatidic acid.

7. All except choline can form hydrogen bonds.

8. No; the two acyl chains of the "head group" are buried in the bilayer interior, leaving a head group of diphosphoglycerol.

9. (a)

   (b) To convert sphingomyelin to sphingosine-1-phosphate, a portion of the head group (typically choline or ethanolamine) must be hydrolytically cleaved, leaving a phosphate group, and the amide linkage to the fatty acyl group must be hydrolyzed.

10. The sulfatide is a galactocerebroside. It differs from other cerebrosides in having a sulfate group covalently linked to C3 of the galactose group.

11. Both DNA and phospholipids have exposed phosphate groups that are recognized by the antibodies.

12. Steroid hormones, which are hydrophobic, can diffuse through the cell membrane to reach their receptors.

13. Eicosanoids synthesized from arachidonic acid are necessary for intercellular communication. Cultured cells do not need such communication and therefore do not require linoleic acid.

14. The branched and ring-containing fatty acids will increase membrane fluidity because they cannot pack next to other lipids as efficiently as straight-chain fatty acids.

15. Triacylglycerols lack polar head groups, so they do not orient themselves in a bilayer with their acyl chains inward and their glycerol moiety toward the surface.

16. The large oligosaccharide head groups of gangliosides would prevent the necessary close packing of the lipids in a bilayer.

17. (a) Saturated; (b) long-chain. By increasing the proportion of saturated and long-chain fatty acids, which have higher melting points, the bacteria can maintain constant membrane fluidity at the higher temperature.

18. The icefish has shorter fatty acyl chains and more unsaturated fatty acyl chains, compared to a tropical fish, in order to maintain membrane fluidity at the low temperatures at which it lives.

19. (a) $(1 \text{ turn}/5.4 \text{ Å})(30 \text{ Å}) = 5.6 \text{ turns}$

    (b) $(3.6 \text{ residues/turn})(5.6 \text{ turns}) = 20 \text{ residues}$

    (c) The additional residues form a helix, which partially satisfies backbone hydrogen bonding requirements, where the lipid head groups do not offer hydrogen bonding partners.

20. No. Although the β strand could span the bilayer, a single strand would be unstable because its backbone could not form the hydrogen bonds it would form with water in aqueous solution.

21. (a) Inner; (b) outer. See Fig. 9-32.

22. (a) Both the intra- and extracellular portions will be labeled. (b) Only the extracellular portion will be labeled. (c) Only the intracellular portion will be labeled.

23. The mutant signal peptidase would cleave many preproteins within their signal peptides, which often contain Leu–Leu sequences. This would not affect translocation into the ER, since signal peptidase acts after the signal peptide enters the ER lumen. Proteins lacking the Leu–Leu sequence would retain their signal peptides. These proteins, and those with abnormally cleaved signal sequences, would be more likely to fold abnormally and therefore function abnormally.

24. In order for a neuron to repeatedly release neurotransmitters, the components of its exocytotic machinery must be recycled. Following the fusion of synaptic vesicles with the plasma membrane, the four-helix SNARE complex is disassembled so that the Q-SNAREs remain in the plasma membrane while portions of the membrane containing R-SNAREs can be used to re-form synaptic vesicles. This recycling process would not be possible if the R- and Q-SNAREs remained associated, and the neuron would eventually be unable to release neurotransmitters.

## Chapter 10

1. $\Delta G = RT \ln \dfrac{[\text{glucose}]_{in}}{[\text{glucose}]_{out}}$

$= (8.3145 \text{ J} \cdot \text{K}^{-1} \cdot \text{mol}^{-1})(298 \text{ K}) \ln \dfrac{(0.003)}{(0.005)}$

$= -1270 \text{ J} \cdot \text{mol}^{-1} = -1.27 \text{ kJ} \cdot \text{mol}^{-1}$

2. (a) $\Delta G = RT \ln([\text{Na}^+]_{in}/[\text{Na}^+]_{out})$

$= (8.314 \text{ J} \cdot \text{K}^{-1} \cdot \text{mol}^{-1})(310 \text{ K})(\ln[10 \text{ mM}/150 \text{ mM}])$

$= (8.314)(310)(-2.71) \text{ J} \cdot \text{mol}^{-1}$

$= -6980 \text{ J} \cdot \text{mol}^{-1} = -7.0 \text{ kJ} \cdot \text{mol}^{-1}$

(b) $\Delta G = RT \ln([\text{Na}^+]_{in}/[\text{Na}^+]_{out}) + Z\mathscr{F}\Delta\Psi$

$= -6980 + (1)(96,485 \text{ J} \cdot \text{V}^{-1}\text{mol}^{-1})(-0.06 \text{ V})$

$= -6980 \text{ J} \cdot \text{mol}^{-1} - 5790 \text{ J} \cdot \text{mol}^{-1}$

$= -12,770 \text{ J} \cdot \text{mol}^{-1} = -12.8 \text{ kJ} \cdot \text{mol}^{-1}$

3. Use Equation 10-3 and let $Z = 2$ and $T = 310$ K:

(a) $\Delta G = RT \ln \dfrac{[\text{Ca}^{2+}]_{in}}{[\text{Ca}^{2+}]_{out}} + Z\mathscr{F}\Delta\Psi$

$= (8.314 \text{ J} \cdot \text{K}^{-1} \cdot \text{mol}^{-1})(310 \text{ K}) \ln(10^{-7})/(10^{-3})$

$+ (2)(96,485 \text{ J} \cdot \text{V}^{-1} \cdot \text{mol}^{-1})(-0.050 \text{ V})$

$= -23,700 \text{ J} \cdot \text{mol}^{-1} - 9600 \text{ J} \cdot \text{mol}^{-1}$

$= -33,300 \text{ J} \cdot \text{mol}^{-1} = -33.3 \text{ kJ} \cdot \text{mol}^{-1}$

The negative value of $\Delta G$ indicates a thermodynamically favorable process.

(b) $\Delta G = RT \ln \dfrac{[\text{Ca}^{2+}]_{in}}{[\text{Ca}^{2+}]_{out}} + Z\mathscr{F}\Delta\Psi$

$= (8.314 \text{ J} \cdot \text{K}^{-1} \cdot \text{mol}^{-1})(310 \text{ K}) \ln(10^{-7})/(10^{-3})$

$+ (2)(96,485 \text{ J} \cdot \text{V}^{-1} \cdot \text{mol}^{-1})( + 0.150 \text{ V})$

$= -23,700 \text{ J} \cdot \text{mol}^{-1} + 28,900 \text{ J} \cdot \text{mol}^{-1}$

$= +5,200 \text{ J} \cdot \text{mol}^{-1} = +5.2 \text{ kJ} \cdot \text{mol}^{-1}$

The positive value of $\Delta G$ indicates a thermodynamically unfavorable process.

4. $\Delta G = RT \ln([\text{K}^+]_{in}/[\text{K}^+]_{out}) + Z\mathscr{F}\Delta\Psi$

$= (8.314 \text{ J} \cdot \text{K}^{-1} \cdot \text{mol}^{-1})(310 \text{ K}) \ln(140\text{mM}/4\text{mM})$

$+ (1)(96,485 \text{ J} \cdot \text{V}^{-1} \cdot \text{mol}^{-1})(-0.070 \text{ V})$

$= 9163 \text{ J} \cdot \text{mol}^{-1} - 6754 \text{ J} \cdot \text{mol}^{-1}$

$= 2409 \text{ J} \cdot \text{mol}^{-1} = 2.4 \text{ kJ} \cdot \text{mol}^{-1}$

5. (a) Nonmediated; (b) mediated; (c) nonmediated; (d) mediated.

6. The less polar a substance, the faster it can diffuse through the lipid bilayer. From slowest to fastest: C, A, B.

7. $\text{K}^+$ transport ceases because the ionophore–$\text{K}^+$ complex cannot diffuse through the membrane when the lipids are immobilized in a gel-like state.

8. The number of ions to be transported is

$(10 \text{ mM})(100 \text{ μm}^3)(N)$

$= (0.010 \text{ mol} \cdot \text{L}^{-1})(10^{-13} \text{ L})(6.02 \times 10^{23} \text{ ions} \cdot \text{mol}^{-1})$

$= 6.02 \times 10^8 \text{ ions}$

Since there are 100 ionophores, each must transport $6.02 \times 10^6$ ions. The time required is $(6.02 \times 10^6 \text{ ions})(1 \text{ s}/10^4 \text{ ions}) = 602 \text{ s} = 10 \text{ min.}$

9. An eight-stranded β barrel has a solid core. A β barrel with 16 or 18 strands has a diameter large enough to accommodate a pore for solute transport.

10. Positively charged Lys and Arg would be relatively abundant at the entrance of the pore, so as to attract the negatively charged phosphate ions and repel cations.

11. The presence of a series of $\text{K}^+$ ions within the channel prevents water molecules from forming a hydrogen-bonded chain.

12. (a) Acetylcholine binding triggers the opening of the channel, an example of a ligand-gated transport protein.

(b) $\text{Na}^+$ ions flow into the muscle cell, where their concentration is low.

(c) The influx of positive charges causes the membrane potential to increase.

13. (a) No; there is no glycerol backbone.

(b) Miltefosine is amphipathic and therefore cannot cross the parasite cell membrane by diffusion. Since it is not a normal cell component, it probably does not have a dedicated active transporter. It most likely enters the cell via a passive transport protein.

(c) This amphipathic molecule most likely accumulates in membranes, with its hydrophobic tail buried in the bilayer and its polar head group exposed to the solvent.

(d) The protein recognizes the phosphocholine head group, which also occurs in some sphingolipids and some glycerophospholipids. Since the protein does not bind all phospholipids or triacylglycerols, it does not recognize the hydrocarbon tail.

14. (a) A transporter similar to a porin would be inadequate since even a large β barrel would be far too small to accommodate the massive ribosome. Likewise, a transport protein with alternating conformations would not be up to the task due to its small size relative to the ribosome. In addition, neither type of protein would be suited for transporting a particle across two membranes. (In fact, ribosomes and other large particles move between the nucleus and cytoplasm via nuclear pores, which are constructed from many different proteins and form a structure that is much larger than the ribosome and spans both nuclear membranes.)

(b) Ribosomal transport might appear to be a thermodynamically favorable process, since the concentration of ribosomes is greater in the nucleus, where they are synthesized. However, free energy would ultimately be required to establish a pore (which would span two membrane thicknesses) for the ribosome to pass through. (In fact, the nucleocytoplasmic transport of all but very small substances requires the activity of GTPases that escort particles through the nuclear pore assembly and help ensure that transport proceeds in one direction.)

15. (a) The data do not indicate the involvement of a transport protein, since the rate of transport does not approach a maximum as [X] increases.

    (b) To verify that a transport protein is involved, increase [X] to demonstrate saturation of the transporter at high [X], or add a structural analog of X to compete with X for binding to the transporter, resulting in a lower flux of X.

16. The hyperbolic curve for glucose transport into pericytes indicates a protein-mediated sodium-dependent process. The transport protein has binding sites for sodium ions. At low [$Na^+$], glucose transport is directly proportional to [$Na^+$]. However, at high [$Na^+$], all $Na^+$ binding sites on the transport protein are occupied, and thus glucose transport reaches a maximum velocity. Glucose transport into endothelial cells is not sodium-dependent and occurs at a high rate whether or not $Na^+$ is present. There is not enough information in the figure to determine whether glucose transport into endothelial cells is protein-mediated.

17. A channel provides an open pore across the membrane, whereas a pump operates by changing its conformation in an ATP-dependent manner. The additional time required for ATP hydrolysis and protein conformation changes causes ion movement through a pump to be slower than through a channel.

18. Overexpression of an MDR transporter would increase the ability of the cancer cell to excrete anticancer drugs. Higher concentrations of the drugs or different drugs would then be required to kill the drug-resistant cells.

19. (a) pH = $-\log[H^+]$ = $-\log(0.15)$ = 0.82

    The pH of the secreted HCl is over 6 pH units lower than the cytosolic pH, which corresponds to a [$H^+$] of $\sim 4 \times 10^{-8}$ M.

    (b) $CO_2 + H_2O \rightleftharpoons HCO_3^- + H^+$

    (c)
    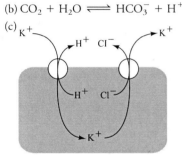

20. (a) Ammonia is small and uncharged and therefore was expected to be able to pass across a membrane by simple diffusion. (b) The ammonia concentration gradient drives ammonia transport (passive-mediated transport). (c) The proton pump is an ATP-requiring active transport system. The combination of $H^+$ and $NH_3$ outside the cell forms $NH_4^+$, which, because it is ionic, is unable to reenter the cell via the ammonia channel.

21. In the absence of ATP, $Na^+$ extrusion by the ($Na^+$–$K^+$)–ATPase would cease, so no glucose could enter the cell by the $Na^+$–glucose symport. The glucose in the cell would then exit via the passive-mediated glucose transporter, and the cellular [glucose] would decrease until it matched the extracellular [glucose] (of course, the cell would probably osmotically burst before this could occur).

22. In order for the protein to function as an antiporter, $H^+$ must move into the cell, down its concentration gradient (which provides the free energy to drive $Na^+$ export). Therefore, the extracellular space has a lower pH (higher $H^+$ concentration).

## Chapter 11

1. b

2. As shown in Table 11-1, the only relationship between the rates of catalyzed and uncatalyzed reactions is that the catalyzed reaction is faster than the uncatalyzed reaction. The absolute rate of an uncatalyzed reaction does not correlate with the degree to which it is accelerated by an enzyme.

3. (a) isomerase (alanine racemase); (b) lyase (pyruvate decarboxylase).

4. (a) oxidoreductase (lactate dehydrogenase); (b) ligase (glutamine synthetase).

5. There are three transition states ($X^\ddagger$) and two intermediates (I). The reaction is not thermodynamically favorable because the free energy of the products is greater than that of the reactants.

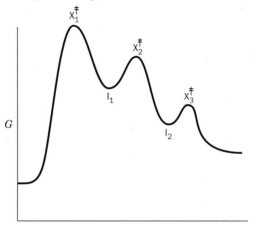

Reaction coordinate

6. The tighter S binds to the enzyme, the greater the value of $\Delta G_E^\ddagger$. As the value of $\Delta G_E^\ddagger$ for reaction c approaches that of $\Delta G_N^\ddagger$, the rate of the enzyme-catalyzed reaction approaches the rate of the nonenzymatic reaction.

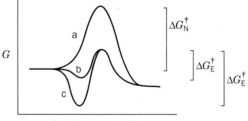

Reaction coordinate

7. At 25°C, every 10-fold increase in rate corresponds to a decrease of about 5.7 kJ · $mol^{-1}$ in $\Delta G^\ddagger$. For the nuclease, with a rate enhancement on the order of $10^{14}$, $\Delta G^\ddagger$ is lowered about $14 \times 5.7$ kJ · $mol^{-1}$, or about 80 kJ · $mol^{-1}$. Alternatively, since the rate enhancement, $k$, is given by $k = e^{\Delta\Delta G_{cat}^\ddagger/RT}$,

    $\ln k = \ln(10^{14}) = \Delta\Delta G_{cat}^\ddagger/8.3145 \times (273 + 25)$

    Hence $\Delta\Delta G_{cat}^\ddagger = 80$ kJ · $mol^{-1}$.

8. Assuming that the free energy of a low-barrier hydrogen bond is $-40$ kJ · $mol^{-1}$, the rate enhancement would be

    Rate enhancement = $e^{\Delta\Delta G_{cat}^\ddagger/RT}$

    $= e^{(40,000 \text{ J} \cdot mol^{-1})/(8.314 \text{ J} \cdot K^{-1} \cdot mol^{-1})(298 \text{ K})}$

    $= e^{16.14}$

    $= 10^7$

9.
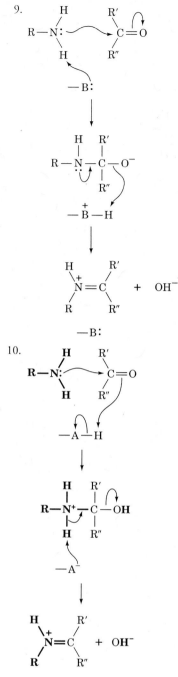

10.

11. As the temperature increases, thermal energy boosts the proportion of reactants that can achieve the transition state per unit time, so the rate increases. Above an optimal temperature, the enzyme becomes denatured and rapidly loses catalytic activity (recall that proteins are typically only marginally stable; Section 6-4).

12. The active form of the enzyme contains the thiolate ion. The increased p$K$ would increase the nucleophilicity of the thiolate and thereby increase the rate of the reaction catalyzed by the active form of the enzyme. However, at physiological pH, there would be less of the active form of the enzyme and therefore the overall rate would be decreased.

13. Glu has a p$K$ of ~4 and, in its ionized form, acts as a base catalyst. Lys has a p$K$ of ~10 and, in its protonated form, acts as an acid catalyst.

14. DNA lacks the 2'-OH group required for the formation of the 2',3'-cyclic reaction intermediate.

15. Inhibition by various metal ions suggests that urease requires a metal ion for catalysis, whose replacement by Hg, Co, or Cd inactivates the enzyme. However, the inhibitory metal ions could also disrupt the enzyme's structure by binding somewhere other than the active site, so this inhibitory effect does not prove that urease acts via metal ion catalysis. (In fact, urease activity requires two catalytic Ni ions.)

16. The ability of RNA molecules to form complex tertiary structures allows them to bind substrates and catalyze reactions through proximity and orientation effects as well as by transition state stabilization, even though RNA lacks a wide variety of functional groups. In addition, the $Mg^{2+}$ ions that RNA normally binds may also have catalytic functions. DNA lacks the 2'-OH functional group that is present in RNA. Moreover, in its double helical form, DNA is conformationally rigid and is therefore unable to form the required complex tertiary structures.

17. The preferential binding of the transition state to an enzyme is an important (often the most important) part of an enzyme's catalytic mechanism. Hence, the substrate binding site is the catalytic site.

18. Two such analogs are

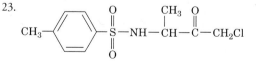

**Furan-2-carboxylate**    **Thiophene-2-carboxylate**

Both of these molecules are planar, particularly at the C atom to which the carboxylate is bonded, as is true of the transition state for the proline racemase reaction.

19. The lysozyme active site is arranged to cleave oligosaccharides between the fourth and fifth residues. Moreover, since the lysozyme active site can bind at least six monosaccharide units, (NAG)$_6$ would be more tightly bound to the enzyme than (NAG)$_4$, and this additional binding free energy would be applied to distorting the D ring to its half-chair conformation, thereby facilitating the reaction.

20. Although the polymeric chains of cellulose, which are β(1→4)-linked glucose residues, have the same overall configuration as the NAG—NAM repeating disaccharide of lysozyme's primary substrate peptidoglycan, cellulose chains typically occur in hydrogen-bonded networks, which would make them unavailable to bind in the surface groove of the enzyme. In addition, the absence of the *N*-acetyl and lactyl groups on the sugar residues would prevent them from fitting snugly into the active site, a prerequisite for efficient catalysis.

21. Asp 101 and Arg 114 form hydrogen bonds with the substrate molecule (Fig. 11-19). Ala cannot form these hydrogen bonds, so the substituted enzyme is less active.

22. The mutations would diminish lysozyme's activity because the removal or addition of a methylene group would move the Glu and Asp carboxyl groups from their catalytically most effective positions.

23.

**Tosyl-L-alanine chloromethylketone**    or

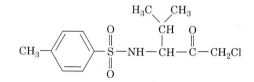

**Tosyl-L-valine chloromethylketone**

24. (a) Little or no effect; (b) catalysis would be much slower because the mutation disrupts the function of the catalytic triad.

25. 

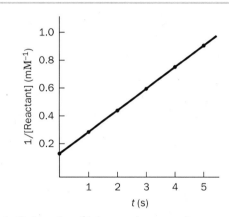

26. The observation that subtilisin and chymotrypsin are genetically unrelated indicates that their active site geometries arose by convergent evolution. Assuming that evolution has optimized the catalytic efficiencies of these enzymes and that there is only one optimal arrangement of catalytic groups, any similarities between the active sites of subtilisin and chymotrypsin must be of catalytic significance. Conversely, any differences are unlikely to be catalytically important.

27. Yes. An enzyme decreases the activation energy barrier for both the forward and the reverse directions of a reaction.

28. As a digestive enzyme, chymotrypsin's function is to indiscriminately degrade a wide variety of ingested proteins, so that their component amino acids can be recovered. Broad substrate specificity would be dangerous for a protease that functions outside of the digestive system, since it might degrade proteins other than its intended target.

29. If the soybean trypsin inhibitor were not removed from tofu, it would inhibit the trypsin in the intestine. At best, this would reduce the nutritional value of the meal by rendering its protein indigestible. It might very well also lead to intestinal upset.

30. The lysosomal enzymes would be expected to have pH optima of around 5, corresponding to their environmental pH. Outside the lysosome, where the pH is closer to neutral, the enzymes would be much less catalytically active and therefore unlikely to carry out hydrolytic reactions in other parts of the cell.

31. Activated factor IXa leads, via several steps, to the activation of the final coagulation protease, thrombin. The absence of factor IX therefore slows the production of thrombin, delaying clot formation, and causing the bleeding of hemophilia. Although activated factor XIa also leads to thrombin production, factor XI plays no role until it is activated by thrombin itself. By this point, coagulation is already well underway, so a deficiency of factor XI does not significantly delay coagulation.

## Chapter 12

1. (a) $v = k[A]$

   $k = v/[A]$

   $k = (5 \, \mu M \cdot min^{-1})/(20 \, mM)$

   $\quad = (0.005 \, mM \cdot min^{-1})/(20 \, mM)$

   $\quad = 2.5 \times 10^{-4} \, min^{-1}$

   (b) The reaction has a molecularity of 1.

2. $v = k[A]^2$

   $v = (10^{-6} \, M^{-1} \cdot s^{-1})(0.010 \, M)(0.010 \, M)$

   $v = 10^{-10} \, M \cdot s^{-1}$

3. From Eq. 12-7, $[A] = [A]_o \, e^{-kt}$. Since $t_{1/2} = 0.693/k$, $k = 0.693/14 \, d$ $= 0.05 \, d^{-1}$. (a) 7 mmol; (b) 5 μmol; (c) 3.5 μmol; (d) 0.3 μmol.

4. For a second-order reaction,

   $t_{1/2} = 1/k[A]_o$

   $$= \frac{1}{(3.6 \times 10^{-3} M^{-1} \cdot s^{-1})(6 \times 10^{-6} M)}$$

   $= 4.63 \times 10^7 \, s$

   $= 4.63 \times 10^7 \, s \, (1 \, h/3600 \, s)(1 \, d/24 \, h)(1 \, yr/365 \, d)$

   $= 1.47 \, yr$

5. Only a plot of 1/[reactant] versus $t$ gives a straight line, so the reaction is second order. The slope, $k$, is $0.15 \, mM^{-1} \cdot s^{-1}$.

| Time (s) | 1/[reactant] (mM$^{-1}$) |
|---|---|
| 0 | 0.16 |
| 1 | 0.32 |
| 2 | 0.48 |
| 3 | 0.62 |
| 4 | 0.78 |
| 5 | 0.91 |

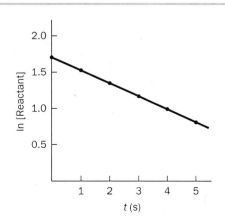

6. Only a plot of ln[reactant] versus $t$ gives a straight line, so the reaction is first order. The negative of the slope, $k$, is $0.17 \, s^{-1}$.

| Time (s) | ln[Reactant] |
|---|---|
| 0 | 1.69 |
| 1 | 1.53 |
| 2 | 1.36 |
| 3 | 1.16 |
| 4 | 0.99 |
| 5 | 0.83 |

7.

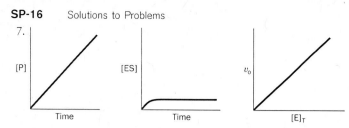

8. Enzyme activity is measured as an initial reaction velocity, the velocity before much substrate has been depleted and before much product has been generated. It is easier to measure the appearance of a small amount of product from a baseline of zero product than to measure the disappearance of a small amount of substrate against a background of a high concentration of substrate.

9. $v_o = V_{max}[S]/(K_M + [S])$
$v_o/V_{max} = [S]/(K_M + [S])$
$0.95 = [S]/(K_M + [S])$
$[S] = 0.95K_M + 0.95[S]$
$0.05[S] = 0.95 K_M$
$[S] = (0.95/0.05)K_M = 19K_M$

10. Acetylcholinesterase, carbonic anhydrase, catalase, and fumarase.

11. Construct a Lineweaver–Burk plot.

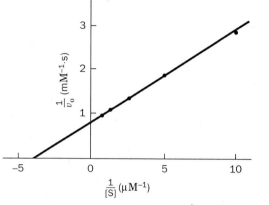

$K_M = -1/x\text{-intercept} = -1/(-4 \ \mu M^{-1}) = 0.25 \ \mu M$
$V_{max} = 1/y\text{-intercept} = 1/(0.8 \ mM^{-1} \cdot s) = 1.25 \ mM \cdot s^{-1}$

12. Set A corresponds to $[S] > K_M$, and set B corresponds to $[S] < K_M$. Ideally, a single data set should include $[S]$ values that are both larger and smaller than $K_M$.

| Set A | | Set B | |
|---|---|---|---|
| **[S] (mM)** | **$v_o$ ($\mu M \cdot s^{-1}$)** | **[S] (mM)** | **$v_o$ ($\mu M \cdot s^{-1}$)** |
| 2 | 0.42 | 0.12 | 0.17 |
| 1 | 0.38 | 0.10 | 0.15 |
| 0.67 | 0.34 | 0.08 | 0.13 |
| 0.50 | 0.32 | 0.07 | 0.11 |

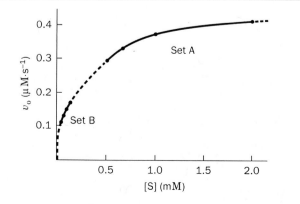

13. Velocity measurements can be made using any convenient unit of change per unit of time. $K_M$ is, by definition, a substrate concentration (the concentration when $v_o = V_{max}/2$), so its value does not reflect how the velocity is measured.

14. (a, b) It is not necessary to know $[E]_T$. The only variables required to determine $K_M$ and $V_{max}$ (for example, by constructing a Lineweaver–Burk plot) are $[S]$ and $v_o$. (c) The value of $[E]_T$ is required to calculate $k_{cat}$ since $k_{cat} = V_{max}/[E]_T$.

15. Comparing the two data points, since a 100-fold increase in substrate concentration only produces a 10-fold increase in reaction velocity, it appears that when $[S] = 100$ mM, the velocity is close to $V_{max}$. Therefore, assume that $V_{max} \approx 50 \ \mu M \cdot s^{-1}$ and use the other data point to estimate $K_M$ using the Michaelis–Menten equation:

$$v_o = \frac{V_{max}[S]}{K_M + [S]}$$

$$K_M + [S] = \frac{V_{max}[S]}{v_o}$$

$$K_M = \frac{V_{max}[S]}{v_o} - [S]$$

$$K_M = \frac{(50 \ \mu M \cdot s^{-1})(1 \ \mu M)}{(5 \ \mu M \cdot s^{-1})} - (1 \ \mu M) = 9 \ \mu M$$

The true $V_{max}$ must be greater than the estimated value, so the value of $K_M$ is an underestimate of the true $K_M$.

16. The experimentally determined $K_M$ would be greater than the true $K_M$ because the actual substrate concentration is less than expected.

17. The enzyme concentration is comparable to the lowest substrate concentration and therefore does not meet the requirement that $[E] \ll [S]$. You could fix this problem by decreasing the amount of enzyme used for each measurement.

18. Enzyme Y is more efficient at low $[S]$; enzyme X is more efficient at high $[S]$.

19. (a) N-Acetyltyrosine ethyl ester, with the lower value of $K_M$, has greater apparent affinity for chymotrypsin. (b) The value of $V_{max}$ is not related to the value of $K_M$, so no conclusion can be drawn.

20. Product P will be more abundant because enzyme A has a much lower $K_M$ for the substrate than enzyme B. Because $V_{max}$ is approximately the same for the two enzymes, the relative efficiency of the enzymes depends almost entirely on their $K_M$ values.

21. A* will appear if the reaction follows a Ping Pong mechanism, since a double-displacement reaction can exchange an isotope from P back to A in the absence of B.

22. In a reaction that has a sequential mechanism, A will not become isotopically labeled, because P cannot be converted back to A in the absence of Q.

23. By irreversibly reacting with chymotrypsin's active site, DIPF would decrease $[E]_T$. The apparent $V_{max}$ would decrease since $V_{max} = k_{cat}[E]_T$. $K_M$ would not be affected since the uninhibited enzyme would bind substrate normally.

24. Molecule B is more likely to be a competitive inhibitor because it more closely resembles the enzyme's substrate (molecule A).

25. If an irreversible inhibitor is present, the enzyme solution's activity would be exactly 100 times lower when the sample is diluted 100-fold. Dilution would not significantly change the enzyme's degree of inhibition.

26. For reversible inhibition, $K_I = [E][I]/[EI]$ so that $[E]/[EI] = K_I/[I]$. Hence, if a reversible inhibitor is present, dilution would lower the

concentrations of both the enzyme and inhibitor so that the degree of dissociation of the inhibitor from the enzyme would increase. The enzyme solution's activity would therefore not be exactly 100 times less than the undiluted sample, but would be somewhat greater than this value because the proportion of uninhibited enzyme would be greater at the lower concentration.

27. The lines of the double-reciprocal plots intersect to the left of the $1/v_o$ axis (on the $1/[S]$ axis). Hence, inhibition is mixed (with $\alpha = \alpha'$).

| [S] | 1/[S] | $1/v_o$ | $1/v_o$ with I |
|---|---|---|---|
| 1 | 1.00 | 0.7692 | 1.2500 |
| 2 | 0.50 | 0.5000 | 0.8333 |
| 4 | 0.25 | 0.3571 | 0.5882 |
| 8 | 0.125 | 0.2778 | 0.4545 |
| 12 | 0.083 | 0.2500 | 0.4167 |

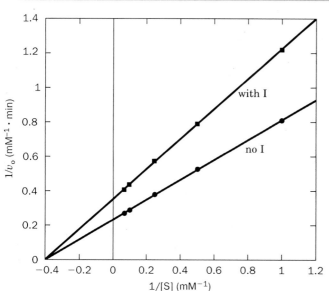

28. From Eq. 12-32, $\alpha$ is 3.

$$\alpha = 3 = 1 + [I]/K_I = 1 + 5 \text{ mM}/K_I$$
$$K_I = 2.5 \text{ mM}$$

29. (a) Inhibition is most likely mixed (noncompetitive) with $\alpha = \alpha'$ since it is reversible and only $V_{max}$ is affected.

(b) Since $V_{max}^{app} = 0.8 V_{max}$, 80% of the enzyme remains uninhibited. Therefore, 20% of the enzyme molecules have bound inhibitor.

(c) As indicated in Table 12-2 for mixed inhibition, $V_{max}^{app} = V_{max}/\alpha'$. Thus,

$$\alpha' = \frac{V_{max}}{V_{max}^{app}} = \frac{1}{0.8} = 1.25$$

From Eq. 12-32,

$$1.25 = 1 + \frac{[I]}{K_I'}$$
$$K_I' = \frac{5 \text{ nM}}{1.25 - 1} = 20 \text{ nM}$$

30.

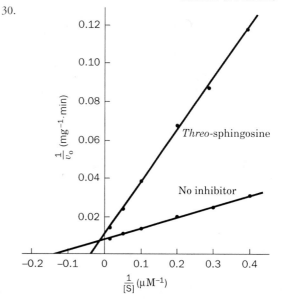

(a) $K_M$ is determined from the $x$-intercept $(= -1/K_M)$. In the absence of inhibitor, $K_M = 1/0.14 \text{ } \mu M^{-1} = 7 \text{ } \mu M$. In the presence of inhibitor, $K_M^{app} = 1/0.04 \text{ } \mu M^{-1} = 25 \text{ } \mu M$. $V_{max}$ is determined from the $y$-intercept $(= 1/V_{max})$. In the absence of inhibitor, $V_{max} = 1/0.008 \text{ mg}^{-1} \cdot \text{min} = 125 \text{ mg} \cdot \text{min}^{-1}$. In the presence of inhibitor, $V_{max}^{app} = 1/0.01 \text{ mg}^{-1} \cdot \text{min} = 100 \text{ mg} \cdot \text{min}^{-1}$.

(b) The lines in the double-reciprocal plots intersect very close to the $1/v_o$ axis. Hence, *threo*-sphingosine is most likely a competitive inhibitor. Competitive inhibition is likely also because of the structural similarity between the inhibitor and the substrate, which allows them to compete for binding to the enzyme active site.

## Chapter 13

1. Somatostatin is a 14-residue peptide hormone and is not lipid-soluble, so it would require a cell-surface receptor.

2. No. Because retinoic acid is a lipid, it can diffuse through the cell membrane to associate with an intracellular receptor.

3. (a) Norepinephrine is synthesized by the decarboxylation of Tyr and by hydroxylation of its $\beta$ carbon and its phenyl group. (b) Epinephrine is derived from norepinephrine by *N*-methylation.

4. (a) Methandrostenolone differs from testosterone by methylation of C17 and desaturation of the C1—C2 bond. (b) Anabolic steroids promote cell growth, a necessary part of wound healing.

5. A plot of fractional saturation ($Y$) versus [L] yields a hyperbola where the estimated value of $0.5 Y$ corresponds to a ligand concentration, or $K_L$, of about 3.5 $\mu M$.

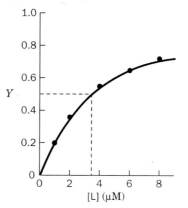

6. A Scatchard plot (B/F versus B) has a slope of $-0.33$ mM$^{-1}$. Since slope $= -1/K_L$, $K_L = 3$ mM.

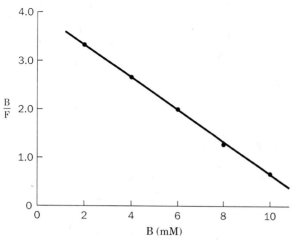

7. Use Equation 13-8, letting $[L] = 1$ μM and $[B] = [I_{50}] = 2.5$ μM,

$$K_I = \frac{[I_{50}]}{\left(1 + \dfrac{[L]}{K_L}\right)}$$

$$= \frac{(2.5 \times 10^{-6})}{\left(1 + \dfrac{(1 \times 10^{-6})}{(5 \times 10^{-6})}\right)}$$

$$= \frac{2.5 \times 10^{-6}}{1.2} = 2.1 \ \mu M$$

8. Rearranging Equation 13-8 gives

$$[I_{50}] = K_I \left(1 + \frac{[L]}{K_L}\right)$$

$$= (8.9 \times 10^{-9}) \left(1 + \frac{1.0 \times 10^{-8}}{4.8 \times 10^{-8}}\right)$$

$$= (8.9 \times 10^{-9})(1 + 0.208) = 1.1 \times 10^{-8} \ M$$

9.

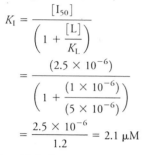

10. ADP is a product of the kinase-catalyzed reaction, so a compound with a similar structure might bind in the kinase active site to act as a competitive inhibitor.

11. The PH domain allows the IRS to be localized to the intracellular leaflet of the cell membrane, close to the insulin receptor and ready to participate in signal transduction. The PTB domain, like an SH2 domain, recognizes phospho-Tyr residues, in this case on the autophosphorylated insulin receptor. As a result, the IRS can be activated following insulin binding to its receptor. The phospho-Tyr residues on the IRS itself are recognition points for SH2-containing proteins, which can thereby become activated.

12. Because Sos functions as a guanine nucleotide exchange factor, it promotes the activity of Ras. The mutation will diminish Ras activity and therefore slow cell growth.

13. (a) Yes, because the binding of Src's SH3 domain to the linker that connects its SH2 domain to the N-terminal lobe of its PTK domain is required for Src to maintain its autoinhibited conformation. Hence, the deletion of this SH3 domain would constitutively activate Src, thereby driving the cell with this mutation to a state of unrestrained proliferation.

(b) No, because the phosphorylation of Tyr 416 is required for the activation of Src and hence the Y416F mutation would inhibit the mutant cell's proliferation.

14. (a) Yes, because the phosphorylation of Tyr 527 is required for the autoinhibition of Src via the binding of its SH2 domain and hence the Y527F mutant Src would be constitutively activated.

(b) No, because replacing the wild-type sequence (one Pro) with a sequence containing two Pro residues would make the 250 to 253 segment of Src a better binding target for the SH3 domain, thereby stabilizing Src's autoinhibited conformation.

15. The SH2 domain allows the enzyme to bind to phospho-Tyr residues on its target proteins. Because targets typically contain more than one phospho-Tyr group, the phosphatase can recognize and bind to one site on the target protein while dephosphorylating another site on the same protein.

16. In the presence of the viral protein, the cell would undergo more cycles of cell division in response to the growth factor.

17. Transformation to the cancerous state results from several genetic changes in a cell. Thus, a single oncogene supplied to an otherwise normal cell will be insufficient to transform it. However, an immortalized cell already has some of the genetic changes necessary for transformation (malignant cells are also immortal). In such cells, the additional oncogene may be all they require to complete their transformation.

18. The antibody might bind to the receptor such that growth factor binding is blocked. Alternatively, antibody binding could generate steric interference that prevents dimerization of the receptor after hormone binding, a necessary step for signal transduction. In either case, growth-promoting signals would be diminished, thereby slowing the growth of cancerous cells.

19. Because the GTP analog cannot be hydrolyzed, $G_\alpha$ remains active. Analog binding to $G_s$ therefore increases cAMP production. Analog binding to $G_i$ decreases cAMP production.

20. In order to survive intracellularly, *M. tuberculosis* must interfere with the host cell's normal activities, for example, by disrupting the cell's signal transduction pathways. In this case, the overproduction of cAMP and the prolonged presence of the second messenger prevents infected cells from responding normally to the presence of the infecting bacteria.

21. Like cholera toxin, *B. anthracis* EF leads to the overproduction of cAMP, which triggers the release of fluid from cells. The result is edema.

22. By preventing MAPK kinases from activating MAPK, LF interferes with the signaling pathway required for white blood cells to become activated and respond to infections.

23. When $G_{s\alpha}$ catalyzes the hydrolysis of its bound GTP to GDP + $P_i$, the Arg side chain that cholera toxin ADP-ribosylates functions to stabilize the transition state's developing negative charge. The ADP-ribosylated $G_{s\alpha}$ therefore hydrolyzes its bound GTP at a greatly reduced rate and hence remains activated far longer than normal $G_{s\alpha}$. Mutating this Arg will have a similar effect. Consequently, in cells in which a $G_{s\alpha} \cdot$ GTP functions to induce cell proliferation, mutating this Arg will drive the cell into a state of unrestrained proliferation, a requirement for the malignant transformation of the cell. The gene encoding such a $G_{s\alpha}$ subunit is a proto-oncogene and the mutation of its Arg residue converts it to an oncogene.

24. Cholera toxin does not cause cancer because the $G_{s\alpha}$ it ADP-ribosylates only mediates the intestinal cell's secretion of digestive fluid, not its rate of proliferation. Moreover, cholera toxin does not pass through the intestine to the other tissues and hence does not affect other cells. However, even if it did so, its effect would only last as long as the cholera infection, whereas a mutation permanently affects the cell in which it has occurred and all its progeny.

25. No. Although the diacylglycerol second messengers are identical, phosphatidylethanolamine does not generate an $IP_3$ second messenger that triggers the release of $Ca^{2+}$, which in turn alters protein kinase C activity.

26. Pertussis toxin ADP ribosylates $G_{i\alpha}$ so as to prevent it from exchanging its bound GDP for GTP and hence from releasing $G_{\beta\gamma}$ on interacting with its cognate activated GPCRs. The resulting decrease in active $G_{\beta\gamma}$ inhibits PLC, which is normally activated through its association with free $G_{\beta\gamma}$s.

27. Diacylglycerol kinase converts DAG to phosphatidic acid (Section 9-1C).

28. Activation of the kinase converts the nonpolar DAG to a more amphiphilic molecule that can no longer activate protein kinase C. In effect, the kinase limits the activity of one of the second messengers produced during signaling by the phosphoinositide pathway.

29. $Li^+$ blocks the conversion of phosphorylated inositol species to inositol, thereby preventing the recycling of $IP_3$ and its degradation products back to inositol. This in turn prevents the synthesis of phosphatidylinositol and $PIP_2$, the precursor of the $IP_3$ second messenger.

30. Insulin binding to its receptor triggers a signaling cascade that begins with the tyrosine kinase activity of the insulin receptor. However, as in all signaling pathways, intracellular responses are eventually shut down. Activation of a phosphatase such as SHP-2, which participates in transducing the insulin signal by activating the MAPK pathway, can also act to limit the cellular response, by dephosphorylating phospho-Tyr groups.

## Chapter 14

1. A heterotroph relies on other organisms for food, which may include substances the heterotroph cannot synthesize (including vitamins). An autotroph can produce all the molecules it needs.

2. (a) The methanogens, which produce methane from inorganic precursors, are autotrophic. The methane consumers rely on the product of the methanogens and so are heterotrophic.

   (b) The organisms associate so that the methanotrophs can consume the methane that is released by the methanogens.

3. Arsenic, which resembles phosphorus, is incorporated into nucleic acids and other compounds that ordinarily contain phosphate groups.

4. Ions of Cd and Hg, which are below Zn in the periodic table, take the place of $Zn^{2+}$ ions that serve as cofactors for essential cellular enzymes.

5. C, D, A, E, B

6. (a) Reduction; (b) reduction.

7. (a) $K = e^{-\Delta G^{\circ\prime}/RT}$

   $K = e^{-(-31,500 \cdot mol^{-1})/(8.3145 J \cdot K^{-1} \cdot mol^{-1})(310\ K)}$

   $K = 2.0 \times 10^5$

   (b) The large change in free energy makes the citrate synthase reaction irreversible, so it could (and does; Section 17-4B) serve as a control point for the citric acid cycle.

8. (a) Since $\Delta G^{\circ\prime} = -RT \ln K$,

   $K = e^{-\Delta G^{\circ\prime}/RT}$

   $K = e^{-(7500\ J \cdot mol^{-1})/(8.3145\ J \cdot K^{-1} \cdot mol^{-1})(298\ K)}$

   $K = 0.048$

   (b) $\Delta G = \Delta G^{\circ\prime} + RT \ln \dfrac{[B]}{[A]}$

   $\Delta G = 7500\ J \cdot mol^{-1}$

   $+ (8.3145\ J \cdot K^{-1} \cdot mol^{-1})(310\ K) \ln \dfrac{(0.0001)}{(0.0005)}$

   $\Delta G = 7500\ J \cdot mol^{-1} - 4150\ J \cdot mol^{-1}$

   $\Delta G = 3350\ J \cdot mol^{-1} = 3.35\ kJ \cdot mol^{-1}$

   The reaction is not spontaneous since $\Delta G > 0$.

   (c) The reaction can proceed in the cell if the product B is the substrate for a second reaction such that the second reaction continually draws off B, causing the first reaction to continually produce more B from A.

9. b

10. (a) For enzymes that catalyze near-equilibrium reactions, $\Delta G \approx 0$ and the direction of flux depends on the relative concentrations of substrates and products. Consequently, these enzymes can catalyze both the forward (e.g., anabolic) and reverse (e.g., catabolic) reactions.

    (b) In order for a metabolic pathway to proceed in the forward direction, $\Delta G$ must be less than zero. The reverse process, involving the same reactions, must therefore have $\Delta G > 0$. However, if the opposing pathways involve different reactions, catalyzed by different enzymes, then both pathways can proceed with favorable changes in free energy.

11. The theoretical maximum yield of ATP is equivalent to ($\Delta G^{\circ\prime}$ for fuel oxidation)/($\Delta G^{\circ\prime}$ for ATP synthesis) = $(-2850\ kJ \cdot mol^{-1})/(-30.5\ kJ \cdot mol^{-1}) \approx$ 93 ATP

12. The theoretical maximum yield of ATP is equivalent to

    ($\Delta G^{\circ\prime}$ for fuel oxidation)/($\Delta G^{\circ\prime}$ for ATP synthesis)

    $= (-9781\ kJ \cdot mol^{-1})/(-30.5\ kJ \cdot mol^{-1}) \approx 320$ ATP

13. At pH 6, the phosphate groups are more ionized than they are at pH 5, which increases their electrostatic repulsion and therefore increases the magnitude of $\Delta G$ for hydrolysis (makes it more negative).

14. Removing a phosphoryl group from ATP has a large change in free energy because there is a large difference in resonance and electrostatic stabilization between the reactants and products. Removing a phosphoryl group from AMP has a smaller change in free energy because the difference in resonance and electrostatic stabilization between AMP (which has only one phosphate group) and $P_i$ is not as large.

15. The exergonic hydrolysis of $PP_i$ by pyrophosphatase ($\Delta G^{\circ\prime} = -19.2$ $kJ \cdot mol^{-1}$) drives fatty acid activation.

16. (a) Because the $\Delta G^{\circ\prime}$ value for the reaction is greater than 0, the reaction will not occur under standard conditions. It could occur under cellular conditions, depending on the actual concentrations of the reactants and products.

    (b) The malate dehydrogenase and citrate synthase reactions are coupled through their common intermediate oxaloacetate

    Malate + $NAD^+ \rightarrow$ oxaloacetate + NADH + $H^+$

    Oxaloacetate + acetyl-CoA $\rightarrow$ citrate + HS-CoA

    so that the overall $\Delta G^{\circ\prime}$ is the sum of the $\Delta G^{\circ\prime}$ values for the two reactions: $29.7\ kJ \cdot mol^{-1} + (-31.5\ kJ \cdot mol^{-1}) = -1.8\ kJ \cdot mol^{-1}$. The citrate synthase reaction helps "pull" the malate dehydrogenase reaction forward by consuming the oxaloacetate produced from malate.

17. Calculating $\Delta G$ for the reaction ATP + creatine $\rightleftharpoons$ phosphocreatine + ADP, using Eq. 14-1:

    $\Delta G = \Delta G^{\circ\prime} + RT \ln \left( \dfrac{[\text{phosphocreatine}][\text{ADP}]}{[\text{creatine}][\text{ATP}]} \right)$

    $= 12.6\ kJ \cdot mol^{-1} + (8.3145\ J \cdot K^{-1} \cdot mol^{-1})(298\ K)$

    $\ln \left( \dfrac{(2.5\ mM)(0.15\ mM)}{(1\ mM)(4\ nM)} \right)$

    $= 12.6\ kJ \cdot mol^{-1} - 5.9\ kJ \cdot mol^{-1} = 6.7\ kJ \cdot mol^{-1}$

    Since $\Delta G > 0$, the reaction will proceed in the opposite direction as written above, that is, in the direction of ATP synthesis.

18. Using the data in Table 14-3, we calculate $\Delta G^{\circ\prime}$ for the adenylate kinase reaction.

|  | $\Delta G^{\circ\prime}$ |
|---|---|
| $ATP + H_2O \rightarrow AMP + PP_i$ | $-45.6$ kJ $\cdot$ mol$^{-1}$ |
| $2\ ADP + 2\ P_i \rightarrow 2\ ATP + 2\ H_2O$ | $2 \times 30.5$ kJ $\cdot$ mol$^{-1}$ |
|  | $= 61.0$ kJ $\cdot$ mol$^{-1}$ |
| $PP_i + H_2O \rightarrow 2\ P_i$ | $-19.2$ kJ $\cdot$ mol$^{-1}$ |
| $2\ ADP \rightarrow ATP + AMP$ | $-3.8$ kJ $\cdot$ mol$^{-1}$ |

Since $\Delta G$ for a reaction at equilibrium is zero, Eq. 14-1 becomes $\Delta G^{\circ\prime} = -RT \ln K_{eq}$ so that $K_{eq} = e^{-\Delta G^{\circ\prime}/RT}$.

$$K_{eq} = \frac{[ATP][AMP]}{[ADP]^2} = e^{-\Delta G^{\circ\prime}/RT}$$

$$[AMP] = \frac{(5 \times 10^{-4}\ M)^2}{(5 \times 10^{-3}\ M)} \times e^{-(-3800\ J \cdot mol^{-1})/(8.3145\ J \cdot K^{-1} \cdot mol^{-1})(298\ K)}$$

$[AMP] = 2.3 \times 10^{-4}\ M = 0.23$ mM

19. The more positive the reduction potential, the greater the oxidizing power. From Table 14-4,

| Compound | $\mathcal{E}^{\circ\prime}$ (V) |
|---|---|
| $SO_4^{2-}$ | $-0.515$ |
| Acetoacetate | $-0.346$ |
| $NAD^+$ | $-0.315$ |
| Pyruvate | $-0.185$ |
| Cytochrome $b$ ($Fe^{3+}$) | $0.077$ |

20. Cytochrome $a$ has a higher standard reduction potential (0.29 V) than cytochrome $c_1$ (0.22 V), so electrons will tend to flow from cytochrome $c_1$ to cytochrome $a$. Under standard conditions, electrons will not flow from cytochrome $c_1$ to cytochrome $b$, whose standard reduction potential (0.077 V) is less than that of cytochrome $c_1$.

21. The ubiquinone half-reaction has a higher reduction potential (0.045 V) than the $NAD^+$ half-reaction ($-0.315$ V). Therefore, electrons will flow from NADH (which becomes oxidized) to ubiquinone (which becomes reduced).

22. The $O_2 \rightleftharpoons H_2O$ half-reaction has a standard reduction potential of 0.815 V, whereas the nitrate $\rightleftharpoons$ nitrite half-reaction has a standard reduction potential of only 0.42 V. Because the free energy change for a redox reaction is proportional to the change in reduction potential, $\Delta G^{\circ\prime} = -n\mathcal{F}\Delta\mathcal{E}^{\circ\prime}$, transferring electrons from the fuel molecule to $O_2$ will have a larger free energy change than transferring the electrons from the fuel molecule to nitrate.

23. Using the data in Table 14-4:

$\Delta\mathcal{E}^{\circ\prime} = \mathcal{E}^{\circ\prime}_{(e^- \text{ acceptor})} - \mathcal{E}^{\circ\prime}_{(e^- \text{ donor})} = \mathcal{E}^{\circ\prime}_{(fumarate)} - \mathcal{E}^{\circ\prime}_{(NAD^+)}$
$= 0.031V\ -(-0.315\ V) = 0.346$ V.

Because $\Delta\mathcal{E}^{\circ\prime} > 0$, $\Delta G^{\circ\prime} < 0$ and the reaction will spontaneously proceed as written.

24. No. Here, $\Delta\mathcal{E}^{\circ\prime} = \mathcal{E}^{\circ\prime}_{(e^- \text{ acceptor})} - \Delta\mathcal{E}^{\circ\prime}_{(e^- \text{ donor})} = \mathcal{E}^{\circ\prime}_{(cyto\ b)} - \mathcal{E}^{\circ\prime}_{(cyto\ a)} = 0.077$ V $- (0.29$ V$) = -0.213$ V. Because $\Delta\mathcal{E}^{\circ\prime} < 0$, $\Delta G^{\circ\prime} > 0$ and the reaction will spontaneously proceed in the opposite direction from that written.

25. The balanced equation is

2 cyto $c$ ($Fe^{3+}$) + ubiquinol $\rightarrow$ 2 cyto $c$ ($Fe^{2+}$) + ubiquinone + 2 $H^+$

Using the data in Table 14-4,

$\Delta\mathcal{E}^{\circ\prime} = \mathcal{E}^{\circ\prime}_{(e^- \text{ acceptor})} - \mathcal{E}^{\circ\prime}_{(e^- \text{ donor})} = 0.235$ V $- 0.045$ V
$= 0.190$ V

$\Delta G^{\circ\prime} = -n\mathcal{F}\ \Delta\mathcal{E}^{\circ\prime} = -(2)(96{,}485\ J \cdot V^{-1} \cdot mol^{-1})(0.190\ V)$
$= -36.7$ kJ $\cdot$ mol$^{-1}$

26. Using the data in Table 14-4, for the oxidation of free $FADH_2$ ($\mathcal{E}^{\circ\prime} = -0.219$ V) by ubiquinone ($\mathcal{E}^{\circ\prime} = 0.045$ V),

$\Delta\mathcal{E}^{\circ\prime}_{(ubiquinone)} - \mathcal{E}^{\circ\prime}_{(FADH_2)} = (0.045\ V) - (-0.219\ V) = 0.264$ V

$\Delta G^{\circ\prime} = -n\mathcal{F}\Delta\mathcal{E}^{\circ\prime} = -(2)(96{,}485\ J \cdot V^{-1} \cdot mol^{-1})(0.264\ V) = -50.9$ kJ $\cdot$ mol

This is more than enough free energy to drive the synthesis of ATP from ADP $+ P_i$ ($\Delta G^{\circ\prime} = +30.5$ kJ $\cdot$ mol$^{-1}$; Table 14-3).

27. $Z \xrightarrow{B} W \xrightarrow{C} Y \xrightarrow{A} X$

28. (a) The step catalyzed by enzyme Y is likely to be the major flux-control point, since this step operates farthest from equilibrium (it is an irreversible step). (b) Inhibition of enzyme Z would cause the concentration of D, the reaction's product, to decrease, and it would cause C, the reaction's substrate, to accumulate. The concentrations of A and B would not change because the steps catalyzed by enzymes X and Y would not be affected. The accumulated C would not be transformed back to B since the step catalyzed by enzyme Y is irreversible.

29. Probably not. Although all cells carry out a similar set of basic metabolic reactions, the enzymes that catalyze the reactions have different amino acid sequences and hence different gene sequences. cDNAs produced from mammalian mRNAs would be unlikely to hybridize with bacterial DNA segments.

30. Only a small portion ($\sim$1.2%) of the human genome consists of protein-coding sequences. Building a gene chip with DNA sequences corresponding to genes that are transcribed increases the likelihood of "capturing" complementary segments derived from the mRNA population of the cells under study.

# Chapter 15

1. (a) Reactions 1, 3, 7, and 10; (b) Reactions 2, 5, and 8; (c) Reaction 6; (d) Reaction 9; (e) Reaction 4.

2. Proteolysis of enzymes such as aldolase and enolase destroys their activity, thereby inhibiting glycolytic flux. As a result, the cell would be unable to generate ATP from glucose and would die. The advantage of this response would be the elimination of the infected cell, which would help limit the growth and dissemination of *Salmonella*.

3. C1 of DHAP and C1 of GAP are achiral but become chiral in FBP (as C3 and C4). There are four stereoisomeric products that differ in configuration at C3 and C4: fructose-1,6-bisphosphate, psicose-1,6-bisphosphate, tagatose-1,6-bisphosphate, and sorbose-1,6-bisphosphate (see Fig. 8-2).

4. The $Zn^{2+}$ polarizes the carbonyl oxygen of the substrate to stabilize the enolate intermediate of the reaction.

5. (a) Glucose $+ 2\ NAD^+ + 2\ ADP + 2\ P_i \rightarrow$
   $2$ pyruvate $+ 2\ NADH + 2\ ATP + 2\ H_2O$

(b) Glucose $+ 2\ NAD^+ + 2\ ADP + 2\ AsO_4^{3-} \rightarrow$
   $2$ pyruvate $+ 2\ NADH + 2\ ADP{-}AsO_3^{2-} + 2\ H_2O$

$\underline{2\ ADP{-}AsO_3^{2-} + 2\ H_2O \rightarrow 2\ ADP + AsO_4^{3-}}$

Overall: Glucose $+ 2\ NAD^+ \rightarrow 2$ pyruvate $+ 2\ NADH$

(c) Arsenate is a poison because it uncouples ATP generation from glycolysis. Consequently, glycolytic energy generation cannot occur.

6. Like phosphoglycerate mutase, phosphoglucomutase catalyzes a phosphoryl group transfer in which a phosphorylated group in the enzyme active site donates its phosphoryl group to the substrate and then receives a second phosphoryl group from the substrate. The active site Ser can undergo reversible phosphorylation.

7. The reaction intermediate is glucose-1,6-bisphosphate (G1,6P).

8. (a) G1,6P inhibits hexokinase. Production of G1,6P indicates that G1P (derived from glycogen or from other sugars such as galactose) is already present and the cell does not need to initiate glycolysis with glucose. (b) G1,6P activates PFK to increase the flux of phosphorylated sugars through the rest of the glycolytic pathway.

9. Alcoholic fermentation, unlike homolactic fermentation, includes a step (the pyruvate decarboxylase reaction) in which a carbon is lost as $CO_2$. Because the $CO_2$ diffuses (bubbles) away, the reaction cannot proceed in reverse.

10. For the coupled reaction

$$\text{Pyruvate} + \text{NADH} + \text{H}^+ \rightarrow \text{lactate} + \text{NAD}^+$$

$\Delta\mathcal{E}°' = (-0.185 \text{ V}) - (-0.315 \text{ V}) = 0.130 \text{ V}$. According to Eq. 14-8,

$$\Delta\mathcal{E} = \Delta\mathcal{E}°' - \frac{RT}{n\mathcal{F}}\ln\left(\frac{[\text{lactate}][\text{NAD}^+]}{[\text{pyruvate}][\text{NADH}]}\right)$$

and $\Delta G = -n\mathcal{F}\Delta\mathcal{E}$ (Eq. 14-7). Since two electrons are transferred in the above reaction, $n = 2$.

(a) $\Delta\mathcal{E} = 0.130 \text{ V} - \frac{RT}{n\mathcal{F}}\ln(1) = 0.130 \text{ V}$

$\Delta G = -(2)(96{,}485 \text{ J} \cdot \text{V}^{-1} \cdot \text{mol}^{-1})(0.130 \text{ V}) = -25.1 \text{ kJ} \cdot \text{mol}^{-1}$

(b) $RT/n\mathcal{F} = (8.3145 \text{ J} \cdot \text{K}^{-1} \cdot \text{mol}^{-1})(298 \text{ K})/$
$\qquad\qquad (2)(96{,}485 \text{ J} \cdot \text{V}^{-1} \cdot \text{mol}^{-1}) = 0.01284 \text{ V}$

$\Delta\mathcal{E} = 0.130 \text{ V} - 0.01284 \text{ V} \ln(160 \times 160)$

$\Delta\mathcal{E} = 0.130 \text{ V} - 0.130 \text{ V} = 0$

$\Delta G = 0$

(c) $\Delta\mathcal{E} = 0.130 \text{ V} - 0.01284 \text{ V} \ln(1000 \times 1000)$

$\Delta\mathcal{E} = 0.130 \text{ V} - 0.177 \text{ V} = -0.047 \text{ V}$

$\Delta G = -(2)(96{,}485 \text{ J} \cdot \text{V}^{-1} \text{ mol}^{-1})(-0.047 \text{ V})$
$\qquad = 9.1 \text{ kJ} \cdot \text{mol}^{-1}$

(d) At the concentration ratios of Part a, $\Delta G$ is negative and the reaction proceeds as written. As [lactate]/[pyruvate] increases, the reaction $\Delta G$ increases even though [NAD$^+$]/[NADH] also increases so that in Part b the reaction is at equilibrium ($\Delta G = 0$) and in Part c $\Delta G$ is positive and the reaction proceeds spontaneously in the opposite direction.

11. $\Delta G$ values differ from $\Delta G°'$ values because $\Delta G = \Delta G°' + RT$ ln[products]/[reactants] and cellular reactants and products are not in their standard states.

12. Yes. The same *in vivo* conditions that decrease the magnitude of $\Delta G$ relative to $\Delta G°'$ may also decrease $\Delta G$ for ATP synthesis.

13. When [GAP] $= 10^{-4}$ M, [DHAP] $= 5.5 \times 10^{-4}$ M. According to Eq. 1-17,

$$K = e^{-\Delta G°'/RT}$$

$$\frac{[\text{GAP}][\text{DHAP}]}{[\text{FBP}]} = e^{-(22{,}800 \text{ J} \cdot \text{mol}^{-1})/(8.3145 \text{ J} \cdot \text{K}^{-1} \cdot \text{mol}^{-1})(310 \text{ K})}$$

$$\frac{(10^{-4})(5.5 \times 10^{-4})}{[\text{FBP}]} = 1.4 \times 10^{-4}$$

$$[\text{FBP}] = 3.8 \times 10^{-4} \text{ M}$$

$$[\text{FBP}]/[\text{GAP}] = (3.8 \times 10^{-4} \text{ M})/(10^{-4} \text{ M}) = 3.8$$

14. The liver enzyme is far more sensitive than the brain enzyme to the three activators. It is possible that liver PFK-1 is subject to a greater degree of regulation than brain PFK-1. Fuel must be supplied to the brain continuously and thus glycolysis is always active, but the liver has a wide variety of physiological roles and is more likely to regulate cellular pathways.

15. Pyruvate kinase regulation is important for controlling the flux of metabolites, such as fructose (in liver), which enter glycolysis after the PFK step.

16. FBP is the product of the third reaction of glycolysis, so it acts as a feed-forward activator of the enzyme that catalyzes Step 10. This regulatory mechanism helps ensure that once metabolites pass the PFK step of glycolysis, they will continue through the pathway.

17. Galactokinase, which catalyzes a highly exergonic reaction, is a possible control point. Because galactose enters the glycolytic pathway as glucose-6-phosphate, the PFK reaction is also likely to be a major control point.

18. The inhibition of phosphoglucomutase, which catalyzes Step 4 of the galactose-metabolizing pathway, would slow the production of G6P from galactose and thereby slow the rate at which galactose is catabolized. Because glucose enters glycolysis without the phosphoglucomutase-catalyzed step, its flux through glycolysis is faster.

19. (a) Glycerol can be converted to the glycolytic intermediate DHAP by the activity of glycerol kinase and glycerol phosphate dehydrogenase (Fig. 15-27). (b) One ATP is consumed by the glycerol kinase reaction, but two ATP are produced (by the PGK and PK reactions), for a net yield of one ATP per glycerol (this does not count the ATP that might be generated through oxidative phosphorylation from the NADH produced in the glycerol phosphate dehydrogenase reaction).

20. Fermentation pathways regenerate one NAD$^+$ needed for the GAPDH reaction of glycolysis. The catabolism of glycerol includes the GAPDH reaction, but it also includes the glycerol phosphate dehydrogenase reaction, which also generates NADH. Therefore, homolactic or alcoholic fermentation could only regenerate half the NAD$^+$ required for glycerol catabolism.

21. The three glucose molecules that proceed through glycolysis yield 6 ATP. The bypass through the pentose phosphate pathway results in a yield of 5 ATP.

22. The label will appear at C1 and C3 of F6P (see Fig. 15-30).

23. 
$$\begin{array}{c}
\text{H} \quad \text{OH} \\
\diagdown \;/ \\
\text{C} \\
\parallel \\
\text{C}-\text{O}^- \\
\mid \\
\text{H}-\text{C}-\text{OH} \\
\mid \\
\text{H}-\text{C}-\text{OH} \\
\mid \\
\text{CH}_2\text{OPO}_3^{2-}
\end{array}$$

**1,2-Enediolate intermediate**

24.
$$\begin{array}{c}
\text{H} \\
\mid \\
\text{H}-\text{C}-\text{OH} \\
\mid \\
\text{C}-\text{O}^- \\
\parallel \\
\text{C}-\text{OH} \\
\mid \\
\text{H}-\text{C}-\text{OH} \\
\mid \\
\text{CH}_2\text{OPO}_3^{2-}
\end{array}$$

**2,3-Enediolate intermediate**

25. Transketolase transfers 2-carbon units from a ketose to an aldose, so the products are a 3-carbon sugar and a 7-carbon sugar.

26. The products are a 4-carbon and a 7-carbon sugar. The order of binding does matter. The ketose binds first and transfers the 2-carbon unit to the TPP on the enzyme. The aldose then binds and accepts the 2-carbon unit.

27. Even when the flux of glucose through glycolysis and hence the citric acid cycle is blocked, glucose can be oxidized by the pentose phosphate pathway, with the generation of $CO_2$.

28. (a)

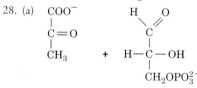

**Pyruvate**              **GAP**

(b) The pyruvate product of the aldolase reaction is not further modified. The GAP product is converted to pyruvate by the actions of the glycolytic enzymes glyceraldehyde-3-phosphate dehydrogenase, phosphoglycerate kinase, phosphoglycerate mutase, enolase, and pyruvate kinase.

(c) One ATP is consumed when glucose is converted to glucose-6-phosphate. One ATP is generated by the phosphoglycerate kinase reaction and one by the pyruvate kinase reaction (these quantities are not doubled because only one three-carbon fragment of glucose follows this route), for a net yield of one ATP per glucose. The standard glycolytic pathway generates two ATP per glucose.

29. (a) Reaction 8, (b) reaction 5, (c) reaction 1, (d) reaction 2, (e) reaction 3.

30. (a) Glucose-6-phosphate dehydrogenase (pentose phosphate pathway, reaction 1 of Fig. 15-30)

(b) UDP–galactose-4-epimerase (galactose metabolism, reaction 3 of Fig. 15-28).

(c) Phosphoglucose isomerase (glycolysis, reaction 2 of Fig. 15-1).

(d) Phosphomannose isomerase (mannose metabolism, Section 15-5C).

(e) Phosphoglycerate mutase (glycolysis, reaction 8 of Fig. 15-1) or phosphoglucomutase (galactose metabolism, reaction 4 of Fig. 15-28).

## Chapter 16

1. Glycogen is broken down when the cell needs to catabolize glucose to produce ATP. The G1P generated by the glycogen phosphorylase reaction is quickly isomerized to G6P and enters glycolysis. The continual consumption of G1P "pulls" the phosphorylase reaction forward, making it thermodynamically favorable.

2. This mechanism allows glycogen phosphorylase activity to be regulated by the concentration of glucose so that glycogen is not broken down when glucose is already plentiful.

3. Phosphoglucokinase activity generates G1,6P, which is necessary to "prime" phosphoglucomutase that has become dephosphorylated and thereby inactivated through the loss of its G1,6P reaction intermediate.

4. A defect in G6P transport would have the symptoms of glucose-6-phosphatase deficiency: accumulation of glycogen and hypoglycemia.

5. The conversion of circulating glucose to lactate in the muscle generates 2 ATP. If muscle glycogen could be mobilized, the energy yield would be 3 ATP, since phosphorolysis of glycogen bypasses the hexokinase-catalyzed step that consumes ATP in the first stage of glycolysis.

6. The deficiency is in branching enzyme (Type IV glycogen storage disease). The high ratio of G1P to glucose indicates abnormally long chains of $\alpha(1\rightarrow4)$-linked residues with few $\alpha(1\rightarrow6)$-linked branch points (the normal ratio is $\sim 10$).

7. In the course of glucose catabolism, a detour through glycogen synthesis and glycogen breakdown begins and ends with G6P. The energy cost of this detour is 1 ATP equivalent, consumed in the UDP–glucose pyrophosphorylase step. The overall energy lost is therefore 1/32 or $\sim 3\%$.

8. The overall free energy change for debranching is

| | |
|---|---|
| Breaking $\alpha(1\rightarrow4)$ bond | $\Delta G^{\circ\prime} = -15.5 \text{ kJ} \cdot \text{mol}^{-1}$ |
| Forming $\alpha(1\rightarrow4)$ bond | $+15.5 \text{ kJ} \cdot \text{mol}^{-1}$ |
| Hydrolyzing $\alpha(1\rightarrow6)$ bond | $-7.1 \text{ kJ} \cdot \text{mol}^{-1}$ |
| Total | $\Delta G^{\circ\prime} = -7.1 \text{ kJ} \cdot \text{mol}^{-1}$ |

The overall free energy change for branching is

| | |
|---|---|
| Breaking $\alpha(1\rightarrow4)$ bond | $\Delta G^{\circ\prime} = -15.5 \text{ kJ} \cdot \text{mol}^{-1}$ |
| Forming $\alpha(1\rightarrow6)$ bond | $+7.1 \text{ kJ} \cdot \text{mol}^{-1}$ |
| Total | $\Delta G^{\circ\prime} = -8.4 \text{ kJ} \cdot \text{mol}^{-1}$ |

Assuming that $\Delta G^{\circ\prime}$ is close to $\Delta G$, the sum of the two reactions of branching has $\Delta G < 0$, but debranching would be endergonic ($\Delta G > 0$) without the additional step of hydrolyzing the $\alpha(1\rightarrow6)$ bond to form glucose.

9. A glycogen molecule with 28 tiers would represent the most efficient arrangement for storing glucose, and its outermost tier would contain considerably more glucose residues than a glycogen molecule with only 12 tiers. However, densely packed glucose residues would be inaccessible to phosphorylase. In fact, such a dense glycogen molecule could not be synthesized because glycogen synthase and branching enzyme would have no room to operate (see Box 16-3 for a discussion of glycogen structure).

10. Amylose is a linear form of starch, so during digestion, glucose monomers can be released only one at a time from the end of the molecule. Amylopectin, a branched form of starch, can release glucose from each of its branches. Consequently, the rate of glucose release from amylose is slower than from amylopectin, and less glucose appears in the blood.

11. (a) Circulating [glucose] is high because cells do not respond to the insulin signal to take up glucose.

(b) Insulin is unable to activate phosphoprotein phosphatase-1 in muscle, so glycogen synthesis is not stimulated. Moreover, glycogen synthesis is much reduced by the lack of available glucose in the cell.

12. The two tissues perform different physiological functions and therefore respond differently to the same hormone. Liver responds by promoting glycogenolysis and gluconeogenesis to produce glucose that can be released for use by other tissues. Muscles respond to the hormone by increasing the flux of glycogen-derived glucose through glycolysis in order to generate ATP to power muscle contraction.

13. The equation for glycolysis is

Glucose + 2 NAD$^+$ + 2 ADP + 2 P$_i \rightarrow$
$$2 \text{ pyruvate} + 2 \text{ NADH} + 4 \text{ H}^+ + 2 \text{ ATP} + 2 \text{ H}_2\text{O}$$

The equation for gluconeogenesis is

2 Pyruvate + 2 NADH + 4 H$^+$ + 4 ATP + 2 GTP + 6 H$_2$O
$$\rightarrow \text{glucose} + 2 \text{ NAD}^+ + 4 \text{ ADP} + 2 \text{ GDP} + 6 \text{ P}_i$$

For the two processes operating sequentially,
$$2 \text{ ATP} + 2 \text{ GTP} + 4 \text{ H}_2\text{O} \rightarrow 2 \text{ ADP} + 2 \text{ GDP} + 4 \text{ P}_i$$

14. The equation for catabolism of 6 G6P by the pentose phosphate pathway is

6 G6P + 12 NADP$^+$ + 6 H$_2$O $\rightarrow$
$$6 \text{ Ru5P} + 12 \text{ NADPH} + 12 \text{ H}^+ + 6 \text{ CO}_2$$

Ru5P can be converted to G6P by transaldolase, transketolase, and gluconeogenesis:

$$6 \text{ Ru5P} + H_2O \rightarrow 5 \text{ G6P} + P_i$$

The net equation is therefore

$$G6P + 12 \text{ NADP}^+ + 7 H_2O \rightarrow$$
$$12 \text{ NADPH} + 12 H^+ + 6 CO_2 + P_i$$

15. (a) +9 ATP, (b) +6 ATP, (c) −18 ATP.

16. (a) Lactate dehydrogenase, pyruvate carboxylase, PEPCK, enolase, phosphoglycerate mutase, phosphoglycerate kinase, GAPDH, triose phosphate isomerase, aldolase, fructose-1,6-bisphosphatase, phosphoglucose isomerase, and glucose-6-phosphatase. (b) Two ATP are produced by glycolysis, and 6 ATP are consumed by gluconeogenesis, so there is a net loss of 4 ATP.

17. (a) Aspartate can be transaminated to produce oxaloacetate, a gluconeogenic precursor. (b) To convert 2 aspartate to glucose, 2 GTP are consumed in the PEPCK reaction and 2 ATP are consumed in the phosphoglycerate kinase reaction, for a total of 4 ATP equivalents.

18. (a) In alcoholic fermentation (Section 15-3B), pyruvate decarboxylase converts pyruvate to acetaldehyde and $CO_2$. In gluconeogenesis (Section 16-4A), pyruvate carboxylase transfers a bicarbonate group to pyruvate to form oxaloacetate. (b) Pyruvate decarboxylase uses a thiamine pyrophosphate cofactor; pyruvate carboxylase uses a biotin cofactor.

19. (a) At the beginning of a fast, blood glucose levels are normal, because dietary sources or glycogenolysis can supply glucose. (b) After a fast, the blood glucose levels are very low, because dietary glucose and glycogen have been depleted and gluconeogenesis is impaired due to the fructose-1,6-bisphosphatase deficiency.

20. Pyruvate, a substrate for gluconeogenesis, cannot be converted to glucose and instead accumulates, because the gluconeogenic enzyme fructose-1,6-bisphosphatase is deficient.

21. A high level of AMP results from a high rate of ATP consumption in the cell, so it would act to promote flux through ATP-generating pathways such as glycolysis. Therefore, AMP would be expected to inhibit the activity of the gluconeogenic enzyme fructose-1,6-bisphosphatase.

22. A high level of acetyl-CoA, the product of fatty acid catabolism (it is also generated from pyruvate and amino acid catabolism), indicates that the cell has adequate metabolic fuel available. The stimulation of pyruvate carboxylase, the first enzyme of gluconeogenesis, allows the cell to direct resources toward glucose synthesis, a mechanism for stockpiling fuel for later use.

23. UDP–Glucose + fructose-6-phosphate

UDP

sucrose-6-phosphate

$H_2O$

$P_i$

sucrose

24. In starch synthesis, α(1→4)-linked glucose residues are added one by one. In cellulose, as indicated in Fig. 8-9, each successive glucose residue is flipped by 180° to form the β(1→4)-linked polymer. This geometry would require an enzyme with a single active site to reorient by 180° with every residue added. By simultaneously accommodating two ADP–glucose substrates, one flipped 180° relative to the other, and catalyzing two glucosyl transfer reactions, the synthase can build cellulose two residues at a time without having to be repositioned.

# Chapter 17

1. The labeled carbon becomes C4 of the succinyl moiety of succinyl-CoA. Because succinate is symmetrical, the label appears at C1 and C4 of succinate. When the resulting oxaloacetate begins the second round, the labeled carbons appear as $^{14}CO_2$ in the isocitrate dehydrogenase and the α-ketoglutarate dehydrogenase reactions (see Fig. 17-2).

2. The labeled carbon becomes C3 of the succinyl moiety of succinyl-CoA and hence appears at C2 and C3 of succinate, fumarate, malate, and oxaloacetate. Neither C2 nor C3 of oxaloacetate is released as $CO_2$ in the second round of the cycle. However, the $^{14}C$ label appears at C1 and C2 of the succinyl moiety of succinyl-CoA in the second round and therefore appears at all four positions of the resulting oxaloacetate. Thus, in the third round, $^{14}C$ is released as $^{14}CO_2$.

3. In mammals, pyruvate can be converted to lactate (reduction), to alanine (transamination), to acetyl-CoA (oxidative decarboxylation), and to oxaloacetate (carboxylation). In yeast, pyruvate is also converted to acetaldehyde (decarboxylation).

4. Citric acid cycle intermediates are all acids and as such represent a source of hydrogen ions that would lead to a decrease in blood pH (acidosis).

5. The decarboxylation step is most likely to be metabolically irreversible since the $CO_2$ product is rapidly hydrated to bicarbonate. The reverse reaction, a carboxylation, requires the input of free energy to become favorable (Section 16-4A). The other four reactions are transfer reactions or oxidation–reduction reactions (transfer of electrons) that are more easily reversed.

6. PDP removes the phosphate group that inactivates the pyruvate dehydrogenase complex. A deficiency of PDP leads to less pyruvate dehydrogenase activity in muscle cells, making it difficult for the muscle to increase flux through the citric acid cycle in order to meet the energy demands of exercise.

7.
$$^-OOC-\overset{\overset{\displaystyle OH}{|}}{C}H-CH_2-CH_2-COO^-$$

8. Inhibition of glutamate transamination to α-ketoglutarate would prevent the increase in citric acid cycle intermediates that would allow increased flux of acetyl carbons through the cycle.

9.
$$\overset{\displaystyle H_3C}{\underset{\displaystyle H_3C}{>}}CH-\overset{\overset{\displaystyle O}{\|}}{C}-S-CoA$$

10. Thiamine pyrophosphate is a cofactor for the pyruvate dehydrogenase and α-ketoglutarate dehydrogenase complexes. In beriberi, the substrates for these enzymes, pyruvate and α-ketoglutarate, would accumulate.

11. $NAD^+$ ($\mathscr{E}°' = -0.315$ V) does not have a high enough reduction potential to support oxidation of succinate to fumarate ($\mathscr{E}°' = +0.031$ V); that is, the succinate dehydrogenase reaction has insufficient free energy to reduce $NAD^+$. Enzyme-bound FAD ($\mathscr{E}°' \approx 0$) is more suitable for oxidizing succinate.

12. The alternate pathway bypasses the succinyl-CoA synthetase reaction of the standard citric acid cycle, a step that is accompanied by the phosphorylation of ADP. The alternate pathway therefore generates one less ATP than the standard citric acid cycle. There is no difference in the number of reduced cofactors generated.

13. Competitive inhibition can be overcome by adding more substrate, in this case succinate. Oxaloacetate overcomes malonate inhibition because it is converted to succinate by the reactions of the citric acid cycle.

14. Malonate inhibits the succinate dehydrogenase reaction, so its substrate succinate would accumulate. Since the succinyl-CoA synthetase reaction operates near equilibrium, its substrate succinyl-CoA would also accumulate.

15. The $\Delta G^{\circ\prime}$ value is the sum of the $\Delta G^{\circ\prime}$ values for the malate dehydrogenase reaction (29. 7 kJ $\cdot$ mol$^{-1}$) and the citrate synthase reaction ($-31.5$ kJ $\cdot$ mol$^{-1}$): $-1.8$ kJ $\cdot$ mol$^{-1}$.

16. For the reaction isocitrate $+$ NAD$^+$ $\rightleftharpoons$ $\alpha$-ketoglutarate $+$ NADH $+$ CO$_2$ $+$ H$^+$, we assume [H$^+$] $= 1$ and [CO$_2$] $= 1$. According to Eq. 14-1,

$$\Delta G = \Delta G^{\circ\prime} + RT\ln\left(\frac{[\text{NADH}][\alpha\text{-ketoglutarate}]}{[\text{NAD}^+][\text{isocitrate}]}\right)$$

$$= -21\ \text{kJ}\cdot\text{mol}^{-1} + (8.3145\ \text{J}\cdot\text{K}\cdot\text{mol}^{-1})(298\ \text{K})\ln\left[\frac{(1)(0.1)}{(8)(0.02)}\right]$$

$$= -21\ \text{kJ}\cdot\text{mol}^{-1} - 1.16\ \text{kJ}\cdot\text{mol}^{-1} = -22.16\ \text{kJ}\ \text{mol}^{-1}$$

With such a large negative free energy of reaction under physiological conditions, isocitrate dehydrogenase is likely to be a metabolic control point.

17. From Table 17-2, for the succinate dehydrogenase reaction, $\Delta G^{\circ\prime} = 6$ kJ $\cdot$ mol$^{-1}$ and $\Delta G = {\sim}0$.

$$\Delta G = \Delta G^{\circ\prime} + RT\ln\left(\frac{[\text{fumarate}]}{[\text{succinate}]}\right)$$

$$\Delta G^{\circ\prime} = -RT\ln\left(\frac{[\text{fumarate}]}{[\text{succinate}]}\right)$$

$$\left(\frac{[\text{fumarate}]}{[\text{succinate}]}\right) = e^{-\Delta G^{\circ\prime}/RT}$$

$$= e^{-(6000\ \text{J}\cdot\text{mol}^{-1})/(8.3145\ \text{J}\cdot\text{K}^{-1}\cdot\text{mol}^{-1})(310\ \text{K})}$$

$$= e^{-2.33} = 0.10$$

18. From Table 17-2, the $\Delta G^{\circ\prime}$ value of the aconitase reaction is ${\sim}5$ kJ $\cdot$ mol$^{-1}$ and the $\Delta G$ value is ${\sim}0$.

$$\Delta G = \Delta G^{\circ\prime} + RT\ln\left(\frac{[\text{isocitrate}]}{[\text{citrate}]}\right)$$

$$\Delta G^{\circ\prime} = -RT\ln\left(\frac{[\text{isocitrate}]}{[\text{citrate}]}\right)$$

$$\left(\frac{[\text{isocitrate}]}{[\text{citrate}]}\right) = e^{-\Delta G^{\circ\prime}/RT}$$

$$= e^{-(5000\ \text{J}\cdot\text{mol}^{-1})/(83145\ \text{J}\cdot\text{K}^{-1}\cdot\text{mol}^{-1})(310\ \text{K})}$$

$$= e^{-1.9} = 0.14$$

19. The phosphofructokinase reaction is the major flux-control point for glycolysis. Inhibiting phosphofructokinase slows the entire pathway, so the production of acetyl-CoA by glycolysis followed by the pyruvate dehydrogenase complex can be decreased when the citric acid cycle is operating at maximum capacity and the citrate concentration is high. As citric acid cycle intermediates are consumed in synthetic pathways, the citrate concentration drops, relieving phosphofructokinase inhibition and allowing glycolysis to proceed in order to replenish the citric acid cycle intermediates.

20. No. When glucose is abundant, a cell generates ATP through glycolysis followed by the citric acid cycle. The insulin-stimulated increase in pyruvate dehydrogenase activity does not function to increase the number of acetyl groups that enter the citric acid cycle, since the cell's energy needs are already being met. Instead, the acetyl groups are destined for fatty acid synthesis, a mechanism that helps the cell store metabolic fuel as triacylglycerols (in addition to glycogen).

21. (a) Because citric acid cycle intermediates such as citrate and succinyl-CoA are precursors for the biosynthesis of other compounds, anaerobes must be able to synthesize them.

(b) These organisms do not need a complete citric acid cycle, which would yield reduced coenzymes that must be reoxidized.

22. Citrate can be cleaved to generate an acetyl group and oxaloacetate. The oxaloacetate can then be converted to succinate to complete the cycle.

23.
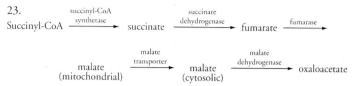

24. (a) The citric acid cycle is a multistep catalyst. Degrading an amino acid to a citric acid cycle intermediate boosts the catalytic activity of the cycle but does not alter the stoichiometry of the overall reaction (acetyl-CoA $\rightarrow$ 2 CO$_2$). In order to undergo oxidation, the citric acid cycle intermediate must exit the cycle and be converted to acetyl-CoA to re-enter the cycle as a substrate.

(b) Pyruvate derived from the degradation of an amino acid can be converted to acetyl-CoA by the pyruvate dehydrogenase complex; these amino acid carbons can then be completely oxidized by the citric acid cycle.

25. To synthesize citrate, pyruvate must be converted to oxaloacetate by pyruvate carboxylase:

Pyruvate $+$ CO$_2$ $+$ ATP $+$ H$_2$O $\rightarrow$ oxaloacetate $+$ ADP $+$ P$_i$

A second pyruvate is converted to acetyl-CoA by pyruvate dehydrogenase:

Pyruvate $+$ CoASH $+$ NAD$^+$ $\rightarrow$ acetyl-CoA $+$ CO$_2$ $+$ NADH

The acetyl-CoA then combines with oxaloacetate to produce citrate:

Oxaloacetate $+$ acetyl-CoA $+$ H$_2$O $\rightarrow$ citrate $+$ CoASH $+$ H$^+$

The net reaction is

2 Pyruvate $+$ ATP $+$ NAD$^+$ $+$ 2 H$_2$O $\rightarrow$

citrate $+$ ADP $+$ P$_i$ $+$ NADH $+$ H$^+$

26. Animals cannot carry out the net synthesis of glucose from acetyl-CoA (to which acetate is converted). However, $^{14}$C-labeled acetyl-CoA enters the citric acid cycle and is converted to oxaloacetate. Some of this oxaloacetate may exchange with the cellular pool of oxaloacetate to be converted to glucose through gluconeogenesis and subsequently taken up by muscle and incorporated into glycogen.

27. Oxaloacetate $+$ FADH$_2$ $+$ NADH $+$ H$^+$ $\rightarrow$

succinate $+$ FAD $+$ NAD$^+$ $+$ H$_2$O

28. (a) $\alpha$-Ketoglutarate $+$ NAD$^+$ $+$ GDP $+$ P$_i$ $\rightarrow$

succinate $+$ CO$_2$ $+$ NADH $+$ H$^+$ $+$ GTP

(b) $\alpha$-Ketoglutarate $+$ CO$_2$ $+$ 2 NADH $+$

2 H$^+$ $+$ CoASH $+$ FADH$_2$ $\rightarrow$

succinate $+$ acetyl-CoA $+$ 2 NAD$^+$ $+$ 2H$_2$O $+$ FAD

# Chapter 18

1. Mitochondria with more cristae have more surface area and therefore more proteins for electron transport and oxidative phosphorylation. Tissues with a high demand for ATP synthesis (such as heart) contain mitochondria with more cristae than tissues with lower demand for oxidative phosphorylation (such as liver).

2. Maternally inherited mitochondrial diseases result from mutations in mitochondrial rather than nuclear DNA. A mother provides mitochondria to her offspring via eggs; a father's mitochondria are not passed to his offspring.

3. When NADH participates in the glycerophosphate shuttle, the electrons of NADH flow to FAD and then to CoQ, bypassing Complex I. Thus, about 1.5 ATP are synthesized per NADH.

4. About 2.5 ATP per NADH are produced when NADH participates in the malate–aspartate shuttle.

5. The relevant half-reactions (Table 14-4) are

$$FAD + 2\,H^+ + 2\,e^- \rightleftharpoons FADH_2 \qquad \mathscr{E}^{o\prime} = -0.219\ V$$

$$\tfrac{1}{2}O_2 + 2\,H^+ + 2\,e^- \rightleftharpoons H_2O \qquad \mathscr{E}^{o\prime} = 0.815\ V$$

Since the $O_2/H_2O$ half-reaction has the more positive $\Delta\mathscr{E}^{o\prime}$, the FAD half-reaction is reversed and the overall reaction is

$$\tfrac{1}{2}O_2 + FADH_2 \rightleftharpoons H_2O + FAD$$

$$\Delta\mathscr{E}^{o\prime} = 0.815\ V - (-0.219\ V) = 1.034\ V$$

Since $\Delta G^{o\prime} = -n\mathscr{F}\,\Delta\mathscr{E}^{o\prime}$,

$$\Delta G^{o\prime} = -(2)(96{,}485\ J\cdot V^{-1}\cdot mol^{-1})(1.034\ V) = -200\ kJ\cdot mol^{-1}$$

The maximum number of ATPs that could be synthesized under standard conditions is therefore $200\ kJ\cdot mol^{-1}/30.5\ kJ\cdot mol^{-1} = 6.6$ mol ATP/mol $FADH_2$ oxidized by $O_2$.

6. The reaction for Complex II is succinate + CoQ $\rightleftharpoons$ fumarate + $CoQH_2$.

The relevant half-reactions and their standard reduction potentials are given in Table 14-4.

$$Fumarate + 2\,H^+ + 2\,e^- \rightleftharpoons succinate \qquad \mathscr{E}^{o\prime} = -0.031\ V$$

$$CoQ + 2\,H^+ + 2\,e^- \rightleftharpoons CoQH_2 \qquad \mathscr{E}^{o\prime} = 0.045\ V$$

$$\Delta\mathscr{E}^{o\prime} = \mathscr{E}^{o\prime}_{(e^-\ acceptor)} - \mathscr{E}^{o\prime}_{(e^-\ donor)}$$

$$= (0.045\ V) - (0.031\ V) = 0.014\ V$$

$$\Delta G = -n\mathscr{F}\Delta\mathscr{E}$$

$$= -(2)(96{,}485\ J\cdot V^{-1}\cdot mol^{-1})(0.014\ V)$$

$$= 2700\ J\cdot mol^{-1} = 2.7\ kJ\cdot mol^{-1}$$

This reaction does not supply enough free energy for the synthesis of ATP, which requires $\sim30.5\ kJ\cdot mol^{-1}$.

7. $NO_3^-$ is the electron acceptor and NADH is the electron donor. Their standard reduction potentials are listed in Table 14-4.

$$\Delta\mathscr{E}^{o\prime} = \mathscr{E}^{o\prime}_{(NO_3^-)} - \mathscr{E}^{o\prime}_{(NADH)}$$

$$= (0.42\ V) - (-0.315\ V) = 0.735\ V$$

$$\Delta G = -n\mathscr{F}\Delta\mathscr{E}$$

$$= -(2)(96{,}485\ J\cdot V^{-1}\cdot mol^{-1})(0.735\ V)$$

$$= -142{,}000\ J\cdot mol^{-1} = -142\ kJ\cdot mol^{-1}$$

Since the standard free energy change for ATP synthesis is $30.5\ kJ\cdot mol^{-1}$, approximately $142/30.5 = 4.6$ ATP could be synthesized.

8. S is the electron acceptor and acetate is the electron donor. Their standard reduction potentials are listed in Table 14-4.

$$\Delta\mathscr{E}^{o\prime} = \mathscr{E}^{o\prime}_{(S)} - \mathscr{E}^{o\prime}_{(acetate)}$$

$$= (-0.23\ V) - (-0.581\ V) = 0.351\ V$$

$$\Delta G = -n\mathscr{F}\Delta\mathscr{E}$$

$$= -(2)(96{,}485\ J\cdot V^{-1}\cdot mol^{-1})(0.351\ V)$$

$$= -68{,}000\ J\cdot mol^{-1} = -68\ kJ\cdot mol^{-1}$$

Since the standard free energy change for ATP synthesis is $30.5\ kJ\cdot mol^{-1}$, approximately $68/30.5 = 2.2$ ATP could be synthesized.

9. *(a)*              *(b)*

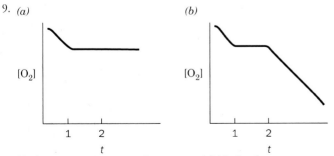

(a) $O_2$ consumption ceases because amytal blocks electron transport in Complex I.

(b) Electrons from succinate bypass the amytal block by entering the electron-transport chain at Complex II and thereby restore electron transport through Complexes III and IV.

10. (a)             (b)

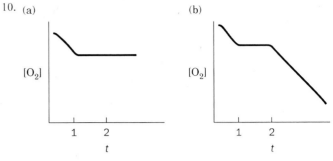

(a) $CN^-$ blocks electron transport in Complex IV, after the point of entry of succinate.

(b) Oligomycin blocks oxidative phosphorylation and hence $O_2$ consumption. DNP uncouples electron transport from oxidative phosphorylation and thereby permits $O_2$ consumption to resume.

11. $\mathscr{E}$ may differ from $\mathscr{E}^{o\prime}$, depending on the redox center's microenvironment and the concentrations of reactants and products. In addition, the tight coupling between successive electron transfers within a complex may "pull" electrons so that the overall process is spontaneous.

12. The range of reduction potentials associated with the Cu ion suggests that the surrounding protein, not just the immediate ligands for the Cu ion, play a major role in determining the affinity of the Cu for electrons.

13.

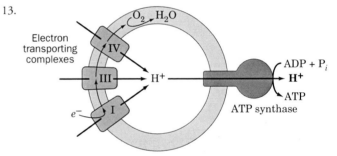

14. An increase in external pH (decrease in $[H^+]$) increases the electrochemical potential across the mitochondrial membrane and therefore leads to an increase in ATP synthesis.

15. For the transport of a proton from outside to inside (Eq. 18-1),

$$\Delta G = 2.3\,RT[\text{pH }(side\ 1) - \text{pH }(side\ 2)] + Z\mathscr{F}\Delta\Psi$$

The difference in pH is $-1.4$. Since an ion is transported from the positive to the negative side of the membrane, $\Delta\Psi$ is negative.

$$\Delta G = (2.3)(8.314\ J\cdot K^{-1}\cdot mol^{-1})(298\ K)(-1.4)$$

$$+ (1)(96{,}485\ J\cdot V^{-1}\cdot mol^{-1})(-0.06\ V)$$

$$\Delta G = -7980\ J\cdot mol^{-1} - 5790\ J\cdot mol^{-1} = -13.8\ kJ\cdot mol^{-1}$$

16. Proton transport has a free energy change of $-13.8\ kJ\cdot mol^{-1}$ (Problem 15). Since $\Delta G^{o\prime}$ for ATP synthesis is $30.5\ kJ\cdot mol^{-1}$ and $30.5/13.8 = 2.2$, between two and three moles of protons must be transported to provide the free energy to synthesize one mole of ATP under standard biochemical conditions.

17. The import of ADP (net charge $-3$) and the export of ATP (net charge $-4$) represents a loss of negative charge from inside the mitochondrion. This decreases the difference in electrical charge across the membrane, since the outside is positive due to the translocation of protons during electron transport. Consequently, the electrochemical gradient is diminished by the activity of the ADP–ATP translocator. The activity of the $P_i$–$H^+$ symport protein diminishes the proton gradient by allowing protons from the intermembrane space to re-enter the matrix.

18. The transport of both ADP and $P_i$ is driven by the free energy of the electrochemical proton gradient, since the transport systems for ADP and $P_i$ both dissipate the proton gradient.

19. In an ATP synthase with more $c$ subunits, more proton translocation events are required to drive one complete rotation of the $c$-ring. Consequently, more substrate oxidation ($O_2$ consumption) is required to synthesize three ATP (the yield of one cycle of the rotary engine), and the P/O ratio is lower.

20. (a) Since one ATP is synthesized for every one-third turn of the $c$-ring, 10/3 or 3.3 protons are required to synthesize 1 ATP.

    (b) 15/3 = 5 protons are required to synthesize 1 ATP.

21. The protonation and subsequent deprotonation of Asp 61 of the $F_1F_0$-ATPase's $c$ subunits induces the rotation of the $c$-ring, which in turn, mechanically drives the synthesis of ATP. DCCD reacts with Asp 61 in a manner that prevents it from binding a proton and thereby prevents the synthesis of ATP.

22. Inhibition of proton transport in ATP synthase prevents ATP production by oxidative phosphorylation. The resulting buildup of the proton gradient causes electron transport to slow, thereby slowing the reoxidation of reduced cofactors produced by processes such as the citric acid cycle. In this situation, continued production of ATP depends on anaerobic glycolysis. Because pyruvate-derived acetyl-CoA cannot be processed by the citric acid cycle and because $NAD^+$ for glycolysis cannot be regenerated by the electron transport chain, homolactic fermentation converts pyruvate to lactate, which accumulates.

23. DNP and related compounds dissipate the proton gradient required for ATP synthesis. The dissipation of the gradient decreases the rate of synthesis of ATP, decreasing the ATP mass action ratio. Decreasing this ratio relieves the inhibition of the electron transport chain, causing an increase in metabolic rate.

24. Hormones stimulate the release of fatty acids from stored triacylglycerols, which activates UCP1 and also provides the fuel whose oxidation yields electrons for the heat-generating electron transfer process. This cascade also amplifies the effect of the hormone.

25. The switch to aerobic metabolism allows ATP to be produced by oxidative phosphorylation. The phosphorylation of ADP increases the [ATP]/[ADP] ratio, which then increases the [NADH]/[NAD$^+$] ratio because a high ATP mass action ratio slows electron transport. The increases in [ATP] and [NADH] inhibit their target enzymes in glycolysis and the citric acid cycle (Fig. 18-30) and thereby slow these processes.

26. (a) If the channels allowed the transit of ions other than $Ca^{2+}$, they would dissipate the proton gradient across the inner mitochondrial membrane and prevent ATP synthesis. (b) Because $Ca^{2+}$ ions stimulate several citric acid cycle enzymes (Fig. 18-30), the effect would be an increase in production of reduced cofactors that would increase electron transport and oxidative phosphorylation.

27. The dead algae are a source of food for aerobic microorganisms lower in the water column. As the growth of these organisms increases, the rate of respiration and $O_2$ consumption increase to the point where the concentration of $O_2$ in the water becomes too low to sustain larger aerobic organisms.

28. The molasses and oil are food for microorganisms. As the food is consumed, the rate of respiration and oxygen consumption increase. Eventually, the depletion of oxygen creates a more reducing environment that favors the reduction of Cr(VI) compounds to Cr(III) compounds.

29. Glucose is shunted through the pentose phosphate pathway to provide NADPH, whose electrons are required to reduce $O_2$ to $O_2^-\cdot$.

30. Because SOD apparently protects cells from oxidative damage, cells with defective SOD would be expected to be more susceptible to such damage. (In fact, the mutant SOD retains its enzymatic activity but may misfold or aggregate so as to disrupt normal cellular activities.)

## Chapter 19

1. $2 H_2O + 2 NADP^+ \rightarrow 2 NADPH + 2 H^+ + O_2$

2. $6 CO_2 + 12 H_2S + light\ energy \rightarrow C_6H_{12}O_6 + 6 S_2 + 6 H_2O$

3. The color of the seawater indicates that the photosynthetic pigments of the algae absorb colors of visible light other than red.

4. The light-harvesting complexes absorb light energy of a variety of wavelengths and then pass the energy to the special pair. In order for the excitation to remain on the special pair, that is, not be transferred back to the light-harvesting complex, the special pair must have a lower excitation energy than the components of the light-harvesting complex. Thus the special pair absorbs light at a longer wavelength than do the antenna chromophores.

5. The energy per photon is $E = hc/\lambda$, so the energy per mole of photons ($\lambda = 700$ nm) is
$$E = Nhc/\lambda$$
$$= (6.022 \times 10^{23}\ mol^{-1})(6.626 \times 10^{-34}\ J\cdot s)$$
$$(2.998 \times 10^8\ m\cdot s^{-1})/(7 \times 10^{-7}\ m)$$
$$= 1.71 \times 10^5\ J\cdot mol^{-1}$$
$$= 171\ kJ\cdot mol^{-1}$$

6. $(171\ kJ\cdot mol^{-1})/(30.5\ kJ\cdot mol^{-1}) = 5.6$

    Thus 5 mol of ATP could theoretically be synthesized.

7. The order of action is water–plastoquinone oxidoreductase (Photosystem II), plastoquinone–plastocyanin oxidoreductase (cytochrome $b_6f$), and plastocyanin–ferredoxin oxidoreductase (Photosystem I).

8. The label appears as $^{18}O_2$:
$$H_2^{18}O + CO_2 \xrightarrow{light} (CH_2O) + {}^{18}O_2$$

9. Both systems mediate cyclic electron flows. The photooxidized bacterial reaction center passes electrons through a series of electron carriers so that electrons return to the reaction center (e.g., P960$^+$) and restore it to its original state. During cyclic electron flow in PSI, electrons from photooxidized P700 are transferred to cytochrome $b_6f$ and, via plastoquinone and plastocyanin, back to P700$^+$. In both cases, there is no net change in the redox state of the reaction center, but the light-driven electron movements are accompanied by the transmembrane movement of protons.

10. Use the data provided in Table 14-4.

    $Q + 2 H^+ + 2 e^- \rightleftharpoons QH_2$  $\mathscr{E}°' = 0.045$ V
    $2\ cyt\ c_2(ox) + 2 H^+ + 2 e^- \rightleftharpoons 2\ cyt\ c_2\ (red)$  $\mathscr{E}°' = 0.230$ V
    The overall reaction is
    $$QH_2 + cyt\ c_2\ (ox) \rightarrow 2\ Q_2 + 2\ cyt\ c_2\ (red)$$
    $\Delta\mathscr{E}°' = 0.230$ V $- (0.045$ V$) = 0.185$ V
    $\Delta G°' = -n\mathscr{F}\Delta\mathscr{E}°'$
    $$= -(2)(96,485\ J\cdot V^{-1}\cdot mol^{-1})(0.185\ V)$$
    $$= 36,000\ J\cdot mol^{-1} = 36\ kJ\cdot mol^{-1}$$

11. The change in reduction potential is about $-1.5$ V (Fig. 19-10). Since $\Delta G°' = -n\mathscr{F}\ \Delta\mathscr{E}°'$,
$$\Delta G = -(1)(96,485\ J\cdot V^{-1}\cdot mol^{-1})(-1.5\ V)$$
$$= 140,000\ J\cdot mol^{-1} = 140\ kJ\cdot mol^{-1}$$

12. The relevant half-reactions are (Table 14-4):

$$O_2 + 4\,H^+ + 4\,e^- \rightleftharpoons H_2O \qquad \mathscr{E}^{\circ\prime} = 0.815\ \text{V}$$
$$NADP^+ + H^+ + 2\,e^- \rightleftharpoons NADPH \qquad \mathscr{E}^{\circ\prime} = -0.320\ \text{V}$$

The overall reaction is

$$2\,NADP^+ + 2\,H_2O \rightarrow 2\,NADPH + O_2 + 2\,H^+$$
$$\Delta\mathscr{E}^{\circ\prime} = -0.320\ \text{V} - (0.815\ \text{V}) = -1.135\ \text{V}$$
$$\Delta G^{\circ\prime} = -n\mathscr{F}\Delta\mathscr{E}^{\circ\prime}$$
$$= -(4)(96{,}485\ \text{J}\cdot\text{V}^{-1}\,\text{mol}^{-1})(-1.135\ \text{V})$$
$$= 438\ \text{kJ}\cdot\text{mol}^{-1}$$

13. One mole of photons of red light ($\lambda = 700$ nm) has an energy of 171 kJ. Therefore, $438/171 = 2.6$ moles of photons are theoretically required to drive the oxidation of $H_2O$ by $NADP^+$ to form one mole of $O_2$.

14. The energy of a mole of photons of UV light ($\lambda = 220$ nm) is

$$E = Nhc/\lambda$$
$$= (6.022 \times 10^{23}\ \text{mol}^{-1})(6.626 \times 10^{-34}\ \text{J}\cdot\text{s})$$
$$(2.998 \times 10^8\ \text{m}\cdot\text{s}^{-1})/(2.2 \times 10^{-7}\ \text{m})$$
$$= 544\ \text{kJ}\cdot\text{mol}^{-1}$$

The number of moles of 220-nm photons required to produce one mole of $O_2$ is $438/544 = 0.8$.

15. The buildup of the proton gradient is indicative of a high level of activity of the photosystems. A steep gradient could therefore trigger photoprotective activity to prevent further photooxidation when the proton-translocating machinery is operating at maximal capacity.

16. Photooxidation would not be a good protective mechanism since it might interfere with the normal redox balance among the electron-carrying groups in the thylakoid membrane. Releasing the energy by exciton transfer or fluorescence (emitting light of a longer wavelength) could potentially funnel light energy back to the overactive photosystems. Dissipation of the excess energy via internal conversion to heat would be the safest mechanism, since the photosystems do not have any way to harvest thermal energy to drive chemical reactions.

17. Because the light-dependent reactions (measured as $O_2$ produced by PSII) and the light-independent reactions (measured as $CO_2$ fixed by the Calvin cycle) are only indirectly linked via ATP and NADPH, they may vary. Cyclic electron flow, which increases ATP production without increasing NADPH production, may increase the amount of $O_2$ produced without increasing $CO_2$ fixation.

18. When cyclic electron flow occurs, photoactivation of PSI drives electron transport independently of the flow of electrons derived from water. Thus, the oxidation of $H_2O$ by PSII is not linked to the number of photons consumed by PSI.

19. Because chloroplast cytochrome $b_6f$ is functionally and structurally similar to mitochondrial Complex III, myxothiazol would be expected to block electron transport in the chloroplast. As a result, the Q cycle would not function and no proton gradient would be generated. No ATP would be produced and no electrons would reach $NADP^+$.

20. At 40°C, membrane fluidity is increased such that protons may leak across the membrane, thereby preventing the synthesis of the ATP required for the Calvin cycle. Without the desaturase, the chloroplast would be unable to synthesize membrane lipids with highly unsaturated tails. As a result, the membrane would be less fluid (Section 9-2) and therefore less likely to become leaky at higher temperatures.

21. Use Equation 18-1,

$$\Delta G = 2.3\ RT[\text{pH}\,(side\,1) - \text{pH}\,(side\,2)]$$
$$\Delta G = 2.3(8.3145\ \text{J}\cdot\text{K}^{-1}\cdot\text{mol}^{-1})(298\ \text{K})(-3.4)$$
$$= -19{,}400\ \text{J}\cdot\text{mol}^{-1} = -19.4\ \text{kJ}\cdot\text{mol}^{-1}$$

22. Because 3 ATP are produced for each complete rotation of the ATP synthase $c$-ring, and one proton is translocated for each $c$ subunit, $14/3 = 4.7$ protons must be translocated in order to synthesis 1 ATP.

23. An uncoupler dissipates the transmembrane proton gradient by providing a route for proton translocation other than ATP synthase. Therefore, chloroplast ATP production would decrease.

24. The uncoupler would not affect $NADP^+$ reduction since light-driven electron transfer reactions would continue regardless of the state of the proton gradient.

25. After the light is turned off, ATP and NADPH levels fall as these substances are used up in the Calvin cycle without being replaced by the light reactions. The RuBP level drops because it is consumed by the RuBP carboxylase reaction (which requires neither ATP nor NADPH) and its replenishment is blocked by the lack of ATP for the phosphoribulokinase reaction.

26. 3PG builds up because it cannot pass through the phosphoglycerate kinase reaction in the absence of ATP.

27. The net synthesis of 2 GAP from 6 $CO_2$ in the initial stage of the Calvin cycle (Fig. 19-26) consumes 18 ATP and 12 NADPH (equivalent to 30 ATP). The conversion of 2 GAP to glucose-6-phosphate (G6P) by gluconeogenesis does not require energy input (Section 16-4B), nor does the isomerization of G6P to glucose-1-phosphate (G1P). The activation of G1P to its nucleotide derivative consumes 2 ATP equivalents (Section 16-5), but ADP is released when the glucose residue is incorporated into starch. These steps represent an overall energy investment of $18 + 30 + 1 = 49$ ATP.

Starch breakdown by phosphorolysis yields G1P, whose subsequent degradation by glycolysis yields 3 ATP, 2 NADH (equivalent to 5 ATP), and 2 pyruvate. Complete oxidation of 2 pyruvate to 6 $CO_2$ by the pyruvate dehydrogenase reaction and the citric acid cycle (Section 17-1) yields 8 NADH (equivalent to 20 ATP), 2 $FADH_2$ (equivalent to 3 ATP), and 2 GTP (equivalent to 2 ATP). The overall ATP yield is therefore $3 + 5 + 20 + 3 + 2 = 33$ ATP.

The ratio of energy spent to energy recovered is $49/33 = 1.5$.

28. (a) Glycolysis generates pyruvate, which is decarboxylated to yield acetyl-CoA for fatty acid synthesis and $CO_2$, which diffuses out of the cell. In order not to waste this $CO_2$, the seed's RuBP carboxylase combines the $CO_2$ with RuBP to incorporate it back into carbohydrates that can be broken down to yield more acetyl-CoA to support fatty acid synthesis.

(b) Photons absorbed by PSII and transferred to cytochrome $b_6f$ could drive the synthesis of ATP to support carbon fixation by RuBP carboxylase. Because the entire Calvin cycle does not function, the RuBP must be regenerated by other mechanisms (in this case, it is derived from glycolytic intermediates).

29. The carbonic anhydrase catalyzes the conversion of bicarbonate to $CO_2$, which is the substrate for RuBP carboxylase.

30. $O_2$ competes with $CO_2$ for the active site of the carboxylase. By minimizing the presence of $O_2$, the carboxysome minimizes the frequency of photooxidation and improves the efficiency of carbon fixation.

31. An increase in $[O_2]$ increases the oxygenase activity of RuBP carboxylase–oxygenase and therefore lowers the efficiency of $CO_2$ fixation.

32. These plants store $CO_2$ by CAM. At night, $CO_2$ reacts with PEP to form malate (malic acid) has accumulated that the leaves have a sour taste. During the day, the malate is converted to pyruvate + $CO_2$. The leaves therefore become less acidic and hence tasteless. Late in the day, when all the malate is consumed, the leaves become slightly basic, that is, bitter.

33. The increased availability of the substrate $CO_2$ would increase the rate of photosynthesis. Because $C_4$ plants spend relatively more energy to acquire $CO_2$ for the Calvin cycle, $C_3$ plants might have the advantage when $CO_2$ is more accessible.

34. The increased concentration of $CO_2$ would mean that plants would need to open their stomata less to obtain the $CO_2$ needed for the Calvin cycle. Consequently, less water would be lost through the stomata and the plants' water consumption would decrease.

# Chapter 20

1.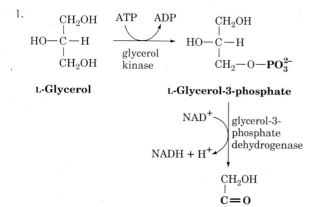

L-Glycerol

L-Glycerol-3-phosphate

Dihydroxyacetone phosphate

2. The reaction products are palmitate, oleate, and 2-oleoylglycerol.

3. Lipoprotein B, with a greater proportion of protein, has a higher density.

4. (a) Perilipin likely resembles the water-soluble apolipoproteins, since it is able to interact with the phospholipid surface of a lipid droplet. Perilipin likely contains amphipathic α helices.

   (b) Phosphorylation of perilipin likely interferes with perilipin–phospholipid interactions, because the negatively charged phosphate groups would repel the phospholipid head groups. As a result, perilipin would associate less tightly with the lipid droplet, allowing access to lipases.

5. The products are heptadecane ($C_{17}H_{36}$) and pentadecane ($C_{15}H_{32}$).

6. The products are one palmitoyl methyl ester, two oleoyl methyl esters, and one glycerol:

$$H_3C-(CH_2)_{14}-\overset{\overset{\displaystyle O}{\|}}{C}-O-CH_3$$

$$H_3C-(CH_2)_7-CH=CH-(CH_2)_7-\overset{\overset{\displaystyle O}{\|}}{C}-O-CH_3$$

$$\underset{HO}{H_2C}-\underset{OH}{CH}-\underset{OH}{CH_2}$$

7. A defect in carnitine palmitoyl transferase II prevents normal transport of activated fatty acids into the mitochondria for β oxidation. Tissues such as muscle that use fatty acids as metabolic fuels therefore cannot generate ATP as needed.

8. The problem is more severe during a fast because other fuels, such as dietary glucose, are not readily available.

9. The first three steps of β oxidation resemble the reactions that convert succinate to oxaloacetate (Sections 17-3F–17-3H).

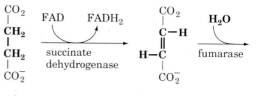

**Succinate**                          **Fumarate**

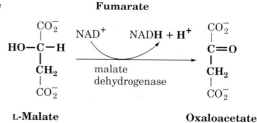

L-Malate                          **Oxaloacetate**

10. Palmitate oxidation produces 106 ATP and glucose catabolism produces 32 ATP (Section 17-4). The standard free energy of ATP synthesis from ADP + $P_i$ is 30.5 kJ · $mol^{-1}$. Palmitate catabolism therefore has an efficiency of

$$106 \times 30.5/9781 \times 100 = 33\%$$

Likewise, glucose catabolism has an efficiency of

$$32 \times 30.5/2850 \times 100 = 34\%$$

Thus, the two processes have very nearly the same overall efficiency.

11. There are not as many usable nutritional calories per gram in unsaturated fatty acids as there are in saturated fatty acids. This is because oxidation of fatty acids containing double bonds yields fewer reduced coenzymes whose oxidation drives the synthesis of ATP. In the oxidation of fatty acids with a double bond at an odd-numbered carbon, the enoyl-CoA isomerase reaction bypasses the acyl-CoA dehydrogenase reaction and therefore does not generate $FADH_2$ (equivalent to 1.5 ATP). A double bond at an even-numbered carbon must be reduced by NADPH (equivalent to the loss of 2.5 ATP).

12. The β oxidation of a saturated $C_{18}$ fatty acid would yield 120 ATP: 8 cycles of β oxidation = 32 ATP; 9 acetyl-CoA yield 90 ATP via the citric acid cycle and oxidative phosphorylation; and 2 ATP are consumed in activating the fatty acid. During β oxidation of oleate, the presence of the double bond allows the $FADH_2$-generating acyl-CoA dehydrogenase step to be skipped, at a cost of 1.5 ATP equivalents. Therefore, the ATP yield from oleate is 118.5.

13. Oxidation of odd-chain fatty acids generates succinyl-CoA, an intermediate of the citric acid cycle. Because the citric acid cycle operates as a multistep catalyst to convert acetyl groups to $CO_2$, increasing the concentration of a cycle intermediate can increase the catalytic activity of the cycle.

14. Conversion of propionyl-CoA to succinyl-CoA consumes 1 ATP. Conversion of succinyl-CoA to malate by the citric acid cycle produces 1 GTP (equivalent to 1 ATP) and 1 $FADH_2$ (equivalent to 1.5 ATP). The conversion of malate to pyruvate produces 1 NADPH (equivalent to 2.5 ATP, assuming NADPH is energetically equivalent to NADH). Conversion of pyruvate to acetyl-CoA produces 1 NADH (equivalent to 2.5 ATP). Each acetyl-CoA that enters the citric acid cycle yields 10 ATP equivalents. Consequently, catabolism of propionyl-CoA yields 16.5 ATP, 6.5 more than for acetyl-CoA.

15. (a) Phytanate can be esterified to CoA, but the methyl group at the β position prevents the dehydrogenation catalyzed by hydroxyacyl-CoA dehydrogenase (reaction 3 of the β oxidation pathway).

(b)

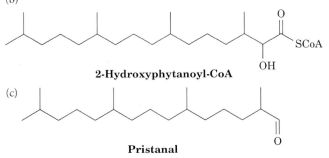

**2-Hydroxyphytanoyl-CoA**

(c)

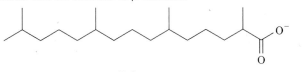

**Pristanal**

16. Pristanate has a methyl group at the α position, which does not interfere with the reactions of β oxidation.

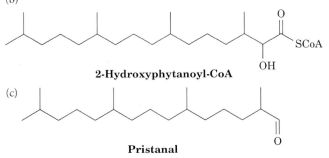

**Pristanate**

The products of β oxidation of pristanate are three acetyl-CoA, three propionyl-CoA, and one methylpropionyl-CoA.

17. 3-Ketoacyl-CoA transferase is required to convert ketone bodies to acetyl-CoA. If the liver contained this enzyme, it would be unable to supply ketone bodies as fuels for other tissues.

18. See Fig. 20-21.

$$H_3C \overset{O}{\underset{||}{\overset{14}{C}}} - CH_2 - \overset{O}{\underset{||}{\overset{14}{C}}} - O^-$$

**Acetoacetate**

19. Palmitate ($C_{16}$) synthesis requires 14 NADPH. The transport of 8 acetyl-CoA to the cytosol by the tricarboxylate transport system supplies 8 NADPH (Fig. 20-24), which represents $8/14 \times 100 = 57\%$ of the required NADPH.

20. This fatty acid (**linolenate**) cannot be synthesized by animals because it contains a double bond closer than 6 carbons from its noncarboxylate end.

21. The label does not appear in palmitate because $^{14}CO_2$ is released in Reaction 2b of fatty acid synthesis (Fig. 20-26).

22. Enoyl-CoA reductase catalyzes step 5 of fatty acid synthesis. Inhibiting this reaction would kill bacteria by preventing them from producing essential lipids.

23. The synthesis of stearate (18:0) from mitochondrial acetyl-CoA requires 9 ATP to transport 9 acetyl-CoA from the mitochondria to the cytosol. Seven rounds of fatty acid synthesis consume 7 ATP (in the acetyl-CoA carboxylase reaction) and 14 NADPH (equivalent to 35 ATP). Elongation of palmitate to stearate requires 1 NADH and 1 NADPH (equivalent to 5 ATP). The energy cost is therefore 9 + 7 + 35 + 5 = 56 ATP.

The degradation of stearate to 9 acetyl-CoA consumes 2 ATP (in the acyl-CoA synthetase reaction) but generates, in eight rounds of β oxidation, 8 $FADH_2$ (equivalent to 12 ATP) and 8 NADH (equivalent to 20 ATP). Thus, the energy yield is 12 + 20 − 2 = 30 ATP. This represents only about half of the energy consumed in synthesizing stearate (30 ATP versus 56 ATP).

24. The synthesis of stearate from acetyl-CoA costs 56 ATP and its β oxidation yields 30 ATP (Problem 23). The complete oxidation of the 9 acetyl-CoA to $CO_2$ by the citric acid cycle yields an additional 9 GTP (equivalent to 9 ATP), 27 NADH (equivalent to 67.5 ATP), and 9 $FADH_2$ (equivalent to 13.5 ATP) for a total of 30 + 9 + 67.5 + 13.5 = 120 ATP. Thus, more than twice the energy investment of synthesizing stearate is recovered (120 ATP versus 56 ATP).

25. Palmitate biosynthesis consumes 7 ATP and 14 NADPH (equivalent to 35 ATP). The addition of four more 2-carbon units as acetyl-CoA in the mitochondrion (Fig. 20-28) consumes 4 NADH (equivalent to 10 ATP) and 4 NADPH (equivalent to 10 ATP), so that a total of 62 ATP are consumed.

26. Palmitate biosynthesis consumes 7 ATP and 14 NADPH (equivalent to 35 ATP). The addition of four more 2-carbon units, initially in the form of acetyl-CoA, in the endoplasmic reticulum consumes 4 ATP (in converting acetyl-CoA to malonyl-CoA). The steps of fatty acid synthesis consume 8 NADPH (equivalent to 20 ATP), so that a total of 66 ATP are consumed.

27. The breakdown of glucose by glycolysis generates the dihydroxyacetone phosphate that becomes the glycerol backbone of triacylglycerols (Fig. 20-29).

28. Glycerol kinase converts glycerol to glycerol-3-phosphate, a precursor for triacylglycerol synthesis. By promoting triacylglycerol synthesis, the drug decreases the concentration of unesterified fatty acids in the body.

29. ACC catalyzes the first committed step of fatty acid synthesis, so blocking this step might decrease the fatty acids available for storage as triacylglycerols (fat). The malonyl-CoA produced in the ACC reaction inhibits import of fatty acyl-CoA into the mitochondria, so lowering the level of malonyl-CoA might help promote fatty acid oxidation and reduce fat accumulation.

30. Dietary fatty acids may be abundant in an obese individual, so that fatty acid synthesis occurs at a low rate. Inhibition of ACC might therefore have little effect on fat metabolism.

31. See Fig. 20-35.

$$\begin{array}{c} OH \\ \overset{14}{|} \\ CH-(CH_2)_{14}-CH_3 \\ | \\ H_2N-C-H \\ | \\ CH_2OH \end{array}$$

**Sphinganine**

32. Statins inhibit the HMG-CoA reductase reaction, which produces mevalonate, a precursor of cholesterol. Although lower cholesterol levels induce the synthesis of HMG-CoA reductase to make up for the loss in activity, some decrease in activity may still be present. Because mevalonate is also the precursor of ubiquinone (coenzyme Q), supplementary ubiquinone may be necessary.

# Chapter 21

1. Proteasome-dependent proteolysis requires ATP to activate ubiquitin in the first step of linking ubiquitin to the target protein (Fig. 21-2) and for denaturing the protein as it enters the proteasome.

2. The structure of the inhibitor suggests that the archaebacterial proteasome cleaves polypeptide substrates at hydrophobic residues such as Leu.

3. The proteasome would facilitate the degradation of intracellular proteins that have been damaged or denatured by heat or oxidation and would therefore help the cell eliminate these nonfunctional and possibly toxic proteins so that they could be replaced by newly synthesized proteins.

4. By interfering with normal cellular protein turnover by the proteasome, ritonavir could promote the accumulation of damaged or unneeded proteins. Such protein accumulation is particularly problematic in long-lived cells such as neurons (see Section 6-5C).

5. A glutamate receptor would have helped human ancestors recognize protein-rich foods, because foods containing significant amounts of protein also contain relatively large amounts of glutamate, one of the most abundant amino acids.

6. Amino acid + $H_2O$ + $O_2$ → α-keto acid + $NH_3$ + $H_2O_2$

7. The urea cycle transforms excess nitrogen from protein breakdown to an excretable form, urea. In a deficiency of a urea cycle enzyme, the preceding urea cycle intermediates may build up to a toxic level. A low-protein diet minimizes the amount of nitrogen that enters the urea cycle and therefore reduces the concentrations of the toxic intermediates.

8. An individual consuming a high-protein diet uses amino acids as metabolic fuels. As the amino acid skeletons are converted to glucogenic or ketogenic compounds, the amino groups are disposed of as urea, leading to increased flux through the urea cycle. During starvation, proteins (primarily from muscle) are degraded to provide precursors for gluconeogenesis. Nitrogen from the protein-derived amino acids must be eliminated, which demands a high level of urea cycle activity.

9. (a) Three ATP are converted to 2 ADP and AMP + $PP_i$, for a total of 4 ATP equivalents.

   (b) The fumarate produced in the urea cycle can be converted to malate and then to pyruvate by malic enzyme, generating NADPH (equivalent to 2.5 ATP). Conversion of pyruvate to acetyl-CoA generates NADH (2.5 ATP equivalents), and the oxidation of the acetyl-CoA by the citric acid cycle yields another 10 ATP, for a total of 15 ATP.

10. (a)

   $$H_2N-\overset{\overset{\displaystyle O}{\|}}{C}-NH_2 + H_2O \rightleftharpoons 2\,NH_3 + CO_2$$

   (b) The $NH_3$ produced by the action of urease can combine with protons in gastric fluid to form $NH_4^+$. This could reduce the concentration of protons and therefore increase the pH.

11. (a) Ala, Arg, Asn, Asp, Cys, Gln, Glu, Gly, His, Met, Pro, Ser, and Val

    (b) Leu and Lys

    (c) Ile, Phe, Thr, Trp, and Tyr

12. Tryptophan can be considered a member of this group since one of its degradation products is alanine, which is converted to pyruvate by deamination.

13. The ε-amino group is removed by the addition of α-ketoglutarate followed by the departure of glutamate (Fig. 21-22, Reactions 1 and 2). The α-amino group is eliminated when α-aminoadipate undergoes transamination with α-ketoglutarate (Fig. 21-22, Reaction 4).

14. Glutamate is converted to α-ketoglutarate by glutamate dehydrogenase, producing 1 NADPH (equivalent to 2.5 ATP). The conversion of α-ketoglutarate to malate by the citric acid cycle produces 1 NADH, 1 GTP, and 1 $FADH_2$ (equivalent to 5 ATP). Malic enzyme converts malate to pyruvate and generates 1 NADH (2.5 ATP equivalents). The conversion of pyruvate to acetyl-CoA also produces 1 NADH (2.5 ATP). The complete oxidation of acetyl-CoA by the citric acid cycle generates 10 ATP, for a total of 22.5 ATP.

   The conversion of methionine to homocysteine costs 3 ATP equivalents. The conversion of homocysteine to propionyl-CoA generates 1 NADH (2.5 ATP equivalents). Conversion of propionyl-CoA to succinyl-CoA consumes 1 ATP. Converting succinyl-CoA to malate by the citric acid cycle generates 1 GTP and 1 $FADH_2$ (1.5 ATP equivalents). The remaining steps are the same as described for glutamate. The net yield of ATP from methionine breakdown is 16 ATP, significantly less than from glutamate.

15. Since the three reactions converting tiglyl-CoA to acetyl-CoA and propionyl-CoA are analogous to those of fatty acid oxidation (β oxidation; Fig. 20-12), the reactions are

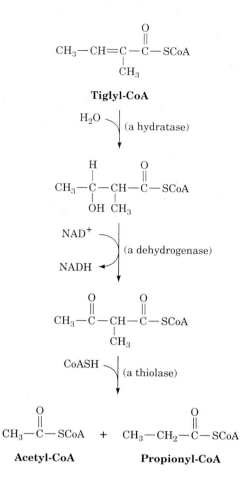

**Tiglyl-CoA**

**Acetyl-CoA**     **Propionyl-CoA**

16.

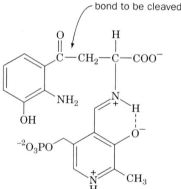

17. Tyrosine is derived from the essential amino acid phenylalanine, and cysteine is derived from the essential amino acid methionine. A diet lacking sufficient phenylalanine and methionine will lead to shortages of tyrosine and cysteine also.

18. Biotin, conversion of pyruvate to oxaloacetate by pyruvate carboxylase (Fig. 16-18); coenzyme $B_{12}$, conversion of (S)-methylmalonyl-CoA to (R)-methylmalonyl-CoA (Fig. 20-19); S-adenosylmethionine, the conversion of norepinephrine to epinephrine by phenylethanolamine N-methyltransferase (Fig. 21-39); tetrahydrofolate, the conversion of glycine to serine by serine hydroxymethyltransferase (Fig. 21-14); thiamine pyrophosphate, the decarboxylation of pyruvate by pyruvate dehydrogenase (Fig. 17-6).

19. Glutamate dehydrogenase, glutamine synthetase, and carbamoyl phosphate synthetase.

20. The γ-carboxylate group of glutamate is reduced to form glutamate-5-semialdehyde. An aminotransferase then transfers an amino group (from glutamate or another amino acid) to yield ornithine.

21. In the absence of uridylyl-removing enzyme, adenylyltransferase · $P_{II}$ will be fully uridylylated, since there is no mechanism for removing the uridylyl groups once they are attached. Uridylylated adenylyltransferase · $P_{II}$ adenylylates glutamine synthetase, which activates it. Hence, the defective *E. coli* cells will have a hyperactive glutamine synthetase and thus a higher than normal glutamine concentration. Reactions requiring glutamine will therefore be accelerated, thereby depleting glutamate and the citric acid cycle intermediate α-ketoglutarate. Consequently, biosynthetic reactions requiring transamination, as well as energy metabolism, will be suppressed.

22. Since only plants and microorganisms synthesize aromatic amino acids, herbicides that inhibit these pathways do not affect amino acid metabolism in animals.

23. Agmatine is derived by decarboxylation from arginine.

24. The compound resembles the urea cycle intermediate ornithine (with a $CHF_2$ group at its $C_\alpha$ atom).

25. The pigment coloring skin and hair is melanin, which is synthesized from tyrosine. When tyrosine is in short supply, as when dietary protein is not available, melanin cannot be synthesized in normal amounts, and the skin and hair become depigmented.

26. Melatonin is derived from tryptophan, which undergoes decarboxylation, *N*-acetylation, hydroxylation, and *O*-methylation.

27. The standard nitrogenase reaction, $N_2 \rightarrow NH_3$, also produces $H_2$. This $H_2$ is used to reduce CO to $C_2H_6$ and $C_3H_8$.

28. The oxidation of ammonia to nitrite is an exergonic process that yields the ATP and reduced NADPH required for the Calvin cycle.

## Chapter 22

1. ATP generating pathways such as glycolysis and fatty acid oxidation require an initial investment of ATP (the hexokinase and phosphofructokinase steps of glycolysis and the acyl-CoA synthetase activation step that precedes β oxidation). This "priming" cannot occur when ATP has been exhausted.

2. (a) In the absence of MCAD, fatty acids cannot be fully oxidized to acetyl-CoA (Section 20-2C). Since ketone bodies are synthesized from acetyl-CoA (Section 20-3), ketogenesis is impaired.

(b) In normal individuals, acetyl-CoA activates pyruvate carboxylase (Section 17-5B), which converts pyruvate to oxaloacetate. This increases the capacity of the citric acid cycle to metabolize acetyl-CoA. When glucose levels are low, the oxaloacetate is used for gluconeogenesis (Section 16-4). In MCAD deficiency, lack of fatty acid–derived acetyl-CoA keeps pyruvate carboxylase activity low, thereby limiting the synthesis of glucose and contributing to hypoglycemia.

3. Rapidly growing cancer cells need ATP as well as the raw materials for synthesizing nucleic acids, proteins, carbohydrates, and lipids. While providing some ATP, catabolism of glucose can also supply ribose (via the pentose phosphate pathway) for nucleotide biosynthesis, pyruvate that can be converted to acetyl-CoA (via the pyruvate dehydrogenase reaction) for fatty acid synthesis, and oxaloacetate (via the pyruvate carboxylase reaction) for amino acid synthesis.

4. The pentose phosphate pathway supplies ribose as well as NADPH to support the biosynthetic processes, including nucleotide synthesis, that are necessary for cell growth and division.

5. At high altitude, less oxygen is available for aerobic metabolism, so glycolysis, an anaerobic pathway, would become relatively more important in active muscles. An increase in GLUT1 would increase the intracellular glucose concentration, and an increase in PFK would increase the flux of glucose through the pathway.

6. Lactate must be converted back to pyruvate, which generates NADH (equivalent to 2.5 ATP). Another NADH (2.5 ATP) is generated by the conversion of pyruvate to acetyl-CoA. Complete oxidation of the acetyl-CoA by the citric acid cycle generates 10 more ATP equivalents, for a total of 15 ATP. The ATP yield from 1 mol of lactate is 15 mol.

7. GLUT2 has a higher $K_M$ than GLUT1 so that the rate of glucose entry into liver cells can vary directly with the concentration of glucose in the blood. A transporter with a high $K_M$ is less likely to be saturated with its ligand and therefore would not limit the rate of transport.

8. Type I glycogen storage disease results from a deficiency of glucose-6-phosphatase so that glucose-6-phosphate produced by glycogenolysis cannot exit the cell as glucose. A defect in the glucose-transport protein GLUT2 would similarly prevent the exit of glucose (a passive transporter can operate in either direction). In both cases, the buildup of intracellular glucose-6-phosphate prevents glycogen breakdown, and glycogen accumulates.

9. The portal vein delivers $NH_4^+$-rich blood from the intestine directly to the liver, which can convert it to urea (only the liver carries out the urea cycle; Section 21-3). The remaining $NH_4^+$ is carried through the circulation to other tissues, where glutamine synthetase converts glutamate to glutamine (Section 21-5A).

10. (a) Glutamate dehydrogenase converts glutamate to α-ketoglutarate and $NH_4^+$ (Section 21-2B). Glutaminase converts glutamine to glutamate and $NH_3$ (Section 21-4C).

(b) α-Ketoglutarate → succinyl-CoA → succinate → fumarate → malate → oxaloacetate → PEP → 2PG → 3PG → 1,3BPG → DHAP/GAP → F1,6BP → F6P → G6P → glucose.

11. Insulin promotes the uptake of glucose via the increase in GLUT4 receptors on the adipocyte surface. A source of glucose is necessary to supply the glycerol-3-phosphate backbone of triacylglycerols.

12. Hyperinsulinemia would result in a decrease in blood glucose. The decrease in [glucose] for the brain would cause loss of brain function (leading to coma and death).

13. Because fatty acids, like glucose, are metabolic fuels, it makes metabolic sense for them to stimulate insulin release, which is a signal of abundant fuel.

14. Elevated levels of circulating fatty acids occur during an extended fast, when dietary glucose and glucose mobilized from glycogen stores are no longer available. Insulin release would be inappropriate for these conditions. A combination of abundant fatty acids and glucose, indicating the fed state, would serve as a better trigger for insulin release.

15. Insulin activates ATP-citrate lyase, which is the enzyme that converts citrate to oxaloacetate and acetyl-CoA (Section 20-4A). The activity of this enzyme is essential for making acetyl units available for fatty acid biosynthesis in the cytosol. The acetyl units, generated from pyruvate in the mitochondria, combine with oxaloacetate to form citrate, which can then be transported from the mitochondria to the cytosol for reconversion to acetyl-CoA.

16. (a) Decrease; (b) decrease; (c) increase; (d) increase.

17. Adipose tissue synthesizes and releases the polypeptide hormones adiponectin, leptin, and resistin.

18. The leptin produced by the normal mouse will enter the circulation of the *ob/ob* mouse, resulting in decreases in its appetite and weight.

19. Since $PYY_{3-36}$ is a peptide hormone, it would be digested if taken orally. Introducing it directly into the bloodstream avoids degradation.

20. The stomach and intestine are endocrine organs, so the reduction in size of the organs decreases the production of hormones that may have been acting in opposition to insulin.

21. Ingesting glucose while in the resting state causes the pancreas to release insulin. This stimulates the liver, muscle, and adipose tissue to synthesize glycogen, fat, and protein from the excess nutrients while inhibiting the breakdown of these metabolic fuels. Hence, ingesting

glucose before a race will gear the runner's metabolism for resting rather than for running.

22. During starvation, the synthesis of glucose from liver oxaloacetate depletes the supply of citric acid cycle intermediates and thus decreases the ability of the liver to metabolize acetyl-CoA via the citric acid cycle.

23. Type I diabetics lack β cells that produce insulin, so providing the hormone is an effective treatment for the disorder. In type II diabetes, cells do not respond efficiently to insulin. Increasing the availability of the hormone may boost its signaling activity in some patients, but in the majority of type II diabetics, insulin levels are already elevated and further increases are ineffective.

24. An intermediate in the biosynthesis of triacylglycerols is diacylglycerol (DAG), a second messenger responsible for activating PKC.

25. PFK-2 catalyzes the production of fructose-2,6-bisphosphate, so increasing PFK-2 activity would increase the concentration of this activator of phosphofructokinase. The effect would be increased flux of glucose through glycolysis, which would help lower the concentration of glucose in the blood.

26. Physical inactivity would lead to a decreased need for ATP in muscle, which would be reflected by a decreased AMP to ATP ratio. A decrease in the ratio would lead to a decrease in AMPK activity. AMPK activity is positively associated with glucose uptake by cells due to an increase in GLUT4 activity. GLUT4 activity is also increased by insulin. A decrease in AMPK activity causes a decrease in GLUT4 activity, making insulin's job more difficult.

# Chapter 23

1. Following aspartate addition to IMP, adenylosuccinate lyase removes fumarate, leaving an amino group. In the urea cycle, following the addition of aspartate to citrulline, argininosuccinase removes fumarate, leaving an amino group.

2. Caffeine is derived by the methylation of xanthine.

3. Amidophosphoribosyl transferase (step 2 of IMP synthesis), FGAM synthetase (step 5 of IMP synthesis), GMP synthetase (GMP synthesis), carbamoyl phosphate synthetase II (step 1 of UMP synthesis), and CTP synthetase (CTP synthesis).

4. PRPP and FGAR accumulate because they are substrates of Reactions 2 and 5 in the IMP biosynthetic pathway (Fig. 23-1). XMP also accumulates because the GMP synthetase reaction is blocked (Fig. 23-3). Although glutamine is a substrate of carbamoyl phosphate synthetase II (the first enzyme of UMP synthesis; Fig. 23-5), the other substrates of this enzyme do not accumulate. UTP, a substrate of the CTP synthetase reaction, also accumulates, although strictly speaking, it is a nucleotide biosynthetic product rather than an intermediate.

5. (a) 7 ATP; (b) 8 ATP; (c) 7 ATP

6. (a) The recovered deoxycytidylate would be equally labeled in its base and ribose components (i.e., the same labeling pattern as in the original cytidine).

   (b) The recovered deoxycytidylate would be unequally labeled in its base and ribose components because the separated $^{14}$C-cytosine and $^{14}$C-ribose would mix with the different-sized pools of unlabeled cellular cytosine and ribose before recombining as the deoxycytidylate that becomes incorporated into DNA. [This experiment established that deoxyribonucleotides, in fact, are synthesized from their corresponding ribonucleotides (alternative a).]

7. UTP functions as a feedback inhibitor of its own synthesis, to prevent the cell from synthesizing too many pyrimidine nucleotides. ATP activates pyrimidine nucleotide synthesis so that when ATP concentrations are high, the production of other nucleotides will increase to match it.

8. Ribose phosphate pyrophosphokinase catalyzes the activation of ribose-5-phosphate to produce PRPP, the substrate for the second reaction of purine nucleotide synthesis and the fifth step of pyrimidine nucleotide synthesis. High concentrations of ADP and GDP signify high metabolic demand and low concentration of nucleoside triphosphates, a situation when the cell's resources should be directed toward energy metabolism rather than the production of nucleotides for the synthesis of DNA or RNA.

9. Hydroxyurea destroys the tyrosyl radical that is essential for the activity of ribonucleotide reductase. Tumor cells are generally fast-growing and cannot survive without this enzyme, which supplies dNTPs for nucleic acid synthesis. In contrast, most normal cells grow slowly, if at all, and hence have less need for nucleic acid synthesis.

10. dATP inhibits ribonucleotide reductase, thereby preventing the synthesis of the deoxynucleotides required for DNA synthesis.

11. Threonine is broken down to acetyl-CoA and glycine, either directly via the serine hydroxymethyltransferase reaction (reaction 5 in Fig. 21-14) or through the intermediacy of α-amino-β-ketobutyrate via the threonine dehydrogenase reaction (reaction 6 of Fig. 21-14) followed by the α-amino-β-ketobutyrate lyase reaction (reaction 7 of Fig. 21-14). The acetyl-CoA can enter the citric acid cycle to produce considerable ATP via oxidative phosphorylation. The glycine is a substrate of the glycine cleavage system, which generates $N^5,N^{10}$-methylene-THF (reaction 3 of Fig. 21-14), the methyl-group donor required for thymidylate synthesis.

12. Serine donates a hydroxymethyl group to THF in order to regenerate the cofactor for the conversion of dUMP to dTMP by thymidylate synthase (Fig. 23-16).

13. FdUMP and methotrexate kill rapidly proliferating cells, such as cancer cells and those of hair follicles. Consequently, hair falls out.

14. The mutant cells grow because the medium contains the thymidine they are unable to make. Normal cells, however, continue to synthesize their own thymidine and thereby convert their limited supply of THF to DHF. The methotrexate inhibits dihydrofolate reductase, so THF cannot be regenerated. Without a supply of THF for the synthesis of nucleotides and amino acids, the cells die.

15. The synthesis of histidine and methionine requires THF. The cell's THF is converted to DHF by the thymidylate synthase reaction, but in the presence of methotrexate, THF cannot be regenerated.

16. The conversion of dUMP to dTMP is a reductive methylation. In the thymidylate synthase reaction shown in Fig. 23-15, THF is oxidized to DHF, so that DHFR must subsequently reduce the DHF to THF. Organisms that lack DHFR use an alternative mechanism for converting dUMP to dTMP in which the FAD cofactor of the enzyme, rather than the folate, undergoes oxidation.

17. Trimethoprim binds to bacterial dihydrofolate reductase but does not permanently inactivate the enzyme. Therefore, it is not a mechanism-based inhibitor.

18. Allopurinol is oxidized by xanthine oxidase to a product that irreversibly binds to the enzyme. It is therefore a mechanism-based inhibitor of xanthine oxidase.

19. In muscles, the purine nucleotide cycle functions to convert aspartate to fumarate to boost the capacity of the citric acid cycle. If glutamate dehydrogenase activity were high, then it would combine the $NH_4^+$ produced by the purine nucleotide cycle with α-ketoglutarate to yield glutamate, a reaction that depletes a citric acid cycle intermediate.

20. In von Gierke's disease (glucose-6-phosphatase deficiency), glucose-6-phosphate accumulates in liver cells, thereby stimulating the pentose phosphate pathway. The resulting increase in ribose-5-phosphate production boosts the concentration of PRPP, which in turn stimulates purine biosynthesis. High levels of uric acid derived from the breakdown of the excess purines causes gout.

21. Fumarate is converted to malate by fumarase; malic enzyme decarboxylates malate to produce pyruvate; pyruvate is converted to acetyl-CoA and $CO_2$ by pyruvate dehydrogenase; and the citric acid cycle oxidizes the acetyl group to 2 $CO_2$.

22. The conversion of thymine to methylmalonyl-CoA consumes NADPH but generates NADH. Methylmalonyl-CoA is converted to succinyl-CoA, which enters the citric acid cycle and is converted to malate, thereby producing one ATP equivalent and $FADH_2$ (equivalent to 1.5 ATP). Malate is converted to pyruvate by malic enzyme, producing NADPH (equivalent to 2.5 ATP). The pyruvate dehydrogenase reaction generates NADH (2.5 ATP equivalents) and acetyl-CoA. Oxidation of acetyl-CoA by the citric acid cycle generates 1 ATP, 3 NADH (7.5 ATP), and $FADH_2$ (1.5 ATP), for a total yield of 17.5 ATP.

23. Uracil and thymine accumulate in the urine because they cannot be further degraded in the absence of the dihydropyrimidine dehydrogenase (Fig. 23-24).

24. 5-Fluorouracil is an anticancer drug because it is converted in the body to FdUMP, an inhibitor of thymidylate synthase. In the absence of adequate dihydropyrimidine dehydrogenase activity, the drug cannot readily be broken down, so it accumulates to toxic levels in the body.

25.

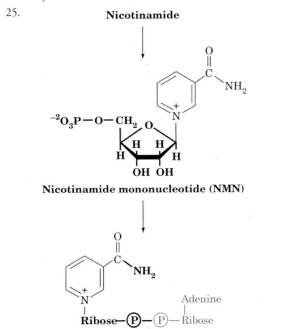

26. The kinase-catalyzed phosphorylation of riboflavin consumes one "high-energy" bond. In the second reaction, an AMP group is transferred from ATP to FMN (thereby breaking a second "high-energy" bond), leaving $PP_i$, whose subsequent hydrolysis breaks a third "high-energy" bond.

## Chapter 24

1. Hypoxanthine pairs with cytosine in much the same way as does guanine.

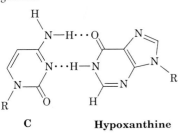

2.

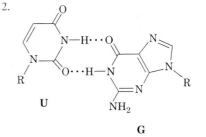

3. Since amino acids have an average molecular mass of ~110 D, the 50-kD protein contains 50,000 D ÷ 110 D/residue = ~455 residues. These residues are encoded by 455 × 3 = 1365 nucleotides. In B-DNA, the rise per base pair is 3.4 Å, so the contour length of 1365 bp is 3.4 Å/bp × 1365 bp = 4641 Å, or 0.46 μm.

4. In A-DNA, the contour length would be 1365 bp × 2.9 Å/bp = 3959 Å, or 0.40 μm.

5.

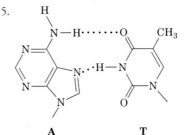

The helix diameter is smaller in this alternate arrangement (Hoogsteen base pairing).

6.

7.

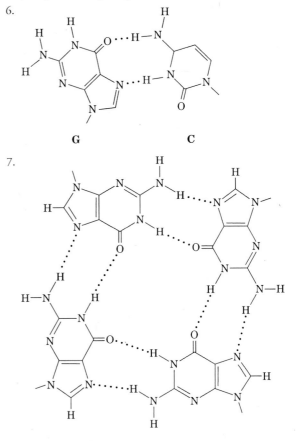

8.

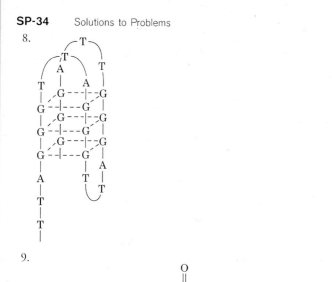

9.

$$^{+}H_3N-CH_2-CH_2-NH-CH_2-\overset{\overset{O}{\|}}{C}-NH-CH_2-CH_2-NH-CH_2-\overset{\overset{O}{\|}}{C}-O^-$$

10. The PNA backbone contains six covalent bonds between each base-attachment site, the same number of bonds as in a DNA backbone (see Fig. 24-5). Consequently, the spacing between bases within each polymer is compatible with base pairing.

11. Assuming all the DNAs in Figure 24-8 contain the same number of base pairs, the most supercoiled structure would move farthest during electrophoresis, because its compact structure allows it to move fastest through the agarose matrix.

12. The enzyme has no effect on the supercoiling of DNA since cleaving the C2'—C3' bond of ribose does not sever the sugar–phosphate chain of DNA.

13. $L = T + W$. For the constrained DNA circle, $W = 0$ so that $L = T = 207$. For the unconstrained DNA circle, $L = 207$ since this quantity is invariant, $T = 2310$ bp/(10.5 bp/turn) = 220, and $W = L - T = 207 - 220 = -13$. For the constrained DNA circle, $\sigma = W/T = 0/207 = 0$. For the unconstrained DNA circle, $\sigma = -13/220 = -0.059$ (a value that is typical of naturally occurring DNA circles *in vivo*).

14. In the B-DNA to Z-DNA transition, a right-handed helix with one turn per 10.5 base pairs converts to a left-handed helix with one turn per 12 base pairs. Since a right-handed duplex helix has a positive twist, the twist decreases:

$$\Delta T = \frac{-100}{12} - \frac{100}{10.5} = -17.9 \text{ turns}$$

The linking number must remain constant ($\Delta L = 0$) since no covalent bonds are broken. Hence, the change in writhing number is $\Delta W = -\Delta T = 17.9$ turns.

15. Its $T_m$ decreases because the charges on the phosphate groups are less shielded from each other at lower ionic strength and hence repel each other more strongly, thereby destabilizing the double helix.

16. The nonpolar solvent diminishes the hydrophobic forces that stabilize double-stranded DNA and hence lowers the $T_m$.

17. The segment with 20% A residues (i.e., 40% A · T base pairs) contains 60% G · C base pairs and therefore melts at a higher temperature than a segment with 30% A residues (i.e., 40% G · C base pairs).

18. (a) As the temperature increases, the stacked bases melt apart so that their ultraviolet absorbance increases (the hyperchromic effect).

(b) The broad shape of the poly(A) melting curve indicates noncooperative changes, as expected for a single-stranded RNA. The sharp melting curve for double-stranded DNA reflects the cooperativity of strand separation.

19.

```
      A U U G G  C
      | | | | |    A
  A A U A G C C  U
```

20.

```
            G
        C       G
        A—T
        C—G
        C—G
        T—A
        G—C
        A—T
5'-T C A   T G C-3'
    | | |   | | |
3'-A G T   A C G-5'
        T—A
        C—G
        A—T
        G—C
        G—C
        T—A
      G       C
            C
```

21.

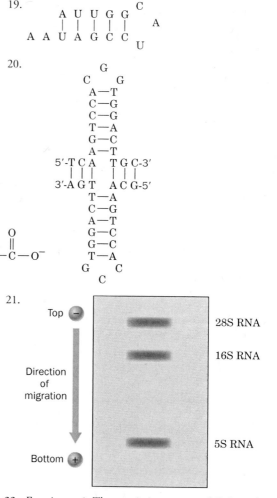

22. *Experiment 1.* The restriction enzyme failed to digest the genomic DNA, leaving the DNA too large to enter the gel during electrophoresis.

*Experiment 2.* The hybridization conditions were too "relaxed," resulting in nonspecific hybridization of the probe to all the DNA fragments. This problem could be corrected by boiling the blot to remove the probe and repeating the hybridization at a higher temperature and/or lower salt concentration.

*Experiment 3.* The probe hybridized with three different mouse genes. The different intensity of each band reflects the relatedness of the sequences. The most intense band is most similar to the human *rxr-1* gene, whereas the least intense band is least similar to the *rxr-1* gene.

23. The target sequence consists of 6 symmetry-related base pairs. Since there are 4 possible base pairs (A · T, T · A, G · C, and C · G), the probability that any two base pairs are randomly related by symmetry is 1/4. Hence, the probability of finding all 6 pairs of base pairs by random chance is $(1/4)^6 = 2.4 \times 10^{-4}$.

24. (a) A 6-nt sequence would be expected to occur, on average, every $4^6 = 4096$ nt in single-stranded DNA. However, in double-stranded DNA, it would be expected to occur at twice this frequency, that is, every 4096/2 = 2048 bp. Thus the expected number of copies of a 6-bp sequence in the *E. coli* genome is 4,639,000 bp/2048 bp = 2265.

(b) A 12-bp sequence would be expected to occur, on average, every $4^{12}/2 = 8,388,608$ bp, which is nearly twice as large as the number of base pairs in the *E. coli* genome. Thus the *trp* repressor is unlikely to bind specifically to any other site in the *E. coli* chromosome.

25. Because they interact closely with DNA, protamines must be rich in basic amino acids. In fact, they are particularly rich in arginine.

26. The decarboxylation of the amino acid ornithine, an intermediate of the urea cycle (Fig. 21-9), generates 1,4-diaminobutane (also known as putrescine):

$$^{+}H_3N—(CH_2)_4—NH_3^{+}$$

This cationic molecule interacts electrostatically with the negatively charged phosphate groups of DNA.

27. (a) The contour length is $5 \times 10^7$ bp $\times$ 3.4 Å/bp $= 1.7 \times 10^8$ Å $= 17$ mm.

(b) A nucleosome, which binds ~200 bp, compresses the DNA to an 80-Å-high supercoil. The length of the DNA is therefore (80 Å/200 bp) $\times$ ($5 \times 10^7$ bp) $= 2 \times 10^7$ Å $= 2$ mm.

28. In the 30-nm fiber, 18.9 nucleosomes cover 316 Å. The length of the DNA is (316 Å/18.9 nucleosomes) $\times$ (1 nucleosome/200 bp) $\times$ ($5 \times 10^7$ bp) $= 4.2 \times 10^6$ Å $= 0.42$ mm.

29. Histones are required in large amounts during a relatively short period when DNA is replicated prior to cell division. The large number of histone genes allows the efficient production of histones.

30. Base methylation is expected to have little or no effect on nucleosomal structure, because there are few contacts between histones and bases and the small, nonpolar methyl group would be unlikely to disrupt the mostly ionic interactions between the histones and the DNA backbone.

## Chapter 25

1. Okazaki fragments are 1000 to 2000 nt long, and the *E. coli* chromosome contains $4.6 \times 10^6$ bp. Therefore, *E. coli* chromosomal replication requires 2300 to 4600 Okazaki fragments.

2. Okazaki fragments are 100 to 200 nt long in humans, and the chromosomes contain $6.0 \times 10^9$ bp (humans are diploid). Therefore, human chromosomal replication requires $6.0 \times 10^7$ to $3.0 \times 10^7$ Okazaki fragments.

3. As indicated in Fig. *a* (below), nucleotides would be added to a polynucleotide strand by attack of the 3'-OH of the incoming nucleotide on the 5' triphosphate group of the growing strand with the elimination of PP$_i$. The hydrolytic removal of a mispaired nucleotide by the 5' → 3' exonuclease activity (Fig. *b*, below) would leave only an OH group or monophosphate group at the 5' end of the DNA chain. This would require an additional activation step before further chain elongation could commence.

*(a)* 3' → 5' Polymerase

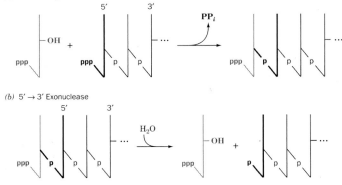

*(b)* 5' → 3' Exonuclease

4. The drug would inhibit DNA synthesis because the polymerization reaction is accompanied by the release and hydrolysis of PP$_i$. Failure to hydrolyze the PP$_i$ would remove the thermodynamic driving force for polymerization, that is, it would be reversible.

5. The 5' → 3' exonuclease activity is essential for DNA replication because it removes RNA primers and replaces them with DNA. Absence of this activity would be lethal.

6. The Klenow fragment, which lacks 5' → 3' exonuclease activity (and therefore cannot catalyze nick translation), is used to ensure that all the replicated DNA chains have the same 5' terminus, a necessity if a sequence is to be assigned according to fragment length.

7. When DNA polymerase begins synthesizing a new strand, it binds the template DNA to which an RNA primer is already base paired. In order to extend the primer, the polymerase active site must accommodate the DNA–RNA hybrid helix, which has an A-DNA-like structure (Fig. 24-4).

8. PP$_i$ is the product of the polymerization reaction catalyzed by DNA polymerase. This reaction also requires a template DNA strand and a primer with a free 3' end.

(a) There is no primer strand, so no PP$_i$ is produced.

(b) There is no primer strand, so no PP$_i$ is produced.

(c) PP$_i$ is produced.

(d) No PP$_i$ is produced because there is no 3' end that can be extended.

(e) PP$_i$ is produced.

(f) PP$_i$ is produced.

9. AT-rich DNA is less stable than GC-rich DNA and therefore would more readily melt apart, a requirement for initiating replication.

10. DNA gyrase adds negative supercoils to relieve the positive supercoiling that helicase-catalyzed unwinding produces ahead of the replication fork.

11. DNA polymerase could extend a primer that entered its active site, but polymerization would not be highly processive unless the sliding clamp was in place. If the primer first associates with the clamp, which then interacts with DNA polymerase, the primer can be extended in a processive and therefore more efficient manner.

12. Mismatch repair and other repair systems correct most of the errors missed by the proofreading functions of DNA polymerases.

13. The *E. coli* replication system can fully replicate only circular DNAs. Bacteria do not have a mechanism (e.g., telomerase-catalyzed extension of telomeres) for replicating the extreme 3' ends of linear template strands.

14. The broken end of a chromosome will not have the characteristic structure of a telomere, which includes repeating DNA sequences plus telomere-binding proteins.

15. After adding the 6-nt telomere sequence, telomerase translocates to the new 3' end to add another repeat. The RNA template includes a region of overlap (~3 nt) that helps position the enzyme and template for the next addition of nucleotides (see Fig. 25-26).

16. (a) DNA polymerase; (b) reverse transcriptase or telomerase; (c) RNA polymerase or primase.

17. (a)

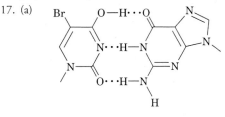

**5BU**
**(enol tautomer)** **Guanine**

(b) When 5BU incorporated into DNA pairs with G, the result is an A · T → G · C transition after two more rounds of DNA replication:

$$A \cdot T \rightarrow A \cdot 5BU \rightarrow G \cdot 5BU \rightarrow G \cdot C$$

18. The cytosine derivative base pairs with adenine, generating a C · G → T · A transition.

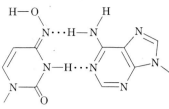

**Adenine**

19. (a) The original A · T base pair becomes an I · T base pair. When the DNA replicates, the template T will pair with A, and the template I will pair with C. Thus, one daughter cell will be normal and one will have an abnormal I · C base pair. (b) After the second round of cell division, two cells will be normal (A · T base pairs). One cell will have an abnormal I · C base pair, and one will have a mutated C · G base pair.

20. (a) After one round of cell division, one daughter cell will contain a normal A · T base pair, and the other cell will contain a C · A mismatch. (b) Two rounds of cell division yield three cells with a normal A · T base pair and one cell with a C · G base pair (a transition mutation).

21. The half-life, $t_{1/2}$ = 0.693/rate = 0.693/(1.7 × 10⁻¹³ s⁻¹) = 4.1 × 10¹² s or 130,000 years.

22. Use Equation 12-6,

$$\ln[A] = \ln[A]_o - kt$$

where $[A]_o = 6.0 \times 10^9$, $[A] = [A]_o - 2 \times 10^4 = 5.99998 \times 10^9$, and $t = 1$ d. Solve the equation for $k$:

$$k = \left(\ln\frac{[A]_o}{[A]}\right)/t = \left(\ln\frac{6 \times 10^9}{5.99998 \times 10^9}\right)/1 \text{ d}$$

$$k = \ln(1.0000033)/1 \text{ d} = 0.0000033 \text{ d}^{-1}$$

Using Equation 12.9,

$$t_{1/2} = 0.693/k = 0.693/0.0000033 \text{ d}^{-1}$$

$$= 2.08 \times 10^5 \text{ d} \times (1 \text{ y}/365 \text{ d}) = 570 \text{ y}$$

23. The triphosphatase destroys nucleotides containing the modified base before they can be incorporated into DNA during replication.

24. (a) Without an active site Cys residue, the alkyl transfer reaction cannot occur.

(b) Although the proteins cannot remove the alkyl group attached to the guanine by direct repair, they can still bind to the modified residue, thereby marking this region of DNA for repair by NER enzymes.

25. When 5-methylcytosine residues deaminate, they form thymine residues.

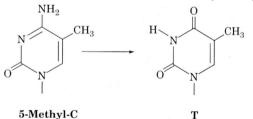

**5-Methyl-C**            **T**

Since thymine is a normal DNA base, the repair systems cannot determine whether such a T or its opposing G is the mutated base. Consequently, only about half of the deaminated 5-methylcytosines are correctly repaired.

26. Mammalian DNA contains 5-methylcytosine residues paired with guanine residues. Oxidative deamination of m⁵C produces thymine. The thymine–DNA glycosylase removes the T in the resulting T · G base pair so that it can be replaced with C to restore the correct C · G base pair.

27. Base excision repair. The deaminated base can be recognized because hypoxanthine does not normally occur in DNA.

28. After the damaged DNA has been removed and replaced by the action of DNA polymerase, the final phosphodiester bond (between the 5′ end of the original DNA and the 3′-OH of the last nucleotide added) must be formed by the action of DNA ligase.

29. *E. coli* contains a low concentration of dUTP, which DNA polymerase incorporates into DNA in place of dTTP. The resulting uracil bases are rapidly excised by uracil–DNA glycosylase followed by nucleotide excision repair (NER), which temporarily causes a break in the DNA chain. DNA that is isolated before DNA polymerase I and DNA ligase can complete the repair process would be fragmented. However, in the absence of a functional uracil–DNA glycosylase, the inappropriate uracil residues would remain in place, and hence leading strand DNA would be free of breaks. The lagging strand, being synthesized discontinuously, would still contain breaks, although fewer than otherwise.

30. Topoisomerases maintain the appropriate degree of supercoiling during DNA replication. These enzymes act by cleaving one or both DNA strands and covalently linking the 3′- or 5′-phosphate to an enzyme Tyr residue (Section 24-1D). If the catalytic cycle is not completed, the enzyme remains associated with the cut DNA and impedes replication and transcription. A tyrosyl–DNA phosphodiesterase frees the trapped topoisomerase so that the DNA can be repaired by other enzymes.

31. DNA polymerase η can synthesize a complementary DNA strand, but the thymine dimer is still present. It can be repaired later by the NER pathway.

32. Pol V is less processive than Pol III. When the progress of Pol III is arrested by the presence of a thymine dimer, Pol V can take over, allowing replication to continue at a high rate, although with a greater incidence of mispairings. The damage is minimal, however, since Pol V soon dissociates from the DNA, allowing the more accurate Pol III to resume replicating DNA.

33. (a) Loss of the helicase DnaB, which unwinds DNA for replication, would be lethal.

(b) Loss of Pol I would prevent the excision of RNA primers and would therefore be lethal.

(c) SSB prevents reannealing of separated single strands. Loss of SSB would be lethal.

(d) RecA protein mediates the SOS response and homologous recombination. Loss of RecA would be harmful but not necessarily lethal.

34. In conjugation, the ssDNA becomes incorporated into the recipient's DNA through homologous recombination. RecBCD includes nuclease and helicase activity, which are necessary to nick and unwind the recipient dsDNA so that the incoming ssDNA can be introduced.

## Chapter 26

1. (a) Cordycepin is the 3′-deoxy analog of adenosine.

(b) Because it lacks a 3′-OH group, the cordycepin incorporated into a growing RNA chain cannot support further chain elongation in the 5′ → 3′ direction.

2. (a) mRNA$_{(n \text{ residues})}$ + P$_i$ → NDP + mRNA$_{(n-1 \text{ residues})}$

(b) The reverse of the phosphorolysis reaction is an RNA polymerization reaction. PNPase uses an NDP substrate to extend the RNA by one nucleotide residue and releases P$_i$ and is template-independent. RNA polymerase uses an NTP substrate, releases PP$_i$, and requires a template DNA.

(c) High processivity would allow the exonuclease to rapidly degrade mRNA molecules. This would be important in cases where the gene product was no longer needed. An mRNA that was degraded more slowly could potentially continue to be translated.

3. The top strand is the sense strand. Its TATGAT segment differs by only one base from the TATAAT consensus sequence of the promoter's −10 sequence; its TTTACA sequence differs by only one base from the TTGACA consensus sequence of the promoter's −35 sequence and is appropriately located ∼25 nt to the 5′ side of the −10 sequence; and the initiating G nucleotide is the only purine that is located ∼10 nt downstream of the −10 sequence.

5′ CAACGTAACACTTTACAGCGGCGCGTCATTTGATATGATGCGCCCCGCTTCCCGATA 3′

           −35                 −10       start
        region          region     point

4. The probe should have a sequence complementary to the consensus sequence of the 6-nt Pribnow box: 5′-ATTATA-3′.

5. G · C base pairs are more stable than A · T base pairs. Hence, the more G · C base pairs that the promoter contains, the more difficult it is to form the open complex during transcription initiation.

6. Promoter elements for RNA polymerase II include sequences at −27 (the TATA box) and between −50 and −100. The insertion of 10 bp would separate the promoter elements by the distance of the turn of the DNA helix, thereby diminishing the binding of proteins required for transcription initiation. However, the protein-binding sites would still be on the same side of the helix. Inserting 5 bp (half of a helical turn) would move the protein-binding sites to opposite sides of the helix, making it even more difficult to initiate transcription.

7. (a) Operons allow cells to turn sets of related genes on and off together, thereby maximizing efficiency, since all the necessary genes are expressed at the same time and in the same amount.

(b) In eukaryotes, genes in different locations can be turned on or off at the same time if they share the same transcriptional regulatory sequences, such as core promoter elements and enhancers that interact with the same transcription factors.

8. Transcription of an rRNA gene yields a single rRNA molecule that is incorporated into a ribosome. In contrast, transcription of a ribosomal protein gene yields an mRNA that can be translated many times to produce many copies of its corresponding protein. The greater number of rRNA genes relative to ribosomal protein genes helps ensure the balanced synthesis of rRNA and proteins necessary for ribosome assembly.

9. Increasing the error rate of transcription increases the chances of introducing a mutation that prevents the virus from completing its life cycle in the host cell.

10.
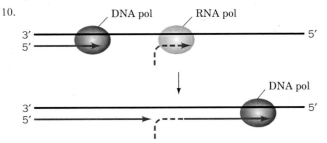

11. In the presence of bicyclomycin, transcription of Rho-dependent genes does not terminate, causing read-through into adjacent coding regions. This results in the transcription of the adjacent gene(s), often causing the inappropriate expression of the corresponding protein(s).

12. (a) Because expression of bacteriophage genes requires the host's RNAP, the increased rate of transcription of bacteriophage genes and decreased rate of transcription termination boost the production of phage-specific mRNAs.

(b) Q must recognize the promoters of bacteriophage genes so that it can bind to the RNAP transcribing those genes. Without this specificity, Q could enhance transcription of all genes in *E. coli.*

13. By introducing a T7 promoter into recombinant DNA and using T7 RNAP, genetic engineers can control the expression of a specific gene without interference from other RNAP enzymes or other promoter sequences that might be present in the experimental system.

14. Because TFIIB can bind directly to DNA at the promoter and indirectly to DNA at the end of the gene, it can cause the intervening DNA to form a loop. As a result, an RNA polymerase that has finished transcribing a gene will be positioned near the promoter so that transcription can be quickly reinitiated.

15. The cell lysates can be applied to a column containing a matrix with immobilized poly(dT). The poly(A) tails of processed mRNAs will bind to the poly(dT) while other cellular components are washed away. The mRNAs can be eluted by decreasing the salt concentration to destabilize the A · T base pairs.

16. The RNA genomes of certain viruses are not processed and hence the first nucleotide includes a triphosphate group. Recognizing this feature allows a cell to detect the presence of an infecting virus. The cell's own RNA molecules (other than mRNA, which is capped) are all processed (hydrolyzed from larger precursors) and therefore contain only a single phosphate at their 5′ ends.

17. (a) The phosphate groups of the phosphodiester backbone of the mRNA will be labeled at all sites where $\alpha$-[$^{32}$P]ATP is used as a substrate by RNA polymerase.

(b) $^{32}$P will appear only at the 5′ end of mRNA molecules that have A as the first residue (this residue retains its $\alpha$ and $\beta$ phosphates). In all other cases where $\beta$-[$^{32}$P]ATP is used as a substrate for RNA synthesis, the $\beta$ and $\gamma$ phosphates are released as $PP_i$ (see Fig. 26-7).

(c) No $^{32}$P will appear in the RNA chain. During polymerization, the $\beta$ and $\gamma$ phosphates are released as $PP_i$. The terminal ($\gamma$) phosphate of an A residue at the 5′ end of an RNA molecule is removed during the capping process.

18. DNA polymerase needs a primer; poly(A) polymerase uses the pre-mRNA as a primer; CCA-adding polymerase uses the immature tRNA as a primer; and RNA polymerase does not require a primer. Both DNA polymerase and RNA polymerase require a DNA template, but neither poly(A) polymerase nor CCA-adding polymerase uses a template. The four polymerases use different sets of nucleotides: DNA polymerase uses all four dNTPS; RNA polymerase uses all four NTPs; poly(A) polymerase uses only ATP; and CCA-adding polymerase uses ATP and CTP.

19. The active site of poly(A) polymerase is narrower because it does not need to accommodate a template strand.

20. The mechanism of RNase hydrolysis requires a free 2′-OH group to form a 2′,3′-cyclic phosphate intermediate (Figure 11-10). Nucleotide residues lacking a 2′-OH group would therefore be resistant to RNase-catalyzed hydrolysis.

21. The mRNA splicing reaction, which requires no free energy input and results in no loss of phosphodiester bonds, is theoretically reversible *in vitro*. However, the degradation of the excised intron makes the reaction irreversible in the cell.

22. The intron must be large enough to include a spliceosome binding site(s).

23. Inhibition of snRNA processing interferes with mRNA splicing. As a result, host mRNA cannot be translated, so the host ribosomes will synthesize only viral proteins.

24.

Alternative mRNA splicing can generate different forms of the protein. If exon 2 encodes a membrane-spanning segment, then joining exon 1 to exon 2 will generate a membrane-bound protein. If exon 3 encodes a soluble segment, then joining exon 1 to exon 3 will generate a soluble protein.

# Chapter 27

1. A 4-nt insertion would add one codon and shift the gene's reading frame by one nucleotide. The proper reading frame could be restored by deleting a nucleotide. Gene function, however, would not be restored if (a) the 4-nt insertion interrupted the codon for a functionally critical amino acid; (b) the 4-nt insertion created a codon for a structure-breaking amino acid; (c) the 4-nt insertion introduced a Stop codon early in the gene; or (d) the 1-nt deletion occurred far from the 4-nt insertion so that even though the reading frame was restored, a long stretch of frame-shifted codons separated the insertion and deletion points.

2. There are two DNA strands corresponding to the sequence: the one whose sequence is given and the one that is complementary to it. Each strand has three possible reading frames, so there are 6 different ways the DNA could be translated.

3. The possible codons are UUU, UUG, UGU, GUU, UGG, GUG, GGU, and GGG. The encoded amino acids are Phe, Leu, Cys, Val, Trp, and Gly (Table 27-1).

4. A UAG Stop codon results from any of the point mutations XAG, UXG, or UAX to UAG. The XAG codons specify Gln, Lys, and Glu; the UXG codons specify Leu, Ser, and Trp; and the UAX codons that are not Stop codons both specify Tyr. Hence some of the codons specifying these amino acids can undergo a point mutation to UAG.

5. There are still two other Stop codons, UAA and UAG, to terminate translation.

6. The likeliest codons to specify Met would be AUU, AUC, or AUA (all of which specify Ile in the standard genetic code), because these codons differ from AUG only at the third position.

7. (a) Each ORF begins with an initiation codon (ATG) and ends with a Stop codon (TGA):

   ATGCTCAACTATATGTGA encodes *vir-2* and ATGCCGCAT-GCTCTGTTAATCACATATAGTTGA on the complementary strand encodes *vir-1*.

   (b) *vir-1*: MPHALLITYS; *vir-2*: MLNYM.

   (c) *vir-1*: MPHALLIPYS; *vir-2*: MLNYMGLTEHAA.

8. There are four exons (the underlined bases)

   TATAATACGCGCAATACAATCTACAGCTTC<u>GCGTAAATCG</u>
   <u>TAG</u>GTAAGTTGTAATAAATATAAGTGAGTATGAT
   ACAG<u>GCTTTGGACCGATAGATGCGACCCTGGAGG</u>TAAG
   TATAGATTAATTAAGCACAG<u>GCATGCAGGGATATCCTC</u>
   <u>CAAAAAG</u>GTAAGTAACCTTACGGTCAATTAAT
   TCAG<u>GCAGTAGATGAATAAACGATATCGATCGGTTAG</u>GTA
   AGTCTGAT

   The mature mRNA, which has a 5′ cap and a 3′ poly(A) tail, therefore has the sequence

   GCGUAAAUCGUAGGCUUUGGACCGAUAG**AUG**CGACC
   CUGGAGGCAUGCAGGGAUAUCCUCCAAAAAGGCAU

GCAGGGAUAUCCUCCAAAUAGGCAGUAGA**UGA**AUAA
ACGAUAUCGAUCGGUUAG

The initiation codon and termination codon are shown in boldface. The encoded protein has the sequence

MRPWRHAGISSKKACRDILQIGSR

9. (a) Like an aaRS, Xpot must recognize features of tRNA structure that are present in all tRNAs, such as the acceptor stem and the TψC loop. (b) Xpot can distinguish mature and pre-tRNAs because mature tRNAs have a processed 5′ end with a single phosphate group, and the 3′ end must be a —CCA sequence (see Fig. 27-3).

10.

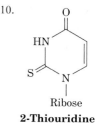

**2-Thiouridine**

11. Gly and Ala; Val and Leu; Ser and Thr; Asn and Gln; and Asp and Glu.

12. A mutation that generates a seldom-used codon that requires a rare tRNA could slow the rate of translation so that although the resulting protein is structurally normal, less of it is synthesized.

13.

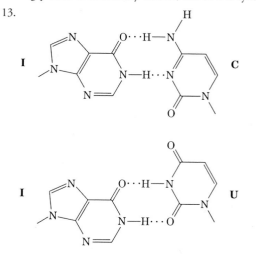

14. This arrangement ensures that the rRNAs will be made in the equal amounts required by functional ribosomes.

15. Only newly synthesized bacterial polypeptides have fMet at their N-terminus. Consequently, the appearance of fMet in a mammalian system signifies the presence of invading bacteria. Leukocytes that recognize the fMet residue can therefore combat these bacteria through phagocytosis.

16. Prokaryotic ribosomes can select an initiation codon located anywhere on the mRNA molecule as long as it lies just downstream of a Shine–Dalgarno sequence. In contrast, eukaryotic ribosomes usually select the AUG closest to the 5′ end of the mRNA. Eukaryotic ribosomes therefore cannot recognize a translation initiation site on a circular mRNA.

17. Ribosomes cannot translate double-stranded RNA, so the base pairing of a complementary antisense RNA to an mRNA prevents its translation.

18. eIF2 is a G protein that delivers the initiator tRNA to the 40S ribosomal subunit and then hydrolyzes its bound GTP to GDP. The GEF eIF2B helps eIF2 release GDP in order to bind GTP so that it can participate in another round of translation initiation.

19. As expected, the correctly charged tRNAs (Ala–tRNA$^{Ala}$ and Gln–tRNA$^{Gln}$) bind to EF-Tu with approximately the same affinity, so they are delivered to the ribosomal A site with the same efficiency. The mischarged Ala–tRNA$^{Gln}$ binds to EF-Tu much more loosely, indicating that it may dissociate from EF-Tu before it reaches the ribosome. The mischarged Gln-tRNA$^{Ala}$ binds to EF-Tu much more tightly, indicating that EF-Tu may not be able to dissociate from it at the ribosome. These results suggest that either a higher or a lower binding affinity could affect the ability of EF-Tu to carry out its function, which would decrease the rate at which mischarged aminoacyl–tRNAs bind to the ribosomal A site during translation.

20. By inducing the same conformational changes that occur during correct tRNA–mRNA pairing, paromomycin can mask the presence of an incorrect codon–anticodon match. Without proofreading at the aminoacyl–tRNA binding step, the ribosome synthesizes a polypeptide with the wrong amino acids, which is likely to be nonfunctional or toxic to the cell.

21. Transpeptidation involves the nucleophilic attack of the amino group of the aminoacyl–tRNA on the carbonyl carbon of the peptidyl–tRNA. As the pH increases, the amino group becomes more nucleophilic (less likely to be protonated).

22. As the pH increases, residue A2486 would be less likely to be protonated and therefore less likely to stabilize the negatively charged oxyanion of the tetrahedral reaction intermediate. Thus, the mechanistic embellishment is inconsistent with the observed effect.

23.
$$^{-}OOC-\underset{\underset{NH_3^+}{|}}{CH}-CH_2-CH_2-\underset{\underset{O}{\|}}{C}-NH-\underset{\underset{\underset{\underset{SH}{|}}{CH_2}}{|}}{CH}-COO^-$$

24. Formation of a peptide bond is an endergonic reaction. In fact, the synthetase requires the free energy of ATP.

25. The enzyme hydrolyzes peptidyl–tRNA molecules that dissociate from a ribosome before normal translation termination takes place. Because peptide synthesis is prematurely halted, the resulting polypeptide, which is still linked to tRNA, is likely to be nonfunctional. Peptidyl–tRNA hydrolase is necessary for recycling the amino acids and the tRNA.

26. When a ribosome translates an mRNA lacking a Stop codon, translation proceeds all the way to the mRNA's poly(A) tail. Since AAA is the codon for Lys, the appearance of a poly(Lys) sequence signals the cell to destroy the newly made polypeptide, which is likely to be nonfunctional.

27. Aminoacylation occurs via pyrophosphate cleavage of ATP, and hence the aminoacylation of 100 tRNAs requires 200 ATP equivalents; translation initiation requires 1 GTP (1 ATP equivalent); 99 cycles of elongation require 99 GTP (99 ATP equivalents) for EF-Tu action; 99 cycles of ribosomal translocation require 99 GTP (99 ATP equivalents) for EF-G action; and translation termination requires 1 GTP (1 ATP equivalent), bringing the total energy cost to 200 + 1 + 99 + 99 + 1 = 400 ATP equivalents.

28.

| | Start | Lys | Pro | Ala |
|---|---|---|---|---|
| 5'-AGGAGCUX$_{\sim4}$ | $^A_G$UG | AA$^A_G$ | CCX | GCX - |

Shine–Dalgarno sequence.
3–10 base pairs with G · U's allowed

| Gly | Thr | Glu | Asn | Ser | Stop |
|---|---|---|---|---|---|
| GGX | ACX | GA$^A_G$ | AA$^U_C$ | UCX | UAA |
| | | | | or | UAG - 3' |
| | | | | AG$^U_C$ | UGA |

# Chapter 28

1. Virtually all the DNA sequences in *E. coli* are present as single copies, so the renaturation of *E. coli* DNA is a straightforward process of each fragment reassociating with its complementary strand. In contrast, the human genome contains many repetitive DNA sequences. The many DNA fragments containing these sequences find each other to form double-stranded regions (renature) much faster than the single-copy DNA sequences that are also present, giving rise to a biphasic renaturation curve.

2. Because genes encoding proteins with related functions often occur in operons, the identification of one or several genes in an operon may suggest functions for the remaining genes in that operon.

3. The *Daphnia* and *Drosophila* genomes are similar in size (200,000 kb versus 180,000 kb), but *Daphnia* contains far more genes (~30,000 versus ~13,000). The *Daphnia* genome is much smaller than the human genome (200,000 kb versus 3,038,000 kb) but appears to contain more genes (~30,000 versus ~23,000).

4. The alga *O. tauri* allots about 13,000 kb/8000 genes or ~1600 bp per gene, which is not much more than *E. coli*, which allots 4639 kb/4289 genes or ~1100 bp per gene.

5. (a) Translation of CAG repeats will yield polypeptides containing polyglutamine (from the CAG codon), polyserine (from the AGC codon), and polyalanine (from the GCA codon). (b) Translation of CTG repeats will yield polypeptides containing polyleucine (from the CUG codon), polycysteine (from the UGC codon), and polyalanine (from the GCU codon).

6. Eleven CNPs of 465 kb each is 5115 kb, or about 5115 kb/3,038,000 kb = 0.0017 of the genome.

7. $O_1$ is the primary repressor-binding site, so *lac* repressor cannot stably bind to the operator in its absence and repression cannot occur.

8. (a) Both $O_2$ and $O_3$ are secondary repressor-binding sequences. If one is absent, the other can still function, resulting in only a small loss of repressor effectiveness.

   (b) In the absence of both $O_2$ and $O_3$, the repressor can bind only to $O_1$, which partially interferes with transcription but does not repress transcription as fully as when a DNA loop forms through the cooperative binding of *lac* repressor to $O_1$ and either $O_2$ or $O_3$.

9. In the absence of β-galactosidase (the product of the *lacZ* gene), lactose is not converted to the inducer allolactose. Consequently, *lac* enzymes, including galactoside permease, are not synthesized.

10. Since operons other than the *lac* operon maintain their sensitivity to the absence of glucose, the defect is probably not in the gene that encodes CAP. Instead, the defect is probably located in the portion of the *lac* operon that binds CAP–cAMP.

11. In eukaryotes, transcription takes place in the nucleus and translation occurs in the cytoplasm. Hence, in eukaryotes, ribosomes are never in contact with nascent mRNAs, an essential aspect of the attenuation mechanism in prokaryotes.

12. Deletion of the leader peptide sequence from *trpL* would eliminate sequence 1 of the attenuator. Consequently, the 2 · 3 hairpin rather than the 3 · 4 terminator hairpin would form. Transcription would therefore continue into the remainder of the *trp* operon, which would then be regulated solely by *trp* repressor.

13. Red–green color blindness is conferred by a mutation in an X-linked gene so that female carriers of the condition, who do not appear to be red–green colorblind, have one wild-type gene and one mutated gene. In placental mammals such as humans, females are mosaics of clones of cells in which only one of their two X chromosomes is transcriptionally active. Hence in a female carrier of red–green color blindness, the transcriptionally active X chromosome in some clones will contain the wild-type gene and the

others will contain the mutated gene. The former type of retinal clone is able to differentiate red and green light, whereas the latter type of retinal clone is unable to do so. Apparently, these retinal clones are small enough so that a narrow beam of light is necessary to separately interrogate them.

14. Transcriptionally active chromatin has a more open structure due to histone modifications that help make the DNA more accessible to transcription factors and RNA polymerase as well as nucleases.

15.

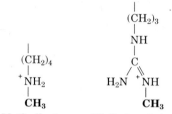

**Acetyllysine**

In acetyllysine, the cationic side chain of Lys has been converted to a polar but uncharged side chain.

16.

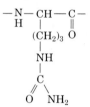

**Methyllysine     Methylarginine**

In methyllysine and methylarginine, the hydrophobic methyl group partially masks the cationic character of the Lys or Arg side chain.

17. Histone and DNA methylation requires *S*-adenosylmethionine (SAM), which becomes *S*-adenosylhomocysteine after it gives up its methyl group (Fig. 21-18). *S*-Adenosylhomocysteine is converted back to methionine, the precursor of SAM, in a reaction in which the methyl group is donated by the folic acid derivative tetrahydrofolate (THF; Fig. 21-18). A shortage of this cofactor could limit cellular production of SAM, which would result in the undermethylation of histones and DNA.

18. The product is a citrulline side chain (Fig. 21-9).

$$-\underset{H}{N}-CH-\overset{\parallel}{C}-$$
$$\quad\quad |\quad\quad O$$
$$\quad(CH_2)_3$$
$$\quad\quad |$$
$$\quad\quad NH$$
$$\quad\quad |$$
$$\quad\quad C$$
$$\quad\quad \overset{\parallel}{O}\quad \diagdown NH_2$$

19. A sequence located downstream of the gene's promoter (i.e., within the coding region) could regulate gene expression if it were recognized by the appropriate transcription factor such that the resulting DNA–protein complex successfully recruited RNA polymerase to the promoter.

20. (a) Protein phosphorylation (Section 13-2B) leads to immediate changes in protein function through allosteric effects, so the release of active NF-κB is rapid.

(b) Phosphorylation of a protein introduces negative charges that might impede the binding of the transcription factor to its recognition sequences in DNA. This potential problem is avoided by the indirect activation of NF-κB through phosphorylation of its inhibitor IκB.

(c) By removing ubiquitin from IκB, the *Yersinia* protein prevents IκB degradation. As a result, IκB is able to bind to NF-κB, preventing the transcription factor from turning on the genes required for lymphocyte proliferation and differentiation. Thus, the bacteria can suppress the immune response.

21. The susceptibility of RNA to degradation *in vivo* makes it possible to regulate gene expression by adjusting the rate of mRNA degradation. If mRNA were very stable, it might continue to direct translation even when the cell no longer needed the encoded protein.

22. A 22-bp segment of RNA, incorporating all four nucleotides, has $4^{22}$ or $1.8 \times 10^{13}$ possible unique sequences. An RNA half this size would have only $4^{11}$ or $4.2 \times 10^6$ possible sequences. The shorter the siRNA, the greater is the probability that it could hybridize with more than one complementary mRNA, thereby making it less efficient in silencing a specific gene. (In the $3.0 \times 10^9$-bp human genome, a sequence of 16 bp has a high probability of randomly occurring at least once.)

23. Since there are 65 $V_H$, 27 $D$, and 6 $J_H$ segments that can be used to assemble the coding sequence of the variable region of the heavy chain, somatic recombination could theoretically generate 65 × 27 × 6 = 10,530 heavy chain genes (junctional flexibility would increase this number). Since each immunoglobulin molecule contains two identical heavy chains and two identical light chains, the possible number of immunoglobulins would be 10530 × 2000 = ~21 million.

24. The imprecise joining of $V$, $D$, and $J$ segments, along with nucleotide addition or removal at the junction, can generate a Stop codon (yielding a truncated and hence nonfunctional immunoglobulin chain) or create a shift in the reading frame (yielding a misfolded and nonfunctional protein).

25. B cells are diploid [have two sets of genes specifying heavy chains and fours sets of genes specifying light chains (two κ's and two λ's)]. Hence, if allelic exclusion were defective, they would continue to rearrange gene segments even after functional heavy and light chain genes had been assembled. Such a B cell could produce more than one type of heavy and light chain. The resulting mixed chain immunoglobulins would be unable to cross-link antigens since the two antigen-binding sites would have different binding specificities.

26. AID, a cytidine deaminase, is required for somatic hypermutation in B cells. If the enzyme were active in another cell, that cell might exhibit a very high rate of mutation.

27. In multicellular organisms, apoptosis of damaged cells minimizes damage to the entire organism. For a single-celled organism, survival of a genetically damaged cell is preferable in a Darwinian sense to its death.

28. Phosphatidylserine is normally present only on the inner leaflet of the plasma membrane (Section 9-4C). The loss of membrane asymmetry in a dying cell would distinguish it from normal cells and facilitate its disposal.

29. The *esc* gene is apparently a maternal-effect gene. Thus, the proper distribution of the *esc* gene product in the fertilized egg, which is maternally specified, is sufficient to permit normal embryonic development regardless of the embryo's genotype.

30. The *knirps* mRNA is expressed in a band posterior to the embryo's midpoint (Fig. 28-55). This band would appear blue due to the action of β-galactosidase on X-gal.